Algebra II
FOR
DUMMIES®

by Mary Jane Sterling

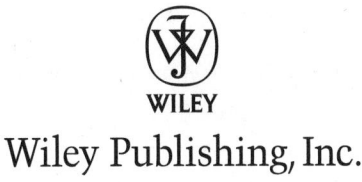

WILEY

Wiley Publishing, Inc.

Algebra II For Dummies®

Published by
Wiley Publishing, Inc.
111 River St.
Hoboken, NJ 07030-5774
www.wiley.com

For general information on our other products and services, please contact our Customer Care Department within the U.S. at 800-762-2974, outside the U.S. at 317-572-3993, or fax 317-572-4002.

For technical support, please visit www.wiley.com/techsupport.

Wiley also publishes its books in a variety of electronic formats. Some content that appears in print may not be available in electronic books.

Library of Congress Control Number: 2006923792

ISBN: 978-0-471-77581-2

Manufactured in the United States of America

10 9 8 7 6 5

1O/TR/QV/QY/IN

WILEY

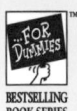

Algebra II For Dummies®

Cheat Sheet

Special Products

Sum and difference: $(a + b)(a - b) = a^2 - b^2$

Binomial squared: $(a + b)^2 = a^2 + 2ab + b^2$

Binomial cubed: $(a + b)^3 = a^3 + 3a^2b + 3ab^2 + b^3$

Rules for Radicals

Root of a product: $\sqrt[n]{a \cdot b} = \sqrt[n]{a} \cdot \sqrt[n]{n}$

Root of a quotient: $\sqrt[n]{\dfrac{a}{b}} = \dfrac{\sqrt[n]{a}}{\sqrt[n]{b}}$

Fractional exponent: $\sqrt[n]{a^m} = a^{m/n}$

Permutations and Combinations:

Number of permutations of n things taken r at a time: $_nP_r = \dfrac{n!}{(n-r)!}$

Number of combinations of n things taken r at a time: $_nC_r = \dfrac{n!}{(n-r)!\,r!}$

Laws of Logarithms

Equivalence: $a^x = y \leftrightarrow \log_a y = x$

$e^x = y \leftrightarrow \ln y = x$

Log of a product: $\log_a x \cdot y = \log_a x + \log_a y$

$\ln x \cdot y = \ln x + \ln y$

Log of a quotient: $\log_a \dfrac{x}{y} = \log_a x - \log_a y$

$\ln \dfrac{x}{y} = \ln x - \ln y$

Log of a power: $\log_a x^n = n\log_a x$

$\ln x^n = n\ln x$

Log of a reciprocal: $\log_a \dfrac{1}{x} = -\log_a x$

$\ln \dfrac{1}{x} = -\ln x$

Log of the base: $\log_a a = 1$

$\ln e = 1$

Log of 1: $\log_a 1 = 0$

$\ln 1 = 0$

Standard Equations of Conics:

Parabolas: $y - k = a(x - h)^2$

$x - h = a(y - k)^2$

Circle: $(x - h)^2 + (y - k)^2 = r^2$

Ellipse: $\dfrac{(x - h)^2}{a^2} + \dfrac{(y - k)^2}{b^2} = 1$

Hyperbola: $\dfrac{(x - h)^2}{a^2} - \dfrac{(y - k)^2}{b^2} = 1$

Quadratic formula

When $ax^2 + bx + c = 0$, then

$x = \dfrac{-b \pm \sqrt{b^2 - 4ac}}{2a}$

Sums of Sequences

Sum of the first n positive integers: $\displaystyle\sum_{k=1}^{n} k = \dfrac{n(n-1)}{2}$

Sum of the first n squares: $\displaystyle\sum_{k=1}^{n} k^2 = \dfrac{n(n+1)(2n+1)}{6}$

Sum of the first n cubes: $\displaystyle\sum_{k=1}^{n} k^3 = \dfrac{n^2(n-1)^2}{4}$

Sum of the first n terms of an arithmetic sequence: $S_n = \dfrac{n}{2}\left[2a_1 + (n-1)d\right] = \dfrac{n}{2}(a_1 + a_n)$

Sum of the first n terms of a geometric sequence: $S_n = \dfrac{g_1(1 - r^n)}{1 - r}$

Sum of all of the terms of a geometric sequence with $|r| < 1$: $S_n \to \dfrac{g_1}{1 - r}$

Algebra II For Dummies®

Cramer's Rule

The solution of the linear system $\begin{cases} ax + by = c \\ dx + ey = f \end{cases}$ is $x = \dfrac{ce - bf}{ae - bd}$, $\quad y = \dfrac{af - cd}{ae - bd}$

Eight Basic Curves:

$y = x$

$y = x^2$

$y = x^3$

$y = |x|$

$y = \dfrac{1}{x}$

$y = \sqrt{x}$

$y = b^x$

$x^2 + y^2 = r^2$

For Dummies: Bestselling Book Series for Beginners

About the Author

Mary Jane Sterling has authored *Algebra For Dummies*, *Trigonometry For Dummies*, *Algebra Workbook For Dummies*, *Trigonometry Workbook For Dummies*, *Algebra I CliffsStudySolver*, and *Algebra II CliffsStudySolver*. She taught junior high and high school math for many years before beginning her current 25-year-and-counting career at Bradley University in Peoria, Illinois. Mary Jane enjoys working with her students both in the classroom and outside the classroom, where they do various community service projects.

Dedication

The author dedicates this book to some of the men in her life. Her husband, Ted Sterling, is especially patient and understanding when her behavior becomes erratic while working on her various projects — his support is greatly appreciated. Her brothers Tom, Don, and Doug knew her "back when." Don, in particular, had an effect on her teaching career when he threw a pencil across the room during a tutoring session. It was then that she rethought her approach — and look what happened! And brother-in-law Jeff is an ongoing inspiration with his miracle comeback and continued recovery.

Author's Acknowledgments

The author wants to thank Mike Baker for being a great project editor — good natured (very important) and thorough. He took the many challenges with grace and handled them with diplomacy. Also, thank you to Josh Dials, a wonderful editor who straightened out her circuitous explanations and made them understandable. A big thank you to the technical editor, Alexsis Venter, who helped her on an earlier project — and still agreed to sign on! Also, thanks to Kathy Cox for keeping the projects coming; she can be counted on to keep life interesting.

Publisher's Acknowledgments

We're proud of this book; please send us your comments through our Dummies online registration form located at www.dummies.com/register/.

Some of the people who helped bring this book to market include the following:

Acquisitions, Editorial, and Media Development

Project Editor: Mike Baker

Acquisitions Editor: Kathy Cox

Copy Editor: Josh Dials

Editorial Program Coordinator: Hanna K. Scott

Technical Editor: Alexsis Venter

Editorial Manager: Christine Meloy Beck

Editorial Assistants: Erin Calligan, David Lutton

Cover Photos: © Wiley Publishing, Inc.

Cartoons: Rich Tennant (www.the5thwave.com)

Composition Services

Project Coordinator: Jennifer Theriot

Layout and Graphics: Lauren Goddard, Denny Hager, Barry Offringa, Heather Ryan

Proofreader: Betty Kish

Indexer: Slivoskey Indexing Services

Publishing and Editorial for Consumer Dummies

Diane Graves Steele, Vice President and Publisher, Consumer Dummies

Joyce Pepple, Acquisitions Director, Consumer Dummies

Kristin A. Cocks, Product Development Director, Consumer Dummies

Michael Spring, Vice President and Publisher, Travel

Kelly Regan, Editorial Director, Travel

Publishing for Technology Dummies

Andy Cummings, Vice President and Publisher, Dummies Technology/General User

Composition Services

Gerry Fahey, Vice President of Production Services

Debbie Stailey, Director of Composition Services

Contents at a Glance

Table of Contents

Introduction

*H*ere you are, contemplating reading a book on Algebra II. It isn't a mystery novel, although you can find people who think mathematics in general is a mystery. It isn't a historical account, even though you find some historical tidbits scattered here and there. Science fiction it isn't; mathematics is a science, but you find more fact than fiction. As Joe Friday (star of the old *Dragnet* series) says, "The facts, ma'am, just the facts." This book isn't light reading, although I attempt to interject humor whenever possible. What you find in this book is a glimpse into the way I teach: uncovering mysteries, working in historical perspectives, providing information, and introducing the topic of Algebra II with good-natured humor. This book has the best of all literary types! Over the years, I've tried many approaches to teaching algebra, and I hope that with this book I'm helping you cope with other teaching methods.

About This Book

Because you're interested in this book, you probably fall into one of four categories:

- ✔ You're fresh off Algebra I and feel eager to start on this new venture.
- ✔ You've been away from algebra for a while, but math has always been your strength, so you don't want to start too far back.
- ✔ You're a parent of a student embarking on or having some trouble with an Algebra II class and you want to help.
- ✔ You're just naturally curious about science and mathematics and you want to get to the good stuff that's in Algebra II.

Whichever category you represent (and I may have missed one or two), you'll find what you need in this book. You can find some advanced algebraic topics, but I also cover the necessary basics, too. You can also find plenty of connections — the ways different algebraic topics connect with each other and the ways the algebra connects with other areas of mathematics.

After all, the many other math areas drive Algebra II. Algebra is the passport to studying calculus, trigonometry, number theory, geometry, and all sorts of good mathematics. Algebra is basic, and the algebra you find here will help you grow your skills and knowledge so you can do well in math courses and possibly pursue other math topics.

Conventions Used in This Book

To help you navigate this book, I use the following conventions:

- ✔ I *italicize* special mathematical terms and define them right then and there so you don't have to search around.

- ✔ I use **boldface** text to indicate keywords in bulleted lists or the action parts of numbered steps. I describe many algebraic procedures in a step-by-step format and then use those steps in an example or two.

- ✔ Sidebars are shaded gray boxes that contain text you may find interesting, but this text isn't necessarily critical to your understanding of the chapter or topic.

Foolish Assumptions

Algebra II is essentially a continuation of Algebra I, so I have some assumptions I need to make about anyone who wants (or has) to take algebra one step further.

I assume that a person reading about Algebra II has a grasp of the arithmetic of signed numbers — how to combine positive and negative numbers and come out with the correct sign. Another assumption I make is that your order of operations is in order. Working your way through algebraic equations and expressions requires that you know the rules of order. Imagine yourself at a meeting or in a courtroom. You don't want to be called out of order!

I assume that people who complete Algebra I successfully know how to solve equations and do basic graphs. Even though I lightly review these topics in this book, I assume that you have a general knowledge of the necessary procedures. I also assume that you have a handle on the basic terms you run across in Algebra I, such as

- ✔ **binomial:** An expression with two terms.

- ✔ **coefficient:** The multiplier or factor of a variable.

- ✔ **constant:** A number that doesn't change in value.

- ✔ **expression:** Combination of numbers and variables grouped together — not an equation or inequality.

- ✔ **factor** (*n.*): Something multiplying something else.

- ✔ **factor** (*v.*): To change the format of several terms added together into a product.

✔ **linear:** An expression in which the highest power of any variable term is one.

✔ **monomial:** An expression with only one term.

✔ **polynomial:** An expression with several terms.

✔ **quadratic:** An expression in which the highest power of any variable term is two.

✔ **simplify:** To change an expression into an equivalent form that you combined, reduced, factored, or otherwise made more useable.

✔ **solve:** To find the value or values of the variable that makes a statement true.

✔ **term:** A grouping of constants and variables connected by multiplication, division, or grouping symbols and separated from other constants and variables by addition or subtraction.

✔ **trinomial:** An expression with three terms.

✔ **variable:** Something that can have many values (usually represented by a letter to indicate that you have many choices for its value).

If you feel a bit over your head after reading through some chapters, you may want to refer to *Algebra For Dummies* (Wiley) for a more complete explanation of the basics. My feelings won't be hurt; I wrote that one, too!

How This Book Is Organized

This book is divided into parts that cover the basics, followed by parts that cover equation solving skills and functions and parts that have some applications of this knowledge. The chapters in each part share a common thread that helps you keep everything straight.

Part I: Homing in on Basic Solutions

Part I focuses on the basics of algebra and on solving equations and factoring expressions quickly and effectively — skills that you use throughout the book. For this reason, I make this material quick and easy to reference.

The first four chapters deal with solving equations and inequalities. The techniques I cover in these chapters not only show you how to find the solutions, but also how to write them so anyone reading your work understands what you've found. I start with linear equations and inequalities and then move to quadratics, rational equations, and radical equations.

The final chapter provides an introduction (or refresher, as the case may be) to the coordinate system — the standard medium used to graph functions and mathematical expressions. Using the coordinate system is sort of like reading a road map where you line up the letter and number to find a city. Graphs make algebraic processes clearer, and graphing is a good way to deal with systems of equations — looking for spots where curves intersect.

Part II: Facing Off with Functions

Part II deals with many of the types of functions you encounter in Algebra II: algebraic, exponential, and logarithmic.

A *function* is a very special type of relationship that you can define with numbers and letters. The mystery involving some mathematical expressions and functions clears up when you apply the basic function properties, which I introduce in this part. For instance, a function's domain is linked to a rational function's asymptotes, and a function's inverse is essential to exponential and logarithmic functions. You can find plenty of links.

Do some of these terms sound a bit overwhelming (asymptote, domain, rational, and so on)? Don't worry. I completely explain them all in the chapters of Part II.

Part III: Conquering Conics and Systems of Equations

Part III focuses on graphing and systems of equations — topics that go together because of their overlapping properties and methods. Graphing is sort of like painting a picture; you see what the creator wants you to see, but you can also look for the hidden meanings.

In this part, you discover ways to picture mathematical curves and systems of equations, and you find alternative methods for solving those systems. Systems of equations can contain linear equations with two, three, and even more variables. Nonlinear systems have curves intersecting with lines, circles intersecting with one another, and all manner of combinations of curves and lines crossing and re-crossing one another. You also find out how to solve systems of inequalities. This takes some shady work — oops, no, that's *shading* work. The solutions are whole sections of a graph.

Part IV: Shifting into High Gear with Advanced Concepts

I find it hard to classify the chapters in Part IV with a single word or phrase. You can just call them special or consequential. Among the topics I cover are matrices, which provide ways to organize numbers and then perform operations on them; sequences and series, which provide other ways to organize numbers but with more nice, neat rules to talk about those numbers; and the set, an organizational method with its own, special arithmetic. The topics here all seem to have a common thread of organization, but they're really quite different and very interesting to read about and work with. After you're finished with this part, you'll be in prime shape for higher-level math courses.

Part V: The Part of Tens

The Part of Tens gives you lists of goodies. Plenty of good things come in tens: fingers and toes, dollars, and the stuff in my lists! Everyone has a unique way of thinking about numbers and operations on numbers; in this part, you find ten special ways to multiply numbers in your head. Bet you haven't seen all these tricks before! You also have plenty of ways to categorize the same number. The number nine is odd, a multiple of three, and a square number, just for starters. Therefore, I also present a list of ten unique ways you can categorize numbers.

Icons Used in This Book

The icons that appear in this book are great for calling attention to what you need to remember or what you need to avoid when doing algebra. Think of the icons as signs along the Algebra II Highway; you pay attention to signs — you don't run them over!

This icon provides you with the rules of the road. You can't go anywhere without road signs — and in algebra, you can't get anywhere without following the rules that govern how you deal with operations. In place of "Don't cross the solid yellow line," you see "Reverse the sign when multiplying by a negative." Not following the rules gets you into all sorts of predicaments with the Algebra Police (namely, your instructor).

This icon is like the sign alerting you to the presence of a sports arena, museum, or historical marker. Use this information to improve your mind, and put the information to work to improve your algebra problem-solving skills.

 This icon lets you know when you've come to a point in the road where you should soak in the information before you proceed. Think of it as stopping to watch an informative sunset. Don't forget that you have another 30 miles to Chicago. Remember to check your answers when working with rational equations.

 This icon alerts you to common hazards and stumbling blocks that could trip you up — much like "Watch for Falling Rock" or "Railroad Crossing." Those who have gone before you have found that these items can cause a huge failure in the future if you aren't careful.

 Yes, Algebra II does present some technical items that you may be interested to know. Think of the temperature or odometer gauges on your dashboard. The information they present is helpful, but you can drive without it, so you can simply glance at it and move on if everything is in order.

Where to Go from Here

I'm so pleased that you're willing, able, and ready to begin an investigation of Algebra II. If you're so pumped up that you want to tackle the material cover to cover, great! But you don't have to read the material from page one to page two and so on. You can go straight to the topic or topics you want or need and refer to earlier material if necessary. You can also jump ahead if so inclined. I include clear cross-references in chapters that point you to the chapter or section where you can find a particular topic — especially if it's something you need for the material you're looking at or if it extends or furthers the discussion at hand.

You can use the table of contents at the beginning of the book and the index in the back to navigate your way to the topic that you need to brush up on. Or, if you're more of a freewheeling type of guy or gal, take your finger, flip open the book, and mark a spot. No matter your motivation or what technique you use to jump into the book, you won't get lost because you can go in any direction from there.

Enjoy!

Part I

Homing in on Basic Solutions

In this part . . .

The chapters in Part I take you through the basics of solving algebraic equations. I start off with a refresher on some of the key points you covered in Algebra I oh so long ago (or what seemed like oh so long ago, even if you've only taken a semester or summer break in between). You'll soon remember that solving an equation for an answer or answers is just peachy keen, hunky dory, far out, or whatever your favorite expression of delight.

After I rehash the basics, I introduce you to a number of new concepts. You discover how to solve linear equations; quadratic equations; and equations with fractions, radicals, and complicated exponents. I wrap the part up with a quick course in Graphing 101. All that for the price of admission.

Chapter 1

Going Beyond Beginning Algebra

In This Chapter

▶ Abiding by (and using) the rules of algebra

▶ Adding the multiplication property of zero to your repertoire

▶ Raising your exponential power

▶ Looking at special products and factoring

*A*lgebra is a branch of mathematics that people study before they move on to other areas or branches in mathematics and science. For example, you use the processes and mechanics of algebra in calculus to complete the study of change; you use algebra in probability and statistics to study averages and expectations; and you use algebra in chemistry to work out the balance between chemicals. Algebra all by itself is esthetically pleasing, but it springs to life when used in other applications.

Any study of science or mathematics involves rules and patterns. You approach the subject with the rules and patterns you already know, and you build on those rules with further study. The reward is all the new horizons that open up to you.

Any discussion of algebra presumes that you're using the correct notation and terminology. Algebra I (check out *Algebra For Dummies* [Wiley]) begins with combining terms correctly, performing operations on signed numbers, and dealing with exponents in an orderly fashion. You also solve the basic types of linear and quadratic equations. Algebra II gets into other types of functions, such as exponential and logarithmic functions, and topics that serve as launching spots for other math courses.

You can characterize any discussion of algebra — at any level — as follows: simplify, solve, and communicate.

Going into a bit more detail, the basics of algebra include rules for dealing with equations, rules for using and combining terms with exponents, patterns to use when factoring expressions, and a general order for combining all the above. In this chapter, I present these basics so you can further your study of algebra and feel confident in your algebraic ability. Refer to these rules whenever needed as you investigate the many advanced topics in algebra.

Outlining Algebra Properties

Mathematicians developed the rules and properties you use in algebra so that every student, researcher, curious scholar, and bored geek working on the same problem would get the same answer — no matter the time or place. You don't want the rules changing on you every day (and I don't want to have to write a new book every year!); you want consistency and security, which you get from the strong algebra rules and properties that I present in this section.

Keeping order with the commutative property

The *commutative property* applies to the operations of addition and multiplication. It states that you can change the order of the values in an operation without changing the final result:

$a + b = b + a$	Commutative property of addition
$a \cdot b = b \cdot a$	Commutative property of multiplication

If you add 2 and 3, you get 5. If you add 3 and 2, you still get 5. If you multiply 2 times 3, you get 6. If you multiply 3 times 2, you still get 6.

Algebraic expressions usually appear in a particular order, which comes in handy when you have to deal with variables and coefficients (multipliers of variables). The number part comes first, followed by the letters, in alphabetical order. But the beauty of the commutative property is that $2xyz$ is the same as $x2zy$. You have no good reason to write the expression in that second, jumbled order, but it's helpful to know that you can change the order around when you need to.

Maintaining group harmony with the associative property

Like the commutative property (see the previous section), the associative property applies only to the operations of addition and multiplication. The *associative property* states that you can change the grouping of operations without changing the result:

$a + (b + c) = (a + b) + c$	Associative property of addition
$a(b \cdot c) = (a \cdot b)c$	Associative property of multiplication

You can use the associative property of addition or multiplication to your advantage when simplifying expressions. And if you throw in the commutative property when necessary, you have a powerful combination. For instance, when simplifying $(x + 14) + (3x + 6)$, you can start by dropping the parentheses (thanks to the associative property). You then switch the middle two terms around, using the commutative property of addition. You finish by reassociating the terms with parentheses and combining the like terms:

$$(x + 14) + (3x + 6)$$
$$= x + 14 + 3x + 6$$
$$= x + 3x + 14 + 6$$
$$= (x + 3x) + (14 + 6)$$
$$= 4x + 20$$

The steps in the previous process involve a lot more detail than you really need. You probably did the problem, as I first stated it, in your head. I provide the steps to illustrate how the commutative and associative properties work together; now you can apply them to more complex situations.

Distributing a wealth of values

The *distributive property* states that you can multiply each term in an expression within a parenthesis by the coefficient outside the parenthesis and not change the value of the expression. It takes one operation, multiplication, and spreads it out over terms that you add to and subtract from one another:

$$a(b + c) = a \cdot b + a \cdot c \qquad \text{Distributing multiplication over addition}$$
$$a(b - c) = a \cdot b - a \cdot c \qquad \text{Distributing multiplication over subtraction}$$

For instance, you can use the distributive property on the problem $12\left(\frac{1}{2} + \frac{2}{3} - \frac{3}{4}\right)$ to make your life easier. You distribute the 12 over the fractions by multiplying each fraction by 12 and then combining the results:

$$12\left(\frac{1}{2} + \frac{2}{3} - \frac{3}{4}\right)$$
$$= 12 \cdot \frac{1}{2} + 12 \cdot \frac{2}{3} - 12 \cdot \frac{3}{4}$$
$$= \overset{6}{\cancel{12}} \cdot \frac{1}{\cancel{2}_1} + \overset{4}{\cancel{12}} \cdot \frac{2}{\cancel{3}_1} + \overset{3}{\cancel{12}} \cdot \frac{3}{\cancel{4}_1}$$
$$= 6 + 8 - 9$$
$$= 5$$

Finding the answer with the distributive property is much easier than changing all the fractions to equivalent fractions with common denominators of 12, combining them, and then multiplying by 12.

You can use the distributive property to simplify equations — in other words, you can prepare them to be solved. You also do the opposite of the distributive property when you *factor* expressions; see the section "Implementing Factoring Techniques" later in this chapter.

Checking out an algebraic 1D

The numbers zero and one have special roles in algebra — as identities. You use *identities* in algebra when solving equations and simplifying expressions. You need to keep an expression equal to the same value, but you want to change its format, so you use an identity in one way or another:

$a + 0 = 0 + a = a$ The *additive identity* is zero. Adding zero to a number doesn't change that number; it keeps its identity.

$a \cdot 1 = 1 \cdot a = a$ The *multiplicative identity* is one. Multiplying a number by one doesn't change that number; it keeps its identity.

Applying the additive identity

One situation that calls for the use of the additive identity is when you want to change the format of an expression so you can factor it. For instance, take the expression $x^2 + 6x$ and add 0 to it. You get $x^2 + 6x + 0$, which doesn't do much for you (or me, for that matter). But how about replacing that 0 with both 9 and –9? You now have $x^2 + 6x + 9 - 9$, which you can write as $(x^2 + 6x + 9) - 9$ and factor into $(x + 3)^2 - 9$. Why in the world do you want to do this? Go to Chapter 11 and read up on conic sections to see why. By both adding and subtracting 9, you add 0 — the additive identity.

Making multiple identity decisions

You use the multiplicative identity extensively when you work with fractions. Whenever you rewrite fractions with a common denominator, you actually multiply by one. If you want the fraction $\frac{7}{2x}$ to have a denominator of $6x$, for example, you multiply both the numerator and denominator by 3:

$$\frac{7}{2x} \cdot \frac{3}{3} = \frac{21}{6x}$$

Now you're ready to rock and roll with a fraction to your liking.

Singing along in-verses

You face two types of *inverses* in algebra: additive inverses and multiplicative inverses. The additive inverse matches up with the additive identity and the multiplicative inverse matches up with the multiplicative identity. The additive inverse is connected to zero, and the multiplicative inverse is connected to one.

A number and its *additive inverse* add up to zero. A number and its *multiplicative inverse* have a product of one. For example, –3 and 3 are additive inverses; the multiplicative inverse of –3 is $-\frac{1}{3}$. Inverses come into play big-time when you're solving equations and want to isolate the variable. You use inverses by adding them to get zero next to the variable or by multiplying them to get one as a multiplier (or coefficient) of the variable.

Ordering Your Operations

When mathematicians switched from words to symbols to describe mathematical processes, their goal was to make dealing with problems as simple as possible; however, at the same time, they wanted everyone to know what was meant by an expression and for everyone to get the same answer to a problem. Along with the special notation came a special set of rules on how to handle more than one operation in an expression. For instance, if you do the problem $4 + 3^2 - 5 \cdot 6 + \sqrt{23 - 7} + \frac{14}{2}$, you have to decide when to add, subtract, multiply, divide, take the root, and deal with the exponent.

The *order of operations* dictates that you follow this sequence:

1. Raise to powers or find roots.

2. Multiply or divide.

3. Add or subtract.

If you have to perform more than one operation from the same level, work those operations moving from left to right. If any grouping symbols appear, perform the operation inside the grouping symbols first.

So, to do the previous example problem, follow the order of operations:

1. The radical acts like a grouping symbol, so you subtract what's in the radical first: $4 + 3^2 - 5 \cdot 6 + \sqrt{16} + \frac{14}{2}$.

2. Raise the power and find the root: $4 + 9 - 5 \cdot 6 + 4 + \frac{14}{2}$.

3. Do the multiplication and division: $4 + 9 - 30 + 4 + 7$.

4. Add and subtract, moving from left to right: $4 + 9 - 30 + 4 + 7 = -6$.

Equipping Yourself with the Multiplication Property of Zero

You may be thinking that multiplying by zero is no big deal. After all, zero times anything is zero, right? Yes, and *that's* the big deal. You can use the multiplication property of zero when solving equations. If you can factor an equation — in other words, write it as the product of two or more multipliers — you can apply the multiplication property of zero to solve the equation. The *multiplication property of zero* states that

If the product of $a \cdot b \cdot c \cdot d \cdot e \cdot f = 0$, at least one of the factors has to represent the number 0.

The only way the product of two or more values can be zero is for at least one of the values to actually be zero. If you multiply $(16)(467)(11)(9)(0)$, the result is 0. It doesn't really matter what the other numbers are — the zero always wins.

The reason this property is so useful when solving equations is that if you want to solve the equation $x^7 - 16x^5 + 5x^4 - 80x^2 = 0$, for instance, you need the numbers that replace the x's to make the equation a true statement. This particular equation factors into $x^2(x^3 + 5)(x - 4)(x + 4) = 0$. The product of the four factors shown here is zero. The only way the product can be zero is if one or more of the factors is zero. For instance, if $x = 4$, the third factor is zero, and the whole product is zero. Also, if x is zero, the whole product is zero. (Head to Chapters 3 and 8 for more info on factoring and using the multiplication property of zero to solve equations.)

The birth of negative numbers

In the early days of algebra, negative numbers weren't an accepted entity. Mathematicians had a hard time explaining exactly what the numbers illustrated; it was too tough to come up with concrete examples. One of the first mathematicians to accept negative numbers was Fibonacci, an Italian mathematician. When he was working on a financial problem, he saw that he needed what amounted to a negative number to finish the problem. He described it as a loss and proclaimed, "I have shown this to be insoluble unless it is conceded that the man had a debt."

Expounding on Exponential Rules

Several hundred years ago, mathematicians introduced powers of variables and numbers called *exponents*. The use of exponents wasn't immediately popular, however. Scholars around the world had to be convinced; eventually, the quick, slick notation of exponents won over, and we benefit from the use today. Instead of writing *xxxxxxxx,* you use the exponent 8 by writing x^8. This form is easier to read and much quicker.

The expression a^n is an exponential expression with a *base* of a and an *exponent* of n. The n tells you how many times you multiply the a times itself.

You use *radicals* to show roots. When you see $\sqrt{16}$, you know that you're looking for the number that multiplies itself to give you 16. The answer? Four, of course. If you put a small superscript in front of the radical, you denote a cube root, a fourth root, and so on. For instance, $\sqrt[4]{81} = 3$, because the number 3 multiplied by itself four times is 81. You can also replace radicals with fractional exponents — terms that make them easier to combine. This system of exponents is very systematic and workable — thanks to the mathematicians that came before us.

Multiplying and dividing exponents

When two numbers or variables have the same base, you can multiply or divide those numbers or variables by adding or subtracting their exponents:

- ✔ $a^n \cdot a^m = a^{m+n}$: When multiplying numbers with the same base, you add the exponents.

- ✔ $\dfrac{a^m}{a^n} = a^{m-n}$: When dividing numbers with the same base, you subtract the exponents (numerator – denominator).

To multiply $x^4 \cdot x^5$, for example, you add: $x^{4+5} = x^9$. When dividing x^8 by x^5, you subtract: $\dfrac{x^8}{x^5} = x^{8-5} = x^3$.

You must be sure that the bases of the expressions are the same. You can combine 3^2 and 3^4, but you can't use the rules for exponents on 3^2 and 4^3.

Getting to the roots of exponents

Radical expressions — such as square roots, cube roots, fourth roots, and so on — appear with a radical to show the root. Another way you can write these values is by using fractional exponents. You'll have an easier time combining variables with the same base if they have fractional exponents in place of radical forms:

- $\sqrt[n]{x} = x^{1/n}$: The root goes in the denominator of the fractional exponent.
- $\sqrt[n]{x^m} = x^{m/n}$: The root goes in the denominator of the fractional exponent, and the power goes in the numerator.

So, you can say $\sqrt{x} = x^{1/2}, \sqrt[3]{x} = x^{1/3}, \sqrt[4]{x} = x^{1/4}$, and so on, along with $\sqrt[5]{x^3} = x^{3/5}$. To simplify a radical expression such as $\dfrac{\sqrt[4]{x}\,\sqrt[6]{x^{11}}}{\sqrt[2]{x^3}}$, you change the radicals to exponents and apply the rules for multiplication and division of values with the same base (see the previous section):

$$\frac{\sqrt[4]{x}\,\sqrt[6]{x^{11}}}{\sqrt[2]{x^3}} = \frac{x^{1/4} \cdot x^{11/6}}{x^{3/2}}$$

$$= \frac{x^{1/4 + 11/6}}{x^{3/2}} = \frac{x^{3/12 + 22/12}}{x^{18/12}}$$

$$= \frac{x^{25/12}}{x^{18/12}} = x^{25/12 - 18/12}$$

$$= x^{7/12}$$

Raising or lowering the roof with exponents

You can raise numbers or variables with exponents to higher powers or reduce them to lower powers by taking roots. When raising a power to a power, you multiply the exponents. When taking the root of a power, you divide the exponents:

- $(a^m)^n = a^{m \cdot n}$: Raise a power to a power by multiplying the exponents.
- $\sqrt[m]{a^n} = \left(a^n\right)^{1/m} = a^{n/m}$: Reduce the power when taking a root by dividing the exponents.

The second rule may look familiar — it's one of the rules that govern changing from radicals to fractional exponents (see Chapter 4 for more on dealing with radicals and fractional exponents).

Here's an example of how you apply the two rules when simplifying an expression:

$$\sqrt[3]{\left(x^4\right)^6 \cdot x^9} = \sqrt[3]{x^{24} \cdot x^9} = \sqrt[3]{x^{33}} = x^{33/3} = x^{11}$$

Making nice with negative exponents

You use negative exponents to indicate that a number or variable belongs in the denominator of the term:

$$a^{-1} = \frac{1}{a}$$

$$a^{-n} = \frac{1}{a^n}$$

Writing variables with negative exponents allows you to combine those variables with other factors that share the same base. For instance, if you have the expression $\frac{1}{x^4} \cdot x^7 \cdot \frac{3}{x}$, you can rewrite the fractions by using negative exponents and then simplify by using the rules for multiplying factors with the same base (see "Multiplying and dividing exponents"):

$$\frac{1}{x^4} \cdot x^7 \cdot \frac{3}{x} = x^{-4} \cdot x^7 \cdot 3x^{-1} = 3x^{-4+7-1} = 3x^2$$

Implementing Factoring Techniques

When you *factor* an algebraic expression, you rewrite the sums and differences of the terms as a product. For instance, you write the three terms $x^2 - x - 42$ in factored form as $(x - 7)(x + 6)$. The expression changes from three terms to one big, multiplied-together term. You can factor two terms, three terms, four terms, and so on for many different purposes. The factorization comes in handy when you set the factored forms equal to zero to solve an equation. Factored numerators and denominators in fractions also make it possible to reduce the fractions.

You can think of factoring as the opposite of distributing. You have good reasons to distribute or multiply through by a value — the process allows you to combine like terms and simplify expressions. Factoring out a common factor also has its purposes for solving equations and combining fractions. The different formats are equivalent — they just have different uses.

Factoring two terms

When an algebraic expression has two terms, you have four different choices for its factorization — if you can factor the expression at all. If you try the following four methods and none of them work, you can stop your attempt; you just can't factor the expression:

$$ax + ay = a(x + y) \qquad \text{Greatest common factor}$$

$$x^2 - a^2 = (x - a)(x + a) \qquad \text{Difference of two perfect squares}$$

$$x^3 - a^3 = (x - a)(x^2 + ax + a^2) \qquad \text{Difference of two perfect cubes}$$

$$x^3 + a^3 = (x + a)(x^2 - ax + a^2) \qquad \text{Sum of two perfect cubes}$$

In general, you check for a greatest common factor before attempting any of the other methods. By taking out the common factor, you often make the numbers smaller and more manageable, which helps you see clearly whether any other factoring is necessary.

To factor the expression $6x^4 - 6x$, for example, you first factor out the common factor, $6x$, and then you use the pattern for the difference of two perfect cubes:

$$6x^4 - 6x = 6x(x^3 - 1)$$
$$= 6x(x - 1)(x^2 + x + 1)$$

A *quadratic trinomial* is a three-term polynomial with a term raised to the second power. When you see something like $x^2 + x + 1$ (as in this case), you immediately run through the possibilities of factoring it into the product of two binomials. You can just stop. These trinomials that crop up with factoring cubes just don't cooperate.

Keeping in mind my tip to start a problem off by looking for the greatest common factor, look at the example expression $48x^3y^2 - 300x^3$. When you factor the expression, you first divide out the common factor, $12x^3$, to get $12x^3(4y^2 - 25)$. You then factor the difference of perfect squares in the parenthesis: $48x^3y^2 - 300x^3 = 12x^3(2y - 5)(2y + 5)$.

Here's one more: The expression $z^4 - 81$ is the difference of two perfect squares. When you factor it, you get $z^4 - 81 = (z^2 - 9)(z^2 + 9)$. Notice that the first factor is also the difference of two squares — you can factor again. The second term, however, is the sum of squares — you can't factor it. With perfect cubes, you can factor both differences and sums, but not with the squares. So, the factorization of $z^4 - 81$ is $(z - 3)(z + 3)(z^2 + 9)$.

Taking on three terms

When a quadratic expression has three terms, making it a *trinomial*, you have two different ways to factor it. One method is factoring out a greatest common factor, and the other is finding two binomials whose product is identical to those three terms:

$$ax + ay + az = a(x + y + z) \qquad \text{Greatest common factor}$$

$$x^{2n} + (a+b)x^n + ab = (x^n + a)(x^n + b) \qquad \text{Two binomials}$$

You can often spot the greatest common factor with ease; you see a multiple of some number or variable in each term. With the product of two binomials, you just have to try until you find the product or become satisfied that it doesn't exist.

For example, you can perform the factorization of $6x^3 - 15x^2y + 24xy^2$ by dividing each term by the common factor, $3x$: $6x^3 - 15x^2y + 24xy^2 = 3x(2x^2 - 5xy + 8y^2)$.

You want to look for the common factor first; it's usually easier to factor expressions when the numbers are smaller. In the previous example, all you can do is pull out that common factor — the trinomial is *prime* (you can't factor it any more).

Trinomials that factor into the product of two binomials have related powers on the variables in two of the terms. The relationship between the powers is that one is twice the other. When factoring a trinomial into the product of two binomials, you first look to see if you have a special product: a perfect square trinomial. If you don't, you can proceed to *unFOIL*. The acronym FOIL helps you multiply two binomials (First, Outer, Inner, Last); unFOIL helps you factor the product of those binomials.

Finding perfect square trinomials

A *perfect square trinomial* is an expression of three terms that results from the squaring of a binomial — multiplying it times itself. Perfect square trinomials are fairly easy to spot — their first and last terms are perfect squares, and the middle term is twice the product of the roots of the first and last terms:

$$a^2 + 2ab + b^2 = (a + b)^2$$
$$a^2 - 2ab + b^2 = (a - b)^2$$

To factor $x^2 - 20x + 100$, for example, you should first recognize that $20x$ is twice the product of the root of x^2 and the root of 100; therefore, the factorization is $(x - 10)^2$. An expression that isn't quite as obvious is $25y^2 + 30y + 9$. You can see that the first and last terms are perfect squares. The root of $25y^2$ is $5y$, and the root of 9 is 3. The middle term, $30y$, is twice the product of $5y$ and 3, so you have a perfect square trinomial that factors into $(5y + 3)^2$.

Resorting to unFOIL

When you factor a trinomial that results from multiplying two binomials, you have to play detective and piece together the parts of the puzzle. Look at the following generalized product of binomials and the pattern that appears:

$$(ax + b)(cx + d) = acx^2 + adx + bcx + bd = acx^2 + (ad + bc)x + bd$$

So, where does FOIL come in? You need to FOIL before you can unFOIL, don't ya think?

The F in FOIL stands for "First." In the previous problem, the First terms are the ax and cx. You multiply these terms together to get acx^2. The Outer terms are ax and d. Yes, you already used the ax, but each of the terms will have two different names. The Inner terms are b and cx; the Outer and Inner products are, respectively, adx and bcx. You add these two values. (Don't worry; when you're working with numbers, they combine nicely.) The Last terms, b and d, have a product of bd. Here's an actual example that uses FOIL to multiply — working with numbers for the coefficients rather than letters:

$$(4x + 3)(5x - 2) = 20x^2 - 8x + 15x - 6 = 20x^2 + 7x - 6$$

Now, think of every quadratic trinomial as being of the form $acx^2 + (ad + bc)x + bd$. The coefficient of the x^2 term, ac, is the product of the coefficients of the two x terms in the parenthesis; the last term, bd, is the product of the two second terms in the parenthesis; and the coefficient of the middle term is the sum of the outer and inner products. To factor these trinomials into the product of two binomials, you have to use the opposite of the FOIL.

Here are the basic steps you take to unFOIL a trinomial:

1. Determine all the ways you can multiply two numbers to get ac, the coefficient of the squared term.

2. Determine all the ways you can multiply two numbers to get bd, the constant term.

3. If the last term is positive, find the combination of factors from Steps 1 and 2 whose *sum* is that middle term; if the last term is negative, you want the combination of factors to be a difference.

4. Arrange your choices as binomials so that the factors line up correctly.

5. Insert the + and – signs to finish off the factoring and make the sign of the middle term come out right.

Arranging the factors in the binomials provides no provisions for positive or negative signs in the unFOIL pattern — you account for the sign part differently. The possible arrangements of signs are shown in the sections that follow. (For a more thorough explanation of FOILing and unFOILing, check out *Algebra For Dummies* [Wiley].)

UnFOILing + +

One of the arrangements of signs you see when factoring trinomials has all the terms separated by positive (+) signs.

Because the last term in the example trinomial, bd, is positive, the two binomials will contain the same operation — the product of two positives is positive, and the product of two negatives is positive.

To factor $x^2 + 9x + 20$, for example, you need to find two terms whose product is 20 and whose sum is 9. The coefficient of the squared term is 1, so you don't have to take any other factors into consideration. You can produce the number 20 with $1 \cdot 20$, $2 \cdot 10$, or $4 \cdot 5$. The last pair is your choice, because $4 + 5 = 9$. Arranging the factors and x's into two binomials, you get $x^2 + 9x + 20 = (x + 4)(x + 5)$.

UnFOILing – +

A second arrangement in a trinomial has a subtraction operation or negative sign in front of the middle term and a positive last term. The two binomials in the factorization of such a trinomial each have subtraction as their operation.

The key you're looking for is the sum of the Outer and Inner products, because the signs need to be the same.

Say that you want to factor the trinomial $3x^2 - 25x + 8$, for example. You start by looking at the factors of 3; you find only one, $1 \cdot 3$. You also look at the factors of 8, which are $1 \cdot 8$ or $2 \cdot 4$. Your only choice for the first terms in the binomials is $(1x \quad)(3x \quad)$. Now you pick either the 1 and 8 or the 2 and 4 so that, when you place the numbers in the second positions in the binomials, the Outer and Inner products have a sum of 25. Using the 1 and 8, you let $3x$ multiply the 8 and $1x$ multiply the 1 — giving you your sum of 25. So, $3x^2 - 25x + 8 = (x - 8)(3x - 1)$. You don't need to write the coefficient 1 on the first x — the 1 is understood.

UnFOILing + – or – –

When the last term in a trinomial is negative, you need to look for a difference between the products. When factoring $x^2 + 2x - 24$ or $6x^2 - x - 12$, for example, the operations in the two binomials have to be one positive and the other negative. Having opposite signs is what creates a negative last term.

To factor $x^2 + 2x - 24$, you need two numbers whose product is 24 and whose difference is 2. The factors of 24 are $1 \cdot 24$, $2 \cdot 12$, $3 \cdot 8$, or $4 \cdot 6$. The first term has a coefficient of 1, so you can concentrate only on the factors of 24. The pair you want is $4 \cdot 6$. Write the binomials with the x's and the 4 and 6; you can wait until the end of the process to put the signs in. You decide that $(x \quad 4)(x \quad 6)$ is the arrangement. You want the difference between the Outer and Inner products to be positive, so let the 6 be positive and the 4 be negative. Writing out the factorization, you have $x^2 + 2x - 24 = (x - 4)(x + 6)$.

The factorization of $6x^2 - x - 12$ is a little more challenging because you have to consider both the factors of 6 and the factors of 12. The factors of 6 are $1 \cdot 6$ or $2 \cdot 3$, and the factors of 12 are $1 \cdot 12$, $2 \cdot 6$, or $3 \cdot 4$. As wizardlike as I may seem, I can't give you a magic way to choose the best combination. It takes practice and luck. But, if you write down all the possible choices, you can scratch them off as you determine which ones don't work. You may start with the factor 2 and 3 for the 6. The binomials are $(2x\ )(3x\ )$. Don't insert any signs until the end of the process. Now, using the factors of 12, you look for a pairing that gives you a difference of 1 between the Outer and Inner products. Try the product of $3 \cdot 4$, matching the 3 with the $3x$ and the 4 with the $2x$. Bingo! You have it. You want $(2x\ 3)(3x\ 4)$. You will multiply the 3 and $3x$ because they're in different parentheses — not the same one. The difference has to be negative, so you can put the negative sign in front of the 3 in the first binomial: $6x^2 - x - 12 = (2x - 3)(3x + 4)$.

Factoring four or more terms by grouping

When four or more terms come together to form an expression, you have bigger challenges in the factoring. As with an expression with fewer terms, you always look for a greatest common factor first. If you can't find a factor common to all the terms at the same time, your other option is *grouping*. To group, you take the terms two at a time and look for common factors for each of the pairs on an individual basis. After factoring, you see if the new groupings have a common factor. The best way to explain this is to demonstrate the factoring by grouping on $x^3 - 4x^2 + 3x - 12$ and then on $xy^2 - 2y^2 - 5xy + 10y - 6x + 12$.

The four terms $x^3 - 4x^2 + 3x - 12$ don't have any common factor. However, the first two terms have a common factor of x^2, and the last two terms have a common factor of 3:

$$x^3 - 4x^2 + 3x - 12 = x^2(x - 4) + 3(x - 4)$$

Notice that you now have two terms, not four, and they both have the factor $(x - 4)$. Now, factoring $(x - 4)$ out of each term, you have $(x - 4)(x^2 + 3)$.

Factoring by grouping only works if a new common factor appears — the exact same one in each term.

The six terms $xy^2 - 2y^2 - 5xy + 10y - 6x + 12$ don't have a common factor, but, taking them two at a time, you can pull out the factors y^2, $-5y$, and -6. Factoring by grouping, you get the following:

$$xy^2 - 2y^2 - 5xy + 10y - 6x + 12 = y^2(x - 2) - 5y(x - 2) - 6(x - 2)$$

The three new terms have a common factor of $(x - 2)$, so the factorization becomes $(x - 2)(y^2 - 5y - 6)$. The trinomial that you create lends itself to the unFOIL factoring method (see the previous section):

$$(x - 2)(y^2 - 5y - 6) = (x - 2)(y - 6)(y + 1)$$

Factored, and ready to go!

Chapter 2

Toeing the Straight Line: Linear Equations

In This Chapter

▶ Isolating values of x in linear equations

▶ Comparing variable values with inequalities

▶ Assessing absolute value in equations and inequalities

The term *linear* has the word *line* buried in it, and the obvious connection is that you can graph many linear equations as lines. But linear expressions can come in many types of packages, not just equations or lines. Add an interesting operation or two, put several first-degree terms together, throw in a funny connective, and you can construct all sorts of creative mathematical challenges. In this chapter, you find out how to deal with linear equations, what to do with the answers in linear inequalities, and how to rewrite linear absolute-value equations and inequalities so that you can solve them.

Linear Equations: Handling the First Degree

Linear equations feature variables that reach only the first degree, meaning that the highest power of any variable you solve for is one. The general form of a linear equation with one variable is

$$ax + b = c$$

The one variable is the x. (If you go to Chapter 12, you can see linear equations with two or three variables.) But, no matter how many variables you see, the common theme to linear equations is that each variable has only one solution or value that works in the equation.

The graph of the single solution, if you really want to graph it, is one point on the number line — the answer to the equation. When you up the ante to two variables in a linear equation, the graph of all the solutions (there are infinitely many) is a straight line. Any point on the line is a solution. Three variables means you have a plane — a flat surface.

Generally, algebra uses the letters at the end of the alphabet for variables; the letters at the beginning of the alphabet are reserved for coefficients and constants.

Tackling basic linear equations

To solve a linear equation, you isolate the variable on one side of the equation by adding the same number to both sides — or you can subtract, multiply, or divide the same number on both sides.

For example, you solve the equation $4x - 7 = 21$ by adding 7 to each side of the equation, to isolate the variable and the multiplier, and then dividing each side by 4, to leave the variable on its own:

$$4x - 7 + 7 = 21 + 7 \rightarrow 4x = 28$$

$$4x \div 4 = 28 \div 4 \rightarrow x = 7$$

When a linear equation has grouping symbols such as parentheses, brackets, or braces, you deal with any distributing across and simplifying within the grouping symbols before you isolate the variable. For instance, to solve the equation $5x - [3(x + 2) - 4(5 - 2x) + 6] = 20$, you first distribute the 3 and –4 inside the brackets:

$$5x - \left[3(x + 2) - 4(5 - 2x) + 6\right] = 20$$
$$5x - \left[3x + 6 - 20 + 8x + 6\right] = 20$$

You then combine the terms that combine and distribute the negative sign (–) in front of the bracket; it's like multiplying through by –1:

$$5x - \left[11x - 8\right] = 20$$
$$5x - 11x + 8 = 20$$

Simplify again, and you can solve for *x:*

$$-6x + 8 = 20$$
$$-6x = 12$$
$$x = -2$$

When distributing a number or negative sign over terms within a grouping symbol, make sure you multiply *every* term by that value or sign. If you don't multiply each and every term, the new expression won't be equivalent to the original.

To check your answer from the previous example problem, replace every x in the original equation with –2. If you do so, you get a true statement. In this case, you get 20 = 20. The solution –2 is the only answer that works — focusing your work on just one answer is what's nice about linear equations.

Clearing out fractions

The problem with fractions, like cats, is that they aren't particularly easy to deal with. They always insist on having their own way — in the form of common denominators before you can add or subtract. And division? Don't get me started!

Seriously, though, the best way to deal with linear equations that involve variables tangled up with fractions is to get rid of the fractions. Your game plan is to multiply both sides of the equation by the least common denominator of all the fractions in the equation.

To solve $\frac{x+2}{5} + \frac{4x+2}{7} = \frac{9-x}{2}$, for example, you multiply each term in the equation by 70 — the least common denominator (also known as the *least common multiple*) for fractions with the denominators 5, 7, and 2:

$$^{14}\cancel{70}\left(\frac{x+2}{\cancel{5}_1}\right) + {}^{10}\cancel{70}\left(\frac{4x+2}{\cancel{7}_1}\right) = {}^{35}\cancel{70}\left(\frac{9-x}{\cancel{2}_1}\right)$$

Now you distribute the reduced numbers over each parenthesis, combine the like terms, and solve for x:

$$14(x+2) + 10(4x+2) = 35(9-x)$$
$$14x + 28 + 40x + 20 = 315 - 35x$$
$$54x + 48 = 315 - 35x$$
$$89x = 267$$
$$x = 3$$

Extraneous (false) solutions can occur when you alter the original format of an equation. When working with fractions and changing the form of an equation to a more easily solved form, always check your answer in the original equation. For the previous example problem, you insert $x = 3$ into $\frac{x+2}{5} + \frac{4x+2}{7} = \frac{9-x}{2}$ and get $3 = 3$.

Isolating different unknowns

When you see only one variable in an equation, you have a pretty clear idea what you're solving for. When you have an equation like $4x + 2 = 11$ or $5(3z - 11) + 4z = 15(8 + z)$, you identify the one variable and start solving for it.

Life isn't always as easy as one-variable equations, however. Being able to solve an equation for some variable when it contains more than one unknown can be helpful in many situations. If you're repeating a task over and over — such as trying different widths of gardens or diameters of pools to find the best size — you can solve for one of the variables in the equation in terms of the others.

The equation $A = \frac{1}{2}h(b_1 + b_2)$, for example, is the formula you use to find the area of a trapezoid. The letter A represents area, h stands for height (the distance between the two parallel bases), and the two b's are the two parallel sides called the *bases* of the trapezoid.

If you want to construct a trapezoid that has a set area, you need to figure out what dimensions give you that area. You'll find it easier to do the many computations if you solve for one of the components of the formula first — for h, b_1, or b_2.

To solve for h in terms of the rest of the unknowns or letters, you multiply each side by two, which clears out the fraction, and then divide by the entire expression in the parenthesis:

$$A = \frac{1}{2}h(b_1 + b_2)$$
$$2A = \cancel{2} \cdot \frac{1}{\cancel{2}}h(b_1 + b_2)$$
$$\frac{2A}{(b_1 + b_2)} = \frac{h\cancel{(b_1 + b_2)}}{\cancel{(b_1 + b_2)}}$$
$$\frac{2A}{(b_1 + b_2)} = h$$

Paying off your mortgage with algebra

A few years ago, one of my mathematically challenged friends asked me if I could help her figure out what would happen to her house payments if she paid $100 more each month on her mortgage. She knew that she'd pay off her house faster, and she'd pay less in interest. But how long would it take and how much would she save? I created a spreadsheet and used the formula for an amortized loan (mortgage). I made different columns showing the principal balance that remained (solved for P) and the amount of the payment going toward interest (solved for the difference), and I extended the spreadsheet down for the number of months of the loan. We put the different payment amounts into the original formula to see how they changed the total number of payments and the total amount paid. She was amazed. I was even amazed! She's paying off her mortgage much sooner than expected!

You can also solve for b_2, the measure of the longer base of the trapezoid. To do so, you multiply each side of the equation by two, divide each side by h, and then subtract b_1 from each side:

$$A = \frac{1}{2}h(b_1 + b_2)$$

$$2A = \cancel{2} \cdot \frac{1}{\cancel{2}}h(b_1 + b_2)$$

$$\frac{2A}{h} = \frac{\cancel{h}(b_1 + b_2)}{\cancel{h}}$$

$$\frac{2A}{h} = b_1 + b_2$$

$$\frac{2A}{h} - b_1 = b_2$$

You can leave the equation in that form, with two terms, or you can find a common denominator and combine the terms on the left:

$$\frac{2A - b_1 h}{h} = b_2$$

When you rewrite a formula aimed at solving for a particular unknown, you can put the formula into a graphing calculator or spreadsheet to do some investigating into how changes in the individual values change the variable that you solve for (see a spreadsheet example of this in the "Paying off your mortgage with algebra" sidebar).

Linear Inequalities: Algebraic Relationship Therapy

Equations — statements with equal signs — are one type of relationship or comparison between things; they say that terms, expressions, or other entities are exactly the same. An inequality is a bit less precise. *Algebraic inequalities* show relationships between a number and an expression or between two expressions. In other words, you use inequalities for comparisons.

Inequalities in algebra are *less than* (<), *greater than* (>), *less than or equal to* (≤), and *greater than or equal to* (≥). A linear equation has only one solution, but a linear inequality has an infinite number of solutions. When you write $x \le 7$, for example, you can replace x with 6, 5, 4, –3, –100, and so on, including all the fractions that fall between the integers that work in the inequality.

Here are the rules for operating on inequalities (you can replace the < symbol with any of the inequality symbols, and the rule will still hold):

- ✔ If $a < b$, $a + c < b + c$ (adding any number).

- ✔ If $a < b$, $a - c < b - c$ (subtracting any number).

- ✔ If $a < b$, $a \cdot c < b \cdot c$ (multiplying by any *positive* number).

- ✔ If $a < b$, $a \cdot c > b \cdot c$ (multiplying by any *negative* number).

- ✔ If $a < b$, $\dfrac{a}{c} < \dfrac{b}{c}$ (dividing by any *positive* number).

- ✔ If $a < b$, $\dfrac{a}{c} > \dfrac{b}{c}$ (dividing by any *negative* number).

- ✔ If $\dfrac{a}{c} < \dfrac{b}{d}$, $\dfrac{c}{a} > \dfrac{d}{b}$ (reciprocating fractions).

You must not multiply or divide each side of an inequality by zero. If you do so, you create an incorrect statement. Multiplying each side of 3 < 4 by 0, you get 0 < 0, which is clearly a false statement. You can't divide each side by 0, because you can never divide anything by 0 — no such number with 0 in the denominator exists.

Solving basic inequalities

To solve a basic inequality, you first move all the variable terms to one side of the inequality and the numbers to the other. After you simplify the inequality down to a variable and a number, you can find out what values of the variable will make the inequality into a true statement. For example, to solve $3x + 4 > 11 - 4x$, you add $4x$ to each side and subtract 4 from each side. The inequality sign stays the same because no multiplication or division by negative numbers is involved. Now you have $7x > 7$. Dividing each side by 7 also leaves the *sense* (direction of the inequality) untouched because 7 is a positive number. Your final solution is $x > 1$. The answer says that any number larger than one can replace the x's in the original inequality and make the inequality into a true statement.

The rules for solving linear equations (see the section "Linear Equations: Handling the First Degree") also work with inequalities — somewhat. Everything goes smoothly until you try to multiply or divide each side of an inequality by a negative number.

When you multiply or divide each side of an inequality by a negative number, you have to *reverse the sense* (change < to >, or vice versa) to keep the inequality true.

The inequality $4(x - 3) - 2 \geq 3(2x + 1) + 7$, for example, has grouping symbols that you have to deal with. Distribute the 4 and 3 through their respective multipliers to make the inequality into $4x - 12 - 2 \geq 6x + 3 + 7$. Simplify the terms on each side to get $4x - 14 \geq 6x + 10$. Now you put your inequality skills to work. Subtract $6x$ from each side and add 14 to each side; the inequality becomes $-2x \geq 24$. When you divide each side by -2, you have to reverse the sense; you get the answer $x \leq -12$. Only numbers smaller than -12 or exactly equal to -12 work in the original inequality.

When solving the previous example, you have two choices when you get to the step $4x - 14 \geq 6x + 10$, based on the fact that the inequality $a < b$ is equivalent to $b > a$. If you subtract $6x$ from both sides, you end up dividing by a negative number. If you move the variables to the right and the numbers to the left, you don't have to divide by a negative number, but the answer looks a bit different. If you subtract $4x$ from each side and subtract 10 from each side, you get $-24 \geq 2x$. When you divide each side by 2, you don't change the sense, and you get $-12 \geq x$. You read the answer as "-12 is greater than or equal to x." This inequality has the same solutions as $x \leq -12$, but stating the inequality with the number coming first is a bit more awkward.

Introducing interval notation

You can alleviate the awkwardness of writing answers with inequality notation by using another format called *interval notation*. You use interval notation extensively in calculus, where you're constantly looking at different intervals involving the same function. Much of higher mathematics uses interval notation, although I really suspect that book publishers pushed its use because it's quicker and neater than inequality notation. Interval notation uses parentheses, brackets, commas, and the infinity symbol to bring clarity to the murky inequality waters.

And, surprise surprise, the interval-notation system has some rules:

✔ You order any numbers used in the notation with the smaller number to the left of the larger number.

✔ You indicate "or equal to" by using a bracket.

✔ If the solution doesn't include the end number, you use a parenthesis.

✔ When the interval doesn't end (it goes up to positive infinity or down to negative infinity), use $+\infty$ or $-\infty$, whichever is appropriate, and a parenthesis.

Here are some examples of inequality notation and the corresponding interval notation:

$$x < 3 \rightarrow (-\infty, 3)$$
$$x \geq -2 \rightarrow [-2, \infty)$$
$$4 \leq x < 9 \rightarrow [4, 9)$$
$$-3 < x < 7 \rightarrow (-3, 7)$$

Notice that the second example has a bracket by the –2, because the "greater than or equal to" indicates that you include the –2, also. The same is true of the 4 in the third example. The last example shows you why interval notation can be a problem at times. Taken out of context, how do you know if (–3, 7) represents the interval containing all the numbers between –3 and 7 or if it represents the point (–3, 7) on the coordinate plane? You can't tell. A problem containing such notation has to give you some sort of hint.

Compounding inequality issues

A *compound inequality* is an inequality with more than one comparison or inequality symbol — for instance, $-2 < x \leq 5$. To solve compound inequalities for the value of the variables, you use the same inequality rules (see the intro to this section), and you expand the rules to apply to each section (intervals separated by inequality symbols).

To solve the inequality $-8 \leq 3x - 5 < 10$, for example, you add 5 to each of the three sections and then divide each section by 3:

$$-8 \leq 3x - 5 < 10$$
$$\underline{+5 \qquad\quad +5 \quad +5}$$
$$-3 \leq 3x \qquad < 15$$
$$\frac{-3}{3} \leq \frac{3x}{3} \qquad < \frac{15}{3}$$
$$-1 \leq x \qquad < 5$$

Ancient symbols for timeless operations

Many ancient cultures used their own symbols for mathematical operations, and the cultures that followed altered or modernized the symbols for their own use. You can see one of the first symbols used for addition in the following figure, located on the far left — a version of the Italian capital P for the word *piu*, meaning *plus*. Tartaglia, a self-taught 16th century Italian mathematician, used this symbol for addition regularly. The modern plus symbol, +, is probably a shortened form of the Latin word *et*, meaning *and*.

The second figure from the left is what Greek mathematician Diophantes liked to use in ancient

Greek times for subtraction. The modern subtraction symbol, −, may be a leftover from what the traders in medieval times used to indicate differences in product weights.

Leibniz, a child prodigy from the 17th century who taught himself Latin, preferred the third symbol from the left for multiplication. One modern multiplication symbol, x or ·, is based on St. Andrew's Cross, but Leibniz used the open circle because he thought that the modern symbol looked too much like the unknown *x*.

The symbol on the far right is a somewhat backward D, used in the 18th century by French mathematician Gallimard for division. The modern division symbol, ÷, may come from a fraction line with dots added above and below.

You write the answer, $-1 \le x < 5$, in interval notation as $[-1, 5)$.

Here's a more complicated example. You solve the problem $-1 < 5 - 2x \le 7$ by subtracting 5 from each section and then dividing each section by −2. Of course, dividing by a negative means that you turn the senses around:

$$-1 < 5 - 2x \le 7$$
$$\underline{-5 \quad -5 \qquad -5}$$
$$-6 < \quad -2x \le 2$$
$$\frac{-6}{-2} > \quad \frac{-2x}{-2} \ge \frac{2}{-2}$$
$$3 > x \qquad \ge -1$$

You write the answer, $3 > x \ge -1$, backward as far as the order of the numbers on the number line; the number −1 is smaller than 3. To flip the inequality in the opposite direction, you reverse the inequalities, too: $-1 \le x < 3$. In interval notation, you write the answer as $[-1, 3)$.

Absolute Value: Keeping Everything in Line

When you perform an *absolute value operation,* you're not performing surgery at bargain-basement prices; you're taking a number inserted between the absolute value bars, $|a|$, and recording the distance of that number from zero on the number line. For instance $|3| = 3$, because 3 is three units away from zero. On the other hand, $|-4| = 4$, because -4 is four units away from zero.

The absolute value of a is defined as $|a| = \begin{cases} a \text{ if } a \geq 0 \\ -a \text{ if } a < 0 \end{cases}$.

You read the definition as follows: "The absolute value of a is equal to a, itself, if a is positive or zero; the absolute value of a is equal to the *opposite* of a if a is negative."

Solving absolute-value equations

A linear absolute value equation is an equation that takes the form $|ax + b| = c$. You don't know, taking the equation at face value, if you should change what's in between the bars to its opposite, because you don't know if the expression is positive or negative. The sign of the expression inside the absolute value bars all depends on the sign of the variable x. To solve an absolute value equation in this linear form, you have to consider both possibilities: $ax + b$ may be positive, or it may be negative.

To solve for the variable x in $|ax + b| = c$, you solve both $ax + b = c$ and $ax + b = -c$.

For example, to solve the absolute value equation $|4x + 5| = 13$, you write the two linear equations and solve each for x:

$$4x + 5 = 13 \qquad 4x + 5 = -13$$
$$4x = 8 \qquad\quad 4x = -18$$
$$x = 2 \qquad\qquad x = -\frac{18}{4} = -\frac{9}{2}$$

You have two solutions: 2 and $-\frac{9}{2}$. Both solutions work when you replace the x in the original equation with their values.

One restriction you should be aware of when applying the rule for changing from absolute value to individual linear equations is that the absolute value term has to be alone on one side of the equation.

Cracking the ISBN check code

Have you ever noticed the bar codes and ISBN numbers that appear on the backs of the books you buy (or, ahem, borrow from friends)? Actually, the International Standard Book Number (ISBN) has been around for only about 30 years. The individual numbers tell those in the know what the number as a whole means: the language the book is printed in, who the publisher is, and what specific number was assigned to that particular book. You can imagine how easy it is to miscopy this long string of numbers — or you can try it with this book's ISBN if you're not into imagination. If you write them down, you could reverse a pair of numbers, skip a number, or just write the number down wrong. For this reason, publishers assign a *check digit* for the ISBN number — the last digit. UPC codes and bank checks have the same feature: a check digit to try to help catch most errors.

To form the check digit on ISBN numbers, you take the first digit of the ISBN number and multiply it by 10, the second by 9, the third by 8, and so on until you multiply the last digit by 2. (Don't do anything with the check digit.) You then add up all the products and change the sum to its opposite — you should now have a negative number. Next, you add 11 to the negative number, and add 11 again, and again, and again, until you finally get a positive number. That number should be the same as the check digit.

For instance, the ISBN number for *Algebra For Dummies* (Wiley), my original masterpiece, is 0-7645-5325-9. Here's the sum you get by performing all the multiplication: $10(0) + 9(7) + 8(6) + 7(4) + 6(5) + 5(5) + 4(3) + 3(2) + 2(5) = 222$.

You change 222 to its opposite, –222. Add 11 to get –211; add 11 again to get –200; add 11 again, and again. Actually, you add the number 21 times — $11(21) = 231$. So, the first positive number you come to after repeatedly adding 11s is 9. That's the check digit! Because the check digit is the same as the number you get by using the process, you wrote down the number correctly. Of course, this checking method isn't foolproof. You could make an error that gives you the same check digit, but this method usually finds most of the errors.

For instance, to solve $3|4 - 3x| + 7 = 25$, you have to subtract 7 from each side of the equation and then divide each side by 3:

$$3|4 - 3x| + 7 = 25$$
$$3|4 - 3x| = 18$$
$$|4 - 3x| = 6$$

Now you can write the two linear equations and solve them for x:

$$4 - 3x = 6 \qquad 4 - 3x = -6$$
$$-3x = 2 \qquad -3x = -10$$
$$x = -\frac{2}{3} \qquad x = \frac{10}{3}$$

Seeing through absolute-value inequality

An absolute value inequality contains an absolute value — |a| — and an inequality — <, >, ≤, or ≥. Surprise, surprise! We're talking algebra, not rocket science.

To solve an absolute value inequality, you have to change the form from absolute value to just plain inequality. The way to handle the change from absolute value notation to inequality notation depends on which direction the inequality points with respect to the absolute-value term. The methods, depending on the direction, are quite different:

- ✔ To solve for x in $|ax + b| < c$, you solve $-c < ax + b < c$.
- ✔ To solve for x in $|ax + b| > c$, you solve $ax + b > c$ and $ax + b < -c$.

The first change sandwiches the $ax + b$ between c and its opposite. The second change examines values greater than c (toward positive infinity) and smaller than $-c$ (toward negative infinity).

Sandwiching the values in inequalities

You apply the first rule of solving absolute-value inequalities to the inequality $|2x - 1| \leq 5$, because of the less-than direction of the inequality. You rewrite the inequality, using the rule for changing the format: $-5 \leq 2x - 1 \leq 5$. Next, you add one to each section to isolate the variable; you get the inequality $-4 \leq 2x \leq 6$. Divide each section by two to get $-2 \leq x \leq 3$. You can write the solution in interval notation as $[-2, 3]$.

Be sure that the absolute-value inequality is in the correct format before you apply the rule. The absolute-value portion should be alone on its side of the inequality sign. If you have $2|3x + 5| -7 < 11$, for example, you need to add 7 to each side and divide each side by 2 before changing the form:

$$2|3x + 5| - 7 < 11$$
$$2|3x + 5| < 18$$
$$|3x + 5| < 9$$
$$-9 < 3x + 5 < 9$$
$$\underline{-5 \qquad -5 \ -5}$$
$$-14 < 3x \qquad < 4$$
$$\frac{-14}{3} < \frac{3x}{3} \qquad < \frac{4}{3}$$
$$\frac{-14}{3} < x \qquad < \frac{4}{3}$$

Use interval notation to write the solution as $\left(-\frac{14}{3}, \frac{4}{3}\right)$.

Harnessing inequalities moving in opposite directions

An absolute-value inequality with a greater-than sign, such as $|7 - 2x| > 11$, has solutions that go infinitely high to the right and infinitely low to the left on the number line. To solve for the values that work, you rewrite the absolute value, using the rule for greater-than inequalities; you get two completely separate inequalities to solve. The solutions relate to the inequality $7 - 2x > 11$ or to the inequality $7 - 2x < -11$. Notice that when the sign of the value 11 changes from positive to negative, the inequality symbol switches direction.

When solving the two inequalities, be sure to remember to switch the sign when you divide by -2:

$$7 - 2x > 11 \qquad 7 - 2x < -11$$
$$-2x > 4 \qquad\quad -2x < -18$$
$$x < -2 \qquad\qquad x > 9$$

The solution $x < -2$ or $x > 9$, in interval notation, is $(-\infty, -2)$ or $(9, \infty)$.

Don't write the solution $x < -2$ or $x > 9$ as $9 < x < -2$. If you do, you indicate that some numbers can be bigger than 9 *and* smaller than -2 at the same time. It just isn't so.

Exposing an impossible inequality imposter

The rules for solving absolute-value inequalities are relatively straightforward. You change the format of the inequality and solve for the values of the variable that work in the problem. Sometimes, however, amid the flurry of following the rules, an impossible situation works its way in to try to catch you off guard.

For example, say you have to solve the absolute-value inequality $2|3x - 7| + 8 < 6$. It doesn't look like such a big deal; you just subtract 8 from each side and then divide each side by 2. The dividing value is positive, so you don't reverse the sense. After performing the initial steps, you use the rule where you change from an absolute-value inequality to an inequality with the variable term sandwiched between inequalities. So, what's wrong with that? Here are the steps:

$$2|3x - 7| + 8 < 6$$
$$\underline{\qquad -8 \quad -8 \qquad}$$
$$2|3x - 7| \qquad < -2$$
$$|3x - 7| \qquad < -1$$

Under the format $-c < ax + b < c$, the inequality looks curious. Do you sandwich the variable term between –1 and 1 or between 1 and –1 (the first number on the left, and the second number on the right)? It turns out that neither works. First of all, you can throw out the option of writing $1 < 3x - 7 < -1$. Nothing is bigger than 1 and smaller than –1 at the same time. The other version seems, at first, to have possibilities:

$$-1 < 3x - 7 < 1$$
$$\underline{+7 \qquad\quad +7 \ +7}$$
$$6 < 3x \qquad\ < 8$$
$$2 < x \qquad\quad < \frac{8}{3}$$

The solution says that x is a number between 2 and $2\frac{2}{3}$. If you check the solution by trying a number — say, 2.1 — in the original inequality, you get the following:

$$2\left|3(2.1) - 7\right| + 8 < 6$$
$$2\left|6.3 - 7\right| + 8 < 6$$
$$2\left|-0.7\right| + 8 < 6$$
$$2(0.7) + 8 < 6$$
$$1.4 + 8 < 6$$
$$9.4 < 6$$

Because 9.4 isn't less than 6, you know the number 2.1 doesn't work. You won't find *any* number that works. So, you can't find an answer to this problem. Did you miss a hint of the situation before you dove into all the work? Yes. (Sorry for the tough love!)

You want to save yourself some time and work? You can do that in this case by picking up on the pesky negative number. When you subtract 8 from each side of the original problem and get $2\left|3x - 7\right| < -2$, the bells should be ringing and the lights flashing. This statement says that 2 times the absolute value of a number is smaller than –2, which is impossible. Absolute value is either positive or zero — it can't be negative — so this expression can't be smaller than –2. If you caught the problem before doing all the work, hurrah for you! Good eye. Often, though, you can get caught up in the process and not notice the impossibility until the end — when you check your answer.

Chapter 3

Cracking Quadratic Equations

. . . .

In This Chapter

▶ Rooting and factoring to solve quadratic equations

▶ Breaking down equations with the quadratic formula

▶ Squaring to prepare for conics

▶ Conquering advanced quadratics

▶ Taking on the inequalities challenge

. . . .

*Q*uadratic equations are some of the more common equations you see in the mathematics classroom. A *quadratic equation* contains a term with an exponent of two, and no term with a higher power. The standard form is $ax^2 + bx + c = 0$.

In other words, the equation is a quadratic expression with an equal sign (see Chapter 1 for a short-and-sweet discussion on quadratic expressions). Quadratic equations potentially have two solutions. You may not find two, but you start out assuming that you'll find two and then proceed to prove or disprove your assumption.

Quadratic equations are not only very manageable — because you can always find ways to tackle them — but also serve as good role models, playing parts in many practical applications. If you want to track the height of an arrow you shoot into the air, for example, you can find your answer with a quadratic equation. The area of a circle is technically a quadratic equation. The profit (or loss) from the production and sales of items often follows a quadratic pattern.

In this chapter, you discover many ways to approach both simple and advanced quadratic equations. You can solve some quadratic equations in only one way, and you can solve others by readers' choice — whatever your preference. It's nice to be able to choose. But if you have a choice, I hope you choose the quickest and easiest way possible, so I cover these first in this chapter. But sometimes the quick-and-easy way doesn't work or doesn't come to you as an inspiration. Read on for other options. You also tackle quadratic inequalities in this chapter — less than jaw dropping, but greater than boring!

Solving Simple Quadratics with the Square Root Rule

Some quadratic equations are easier to solve than others; half the battle is recognizing which equations are easy and which are more challenging.

The simplest quadratic equations that you can solve quickly are those that allow you to take the square root of both sides. These lovely equations are made up of a squared term and a number, written in the form $x^2 = k$. You solve equations written this way by using the *square root rule:* If $x^2 = k$, $x = \pm\sqrt{k}$.

Notice that by using the square root rule, you come up with two solutions: both the positive and the negative. When you square a positive number, you get a positive result, and when you square a negative number, you also get a positive result.

The number represented by k has to be positive if you want to find real answers with this rule. If k is negative, you get an imaginary answer, such as $3i$ or $5 - 4i$. (For more on imaginary numbers, check out Chapter 14.)

Finding simple square-root solutions

You can use the square root rule to solve equations that take the simple form $x^2 = k$, such as $x^2 = 25$: $x = \pm\sqrt{25} = \pm5$.

The solution for this problem is simple enough, but there's more to the square root rule than meets the eye. For instance, what if you have a coefficient on the x term? The equation $6x^2 = 96$ doesn't strictly follow the format for the square root rule because of the coefficient 6, but you can get to the proper form pretty quickly. You divide each side of the equation by the coefficient; in this case, you get $x^2 = 16$; now you're in business. Taking the square root of each side, you get $x = \pm 4$.

Dealing with radical square-root solutions

The choice to use the square root rule is pretty obvious when you have an equation with a squared variable and a perfect square number. The decision may seem a bit murkier when the number involved isn't a perfect square. But no need to fret; you can still use the square root rule in these situations.

If you want to solve $y^2 = 40$, for example, you can take the square root of each side and then simplify the radical term:

$$y = \pm\sqrt{40} = \pm\sqrt{4}\sqrt{10} = \pm2\sqrt{10}$$

You use the law of radicals to separate a number under a radical into two factors — one of which is a perfect square: $\sqrt{a \cdot b} = \sqrt{a}\sqrt{b}$.

At this point, you're pretty much done. The number $2\sqrt{10}$ is an exact number or value, and you can't simplify the radical part of the term any more because the value 10 doesn't have any factors that double as perfect squares.

If a number under a square root isn't a perfect square, as with $\sqrt{10}$, you say that radical value is *irrational,* meaning that the decimal value never ends. A radical portion with an *exact value* is simplified as far as it can go.

Depending on the instructions you receive for an exercise, you can leave your answer as a simplified radical value, or you can round your answer to a certain number of decimal points. The decimal value of $\sqrt{10}$ rounded to the first eight decimal places is 3.16227766. You can round to fewer decimal places if you want to (or need to). For instance, rounding 3.16227766 to the first five decimal places gives you 3.16228. You can then estimate $2\sqrt{10}$ to be $2(3.16228) = 6.32456$.

Dismantling Quadratic Equations into Factors

You can *factor* many quadratic expressions — one side of a quadratic equation — by rewriting them as products of two or more numbers, variables, terms in parentheses, and so on. The advantage of the factored form is that you can solve quadratic equations by setting the factored expression equal to zero (making it an equation) and then using the multiplication property of zero (described in detail in Chapter 1). How you factor the expression depends on the number of terms in the quadratic and how those terms are related.

Factoring binomials

You can factor a quadratic binomial (which contains two terms; one of them with a variable raised to the power 2) in one of two ways — if you can factor it at all (you may find no common factor, or the two terms may not both be squares):

✔ Divide out a common factor from each of the terms.

✔ Write the quadratic as the product of two binomials, if the quadratic is the difference of perfect squares.

Taking out a greatest common factor

The *greatest common factor* (GCF) of two or more terms is the largest number (and variable combination) that divides each of the terms evenly. To solve the equation $4x^2 + 8x = 0$, for example, you factor out the greatest common factor, which is $4x$. After dividing, you get $4x(x + 2) = 0$. Using the multiplication property of zero (see Chapter 1), you can now state three facts about this equation:

✔ $4 = 0$, which is false — this isn't a solution

✔ $x = 0$

✔ $x + 2 = 0$, which means that $x = -2$

You find two solutions for the original equation $4x^2 + 8x = 0$: $x = 0$ or $x = -2$. If you replace the x's with either of these solutions, you create a true statement.

Here's another example, with a twist: The quadratic equation $6y^2 + 18 = 0$. You can factor this equation by dividing out the 6 from each term:

$$6y^2 + 18 = 6(y^2 + 3) = 0$$

Unfortunately, this factored form doesn't yield any real solutions for the equation, because it doesn't have any. Applying the multiplication property of zero, you first get $6 = 0$. No help there. Setting $y^2 + 3 = 0$, you can subtract 3 from each side to get $y^2 = -3$. The number -3 isn't positive, so you can't apply the square root rule, because you get $y = \pm\sqrt{-3}$. No real number's square is -3. So, you can't find an answer to this problem among real numbers. (For information on complex or imaginary answer, head to Chapter 14.)

Be careful when the GCF of an expression is just x, and always remember to set that front factor, x, equal to zero so you don't lose one of your solutions. A really common error in algebra is to take a perfectly nice equation such as $x^2 + 5x = 0$, factor it into $x(x + 5) = 0$, and give the answer $x = -5$. For some reason, people often ignore that lonely x in front of the parenthesis. Don't forget the solution $x = 0$!

Factoring the difference of squares

If you run across a binomial that you don't think calls for the application of the square root rule, you can factor the difference of the two squares and solve for the solution by using the multiplication property of zero (see Chapter 1). If any solutions exist, and you can find them with the square root rule, you can also find them by using the difference of two squares method.

This method states that if $x^2 - a^2 = 0$, $(x - a)(x + a) = 0$, and $x = a$ or $x = -a$. Generally, if $k^2x^2 - a^2 = 0$, $(kx - a)(kx + a) = 0$, and $x = \frac{a}{k}$ or $x = -\frac{a}{k}$.

To solve $x^2 - 25 = 0$, for example, you factor the equation into $(x - 5)(x + 5) = 0$, and the rule (or the multiplication property of zero) tells you that $x = 5$ or $x = -5$.

When you have a perfect square multiplier of the variable, you can still factor into the difference and sum of the square roots. To solve $49y^2 - 64 = 0$, for example, you factor the terms on the left into $(7y - 8)(7y + 8) = 0$, and the two solutions are $y = \frac{8}{7}$, $y = -\frac{8}{7}$.

Factoring trinomials

Like quadratic binomials (see the previous pages), a quadratic trinomial can have as many as two solutions — or it may have one solution or no solution at all. If you can factor the trinomial and use the multiplication property of zero (see Chapter 1) to solve for the roots, you're home free. If the trinomial doesn't factor, or if you can't figure out how to factor it, you can utilize the *quadratic formula* (see the section "Resorting to the Quadratic Formula" later in this chapter). The rest of this section deals with the trinomials that you *can* factor.

Finding two solutions in a trinomial

The trinomial $x^2 - 2x - 15 = 0$, for example, has two solutions. You can factor the left side of the equation into $(x - 5)(x + 3) = 0$ and then set each factor equal to zero. When $x - 5 = 0$, $x = 5$, and when $x + 3 = 0$, $x = -3$. (If you can't remember how to factor these trinomials, refer to Chapter 1, or, for even more detail, see *Algebra For Dummies* [Wiley].)

It may not be immediately apparent how you should factor a seemingly complicated trinomial like $24x^2 + 52x - 112 = 0$. Before you bail out and go to the quadratic formula, consider factoring 4 out of each term to simplify the picture a bit; you get $4(6x^2 + 13x - 28) = 0$. The quadratic in the parenthesis factors into the product of two binomials (with a little trial and error and educated guessing), giving you $4(3x - 4)(2x + 7) = 0$. Setting $3x - 4$ equal to 0, you get $x = \frac{4}{3}$, and setting $2x + 7$ equal to 0, you get $x = -\frac{7}{2}$. How about the factor of 4? If you set 4 equal to 0, you get a false statement, which is fine; you already have the two numbers that make the equation a true statement.

Doubling up on a trinomial solution

The equation $x^2 - 12x + 36 = 0$ is a *perfect square trinomial,* which simply means that it's the square of a single binomial. Assigning this equation that special name points out why the two solutions you find are actually just one. Look at the factoring: $x^2 - 12x + 36 = (x - 6)(x - 6) = (x - 6)^2 = 0$. The two different factors give the same solution: $x = 6$. A quadratic trinomial can have as many as two solutions or roots. This trinomial technically does have two roots, 6 and 6. You can say that the equation has a *double root*.

Pay attention to double roots when you're graphing, because they act differently on the axes. This distinction is important when you're graphing any polynomial. Graphs at double roots don't cross the axis — they just touch. You also see these entities when solving inequalities; see the "Solving Quadratic Inequalities" section later in the chapter for more on how they affect those problems. (Chapter 5 gives you some pointers on graphing.)

Factoring by grouping

Factoring by grouping is a great method to use to rewrite a quadratic equation so that you can use the multiplication property of zero (see Chapter 1) and find all the solutions. The main idea behind factoring by grouping is to arrange the terms into smaller groupings that have a common factor. You go to little groupings because you can't find a greatest common factor for all the terms; however, by taking two terms at a time, you can find something to divide them by.

Grouping terms in a quadratic

A quadratic equation such as $2x^2 + 8x - 5x - 20 = 0$ has four terms. Yes, you can combine the two middle terms on the left, but leave them as is for the sake of the grouping process. The four terms in the equation don't share a greatest common factor. You can divide the first, second, and fourth terms evenly by 2, but the third term doesn't comply. The first three terms all have a factor of x, but the last term doesn't. So, you group the first two terms together and take out their common factor, $2x$. The last two terms have a common factor of -5. The factored form, therefore, is $2x(x + 4) - 5(x + 4) = 0$.

The new, factored form has two terms. Each of the terms has an $(x + 4)$ factor, so you can divide that factor out of each term. When you divide the first term, you have $2x$ left. When you divide the second term, you have -5 left. Your new factored form is $(x + 4)(2x - 5) = 0$. Now you can set each factor equal to zero to get $x = -4$ and $x = \frac{5}{2}$.

Factoring by grouping works only when you can create a new form of the quadratic equation that has fewer terms and a common factor. If the factor $(x + 4)$ hadn't shown up in both of the factored terms in the previous example, you would've gone in a different direction.

Finding quadratic factors in a grouping

Solving quadratic equations by grouping and factoring is even more important when the exponents in the equations get larger. The equation $5x^3 + x^2 - 45x - 9 = 0$, for example, is a third-degree equation (the highest power on any of the variables is 3), so it has the potential for three different solutions. You can't find a factor common to all four terms, so you group the first two terms, factor out x^2, group the last two terms, and factor out -9. The factored equation is as follows: $x^2(5x + 1) - 9(5x + 1) = 0$.

The common factor of the two terms in the new equation is $(5x + 1)$, so you divide it out of the two terms to get $(5x + 1)(x^2 - 9) = 0$. The second factor is the difference of squares, so you can rewrite the equation as $(5x + 1)(x - 3)(x + 3) = 0$. The three solutions are $x = -\frac{1}{5}$, $x = 3$, and $x = -3$.

Resorting to the Quadratic Formula

The quadratic formula is a wonderful tool to use when other factoring methods fail (see the previous section) — an algebraic vending machine, of sorts. You take the numbers from a quadratic equation, plug them into the formula, and out come the solutions of the equation. You can even use the formula when the equation does factor, but you don't see how.

The *quadratic formula* states that when you have a quadratic equation in the form $ax^2 + bx + c = 0$ (with the a as the coefficient of the squared term, the b as the coefficient of the first-degree term, and c as the constant), the equation has the solutions

$$x = \frac{-b \pm \sqrt{b^2 - 4ac}}{2a}$$

The process of solving quadratic equations is almost always faster and more accurate if you can factor the equations. The quadratic formula is wonderful, but like a vending machine that eats your quarters, it has some built-in troublesome parts:

- ✔ You have to remember to find the opposite of b.
- ✔ You have to simplify the numbers under the radical correctly.
- ✔ You have to divide the whole equation by the denominator.

Don't get me wrong, you shouldn't hesitate to use the quadratic formula whenever necessary! It's great. But factoring is usually better, faster, and more accurate (to find out when it isn't, check out the section "Formulating huge quadratic results" later in the chapter).

Finding rational solutions

You can factor quadratic equations such as $3x^2 + 11x + 10 = 0$ to find their solutions, but the factorization may not leap right out at you. Using the quadratic formula for this example, you let $a = 3$, $b = 11$, and $c = 10$. Filling in the values and solving for x, you get

$$x = \frac{-11 \pm \sqrt{11^2 - 4(3)(10)}}{2(3)}$$
$$= \frac{-11 \pm \sqrt{121 - 120}}{6}$$
$$= \frac{-11 \pm \sqrt{1}}{6}$$
$$= \frac{-11 \pm 1}{6}$$

You finish up the answers by dealing with the ± symbol one sign at a time. First, you address the +:

$$x = \frac{-11 + 1}{6} = \frac{-10}{6} = -\frac{5}{3}$$

Now you address the –:

$$x = \frac{-11 - 1}{6} = \frac{-12}{6} = -2$$

You find two different solutions. The fact that the solutions are *rational numbers* (numbers that you can write as fractions) tells you that you could've factored the equation. If you end up with a radical in your answer, you know that factorization isn't possible for that equation.

Hints to the actual factorization of the equation $3x^2 + 11x + 10 = 0$ come from the solutions: the 5 divided by the 3 and the –2 by itself help you factor the original equation as $3x^2 + 11x + 10 = (3x + 5)(x + 2) = 0$.

Straightening out irrational solutions

The quadratic formula is especially valuable for solving quadratic equations that don't factor. Unfactorable equations, when they do have solutions, have irrational numbers in their answers. *Irrational numbers* have no fractional equivalent; they feature decimal values that go on forever and never have patterns that repeat.

You have to solve the quadratic equation $2x^2 + 5x - 6 = 0$, for example, with the quadratic formula. Letting $a = 2$, $b = 5$, and $c = -6$, you get

$$x = \frac{-5 \pm \sqrt{5^2 - 4(2)(-6)}}{2(2)}$$
$$= \frac{-5 \pm \sqrt{25 + 48}}{4}$$
$$= \frac{-5 \pm \sqrt{73}}{4}$$

The answer $\frac{-5 + \sqrt{73}}{4}$ is approximately 0.886, and $\frac{-5 - \sqrt{73}}{4}$ is approximately −3.386. You find perfectly good answers, rounded off to the nearest thousandth. The fact that the number under the radical isn't a perfect square tells you something else: You couldn't have factored the quadratic, no matter how hard you tried.

If you get a negative number under the radical when using the quadratic formula, you know that the problem has no real answer. Chapter 14 explains how to deal with these imaginary/complex answers.

Formulating huge quadratic results

Factoring a quadratic equation is almost always preferable to using the quadratic formula. But at certain times, you're better off opting for the quadratic formula, even when you can factor the equation. In cases where the numbers are huge and have many multiplication possibilities, I suggest that you bite the bullet, haul out your calculator, and go for it.

For instance, a great problem in calculus (known as "Finding the largest box that can be formed from a rectangular piece of cardboard" for the curious among you) has an answer that you find when you solve the quadratic equation $48x^2 - 155x + 125 = 0$. The factors of 48 are 1, 2, 3, 4, 6, 8, 12, 16, 24, and 48. You can find that 125 has only four factors: 1, 5, 25, and 125. But, before you continue, imagine having to decide how to line up these numbers or their multiples in parentheses to create the factorization. Instead, using the quadratic formula and a handy-dandy calculator, you find the following:

$$x = \frac{-(-155) \pm \sqrt{(-155)^2 - 4(48)(125)}}{2(48)}$$
$$= \frac{155 \pm \sqrt{24,025 - 24,000}}{96}$$
$$= \frac{155 \pm \sqrt{25}}{96}$$
$$= \frac{155 \pm 5}{96}$$

Starting with the plus sign, you get $\frac{155+5}{96} = \frac{160}{96} = \frac{5}{3}$. For the minus sign, you get $\frac{155-5}{96} = \frac{150}{96} = \frac{25}{16}$. The fact that you get fractions tells you that you could've factored the quadratic: $48x^2 - 155x + 125 = (3x - 5)(16x - 25) = 0$. Do you see where the 3 and 5 and the 16 and 25 come from in the answers?

Completing the Square: Warming Up for Conics

Of all the choices you have for solving a quadratic equation (factoring and the quadratic formula, to name a couple; see the previous sections in this chapter), completing the square should be your last resort. *Completing the square* means to form a perfect square trinomial, which factors into a binomial squared. The binomial-squared format is very nice to have when you're working with conic sections (circles, ellipses, hyperbolas, and parabolas) and writing their standard forms (as you can see in Chapter 11).

Completing the square isn't as quick and easy as factoring, and it's more complicated than the quadratic formula. So, when should you consider it, if you have better choices?

You need to use completing the square when you're told that the method is "good for you" — sort of like bran cereal. But that reason isn't really compelling. The next best reason is to put an equation into a standard form so you can get certain information from it. For instance, using completing the square on the equation of a parabola gives you a visual answer to questions about where the vertex is and how it opens (to the side, up, or down; see Chapter 7). The big payoff is that you have a result for all your work. After all, you *do* get answers to the quadratic equations by using this process.

Completing the square is a great skill to have and will come in handy in later chapters of this book and in other mathematics courses such as analytic geometry and calculus. Plus, it's good for you.

Squaring up to solve a quadratic equation

To solve a quadratic equation — such as $3x^2 + 10x - 8 = 0$ — by completing the square, follow these steps:

1. **Divide every term in the equation by the coefficient of the squared term.**

   For the example problem, you divide each term by the coefficient 3:

   $$3x^2 + 10x - 8 = 0$$

   $$x^2 + \frac{10}{3}x - \frac{8}{3} = 0$$

2. **Move the *constant term* (the term without a variable) to the opposite side of the equation by adding or subtracting.**

   Add $\frac{8}{3}$ to each side:

   $$x^2 + \frac{10}{3}x - \frac{8}{3} + \frac{8}{3} = 0 + \frac{8}{3}$$

   $$x^2 + \frac{10}{3}x = \frac{8}{3}$$

3. **Find half the value of the coefficient on the first-degree term of the variable; square the result of the halving; and add that amount to each side of the equation.**

   Find half of $\frac{10}{3}$, which is $\frac{\overset{5}{\cancel{10}}}{3} \cdot \frac{1}{\underset{1}{\cancel{2}}} = \frac{5}{3}$. Square the fraction, and add the square to each side of the equation:

   $$\left(\frac{5}{3}\right)^2 = \frac{25}{9}$$

   $$x^2 + \frac{10}{3}x + \frac{25}{9} = \frac{8}{3} + \frac{25}{9}$$

   $$= \frac{24}{9} + \frac{25}{9}$$

   $$= \frac{49}{9}$$

4. **Factor the side of the equation that's a perfect square trinomial (you just created it) into the square of a binomial.**

   Factor the left side of the equation:

   $$x^2 + \frac{10}{3}x + \frac{25}{9} = \frac{49}{9}$$

   $$\left(x + \frac{5}{3}\right)^2 = \frac{49}{9}$$

5. **Find the square root of each side of the equation.**

   $$\sqrt{\left(x + \frac{5}{3}\right)^2} = \pm\sqrt{\frac{49}{9}}$$

   $$x + \frac{5}{3} = \pm\frac{7}{3}$$

6. **Isolate the variable term by adding or subtracting to move the constant to the other side.**

   Subtract $\frac{5}{3}$ from each side and solve for the value of x:

   $$x + \frac{5}{3} - \frac{5}{3} = -\frac{5}{3} \pm \frac{7}{3}$$
   $$x = -\frac{5}{3} \pm \frac{7}{3}$$
   $$x = -\frac{5}{3} + \frac{7}{3} = \frac{2}{3} \text{ or } x = -\frac{5}{3} - \frac{7}{3} = -4$$

   You can verify the two example solutions by factoring the original equation: $3x^2 + 10x - 8 = (3x - 2)(x + 4) = 0$.

Completing the square twice over

Completing the square on an equation with both x's and y's leaves you just one step away from what you need to work with conics. *Conic sections* (circles, ellipses, hyperbolas, and parabolas) have standard equations that give you plenty of information about individual curves — where their centers are, which direction they go in, and so on. Chapter 11 covers this information in detail. In the meantime, you practice completing the square twice over in this section.

For example, you can write the equation $x^2 + 6x + 2y^2 - 8y + 13 = 0$ as the sum of two binomials squared and a constant. Think of the equation as having two separate completing the square problems to complete. Follow these steps to give the equation a twice-over:

1. **To handle the two completions more efficiently, rewrite the equation with a space between the x terms and the y terms and with the constant on the other side:**

   $$x^2 + 6x \qquad + 2y^2 - 8y \qquad = -13$$

   You don't divide through by the 2 on the y^2 term because that would leave you with a fractional coefficient on the x^2 term.

2. **Find numerical factors for each grouping — you want the coefficient of the squared term to be one. Write the factor outside a parenthesis with the variables inside.**

   In this case, you factor the 2 out of the two y terms and leave it outside the parenthesis:

   $$x^2 + 6x \qquad + 2(y^2 - 4y \quad) \qquad = -13$$

3. **Complete the square on the *x*'s, and add whatever you used to complete the square to the other side of the equation, too, to keep the equation balanced.**

   Here, you take half of 6, square the 3 to get 9, and then add 9 to each side of the equation:

   $$x^2 + 6x + 9 \qquad + 2(y^2 - 4y \quad) \qquad = -13 + 9$$

4. **Complete the square on the *y*'s, and add whatever you used to complete the square to the other side of the equation, too.**

   If the trinomial is inside a parenthesis, be sure you multiply what you added by the factor outside the parenthesis before adding to the other side.

   Completing the square on the *y*'s means you need to take half the value (–4), square the –2 to get +4, and then add 8 to each side:

   $$x^2 + 6x + 9 \qquad + 2(y^2 - 4y + 4) \qquad = -13 + 9 + 8$$

   Why do you add 8? Because when you put the 4 inside the parenthesis with the *y*'s, you multiply by the 2. To keep the equation balanced, you put 4 inside the parenthesis and 8 on the other side of the equation.

5. **Simplify each side of the equation by writing the trinomials on the left as binomials squared and by combining the terms on the right.**

   For this example, you get $(x + 3)^2 + 2(y - 2)^2 = 4$.

You're done — that is, until you get to Chapter 11, where you find out that you have an ellipse. How's that for a cliffhanger?

Getting Promoted to High-Powered Quadratics (without the Raise)

A *polynomial* is an algebraic expression with one, two, three, or however many terms. The degree (power) of the polynomial is determined by whatever the highest power is that appears in the expression. Polynomials have powers that are whole numbers — no fractions or negatives. Put a polynomial expression up next to "= 0," and you have a polynomial equation.

Solving polynomial equations requires that you know how to count and plan. Okay, so it isn't really *that* simple. But if you can count up to the number that represents the degree (highest power) of the equation, you can account for the solutions you find and determine if you have them all. And if you can make a plan to use patterns in binomials or techniques from quadratics, you're well on your way to a solution.

Handling the sum or difference of cubes

As I explain in Chapter 1, you factor an expression that's the difference between two perfect squares into the difference and sum of the roots, $a^2 - b^2 = (a - b)(a + b)$. If you're solving an equation that represents the difference of two squares, you can apply the multiplication property of zero and solve. However, you can't factor the sum of two squares this way, so you're usually out of luck when it comes to finding any real solution.

In the case of the difference or sum of two cubes, you can factor the binomial, and you do find a solution.

Here are the factorizations of the difference and sum of cubes:

$$a^3 - b^3 = (a - b)(a^2 + ab + b^2)$$
$$a^3 + b^3 = (a + b)(a^2 - ab + b^2)$$

Solving cubes by factoring

If you want to solve a cubed equation such as $x^3 - 64 = 0$ by using the factorization of the difference of cubes, you get $(x - 4)(x^2 + 4x + 16) = 0$. Using the multiplication property of zero (see Chapter 1), when $x - 4$ equals 0, x equals 4. But when $x^2 + 4x + 16 = 0$, you need to use the quadratic formula, and you won't be too happy with the results.

Applying the quadratic formula, you get

$$x = \frac{-4 \pm \sqrt{4^2 - 4(1)(16)}}{2(1)} = \frac{-4 \pm \sqrt{16 - 64}}{2} = \frac{-4 \pm \sqrt{-48}}{2}$$

You have a negative number under the radical, which means you find no real root. (Chapter 14 discusses what to do about negatives under radicals.) You deduce from this that the equation $x^3 - 64 = 0$ has just one solution, $x = 4$.

Don't misinterpret the function of the power in a quadratic equation. In the previous example, the power three suggests that you may find as many as three solutions. But in actuality, the exponent just tells you that you can find *no more than* three solutions.

Solving cubes by taking the cube root

You may wonder if the factorization of the difference or sum of two perfect cubes always results in a quadratic factor that has no real roots (like the example I show in the previous section, "Solving cubes by factoring"). Well, wonder no longer. The answer is a resounding "Yes!" When you factor a difference — $a^3 - b^3 = (a - b)(a^2 + ab + b^2)$ — the quadratic $a^2 + ab + b^2 = 0$ doesn't have a real root; when you factor a sum — $a^3 + b^3 = (a + b)(a^2 - ab + b^2)$ — the equation $a^2 - ab + b^2 = 0$ doesn't have a real root, either.

You can make the best of the fact that you find only one real root for equations in the cube form by deciding how to deal with solving equations of that form. I suggest changing the forms to $x^3 = b^3$ and $x^3 = -b^3$, respectively, and taking the cube root of each side.

To solve $x^3 - 27 = 0$, for example, rewrite it as $x^3 = 27$ and then find the cube root, $\sqrt[3]{x^3} = \sqrt[3]{27}$, $x = 3$. With the equation $8a^3 + 125 = 0$, you first subtract 125 from each side and then divide each side by 8 to get $x^3 = -\dfrac{125}{8}$.

When you take the cube root of a negative number, you get a negative root. A cube root is an odd root, so you can find cube roots of negatives. You can't do roots of negative numbers if the roots are even (square root, fourth root, and so on). For the previous example, you find $\sqrt[3]{x^3} = \sqrt[3]{-\dfrac{125}{8}}$, $x = -\dfrac{5}{2}$.

Tackling quadratic-like trinomials

A *quadratic-like trinomial* is a trinomial of the form $ax^{2n} + bx^n + c = 0$. The power on one variable term is twice that of the other variable term, and a constant term completes the picture. The good thing about quadratic-like trinomials is that they're candidates for factoring and then for the application of the multiplication property of zero (see Chapter 1). One such trinomial is $z^6 - 26z^3 - 27 = 0$.

You can think of this equation as being like the quadratic $x^2 - 26x - 27$, which factors into $(x - 27)(x + 1)$. If you replace the x's in the factorization with z^3, you have the factorization for the equation with the z's. You then set each factor equal to zero:

$$z^6 - 26z^3 - 27 = \left(z^3 - 27\right)\left(z^3 + 1\right) = 0$$
$$z^3 - 27 = 0,\ z^3 = 27,\ z = 3$$
$$z^3 + 1 = 0,\ z^3 = -1,\ z = -1$$

You can just take the cube roots of each side of the equations you form (see the previous section), because when you take that odd root, you know you can find only one solution.

Here's another example. When solving the quadratic-like trinomial $y^4 - 17y^2 + 16 = 0$, you can factor the left side and then factor the factors again:

$$y^4 - 17y^2 + 16 = \left(y^2 - 16\right)\left(y^2 - 1\right)$$
$$= \left(y - 4\right)\left(y + 4\right)\left(y - 1\right)\left(y + 1\right)$$

Setting the individual factors equal to zero, you get $y = 4$, $y = -4$, $y = 1$, $y = -1$.

Solving Quadratic Inequalities

A *quadratic inequality* is just what it says: an inequality ($<$, $>$, $\leq$, or $\geq$) that involves a quadratic expression. You can employ the same method you use to solve a quadratic inequality to solve high-degree inequalities and rational inequalities (which contain variables in fractions).

You need to be able to solve quadratic equations in order to solve quadratic inequalities. With quadratic equations, you set the expressions equal to zero; inequalities deal with what's on either side of the zero (positives and negatives).

To solve a quadratic inequality, follow these steps:

1. **Move all the terms to one side of the inequality sign.**

2. **Factor, if possible.**

3. **Determine all zeros (roots, or solutions).**

   Zeros are the values of x that make each factored expression equal to zero. (Check out "The name game: Solutions, roots, and zeros" sidebar in this chapter for information on these terms.)

4. **Put the zeros in order on a number line.**

5. **Create a sign line to show where the expression in the inequality is positive or negative.**

   A *sign line* shows the signs of the different factors in each interval. If the expression is factored, show the signs of the individual factors.

6. **Determine the solution, writing it in inequality notation or interval notation (I cover interval notation in Chapter 1).**

The name game: Solutions, roots, and zeros

Algebra lets you describe the x-values that you find when an equation is set equal to 0 in several different ways. For instance, when $(x - 3)(x + 4) = 0$, you have two:

✔ **Solutions** to the equation, $x = 3$ and $x = -4$.

✔ **Roots** of the equations, 3 and –4, because they make the equation true.

✔ **Zeros** for the equation (values that make the equation equal to 0) that occur when $x = 3$ and $x = -4$.

✔ **x-intercepts** at (3, 0) and (–4, 0).

The descriptions are often used interchangeably, because you determine these values in exactly the same way.

Keeping it strictly quadratic

The techniques you use to solve the inequalities in this section are also applicable for solving higher degree polynomial inequalities and rational inequalities. If you can factor a third- or fourth-degree polynomial (see the previous section to get started), you can handily solve an inequality where the polynomial is set less than zero or greater than zero. You can also use the sign-line method to look at factors of rational (fractional) expressions. For now, however, consider sticking to the quadratic inequalities.

To solve the inequality $x^2 - x > 12$, for example, you need to determine what values of x you can square so that when you subtract the original number, your answer will be bigger than 12. For instance, when $x = 5$, you get $25 - 5 = 20$. That's certainly bigger than 12, so the number 5 works; $x = 5$ is a solution. How about the number 2? When $x = 2$, you get $4 - 2 = 2$, which isn't bigger than 12. You can't use $x = 2$ in the solution. Do you then conclude that smaller numbers don't work? Not so. When you try $x = -10$, you get $100 + 10 = 110$, which is most definitely bigger than 12. You can actually find an infinite amount of numbers that make this inequality a true statement.

Therefore, you need to solve the inequality by using the steps I outline in the introduction to this section:

1. **Subtract 12 from each side of the inequality $x^2 - x > 12$ to move all the terms to one side.**

   You end up with $x^2 - x - 12 > 0$.

2. **Factoring on the left side of the inequality, you get $(x - 4)(x + 3) > 0$.**

3. **Determine that all the zeroes for the inequality are $x = 4$ and $x = -3$.**

4. **Put the zeros in order on a number line, shown in the following figure.**

5. **Create a sign line to show the signs of the different factors in each interval.**

   Between -3 and 4, try letting $x = 0$ (you can use any number between -3 and 4). When $x = 0$, the factor $(x - 4)$ is negative, and the factor $(x + 3)$ is positive. Put those signs on the sign line to correspond to the factors. Do the same for the interval of numbers to the left of -3 and to the right of 4 (see the following illustration).

$$(x-4)(x+3)$$
$$x = -5$$
$$(-)(-)$$

$$(x-4)(x+3)$$
$$x = 0$$
$$(-)(+)$$

$$(x-4)(x+3)$$
$$x = 10$$
$$(+)(+)$$

REMEMBER

The x values in each interval are really random choices (as you can see from my choice of $x = -5$ and $x = 10$). Any number in each of the intervals gives you the same positive or negative value to the factor.

6. **To determine the solution, look at the signs of the factors; you want the expression to be positive, corresponding to the inequality *greater than zero*.**

The interval to the left of –3 has a negative times a negative, which is positive. So, any number to the left of –3 works. You can write that part of the solution as $x < -3$ or, in interval notation (see Chapter 1), $(-\infty, -3)$. The interval to the right of 4 has a positive times a positive, which is positive. So, $x > 4$ is a solution; you can write it as $(4, \infty)$. The interval between –3 and 4 is always negative; you have a negative times a positive. The complete solution lists both intervals that have working values in the inequality.

The solution of the inequality $x^2 - x > 12$, therefore, is $x < -3$ or $x > 4$.

Signing up for fractions

The sign-line process (see the introduction to this section and the previous example problem) is great for solving rational inequalities, such as $\frac{x-2}{x+6} \le 0$. The signs of the results of multiplication and division use the same rules, so to determine your answer, you can treat the numerator and denominator the same way you treat two different factors in multiplication.

Using the steps from the list I present in the introduction to this section, determine the solution for a rational inequality:

1. Every term in $\frac{x-2}{x+6} \le 0$ is to the left of the inequality sign.

2. Neither the numerator nor the denominator factors any further.

3. The two zeros are $x = 2$ and $x = -6$.

4. You can see the two numbers on a number line in the following illustration.

5. Create a sign line for the two zeros; you can see in the following figure that the numerator is positive when x is greater than 2, and the denominator is positive when x is greater than –6.

6. When determining the solution, keep in mind that the inequality calls for something less than or equal to zero.

   The fraction is a negative number when you choose an x between –6 and 2. You get a negative numerator and a positive denominator, which gives a negative result. Another solution to the original inequality is the number 2. Letting $x = 2$, you get a numerator equal to 0, which you want because the inequality is less than or equal to zero. You can't let the denominator be zero, though. Having a zero in the denominator isn't allowed because no such number exists. So, the solution of $\frac{x-2}{x+6} \le 0$ is $-6 < x \le 2$. In interval notation, you write the solution as (–6, 2]. (For more on interval notation, see Chapter 1.)

Increasing the number of factors

The method you use to solve a quadratic inequality (see the section "Keeping it strictly quadratic") works nicely with fractions and high-degree expressions. For example, you can solve $(x + 2)(x - 4)(x + 7)(x - 5)^2 \ge 0$ by creating a sign line and checking the products.

The inequality is already factored, so you move to the step (Step 3) where you determine the zeros. The zeros are –2, 4, –7, and 5 (the 5 is a double root and the factor is always positive or 0). The following illustration shows the values in order on the number line.

Now you choose a number in each interval, substitute the numbers into the expression on the left of the inequality, and determine the signs of the four factors in those intervals. You can see from the following figure that the last factor, $(x-5)^2$, is always positive or zero, so that's an easy factor to pinpoint.

You want the expression on the left to be positive or zero, given the original language of the inequality. You find an even number of positive factors between −7 and −2 and for numbers greater than 4. You include the zeros, so the solution you find is $-7 \le x \le -2$ or $x \ge 4$. In interval notation, you write the solution as $[-7, -2]$ or $[4, \infty)$. (For more on interval notation, see Chapter 1.)

Chapter 4

Rooting Out the Rational, Radical, and Negative

Solving an algebraic equation requires some know-how. You need the basic mathematical tools, and you need to know what is and isn't allowed. You don't want to take a perfectly good equation and change it into drivel. You need a game plan to solve equations with fractions, radicals, and negative or fractional exponents — one that involves careful planning and a final check of your answers. In this chapter, you find out how to tackle equations by changing them into new equations that are more familiar and easier to solve. You also see a recurring theme of *check your answers,* because changing equations into different forms can introduce mysterious strangers into the mix — in the form of false answers.

Acting Rationally with Fraction-Filled Equations

A *rational* term in an equation is a fraction — an equation with one or more terms, all of which are rational, must be a *rational equation.* You probably hope that all your problems (and the people you associate with) are rational, but an equation that contains fractions isn't always easy to handle.

A general plan for solving a rational equation is to get rid of the fraction or fractions by changing the equation into an equivalent form with the same answer — a form that makes it easier to solve.

Two of the most common ways to get rid of the fractions are multiplying through by the least common denominator (LCD) and cross-multiplying proportions. I just happen to discuss both of these techniques in the sections that follow.

This mathematical slight of hand — using alternate equations to solve more complicated problems — isn't without its potential problems. At times, the new equation produces an *extraneous solution* (also referred to as an *extraneous root*), a false solution that pops up because you messed around with the original format of the equation. To guard against including extraneous solutions in your answers, you need to check the solutions you come up with in the original equations. Don't worry; I have you covered in the following sections.

Solving rational equations by tuning in your LCD

You can solve rational equations, such as $\frac{3x+2}{2} - \frac{5}{2x-3} = \frac{x+3}{4}$, without as much hassle if you simply get rid of all the denominators. To do so, you work with an old friend, the least common denominator. In case you're not on a first-name basis with her, the *least common denominator* (LCD) is also known as the *least common multiple* — the smallest number that two or more other numbers all divide into evenly (such as 2, 3, and 4 all dividing the LCD 12 evenly). (Head to Chapter 18 for a quick trick on finding an LCD.)

To solve the previous example equation with the LCD, you find a common denominator, write each fraction with that common denominator, and then multiply each side of the equation by that same denominator to get a nice quadratic equation (see Chapter 3 for a full discussion of quadratic equations).

Quadratic equations have two solutions, so they present more opportunities for extraneous solutions. Be on the lookout!

1. Find a common denominator

The first step in solving the rational equation is to find the least common denominator (LCD) for all the terms in the equation.

For example, the common denominator of all three fractions in the equation $\frac{3x+2}{2} - \frac{5}{2x-3} = \frac{x+3}{4}$ consists of all the factors in the three denominators. Each of the denominators has to be able to divide into the common denominator evenly. In other words, the LCD is a multiple of each of the original denominators. To solve this equation, use $4(2x-3)$ as the common denominator, because it's a multiple of 2 — you multiply by $2(2x-3)$ to get it; it's a multiple of $2x - 3$ — you multiply by 4 to get it; and it's a multiple of 4 — you multiply by $(2x - 3)$ to get it. All three denominators divide this product evenly.

2. Write each fraction with the common denominator

Multiply each of the terms in the original equation by some value so that, after multiplying, each resulting term has the same denominator — the LCD you're so fond of:

$$\frac{3x+2}{2} \cdot \frac{2(2x-3)}{2(2x-3)} - \frac{5}{2x-3} \cdot \frac{4}{4} = \frac{x+3}{4} \cdot \frac{2x-3}{2x-3}$$

$$\frac{2(3x+2)(2x-3)}{4(2x-3)} - \frac{20}{4(2x-3)} = \frac{(x+3)(2x-3)}{4(2x-3)}$$

The "some value" I speak of is equal to one, because each of the fractions multiplying the terms is the same in the numerator and denominator. But you carefully select the fractions that serve as multipliers — the numerators and denominators must consist of all the factors necessary to complete the LCD.

You can just divide the LCD by the current denominator to determine what more you need to create the common denominator in that term.

3. Multiply each side of the equation by that same denominator

Multiply each term in the equation by the least common denominator to reduce each term and get rid of the denominators:

$$4(2x-3) \cdot \frac{2(3x+2)(2x-3)}{4(2x-3)} - 4(2x-3) \cdot \frac{20}{4(2x-3)} = 4(2x-3) \cdot \frac{(x+3)(2x-3)}{4(2x-3)}$$

$$2(3x+2)(2x-3) - 20 = (x+3)(2x-3)$$

One pitfall of multiplying both sides of an equation by a variable is that you may have to multiply both sides by zero, which may introduce an extraneous solution. Be sure to check your answer in the *original* equation when you're finished to make sure your answer doesn't make one or more of the denominators equal to zero.

4. Solve the new equation

By completing the previous steps for this example problem, you produce a quadratic equation (if you don't know what to do with those, turn to Chapter 3).

To solve the new quadratic equation, you multiply out the terms, simplify, and set the equation equal to zero:

$$2(3x+2)(2x-3) - 20 = (x+3)(2x-3)$$
$$12x^2 - 10x - 12 - 20 = 2x^2 + 3x - 9$$
$$10x^2 - 13x - 23 = 0$$

Now you find out if the quadratic equation factors. If it doesn't factor, you can resort to the quadratic formula; fortunately, that isn't necessary here. After factoring, you set each factor equal to zero and solve for x:

$$10x^2 - 13x - 23 = 0$$
$$(10x - 23)(x + 1) = 0$$
$$10x - 23 = 0, \ x = \frac{23}{10}$$
$$x + 1 = 0, \ x = -1$$

You find two solutions for the quadratic equation: $x = \frac{23}{10}$ and $x = -1$.

5. Check your answers to avoid extraneous solutions

You now have to check to be sure that both your solutions work in the *original* equation. As I discuss in the introduction to this section, one or both may be extraneous solutions.

The most common indication that you have an extraneous solution is that you end up with a zero in the denominator after replacing all the variables with that answer. Occasionally, you get a "nonsense" equation such as 4 = 7 when checking — and that tells you that the solution is extraneous — but those are very special cases. You should always check your answers after solving equations. Make sure that the value(s) you find create true statements.

Checking the original equation to see if the two solutions work, you first look at $x = \frac{23}{10}$:

$$\frac{3\left(\frac{23}{10}\right) + 2}{2} - \frac{5}{2\left(\frac{23}{10}\right) - 3} = \frac{\left(\frac{23}{10}\right) + 3}{4}$$

$$\frac{\frac{69}{10} + \frac{20}{10}}{2} - \frac{5}{\frac{46}{10} - \frac{30}{10}} = \frac{\frac{23}{10} + \frac{30}{10}}{4}$$

$$\frac{\frac{89}{10}}{2} - \frac{5}{\frac{16}{10}} = \frac{\frac{53}{10}}{4}$$

$$\frac{89}{20} - \frac{50}{16} = \frac{53}{40}$$

Checking this equation with a fractional answer appears to be pretty nasty work, but the work has to be done. Don't consider resorting to a calculator to check your answer as wimping out at this stage. You can always hope for integers so you don't have these complex fractions to simplify, but you don't always get what you wish for.

Your next step is to find a common denominator so you can compare the two sides of the equation:

$$\frac{89}{20}\cdot\frac{4}{4}-\frac{50}{16}\cdot\frac{5}{5}=\frac{53}{40}\cdot\frac{2}{2}$$
$$\frac{356}{80}-\frac{250}{80}=\frac{106}{80}$$
$$\frac{106}{80}=\frac{106}{80}$$

Nice! The first solution works. The next check, to see if $x = -1$ is a solution, should be a piece of cake:

$$\frac{3(-1)+2}{2}-\frac{5}{2(-1)-3}=\frac{(-1)+3}{4}$$
$$\frac{-3+2}{2}-\frac{5}{-5}=\frac{2}{4}$$
$$-\frac{1}{2}+1=\frac{1}{2}$$
$$\frac{1}{2}=\frac{1}{2}$$

Very nice! The solutions of $\frac{3x+2}{2}-\frac{5}{2x-3}=\frac{x+3}{4}$ are $x = {}^{23}\!/_{10}$ and $x = -1$.

Rational equations don't always work out so well, though. Take the equation $\frac{x}{x+2}+\frac{3x+2}{x(x+2)}=\frac{3}{x}$, for example. If you work through Steps 1 through 4, you get a new equation:

$$\frac{x}{x+2}\cdot\frac{x}{x}+\frac{3x+2}{x(x+2)}=\frac{3}{x}\cdot\frac{x+2}{x+2}$$
$$\frac{x^2}{x(x+2)}+\frac{3x+2}{x(x+2)}=\frac{3(x+2)}{x(x+2)}$$
$$x(x+2)\cdot\frac{x^2}{x(x+2)}+x(x+2)\cdot\frac{3x+2}{x(x+2)}=x(x+2)\cdot\frac{3(x+2)}{x(x+2)}$$
$$x^2+3x+2=3(x+2)$$
$$x^2-4=0$$
$$(x+2)(x-2)=0$$

The solutions for this equation are $x = -2$ and $x = 2$.

When you try $x = 2$ in the original equation, it works out:

$$\frac{2}{2+2} + \frac{3(2)+2}{2(2+2)} = \frac{3}{2}$$

$$\frac{2}{4} + \frac{8}{8} = \frac{3}{2}$$

$$\frac{1}{2} + 1 = \frac{3}{2}$$

However, when you substitute $x = -2$ into the original equation, you get

$$\frac{-2}{-2+2} + \frac{3(-2)+2}{-2(-2+2)} = \frac{3}{-2}$$

$$\frac{-2}{0} + \frac{-4}{0} = \frac{3}{-2}$$

Stop right there! You can't have a zero in the denominator. The solution $x = -2$ works just fine in the quadratic equation, but it isn't a solution of the rational equation — -2 is extraneous.

Solving rational equations with proportions

A *proportion* is an equation in which one fraction is set equal to another. For example, the equation $\frac{a}{b} = \frac{c}{d}$ is a proportion. Proportions have several very nice features that make them desirable to work with when you're solving rational equations because you can eliminate the fractions or change them so that they feature better denominators. Also, they factor in four different ways.

When you have the proportion $\frac{a}{b} = \frac{c}{d}$, the following points are also true:

- ✔ ad and bc, the cross-products, are equal, giving you $ad = bc$.
- ✔ $\frac{b}{a} = \frac{d}{c}$, the reciprocals, are equal (you can flip the proportion).

Using cross-products to solve a rational equation

To solve an equation such as $\frac{x+5}{2} - \frac{3}{x} = \frac{9}{x}$, you can find a common denominator and then multiply each side by the common denominator (see the "Solving rational equations by tuning in your LCD" section earlier in this chapter), but here's a quicker, easier way:

1. **Add $\frac{3}{x}$ to each side and add the terms with the same denominator to form a proportion.**

$$\frac{x+5}{2} - \frac{3}{x} + \frac{3}{x} = \frac{9}{x} + \frac{3}{x}$$

$$\frac{x+5}{2} = \frac{12}{x}$$

2. **Cross multiply.**

$$(x+5)\,x = 24$$

This is a quadratic equation. Here's how you solve for the solutions (see Chapter 3):

1. **Simplify the quadratic equation.**

$$(x+5)\,x = 24$$
$$x^2 + 5x = 24$$

2. **Set it equal to zero.**

$$x^2 + 5x - 24 = 0$$

3. **Solve for the solutions by factoring.**

$$(x+8)(x-3) = 0$$
$$x+8 = 0,\ x = -8$$
$$x-3 = 0,\ x = 3$$

You have two solutions, x = –8 and x = 3. You have to check both to be sure that neither is an extraneous solution. When x = –8,

$$\frac{-8+5}{2} - \frac{3}{-8} = \frac{9}{-8}$$
$$-\frac{3}{2} + \frac{3}{8} = -\frac{9}{8}$$
$$-\frac{12}{8} + \frac{3}{8} = -\frac{9}{8}$$
$$-\frac{9}{8} = -\frac{9}{8}$$

The solution x = –8 works. As you can see, so does the solution x = 3:

$$\frac{3+5}{2} - \frac{3}{3} = \frac{9}{3}$$
$$\frac{8}{2} - \frac{3}{3} = 3$$
$$4 - 1 = 3$$

Reducing every which way but loose

TIP

Another wonderful feature of proportions is that you can reduce the fractions in a proportion by finding common factors in four different directions: top, bottom, left, and right. The ability to reduce a proportion comes in handy when you have large numbers in the equation.

Here are some examples of reducing proportions across the top (numerators), bottom (denominators), left, and right:

Numerators	**Denominators**	**Left**	**Right**
$\dfrac{15x}{28} = \dfrac{5}{49}$	$\dfrac{3x}{28} = \dfrac{1}{49}$	$\dfrac{100y}{300(y+1)} = \dfrac{121}{77y}$	$\dfrac{y}{3(y+1)} = \dfrac{121}{77y}$
$\dfrac{\overset{3}{\cancel{15}}x}{28} = \dfrac{\cancel{5}^{\,1}}{49}$	$\dfrac{3x}{\underset{4}{\cancel{28}}} = \dfrac{1}{\cancel{49}_{\,7}}$	$\dfrac{\overset{1}{\cancel{100}}y}{\underset{3}{\cancel{300}}(y+1)} = \dfrac{121}{77y}$	$\dfrac{y}{3(y+1)} = \dfrac{\cancel{121}^{\,11}}{\underset{7}{\cancel{77}}y}$
$\dfrac{3x}{28} = \dfrac{1}{49}$	$\dfrac{3x}{4} = \dfrac{1}{7}$	$\dfrac{y}{3(y+1)} = \dfrac{121}{77y}$	$\dfrac{y}{3(y+1)} = \dfrac{11}{7y}$

The reduced forms of the proportions make cross-multiplication much easier and more manageable. Take the following proportion, for example. You first reduce across the numerators, and then you reduce the left fractions. Finally, you cross-multiply and solve the quadratic equation:

$$\frac{80x}{16} = \frac{30}{x-5}$$

$$\frac{\overset{8}{\cancel{80}}x}{16} = \frac{\cancel{30}^{\,3}}{x-5}$$

$$\frac{8x}{16} = \frac{3}{x-5}$$

$$\frac{\overset{1}{\cancel{8}}x}{\underset{2}{\cancel{16}}} = \frac{3}{x-5}$$

$$\frac{x}{2} = \frac{3}{x-5}$$

$$x(x-5) = 6$$

$$x^2 - 5x = 6$$

$$x^2 - 5x - 6 = 0$$

$$(x-6)(x+1) = 0$$

$$x - 6 = 0,\ x = 6$$

$$x + 1 = 0,\ x = -1$$

The solutions are $x = 6$ and $x = -1$. As usual, you need to check to be sure that you haven't introduced any extraneous roots:

$$\frac{80x}{16} = \frac{30}{x-5}$$

$$x = 6, \frac{80(6)}{16} = \frac{30}{(6)-5}, \frac{480}{16} = \frac{30}{1}, 480 = 30(16)$$

$$x = -1, \frac{80(-1)}{16} = \frac{30}{(-1)-5}, \frac{-80}{16} = \frac{30}{-6}, (-80)(-6) = 30(16)$$

Both of the solutions check out.

Reciprocating rational equations

The property of proportions that states the proportion $\frac{a}{b} = \frac{c}{d}$ is equivalent to its reciprocal $\frac{b}{a} = \frac{d}{c}$ comes in handy in an equation such as $\frac{1}{x-3} = \frac{2}{5}$. After you flip the proportion, you end up with a 1 in the denominator on the left. At that point, all you need to do to solve the equation is to add 3 to each side.

$$\frac{x-3}{1} = \frac{5}{2}$$

$$x - 3 = 2.5$$

$$x = 5.5$$

Ridding Yourself of a Radical

The term *radical* often indicates that you want to find a root — the square root of a number, its cube root, and so on. A radical in an equation gives the same message, but it adds a whole new dimension to what could've been a perfectly nice equation to solve. In general, you deal with radicals in equations the same way you deal with fractions in equations — you get rid of them. But watch out: The extraneous answers that first rear their ugly heads in the "Solving rational equations by tuning in your LCD" section pop up here as well. So — you guessed it — you have to check your answers.

Squaring both sides of a radical equation

If you have an equation in the form $\sqrt{ax+b} = c$, you square both sides of the equation to get rid of the radical. The only problem arises when you end up with an extraneous root.

Consider the non-equation –3 = 3. You know that the equation isn't correct, but what happens when you square both sides of this statement? You get $(-3)^2 = (3)^2$, or 9 = 9. Now you have an equation. Squaring both sides can mask or hide an incorrect statement.

Much like the process of getting rid of fractions in equations, the method of squaring both sides is the easiest way to deal with radicals in equations. You just accept that you always have to watch for extraneous roots when solving equations by squaring.

For example, to solve the equation $\sqrt{4x+21} - 6 = x$, follow these steps:

1. **Change the equation so that the radical term is by itself on the left.**

2. **Square both sides of the equation.**

On paper, the process looks like this:

$$\sqrt{4x+21} = x + 6$$
$$\left(\sqrt{4x+21}\right)^2 = (x+6)^2$$
$$4x + 21 = x^2 + 12x + 36$$

A very common error when squaring problems is to square the binomial on the right incorrectly. Don't forget the middle term — you can't just square the two terms alone $[(a+b)^2 = a^2 + 2ab + b^2]$.

At this point, you have a quadratic equation (see Chapter 3). Set it equal to zero and solve it:

$$4x + 21 = x^2 + 12x + 36$$
$$0 = x^2 + 8x + 15$$
$$0 = (x+3)(x+5)$$
$$x + 3 = 0,\ x = -3$$
$$x + 5 = 0,\ x = -5$$

Now you check to see if your solutions fit with the original equation. When $x = -3$, you get

$$\sqrt{4(-3)+21} - 6 = \sqrt{-12+21} - 6 = \sqrt{9} - 6 = 3 - 6 = -3$$

It works. Checking $x = -5$, you get

$$\sqrt{4(-5)+21} - 6 = \sqrt{-20+21} - 6 = \sqrt{1} - 6 = -5$$

This solution works, too.

Both solutions working out is more the exception rather than the rule when you're dealing with radicals. Most of the time, one solution or the other works, but not both. And, unfortunately, sometimes you go through all the calculations and find that neither solution works in the original equation. You get an answer, of course (that there is no answer), but it isn't very fulfilling.

Calming two radicals

Some equations that contain radicals call for more than one application of squaring both sides. For example, you have to square both sides more than once when you can't isolate a radical term by itself on one side of the equation. And you usually need to square both sides more than once when you have three terms in the equation — two of them with radicals.

For example, say you have to work with the equation $\sqrt{3x+19} - \sqrt{5x-1} = 2$. Here's how you solve the problem:

1. **Move the radicals so that only one appears on each side.**

2. **Square both sides of the equation.**

   After the first two steps, you have the following:

   $$\sqrt{3x+19} = 2 + \sqrt{5x-1}$$
   $$\left(\sqrt{3x+19}\right)^2 = \left(2 + \sqrt{5x-1}\right)^2$$
   $$3x+19 = 4 + 4\sqrt{5x-1} + 5x - 1$$

3. **Move all the nonradical terms to the left and simplify.**

   This gives you

   $$3x+19-4-5x+1 = 4\sqrt{5x-1}$$
   $$-2x+16 = 4\sqrt{5x-1}$$

4. **Make the job of squaring the binomial on the left easier by dividing each term by two — the common factor of all the terms on both sides.**

   You end up with

   $$\frac{-2x}{2} + \frac{16}{2} = \frac{4\sqrt{5x-1}}{2}$$
   $$-x+8 = 2\sqrt{5x-1}$$

5. **Square both sides, simplify, set the quadratic equal to zero, and solve for x.**

   This process gives you the following:

   $$(-x+8)^2 = \left(2\sqrt{5x-1}\right)^2$$
   $$x^2 - 16x + 64 = 4(5x-1)$$
   $$x^2 - 16x + 64 = 20x - 4$$
   $$x^2 - 36x + 68 = 0$$
   $$(x-2)(x-34) = 0$$
   $$x-2 = 0,\ x = 2$$
   $$x-34 = 0,\ x = 34$$

   The two solutions you come up with are $x = 2$ and $x = 34$.

Packing a palindromic punch

What do the phrases "Never Odd Or Even," "I Prefer Pi," and "A Toyota's a Toyota" and the square root of the number 14641 have in common? Here's a hint: The square of the number 111 can also fall in that list. Yes, they're all palindromes. A *palindrome* reads the same way front to back and back to front. A palindromic number is just a number that reads the same both ways. Many numbers are palindromes, but you can also find some special palindromic numbers: those whose squares or

square roots are also palindromes. For instance, the square of 111 is 12,321. In fact, the squares of all numbers made up of all 1s (up to nine of them) are palindromes.

Even though you can find an infinite amount of numbers whose squares are palindromes, none of those numbers begins with a 4, 5, 6, 7, 8, or 9. Did you appreciate this little journey, compliments of Ferdinand de Lesseps? "A man, a plan, a canal, Panama." Didn't he say, "Too hot to hoot?"

Don't forget to check each solution in the original equation:

$$\sqrt{3x + 19} - \sqrt{5x - 1} = 2$$
$$x = 2, \sqrt{3(2) + 19} - \sqrt{5(2) - 1} = \sqrt{25} - \sqrt{9} = 5 - 3 = 2$$
$$x = 34, \sqrt{3(34) + 19} - \sqrt{5(34) - 1} = \sqrt{121} - \sqrt{169} = 11 - 13 = -2$$

The solution $x = 2$ works. The other solution, $x = 34$, doesn't work in the equation. The number 34 is an extraneous solution.

Changing Negative Attitudes about Exponents

Equations with negative exponents offer some unique challenges. The first challenge deals with the fact that you're working with negative numbers and have to keep track of the rules needed to add, subtract, and multiply those negative numbers. Another challenge deals with the solution — if you find one — and checking to see if it works in the original form of the equation. The original form will take you back to those negative exponents, so it's round and round you go with number challenges.

Flipping negative exponents out of the picture

In general, negative exponents are easier to work with if they disappear. Yes, as wonderful as negative exponents are in the world of mathematics, solving equations that contain them is just easier if you change the format to positive exponents and fractions and then deal with solving the fractional equations (see the previous section).

For example, the equation $x^{-1} = 4$ has a fairly straightforward solution. You write the variable x in the denominator of a fraction and then solve for x. A nice way to solve for x is to write the 4 as a fraction, creating a proportion, and then cross-multiply (check out the aptly-named "Using cross-products to solve a rational equation" section earlier in the chapter for the lowdown):

$$x^{-1} = 4$$
$$\frac{1}{x} = 4$$
$$\frac{1}{x} = \frac{4}{1}$$
$$4x = 1$$
$$x = \frac{1}{4}$$

The process can get a bit dicey when you have more than one term with a negative exponent or when the negative exponent applies to more than one term. For instance, in the problem $(x-3)^{-1} - x^{-1} = \frac{3}{10}$, you have to rewrite the equation, changing the terms with negative exponents into rational or fractional terms.

$$\frac{1}{x-3} - \frac{1}{x} = \frac{3}{10}$$

You then find the common denominator for $x - 3$, x, and 10, which is the product of the three different denominators, $10x(x-3)$. Next, you rewrite each fraction as an equivalent fraction with that common denominator, multiply through to get rid of all the denominators (whew!), and solve the resulting equation. Didn't think you'd ever prefer to switch to fractions, did you?

$$\frac{10x}{10x\,(x-3)} - \frac{10\,(x-3)}{10x\,(x-3)} = \frac{3x\,(x-3)}{10x\,(x-3)}$$

$$10x - 10\,(x-3) = 3x\,(x-3)$$

$$10x - 10x + 30 = 3x^2 - 9x$$

$$0 = 3x^2 - 9x - 30$$

$$0 = 3\left(x^2 - 3x - 10\right)$$

$$0 = 3\,(x-5)(x+2)$$

$$x - 5 = 0,\ x = 5$$

$$x + 2 = 0,\ x = -2$$

You can't simplify $(x-3)^{-1}$ by distributing the exponent or multiplying it out in any way. You have to rewrite the term as a fraction to get rid of the negative exponent.

Your answers are $x = 5$ and $x = -2$. You need to check to be sure that they work in the original equation:

$$(x-3)^{-1} - x^{-1} = \frac{3}{10}$$

$$x = 5,\ (5-3)^{-1} - 5^{-1} = 2^{-1} - 5^{-1} = \frac{1}{2} - \frac{1}{5} = \frac{5}{10} - \frac{2}{10} = \frac{3}{10}$$

$$x = -2,\ (-2-3)^{-1} - (-2)^{-1} = (-5)^{-1} - (-2)^{-1}$$

$$= -\frac{1}{5} - \left(-\frac{1}{2}\right) = -\frac{2}{10} + \frac{5}{10} = \frac{3}{10}$$

Good news! Both of the solutions work.

Factoring out negatives to solve equations

Negative exponents don't have to have the same power within a particular equation. In fact, it may be more common to have a mixture of powers in an equation. Two useful methods for solving equations with negative exponents are factoring out a greatest common factor (GCF) and solving the equation as if it's a quadratic (quadratic-like; see Chapter 3).

Factoring out a negative GCF

An equation such as $3x^{-3} - 5x^{-2} = 0$ has a solution that you can find without switching to fractions right away. In general, equations that have no constant terms — all the terms have variables with exponents on them — work best with this technique.

Here are the steps:

1. **Factor out the greatest common factor (GCF).**

   In this case, the GCF is x^{-3}:

   $$3x^{-3} - 5x^{-2} = 0$$
   $$x^{-3}(3 - 5x) = 0$$

   Did you think the exponent of the greatest common factor was –2? Remember, –3 is smaller than –2. When you factor out a greatest common factor, you choose the smallest exponent out of all the choices and then divide each term by that common factor.

   The tricky part of the factoring is dividing $3x^{-3}$ and $5x^{-2}$ by x^{-3}. The rules of exponents say that when you divide two numbers with the same base, you subtract the exponents, so you have

   $$\frac{3x^{-3}}{x^{-3}} = 3$$
   $$\frac{-5x^{-2}}{x^{-3}} = -5x^{-2-(-3)} = -5x^{-2+3} = -5x^{1}$$

2. **Set each term in the factored form equal to zero to solve for x.**

   $$x^{-3}(3 - 5x) = 0$$
   $$x^{-3} = 0, \; \frac{1}{x^{3}} = 0$$
   $$3 - 5x = 0, \; x = \frac{3}{5}$$

3. **Check your answers.**

   The only solution for this equation is $x = \frac{3}{5}$ — a perfectly dandy answer.

   The other factor, with the x^{-3}, doesn't yield an answer at all. The only way a fraction can equal zero is if the numerator is zero. Having a one in the numerator makes it impossible for the term to be equal to zero. And you can't let the x be equal to zero because that would put a zero in the denominator.

Solving quadratic-like trinomials

Trinomials are expressions with three terms, and if the terms are raised to the second degree, the expression is quadratic. You can simplify quadratic trinomials by factoring them into two binomial factors. (See Chapter 3 for details on factoring trinomials.)

Often, you can factor trinomials with negative powers into two binomials if they have the following pattern: $ax^{-2n} + bx^{-n} + c$. The exponents on the variables have to be in pairs, where one of the exponents is twice the other. For instance, the trinomial $3x^{-2} + 5x^{-1} - 2$ fits this description. Make this into an equation, and you can solve it by factoring and setting the two factors equal to zero:

$$3x^{-2} + 5x^{-1} - 2 = 0$$
$$(3x^{-1} - 1)(x^{-1} + 2) = 0$$
$$3x^{-1} - 1 = 0, \frac{3}{x} = 1, x = 3$$
$$x^{-1} + 2 = 0, \frac{1}{x} = -2, x = -\frac{1}{2}$$

You produce two solutions, and both work when substituted into the original equation. You haven't changed the format of the equation, but you still have to be sure that you aren't putting a zero in the denominator for an answer.

REMEMBER

You need to be careful when solving an equation containing negative exponents that involves taking an even root (square root, fourth root, and so on). The following problem starts out behaving very nicely, factoring into two binomials:

$$x^{-4} - 15x^{-2} - 16 = 0$$
$$(x^{-2} - 16)(x^{-2} + 1) = 0$$
$$x^{-2} - 16 = 0 \text{ or } x^{-2} + 1 = 0$$

The first factor offers no big surprise. You get two solutions after changing the negative exponent and solving the equation by using the square root rule (see Chapter 3 for more on this rule):

$$x^{-2} - 16 = 0, \frac{1}{x^2} = 16, x^2 = \frac{1}{16}$$
$$x = \pm\frac{1}{4}$$

The other factor doesn't have a real solution because you can't find a square root of a negative number (in Chapter 3, you find more information on what happens when you try to take the square root of a negative number; in Chapter 14, you find out how to deal with imaginary numbers — those square roots of the negatives):

$$x^{-2} + 1 = 0, \frac{1}{x^2} = -1, x^2 = -1$$

Look at all the possibilities — and ways to trip up with solutions. Watch out for zeros in the denominator, because those numbers don't exist, and be wary of imaginary numbers — they exist somewhere, in some mathematician's imagination. Factoring into binomials is a nifty way for solving equations with negative exponents; just be sure to proceed cautiously.

Fooling Around with Fractional Exponents

You use fractional exponents ($x^{1/2}$, for example) to replace radicals and powers under radicals. Writing terms with fractional exponents allows you to perform operations on terms more easily when they have the same base or variable.

You write the radical expression $\sqrt[3]{x^4}$, for example, as $x^{4/3}$. The power of the variable under the radical goes in the numerator of the fraction, and the root of the radical goes in the denominator of the fraction.

Combining terms with fractional exponents

Fortunately, the rules of exponents stay the same when the exponents are fractional:

- **You can add or subtract terms with the same base and exponent:**
  $4x^{2/3} + 5x^{2/3} - 2x^{2/3} = 7x^{2/3}$.

- **You can multiply terms with the same base by adding their exponents:**
  $\left(5x^{3/4}\right)\left(9x^{2/3}\right) = 45x^{3/4+2/3} = 45x^{9/12+8/12} = 45x^{17/12}$.

- **You can divide terms with the same base by subtracting their exponents:** $\dfrac{45x^{3/5}}{9x^{1/5}} = 5x^{3/5-1/5} = 5x^{2/5}$.

- **You can raise a fractional power to a power by multiplying the two powers:** $\left(x^{3/4}\right)^{-5/2} = x^{3/4\,(-5/2)} = x^{-15/8}$.

Fractional exponents may not look that much better than the radicals they represent, but can you imagine trying to simplify $5\sqrt[4]{x^3} \cdot 9\sqrt[3]{x^2}$? You can always refer to the second entry in the previous list to see how fractional powers make doing the multiplication possible.

Factoring fractional exponents

You can easily factor expressions that contain variables with fractional exponents if you know the rule for dividing numbers with the same base (see the previous section): *Subtract their exponents.* Of course, you have the challenge of finding common denominators when you subtract fractions. Other than that, it's smooth sailing.

To factor the expression $2x^{1/2} - 3x^{1/3}$, for example, you note that the smaller of the two exponents is the fraction ⅓. Factor out x raised to that lower power, changing to a common denominator where necessary:

$$2x^{1/2} - 3x^{1/3} = x^{1/3}\left(2x^{1/2-1/3} - 3x^{1/3-1/3}\right)$$
$$= x^{1/3}\left(2x^{3/6-2/6} - 3x^{0}\right)$$
$$= x^{1/3}\left(2x^{1/6} - 3\right)$$

A good way to check your factoring work is to mentally distribute the first term through the terms in the parenthesis to be sure that the product is what you started with.

Solving equations by working with fractional exponents

Fractional exponents represent radicals and powers, so when you can isolate a term with a fractional exponent, you can raise each side to an appropriate power to get rid of the exponent and eventually solve the equation. When you can't isolate the fractional exponent, you must resort to other methods for solving equations, such as factoring.

Raising each side to a power

When you can isolate a term that has a fractional exponent attached to it, go for it. The goal is to make the exponent equal to one so you can solve the equation for x. You accomplish your goal by raising each side of the equation to a power that's equal to the reciprocal of the fractional exponent.

For example, you solve the equation $x^{4/3} = 16$ by raising each side of the equation to the ¾th power, because multiplying a number and its reciprocal always gives you a product of one:

$$\left(x^{4/3}\right)^{3/4} = (16)^{3/4}$$
$$x^{1} = \sqrt[4]{16^{3}}$$

You finish the problem by evaluating the radical (see the section "Ridding Yourself of a Radical"):

$$x = \sqrt[4]{16^{3}} = \left(\sqrt[4]{16}\right)^{3} = (2)^{3} = 8$$

The evaluation is easier if you take the fourth root first and then raise the answer to the third power. You can do this because powers and roots are on the same level in the *order of operations* (refer to Chapter 1 for more on this topic), so you can calculate them in either order — whatever is most convenient.

Factoring out variables with fractional exponents

You don't always have the luxury of being able to raise each side of an equation to a power to get rid of the fractional exponents. Your next best plan of attack involves factoring out the variable with the smaller exponent and setting the two factors equal to zero.

To solve $x^{5/6} - 3x^{1/2} = 0$, for example, you first factor out an x with an exponent of ½:

$$x^{5/6} - 3x^{1/2} = 0$$
$$x^{1/2}\left(x^{5/6-1/2} - 3\right) = 0$$
$$x^{1/2}\left(x^{5/6-3/6} - 3\right) = 0$$
$$x^{1/2}\left(x^{2/6} - 3\right) = x^{1/2}\left(x^{1/3} - 3\right) = 0$$

Now you can set the two factors equal to zero and solve for x:

$$x^{1/2} = 0, \ x = 0$$
$$x^{1/3} - 3 = 0, \ x^{1/3} = 3, \ \left(x^{1/3}\right)^3 = (3)^3, \ x = 27$$

You come up with two perfectly civilized answers: $x = 0$ and $x = 27$.

Factoring into the product of two binomials

Often, you can factor trinomials with fractional exponents into the product of two binomials. After the factoring, you set the two binomials equal to zero to determine if you can find any solutions.

To solve $x^{1/2} - 6x^{1/4} + 5 = 0$, for example, you first factor into two binomials. The exponent of the first variable is twice that of the second, which should indicate to you that the trinomial has factoring potential. After you factor (see Chapter 3), you set the expression equal to zero and solve for x:

$$\left(x^{1/4} - 1\right)\left(x^{1/4} - 5\right) = 0$$
$$x^{1/4} - 1 = 0, \ x^{1/4} = 1, \ \left(x^{1/4}\right)^4 = (1)^4, \ x = 1$$
$$x^{1/4} - 5 = 0, \ x^{1/4} = 5, \ \left(x^{1/4}\right)^4 = (5)^4, \ x = 625$$

Check your answers in the original equation (see the fifth step in the section "Solving rational equations by tuning in your LCD" earlier in this chapter); you find that both $x = 1$ and $x = 625$ work.

Putting fractional and negative exponents together

This chapter wouldn't be complete without an explanation of how you can combine fractional and negative exponents into one big equation. Creating this mega problem isn't something you do just to see how exciting an equation can be. The following is an example of a situation that occurs in calculus problems. The derivative (a calculus process) has already been performed, and now you have to solve the equation. The hardest part of calculus is often the algebra, so I feel I should address this topic before you get to calculus.

The two terms in the equation $x(x^3+8)^{1/3}(x^2-4)^{-1/2}+x^2(x^3+8)^{-2/3}(x^2-4)^{1/2}=0$ have the common factor of $x(x^3+8)^{-2/3}(x^2-4)^{-1/2}$. Notice that both terms have a power of x, a power of (x^3+8), and a power of (x^2-4) in them. (You have to choose the lower of the powers on a factor and use that on the greatest common factor.)

Dividing each term by the greatest common factor (see the section "Factoring out a negative GCF" for more), the equation becomes

$$x(x^3+8)^{-2/3}(x^2-4)^{-1/2}\left[(x^3+8)^{1/3-(-2/3)}(x^2-4)^{-1/2-(-1/2)}+x(x^3+8)^{-2/3-(-2/3)}(x^2-4)^{1/2-(-1/2)}\right]=0$$

$$x(x^3+8)^{-2/3}(x^2-4)^{-1/2}\left[(x^3+8)^{1/3+2/3}(x^2-4)^{-1/2+1/2}+x(x^3+8)^{-2/3+2/3}(x^2-4)^{1/2+1/2}\right]=0$$

$$x(x^3+8)^{-2/3}(x^2-4)^{-1/2}\left[(x^3+8)^{1}(x^2-4)^{0}+x(x^3+8)^{0}(x^2-4)^{1}\right]=0$$

$$x(x^3+8)^{-2/3}(x^2-4)^{-1/2}\left[(x^3+8)^{1}+x(x^2-4)^{1}\right]=0$$

Now you simplify the terms inside the brackets:

$$x(x^3+8)^{-2/3}(x^2-4)^{-1/2}\left[(x^3+8)^{1}+x(x^2-4)^{1}\right]=0$$

$$x(x^3+8)^{-2/3}(x^2-4)^{-1/2}\left[x^3+8+x^3-4x\right]=0$$

$$x(x^3+8)^{-2/3}(x^2-4)^{-1/2}\left[2x^3-4x+8\right]=0$$

$$2x(x^3+8)^{-2/3}(x^2-4)^{-1/2}\left[x^3-2x+4\right]=0$$

You can set each of the four factors (the three in the greatest common factor and the fourth in the brackets) equal to zero to find any solutions for the equation (see Chapter 3):

$$2x(x^3+8)^{-2/3}(x^2-4)^{-1/2}\left[x^3-2x+4\right]=0$$

$$2x=0,\ x=0$$

$$(x^3+8)^{-2/3}=0,\ x^3+8=0,\ x^3=-8,\ x=-2$$

$$(x^2-4)^{-1/2}=0,\ x^2-4=0,\ x^2=\pm2$$

$$x^3-2x+4=(x+2)(x^2-2x+2)=0,\ x=-2$$

The solutions you find are $x = 0$, $x = 2$, and $x = -2$. You see repeated roots, but I name each of them only once here. The quadratic that remains from factoring $x^3 - 2x + 4$ doesn't have any real solution.

Chapter 5

Graphing Your Way to the Good Life

A *graph* is a drawing that illustrates an algebraic operation or equation in a two-dimensional plane (like a piece of graph paper). A graph allows you to see the characteristics of an algebraic statement immediately.

The graphs in algebra are unique because they reveal relationships that you can use to model a situation. The graph of a mathematical curve has a ton of information crammed into an elegant package. (Okay, not everyone thinks that the graph of a parabola is elegant, but beauty is in the eye of the beholder.) For example, parabolas can model daily temperature, and a flat, S-shaped curve can model the number of people infected with the flu.

In this chapter, you find out how lines, parabolas, other polynomials, and radical curves all fit into the picture. I start off with a quick refresher on graphing basics. The rest of the chapter helps you become a more efficient grapher so you don't have to plot a million points every time you need to create a graph. Along the way, I discuss using intercepts and symmetry and working with linear equations. I also provide you with a quick rundown of ten basic forms and equations you run into again and again in Algebra II. After you recognize a few landmarks, you can quickly sketch a reasonable graph of an equation. And to top it all off, you get some pointers on using a graphing calculator.

Coordinating Your Graphing Efforts

You do most graphing in algebra on the *Cartesian coordinate system* — a grid-like system where you plot points depending on the position and signs of numbers. You may be familiar with the basics of graphing coordinates, (*x, y*) points, from Algebra I or junior-high math. Within the Cartesian coordinate system (which is named for the philosopher and mathematician Rene Descartes), you can plug-and-plot points to draw a curve, or you can take advantage of knowing a little something about what the graphs should look like. In either case, the coordinates and points fit together to give you a picture.

Identifying the parts of the coordinate plane

A *coordinate plane* (shown in Figure 5-1) features two intersecting *axes* — two perpendicular lines that cross at a point called the *origin.* The axes divide up the coordinate plane into four *quadrants,* usually numbered with Roman numerals starting in the upper right-hand corner and moving counterclockwise.

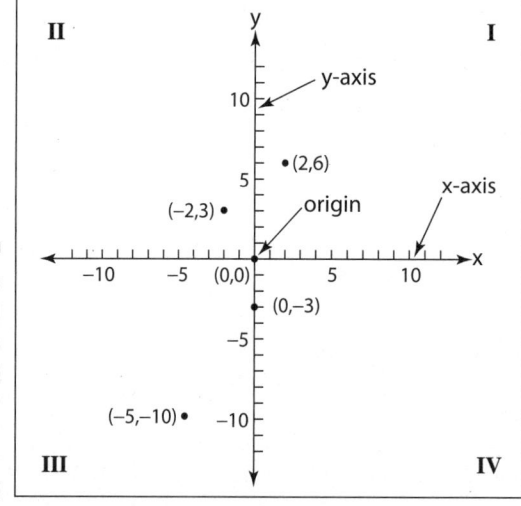

Figure 5-1:
Identifying
all the
players
in the
coordinate
plane.

The horizontal line is called the X-axis, and the vertical line is called the Y-axis. The scales on the axes are shown with little tick marks. Usually, you have the same scale on both axes — each tick mark may represent one unit or five units — but sometimes the plane needs to have two different scales. In any case, the scale on any one axis is the same all along that axis.

You identify points on the coordinate plane with numbers called *coordinates,* which come in *ordered pairs* — where the order matters. In the ordered pair (*x, y*), the first number is the *x*-coordinate; it tells you how far and in which direction the point is from the origin along the X-axis. The second number is the *y*-coordinate; it tells you how far from the origin and in which direction the point is along the Y-axis. In Figure 5-1, you can see several points drawn in with their corresponding coordinates.

Plotting from dot to dot

Graphing in algebra isn't quite as simple as a quick game of connect the dots, but the main concept is the same: You connect the points in the right order, and you see the desired picture.

In the section "Looking at 10 Basic Forms," later in this chapter, you discover the advantage of having at least a hint of what a graph is supposed to look like. But even when you have a pretty good idea of what you're going to get, you still need the down-and-dirty ability to plot points.

For example, the list of points (0, 3), (–2, 4), (–4, 3), (–5, 0), (–2, –3), (0, –2.5), (2, –3), (5, 0), (4, 3), (2, 4), (0, 3), (1.5, 5), (1, 6), (0, 3) doesn't mean too much, and the list doesn't look all that informative when graphed. But if you connect the points in order, you get a picture. You can see the points and the corresponding picture in Figure 5-2.

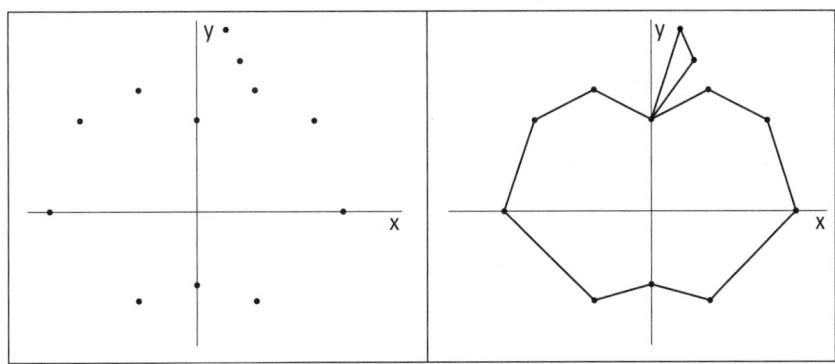

Figure 5-2:
Connecting a list of points in order creates a graphing picture.

In algebra, you usually have to create your set of points to graph. A problem gives you a formula or equation, and you determine the coordinates of the points that work in that equation. Your main goal is to sketch the graph after finding as few points as necessary. If you know what the general shape of the graph should be, all you need are a few anchor points to be on your way.

For instance, if you want to sketch the graph of $y = x^2 - x - 6$, you can find some points $[(x, y)]$ that make the equation true. Some points that work (picked at random, making sure they work in the equation) include: (4, 6), (3, 0), (2, –4), (1, –6), (0, –6), (–1, –4), (–2, 0), and (–3, 6). You can find many, many more points that work, but these points should give you a pretty good idea of what's going on. In Figure 5-3a, you can see the points graphed; I connect them with a smooth curve in Figure 5-3b. The order of connecting the points goes by the order of the x-coordinates.

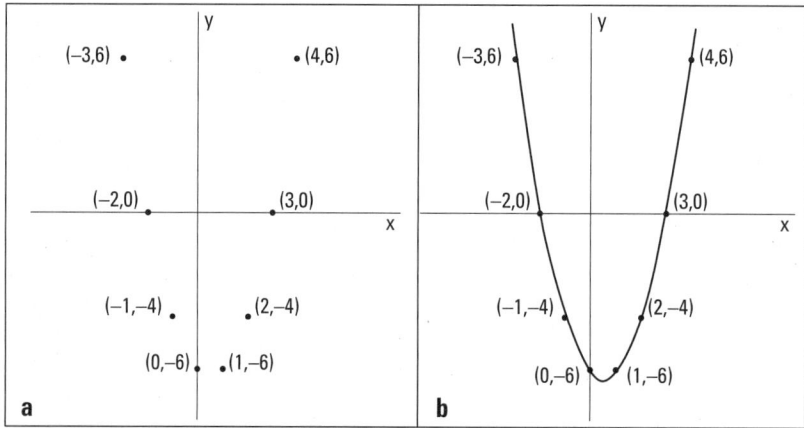

Figure 5-3:
Creating a
set of points
to fit the
graph of an
equation.

Streamlining the Graphing Process with Intercepts and Symmetry

Graphing curves can take as long as you like or be as quick as you like. If you take advantage of the characteristics of the curves you're graphing, you can cut down on the time it takes to graph and improve your accuracy. Two features that you can quickly recognize and solve for are the intercepts and symmetry of the graphs.

Finding x- and y-intercepts

The *intercepts* of a graph appear at the points where the lines of the graph cross the axes. The graph of a curve may never cross an axis, but when it does, knowing the points that represent the intercepts is very helpful.

The x-intercepts always have the format $(h, 0)$ — the y-coordinate is 0 because the point is on the X-axis. The y-intercepts have the form $(0, k)$ — the x-coordinate is zero because the point is on the Y-axis. You find the x- and y-intercepts by letting y and x, respectively, equal zero. To find the x-intercept(s) of a curve, you set y equal to zero and solve a given equation for x. To find the y-intercept(s) of a curve, you set x equal to zero and solve the equation for y.

For example, the graph of $y = -x^2 + x + 6$ has two x-intercepts and one y-intercept:

To find the x-intercepts, let $y = 0$; you then have the quadratic equation $0 = -x^2 + x + 6 = -(x^2 - x - 6)$. You solve this equation by factoring it into $0 = -(x - 3)(x + 2)$. You find two solutions, $x = 3$ and -2, so the two x-intercepts are $(3, 0)$ and $(-2, 0)$. (For more on factoring, see Chapters 1 and 3.)

To find the y-intercept, let $x = 0$. This gives you the equation $y = -0 + 0 + 6 = 6$. The y-intercept, therefore, is $(0, 6)$.

Figure 5-4a shows the intercepts of the previous example problem placed on the axes. The graph doesn't tell you a whole lot unless you're aware that the equation is a parabola. If you know that you have a parabola (which you will if you read Chapter 7), you know that a U-shaped curve goes through the intercepts. For now, you can just plug-and-plot to find a few more points to help you with the graph (see the section "Plotting from dot to dot"). Using the equation to find other points that work with the graph, you get $(1, 6)$, $(2, 4)$, $(4, -6)$ and $(-1, 4)$. Figure 5-4b shows the completed graph.

Figure 5-4:
Plotting the intercepts and calculated points on a graph to get the whole picture.

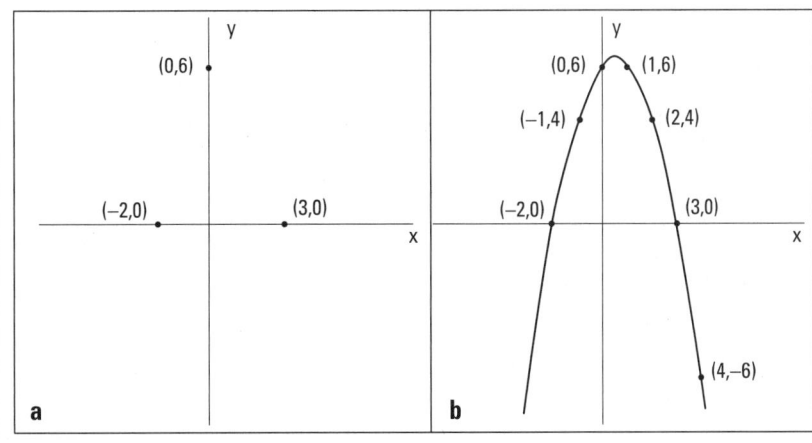

Reflecting on a graph's symmetry

When an item or individual is symmetric, you can see a sameness or pattern to it. A graph that's *symmetric* with respect to one of the axes appears to be a mirror image of itself on either side of the axis. A graph symmetric about the origin appears to be the same after a 180-degree turn. Figure 5-5 shows three curves and three symmetries: symmetry with respect to the Y-axis (a), symmetry with respect to the X-axis (b), and symmetry with respect to the origin (c).

Figure 5-5:
Symmetry
in a graph
makes for
a pretty
picture.

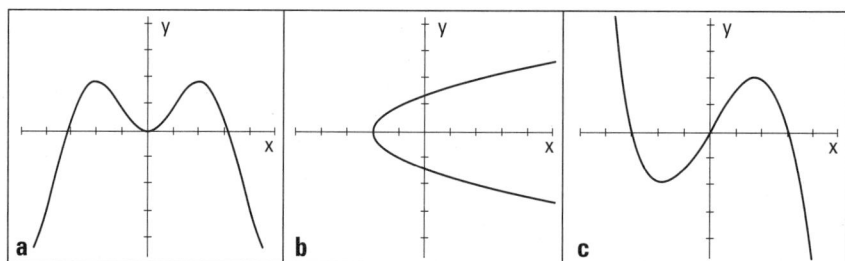

Recognizing that the graph of a curve has symmetry helps you sketch the graph and determine its characteristics. The following sections outline ways to determine, from a graph's equation, if symmetry exists.

With respect to the Y-axis

Consider an equation in this form: y = some expression in x's. If replacing every x with $-x$ doesn't change the value of y, the curve is the mirror image of itself over the Y-axis. The graph contains the points (x, y) and $(-x, y)$.

For example, the graph of the equation $y = x^4 - 3x^2 + 1$ is symmetric with respect to the Y-axis. If you replace each x with $-x$, the equation remains unchanged. Replacing each x with $-x$, $y = (-x)^4 - 3(-x)^2 + 1 = x^4 - 3x^2 + 1$. The y-intercept is $(0, 1)$. Some other points include $(1, -1)$, $(-1, -1)$ and $(2, 5)$, $(-2, 5)$. Notice that the y is the same for both positive and negative x's. With symmetry about the Y-axis, for every point (x, y) on the graph, you also find $(-x, y)$. It's easier to find points when an equation has symmetry because of the pairs. Figure 5-6 shows the graph of $y = x^4 - 3x^2 + 1$.

With respect to the X-axis

Consider an equation in this form: x = some expression in y's. If replacing every y with $-y$ doesn't change the value of x, the curve is the mirror image of itself over the X-axis. The graph contains the points (x, y) and $(x, -y)$.

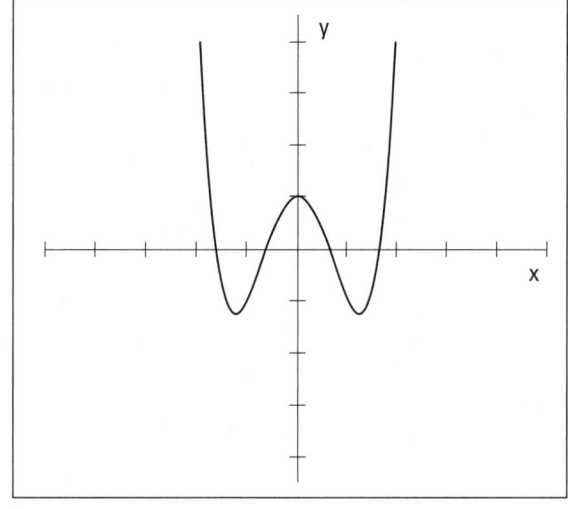

Figure 5-6:
A graph's
reflection
over a
vertical line.

For example, the graph of $x = \dfrac{10}{y^2 + 1}$ is symmetric with respect to the
X-axis. When you replace each *y* with *–y*, the *x* value remains unchanged. The
x-intercept is (10, 0). Some other points on the graph include (5, 1), (5, –1);
(2, 2), (2, –2); and (1, 3), (1, –3). Notice the pairs of points that have positive
and negative values for *y* but the same value for *x*. This is where the symmetry
comes in: The points have both positive and negative values — on both sides
of the X-axis — for each *x* coordinate. Symmetry about the X-axis means that for
every point (*x*, *y*) on the curve, you also find the point (*x*, *–y*). This symmetry
makes it easy to find points and plot the graph. The graph of $x = \dfrac{10}{y^2 + 1}$ is
shown in Figure 5-7.

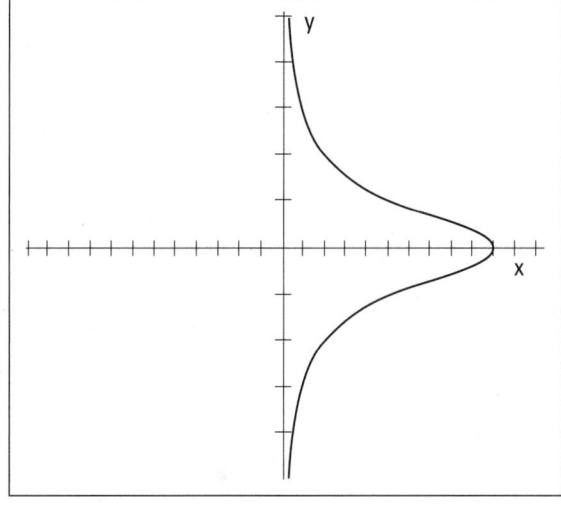

Figure 5-7:
A graph's
reflection
over a
horizontal
line.

With respect to the origin

REMEMBER

Consider an equation in this form: y = some expression in x's or x = some expression in y's. If replacing every variable with its opposite is the same as multiplying the entire equation by –1, the curve can rotate by 180 degrees about the origin and be its own image. The graph contains the points (x, y) and $(-x, -y)$.

For example, the graph of $y = x^5 - 10x^3 + 9x$ is symmetric with respect to the origin. When you replace every x and y with $-x$ and $-y$, you get $-y = -x^5 + 10x^3 - 9x$, which is the same as multiplying everything through by –1. The origin is both an x and a y-intercept. The other x-intercepts are $(1, 0)$, $(-1, 0)$, $(3, 0)$, and $(-3, 0)$. Other points on the graph of the curve include $(2, -30)$, $(-2, 30)$, $(4, 420)$, and $(-4, -420)$. These points illustrate the fact that (x, y) and $(-x, -y)$ are both on the graph. Figure 5-8 shows the graph of $y = x^5 - 10x^3 + 9x$ (I changed the scale on the Y-axis to make each tick-mark equal 10 units).

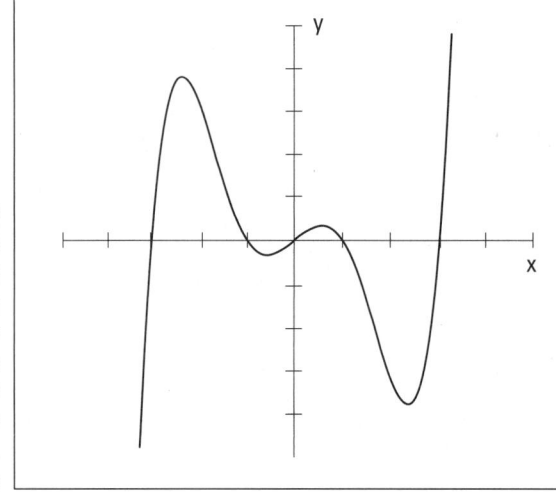

Figure 5-8:
A graph
revolving
180 degrees
about the
origin of the
coordinate
plane.

Graphing Lines

Lines are some of the simplest graphs to sketch. It takes only two points to determine the one and only line that exists in a space, so one simple method for graphing lines is to find two points — any two points — on the line. Another useful method is to use a point and the slope of the line. The method you choose is often just a matter of personal preference.

The slope of a line also plays a big role in comparing it with other lines that run parallel or perpendicular to it. The slopes are closely related to one another.

Finding the slope of a line

The *slope* of a line, *m,* has a complicated math definition, but it's basically a number — positive or negative, large or small — that highlights some of the characteristics of the line, such as its steepness and direction. The numerical value of the slope tells you if the line slowly rises or drops from left to right or dramatically soars or falls from left to right. In order to find the slope and discover the properties of the line, you can use the equation of the line to solve for the information you need, or you can look at the graph of the line to get the general picture. If you opt to concentrate on the equation of the line, you can solve for points on the line, which give you even more information.

Identifying characteristics of a line's slope

A line can have a positive or a negative slope. If a line's slope is positive, the line rises from left to right. If the slope is negative, the line falls from left to right. The greater the *absolute value* (the value of the number without regard to the sign; in other words, the distance of the number from zero) of a line's slope, the steeper the line is. For example, if the slope is a number between –1 and 1, the line is rather flat. A slope of zero means that the line is absolutely horizontal.

A vertical line doesn't have a slope. This is tied to the fact that numbers go infinitely high, and math doesn't have a highest number — you just say *infinity.* Only an infinitely high number can represent a vertical line's slope, but usually, if you're talking about a vertical line, you just say that the slope doesn't exist.

Formulating the value of a line's slope

You can determine the slope of a line, *m,* if you know two points on the line. The formula for finding the slope with this method involves finding the difference between the *y*-coordinates of the points and dividing that difference by the difference of the *x*-coordinates of the points.

You find the slope of the line that goes through the points (x_1, y_1) and (x_2, y_2) with the formula $m = \frac{y_2 - y_1}{x_2 - x_1}$.

For example, to find the slope of the line that crosses the points (–3, 2) and (4, –12), you use the formula to get $m = \frac{y_2 - y_1}{x_2 - x_1} = \frac{-12 - 2}{4 - (-3)} = \frac{-14}{7} = -2$. This line is fairly steep — the absolute value of –2 is 2 — and it falls as it moves from left to right, which makes it negative.

When you use the slope formula, it doesn't matter which point you choose to be (x_1, y_1) — the order of the points doesn't matter — but you can't mix up the order of the two different coordinates. You can run a quick check by seeing if the coordinates of each point are above and below one another. Also, be sure that the y-coordinates are in the numerator; a common error is to have the difference of the coordinates in the denominator.

Facing two types of equations for lines

Algebra II offers two different forms for the equation of a line. The first is the *standard form,* written $Ax + By = C$, with the two variable terms on one side and the constant on the other side. The other form is the *slope-intercept* form, written $y = mx + b$; the y value is set equal to the product of the slope, m, and x added to the y-intercept, b.

Meeting high standards with the standard form

The standard form for the equation of a line is written $Ax + By = C$. One example of such a line is $4x + 3y = 12$. You can find an infinite number of points that satisfy this equation; to name a few, you can use $(0, 4)$, $(-3, 8)$, and $(6, -4)$. It takes only two points to graph a line, so you can choose any two points and plot them on the coordinate plane to create your line.

However, the standard form has more information than may be immediately apparent. You can determine, just by looking at the numbers in the equation, the intercepts and slope of the line. The intercepts, in particular, are great for graphing the line, and you can find them easily because they fall right on the axes.

The line $Ax + By = C$ has

- An x-intercept of $\left(\frac{C}{A}, 0\right)$
- A y-intercept of $\left(0, \frac{C}{B}\right)$
- A slope of $m = -\frac{A}{B}$

To graph the line $4x + 3y = 12$, you find the two intercepts, $\left(\frac{C}{A}, 0\right) = \left(\frac{12}{4}, 0\right) = (3, 0)$ and $\left(0, \frac{C}{B}\right) = \left(0, \frac{12}{3}\right) = (0, 4)$, plot them, and create the line. Figure 5-9 shows the two intercepts and the graph of the line. Note that the line falls as it moves from left to right, confirming the negative value of the slope from the formula $m = -\frac{A}{B} = -\frac{4}{3}$.

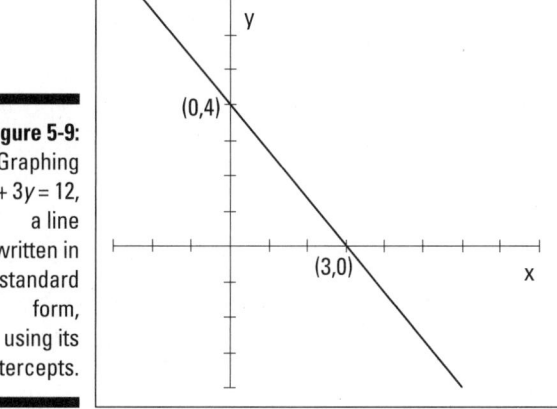

Figure 5-9:
Graphing
$4x + 3y = 12$,
a line
written in
standard
form,
using its
intercepts.

Picking off the slope-intercept form

When the equation of a line is written in the slope-intercept form, $y = mx + b$, you have good information right at your fingertips. The *coefficient* (a number multiplied with a variable or unknown quantity) of the x term, m, is the slope of the line. And the constant, b, is the y value of the y-intercept (the coefficient of the y term must be one). With these two bits of information, you can quickly sketch the line.

If you want to graph the line $y = 2x + 5$, for example, you first plot the y-intercept, $(0, 5)$, and then count off the slope from that point. The slope of the line $y = 2x + 5$ is 2; think of the 2 as the slope fraction, with y-coordinates on top and the x-coordinates on bottom. The slope then becomes $\frac{2}{1}$.

Counting off the slope means to start at the y-intercept, move one unit to the right (for the one in the denominator of the previous fraction), and, from the point one unit to the right, count two units up (for the two in the numerator). This process gets you to another point on the line.

You connect the point you count off with the y-intercept to create your line. Figure 5-10 shows the intercept, an example of counting off, and the new point that appears one unit to the right and two units up from the intercept — the point $(1, 7)$.

Changing from one form to the other

You can graph lines by using the standard form or the slope-intercept form of the equations. If you prefer one form to the other — or if you need a particular form for an application you're working on — you can change the equations to your preferred form by performing simple algebra:

✔ To change the equation $2x - 5y = 8$ to the slope-intercept form, you first subtract $2x$ from each side and then divide by the -5 multiplier on the y term:

$$2x - 5y = 8$$
$$-5y = -2x + 8$$
$$\frac{-5}{-5}\,y = \frac{-2}{-5}\,x + \frac{8}{-5}$$
$$y = \frac{2}{5}\,x - \frac{8}{5}$$

In the slope-intercept form, you can quickly determine the slope and y-intercept, $m = \frac{2}{5}$, $b = -\frac{8}{5}$.

✔ To change the equation $y = -\frac{3}{4}\,x + 5$ to the standard form, you first multiply each term by 4 and then add $3x$ to each side of the equation:

$$4(y) = 4\left(-\frac{3}{4}\,x + 5\right)$$
$$4y = \cancel{4}\left(-\frac{3}{\cancel{4}}\,x\right) + 4(5)$$
$$4y = -3x + 20$$
$$3x + 4y = 20$$

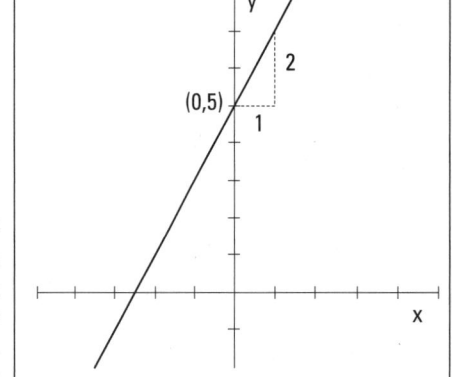

Figure 5-10:
A line with a slope of 2 is fairly steep.

Identifying parallel and perpendicular lines

Lines are *parallel* when they never touch — no matter how far out you draw them. Lines are *perpendicular* when they intersect at a 90-degree angle. Both of these instances are fairly easy to spot when you see the lines graphed, but how can you be sure that the lines are truly parallel or that the angle is really 90 degrees and not 89.9 degrees? The answer lies in the slopes.

Consider two lines, $y = m_1 x + b_1$ and $y = m_2 x + b_2$.

REMEMBER

Two lines are *parallel* when their slopes are equal ($m_1 = m_2$). Two lines are *perpendicular* when their slopes are negative reciprocals of one another $\left(m_2 = -\dfrac{1}{m_1} \right)$.

For example, the lines $y = 3x + 7$ and $y = 3x - 2$ are parallel. Both lines have slopes of 3, but their y-intercepts are different — one crosses the Y-axis at 7 and the other at –2. Figure 5-11a shows these two lines.

Figure 5-11:
Parallel lines have equal slopes, and perpendicular lines have slopes that are negative reciprocals.

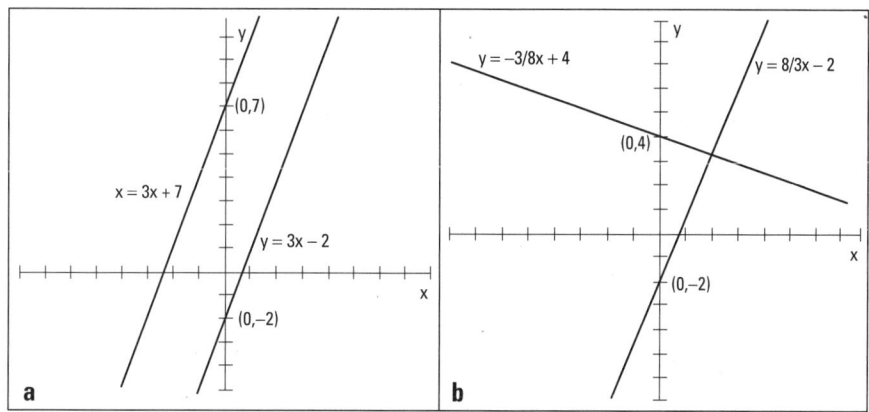

The lines $y = -\dfrac{3}{8} x + 4$ and $y = \dfrac{8}{3} x - 2$ are perpendicular. The slopes are negative reciprocals of one another. You see these two lines graphed in Figure 5-11b.

TIP

A quick way to determine if two numbers are negative reciprocals is to multiply them together; you should get a result of –1. For the previous example problem, you get $-\dfrac{3}{8} \cdot \dfrac{8}{3} = -1$.

Looking at 10 Basic Forms

You study many types of equations and graphs in Algebra II. You find very specific details on graphs of quadratics, polynomials, radicals, rationals, exponentials, and logarithms in the various chapters of this book devoted to the different types. What I present in this section is a general overview of some of the types of graphs, designed to help you distinguish one type from another as you get ready to ferret out all the gory details. Knowing the basic graphs is the starting point for graphing variations on the basic graphs or more complicated curves.

Ten basic graphs seem to occur most often in Algebra II. I cover the first, a line, earlier in this chapter (see the section "Graphing Lines"), but I also include it in the following sections with the other nine basic graphs.

Lines and quadratics

Figure 5-12a shows the graph of a line. Lines can rise or drop as you move from left to right, or they can be horizontal or vertical. The line in Figure 5-12a has a positive slope and is fairly steep. For more details on lines and other figures, see the section "Graphing Lines."

The slope-intercept equation of a line is $y = mx + b$.

Figure 5-12b shows a general quadratic (second-degree polynomial). This curve is called a *parabola*. Parabolas can open upward like this one, downward, to the left, or to the right. Figures 5-3, 5-4, and 5-5b show you some other quadratic curves. (Chapters 3 and 7 go into more quadratic detail.)

Two general equations for quadratics are $y = ax^2 + bx + c$ and $x = ay^2 + by + c$.

Figure 5-12: Graphs of a steep line and an upward-facing quadratic.

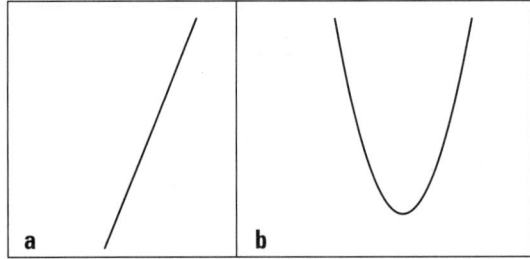

a   b

Cubics and quartics

Figure 5-13a shows the graph of a cubic (third-degree polynomial). A cubic curve (a polynomial with degree 3) can be an S-shaped curve as shown, or it can flatten out in the middle and not feature those turns. A cubic can appear to come from below and end up high and to the right, or it can drop from the left, take some turns, and continue on down (for more on polynomials, check out Chapter 8).

The general equation for a cubic is $y = ax^3 + bx^2 + cx + d$.

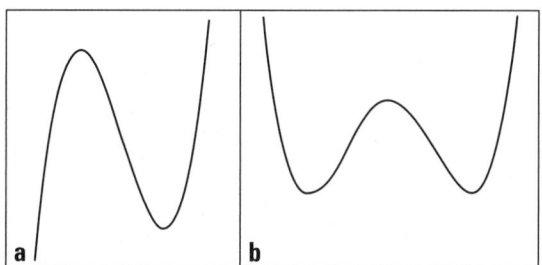

Figure 5-13:
Graphs
of an
S-shaped
cubic and a
W-shaped
quartic.

Figure 5-13b shows the graph of a quartic (fourth-degree polynomial). The graph has a distinctive W-shape, but that shape can flatten out, depending on how many terms appear in its equation. Similar to the quadratic (parabola), the quartic can open downward, in which case it looks like an M rather than a W.

The quartic has the following general equation: $y = ax^4 + bx^3 + cx^2 + dx + e$.

Radicals and rationals

In Figure 5-14a, you see a radical curve, and in Figure 5-14b, you see a rational curve. They seem like opposites, don't they? But a major characteristic radical and rational curves have in common is that you can't draw them everywhere. A radical curve can have an equation such as $y = \sqrt{x-4}$, where the square root doesn't allow you to put values under the radical that give you a negative result. In this case, you can't use any number smaller than four, so you have no graph when x is smaller than four. The radical curve in Figure 5-14a shows an abrupt stop at the left end, which is typical of a radical curve. (You can find more on radicals in Chapter 4.)

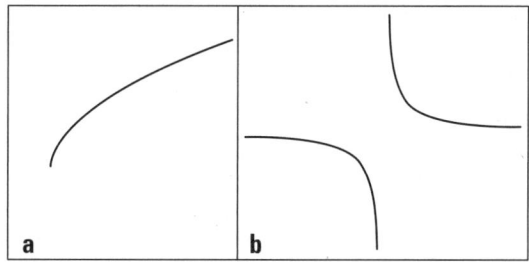

Figure 5-14:
Graphs of
radicals
often have
abrupt
stops, and
graphs of
rationals
have gaps.

A general equation for a radical curve may be $y = \sqrt[n]{x+a}$, where n is an even number.

Rational curves have a different type of restriction. When the equation of a curve involves a fraction, it opens itself up for the possibility of having a zero in the denominator of the fraction — a big no-no. You can't have a zero in the denominator because math offers no such number, which is why the graphs of rational curves have spaces in them — places where the graph doesn't touch. In Figure 5-14b, you see the graph of $y = \frac{1}{x}$. The graph doesn't have any value when $x = 0$. (Refer to Chapter 9 for information on rational functions.)

A general equation for a rational curve may be $y = \frac{a}{x+b}$.

Exponential and logarithmic curves

Exponential and logarithmic curves are sort of opposites because the exponential and logarithmic functions are inverses of one another. Figure 5-15a shows you an exponential curve, and Figure 5-15b shows a logarithmic curve. Exponential curves have a starting point, or initial value, of a, which is actually the y-intercept. The value of b determines whether the curve rises or falls in that gentle C shape. If b is larger than one, the curve rises. If b is between zero and one, the curve falls. They generally rise or fall from left to right, but they face upward (called *concave up*). It helps to think of them collecting snow or sand in their gentle curves. Logarithmic curves also can rise or fall as you go from left to right, but they face downward.

Figure 5-15: The graph of an exponential faces upward, and a graph of a logarithmic faces downward.

Both exponential and logarithmic curves can model growth (or decay), where a previous amount is multiplied by some factor. The change continues on at the same rate throughout, so you produce steady curves up or down. (See Chapter 10 for more info on exponential and logarithmic functions.)

The general form for an exponential curve is $y = ab^x$.

The general form for a logarithmic curve is $y = \log_b x$.

Absolute values and circles

The absolute value of a number is a positive value, which is why you get a distinctive V-shaped curve from absolute-value relations. Figure 5-16a shows a typical absolute value curve.

A general equation for absolute value is $y = |ax + b|$.

The graph in Figure 5-16b is probably the most recognizable shape this side of the line. A circle goes round and round a fixed distance from its center. (I cover circles and other conics in detail in Chapter 11.)

Circles with their centers at the origin have equations such as $x^2 + y^2 = r^2$.

Figure 5-16:
Graphs of
absolute
values show
only positive
values, and
graphs
of circles
go round
and round.

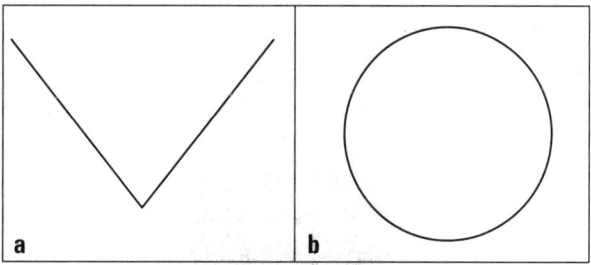

Solving Problems with a Graphing Calculator

You may think that discussions on graphing aren't necessary if you have a calculator that does the graphing for you. Not so, Kemo Sabe. (If you aren't a *Lone Ranger* fan, you can read the last sentence as, "Oh, that's not true, my dear friend.")

Graphing calculators *are* wonderful. They take much of the time-consuming drudgery out of graphing lines, curves, and all sorts of intricate functions. But you have to know how to use the calculator correctly and effectively. Two of the biggest stumbling blocks students and teachers face when using a graphing calculator to solve problems have to do with entering a function correctly and then finding the correct graphing window. You can use your calculator's manual to get all the other nitty-gritty details.

Entering equations into graphing calculators correctly

The trick to entering equations correctly into your graphing calculator is to insert what you mean (and mean what you insert). Your calculator will follow the *order of operations,* so you need to use parentheses to create the equation that you really want. (Refer to Chapter 1 if you need a review of the order of operations.) The four main trouble areas are fractions, exponents, radicals, and negative numbers.

Facing fractions

Say, for example, that you want to graph the equations $y = \dfrac{1}{x+2}$, $y = \dfrac{x}{x+3}$, and $y = \dfrac{1}{4}x + 3$. You may mistakenly enter them into your graphing calculator as $y_1 = 1 \div x + 2$, $y_2 = x \div x + 3$, and $y_3 = 1 \div 4x + 3$. They don't appear at all as you expected. Why is that?

When you enter $1 \div x + 2$, your calculator reads that as "Divide 1 by x and then add 2." Instead of having a gap when $x = -2$, like you expect, the gap or discontinuity appears when $x = 0$ — your hint that you have a problem. If you want the numerator divided by the entire denominator, you use a parenthesis around the terms in the denominator. Enter the first equation as $y_1 = 1 \div (x + 2)$.

Most graphing calculators try to connect the curves, even when you don't want the curves to be connected. You see what appear to be vertical lines in the graph where you shouldn't see anything (see the section "Disconnecting curves" later in this chapter).

If you enter the second equation as $y_2 = x \div x + 3$, the calculator does the division first, as dictated by the order of operations, and then adds 3. You get a horizontal line for the graph rather than the rational curve you're expecting. The correct way to enter this equation is $y_2 = x \div (x + 3)$.

The last equation, if entered as $y_3 = 1 \div 4x + 3$, may or may not come out correctly. Some calculators will grab the x and put it in the denominator of the fraction. Others will read the equation as ¼ times the x. To be sure, you should write the fraction in a parenthesis: $y_3 = (¼)x + 3$.

Expressing exponents

Graphing calculators have some built-in exponents that make it easy to enter the expressions. You have a button for squared, ^{2}, and, if you look hard, you can find one for cubed, ^{3}. Calculators also provide an alternative to these options, and to all the other exponents: You can enter an exponent with a *caret,* $^\wedge$ (pronounced *carrot,* like what Bugs Bunny eats).

You have to be careful when entering exponents with carets. When graphing $y = x^{1/2}$ or $y = 2^{x+1}$, you don't get the correct graphs if you type in $y_1 = x^\wedge \frac{1}{2}$ and $y_2 = 2^\wedge x + 1$. Instead of getting a radical curve (raising to the $\frac{1}{2}$ power is the same as finding the square root) for y_1, you get a line. You need to put the fractional exponent in parenthesis: $y_1 = x^\wedge(\frac{1}{2})$. The same goes for the second curve. When you have more than one term in an exponent, you put the whole exponent in parenthesis: $y_2 = 2^\wedge(x + 1)$.

Raising Cain over radicals

Graphing calculators often have keys for square roots — and even other roots. The main problem is not putting every term that falls under the radical where it belongs.

If you want to graph $y = \sqrt{4 - x}$ and $y = \sqrt[6]{4 + x}$, for example, you use parentheses around what falls under the radical and parentheses around any fractional exponents. Here are two ways to enter each of the equations:

$$y_1 = \sqrt{(4 - x)}$$
$$y_2 = (4 - x)^{(1/2)}$$
$$y_3 = \sqrt[6]{(4 + x)}$$
$$y_4 = (4 + x)^{(1/6)}$$

The entries for y_1 and y_3 depend on your calculator having the appropriate buttons. The other two entries work whether you have those wonderful buttons or not. It may be just as easy to use carets for your powers (see the previous section for more on carets).

Negating or subtracting

In all phases of algebra, you treat *subtract, minus, opposite,* and *negative* the same way, and they all have the same symbol, "–". Graphing calculators aren't quite as forgiving. They have a special button for *negative,* and they have another button that means *subtract.*

If you're performing the operation of subtraction, such as in $4 - 3$, you use the subtract button found between the addition and multiplication buttons. If you want to type in the number –3, you use the button with a parenthesis around the negative sign, (–). You get an error message if you enter the wrong one, but it helps to know why the action is considered an error.

Another problem with negatives has to do with the order of operations (see Chapter 1). If you want to square –4, you can't enter -4^2 into your calculator — if you do, you get the wrong answer. Your calculator squares the 4 and then takes its opposite, giving you –16. To square the –4, you put parentheses around both the negative sign and the 4: $(-4)^2 = 16$.

Looking through the graphing window

Graphing calculators usually have a standard window to graph in — from –10 to +10, both side-to-side and up-and-down. The standard window is a wonderful starting point and does the job for many graphs, but you need to know when to change the window or view of your calculator to solve problems.

Using x-intercepts

The graph of $y = x(x - 11)(x + 12)$ appears only as a set of axes if you use the standard setting. The funky picture shows up because the two intercepts — $(11, 0)$ and $(-12, 0)$ — don't show up if you just go from –10 to 10. (In Chapter 8, you find out how to determine these intercepts, if you're not familiar with that process.) Knowing where the x-intercepts are helps you to set the window properly. You need to change the view or window on your calculator so that it includes all the intercepts. One possibility for the window of the example curve is to have the x values go from –13 to 12 (one lower than the lowest and one higher than the highest).

If you change the window in this fashion, you have to adjust the window upward and downward for the heights of the graph, or you can use *Fit*, a graphing capability of most graphing calculators. After you set how far to the left and right the window must go, you set the calculator to adjust the upward and downward directions with *Fit*. You have to tell the calculator how wide you want the graph to be. After you set the parameters, the *Fit* button automatically adjusts the height (up and down) so the window includes all the graph in that region.

Disconnecting curves

Your calculator, bless its heart, tries to be very helpful. But sometimes helpful isn't accurate. For example, when you graph rational functions that have gaps, or graph piecewise functions that have holes, your calculator tries to connect the two pieces. You can just ignore the connecting pieces of graph, or you can change the mode of your calculator to *dot mode*. Most calculators have a *Mode* menu where you can change decimal values, angle measures, and so on. Go to that general area and toggle from *connected* to *dot* mode.

The downside to using dot mode is that sometimes you lose some of the curve — it won't be nearly as complete as with the connected mode. As long as you recognize what's happening with your calculator, you can adjust for its quirks.

Part II

Facing Off with Functions

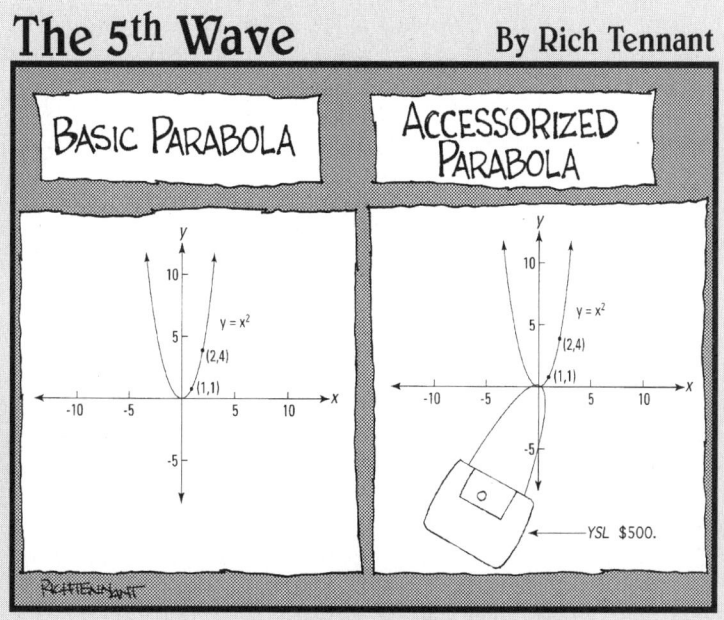

In this part . . .

Solving equations is fundamental to basic algebra, but advanced algebra presents new challenges in the form of functions. Functions are what you solve the equations for. The ability to solve functions is what separates highly trained Algebra II students from the rookie Algebra I crowd. Functions are specific types of equations that have certain properties that set them apart from run-of-the-mill equations.

In Part II, you see how each type of function has its own peculiarities (or characteristics, to be nicer). Quadratic functions have distinctive shapes, exponential functions just go on and on, and rational functions can have asymptotes. Doesn't that sound like fun? You graph and solve functions for zero and group them together on the way to performing all sorts of tasks, and you see the effect of using real versus imaginary numbers with the functions.

Chapter 6

Formulating Function Facts

*I*s your computer functioning well? Are you going to your evening function for work? What *is* all this business about function? In algebra, the word *function* is very specific. You reserve it for certain math expressions that meet the tough standards of input and output values, as well as other mathematical rules of relationships. Therefore, when you hear that a certain relationship is a function, you know that the relationship meets some particular requirements.

In this chapter, you find out more about these requirements. I also cover topics ranging from the domain and range of functions to the inverses of functions, and I show you how to deal with piecewise functions and do composition of functions. After grazing through these topics, you can confront a function equation with great confidence and a plan of attack.

Defining Functions

A *function* is a relationship between two variables that features exactly one output value for every input value — in other words, exactly one answer for every number inserted.

For example, the equation $y = x^2 + 5x - 4$ is a function equation or function rule that uses the variables x and y. The x is the *input variable,* and the y is the *output variable.* (For more on how these designations are determined, flip to the section "Homing In on Domain and Range" later in this chapter.) If you input the number 3 for each of the x's, you get $y = 3^2 + 5(3) - 4 = 9 + 15 - 4 = 20$.

The output is 20, the only possible answer. You won't get another number if you input the 3 again.

The single-output requirement for a function may seem like an easy requirement to meet, but you encounter plenty of strange math equations out there. You have to watch out.

Introducing function notation

Functions feature some special notation that makes working with them much easier. The notation doesn't change any of the properties, it just allows you to identify different functions quickly and indicate various operations and processes more efficiently.

The variables x and y are pretty standard in functions and come in handy when you're graphing functions. But mathematicians also use another format called *function notation*. For instance, say I have these three functions:

$$y = x^2 + 5x - 4$$
$$y = \sqrt{3x - 8}$$
$$y = 6xe^x - 2e^{2x}$$

Assume you want to call them by name. No, I don't mean you should say, "Hey, you over there, Clarence!" Instead, you should say

$$f(x) = x^2 + 5x - 4$$
$$g(x) = \sqrt{3x - 8}$$
$$h(x) = 6xe^x - 2e^{2x}$$

The names of these functions are f, g, and h. (How boring!) You read them as follows: "f of x is x squared plus $5x$ minus 4," and so on. When you see a bunch of functions written together, you can be efficient by referring to individual functions as f or g or h so listeners don't have to question what you're referring to.

Evaluating functions

When you see a written function that uses function notation, you can easily identify the input variable, the output variable, and what you have to do to *evaluate* the function for some input (or replace the variables with numbers and simplify). You can do so because the input value is placed in the parenthesis right after the function name or output value.

If you see $g(x) = \sqrt{3x-8}$, for example, and you want to evaluate it for $x = 3$, you write down $g(3)$. This means you substitute a 3 for every x in the function expression and perform the operations to get the output answer: $g(3) = \sqrt{3(3)-8} = \sqrt{9-8} = \sqrt{1} = 1$. Now you can say that $g(3) = 1$, or "g of 3 equals 1." The output of the function g is 1 if the input is 3.

Homing In on Domain and Range

The input and output values of a function are of major interest to people working in algebra. These terms don't yet strum your guitar? Well, allow me to pique your interest. The words *input* and *output* describe what's happening in the function (namely what number you put in and what result comes out), but the official designations for these sets of values are *domain* and *range*.

Determining a function's domain

The *domain* of a function consists of all the input values of the function (think of a king's domain of all his servants entering his kingdom). In other words, the domain is the set of all numbers that you can input without creating an unwanted or impossible situation. Such situations can occur when operations appear in the definition of the function, such as fractions, radicals, logarithms, and so on.

Many functions have no exclusions of values, but fractions are notorious for causing trouble when zeros appear in the denominators. Radicals have restrictions as to what you can find roots of, and logarithms can deal only with positive numbers.

You need to be prepared to determine the domain of a function so that you can tell where you can use the function — in other words, for what input values it does any good. You can determine the domain of a function from its equation or function definition. You look at the domain in terms of which real numbers you can use for input and which ones you have to eliminate. You can express the domain by using the following:

- ✔ **Words:** The domain of $f(x) = x^2 + 2$ is all real numbers (anything works).
- ✔ **Inequalities:** The domain of $g(x) = \sqrt{x}$ is $x \geq 0$.
- ✔ **Interval notation:** The domain of $h(x) = ln(x-1)$ is $(1, \infty)$. (Check out Chapter 2 for information on interval notation.)

The way you express domain depends on what's required of the task you're working on — evaluating functions, graphing, or determining a good fit as a model, to name a few. Here are some examples of functions and their respective domains:

- $f(x) = \sqrt{x - 11}$. The domain consists of the number 11 and every greater number thereafter. You write this as $x \geq 11$ or, in interval notation, $[11, \infty)$. You can't use numbers smaller than 11 because you'd be taking the square root of a negative number, which isn't a real number.

- $g(x) = \dfrac{x}{x^2 - 4x - 12} = \dfrac{x}{(x - 6)(x + 2)}$. The domain consists of all real numbers except 6 and –2. You write this domain as $x < -2$ or $-2 < x < 6$ or $x > 6$, or, in interval notation, as $(-\infty, -2) \cap (-2, 6) \cap (6, \infty)$. It may be easier to simply write "All real numbers except $x = -2$ and $x = 6$." The reason you can't use –2 or 6 is because these numbers result in a 0 in the denominator of the fraction, and a fraction with 0 in the denominator creates a number that doesn't exist.

- $h(x) = x^3 - 3x^2 + 2x - 1$. The domain of this function is all real numbers. You don't have to eliminate anything, because you can't find a fraction with the potential of a zero in the denominator, and you have no radical to put a negative value into. You write this domain with a fancy R, $\Re$, or with interval notation as $(-\infty, \infty)$.

Describing a function's range

The *range* of a function is all its output values — every value you get by inputting the domain values into the *rule* (the function equation) for the function. You may be able to determine the range of a function from its equation, but sometimes you have to graph it to get a good idea of what's going on.

A range may consist of all real numbers, or it may be restricted because of the way a function equation is constructed. You have no easy way to describe ranges — at least, not as easy as describing domains — but you can discover clues with some functions by looking at their graphs and with others by knowing the characteristics of those kinds of curves.

The following are some examples of functions and their ranges. Like domains, you can express ranges in words, inequalities, or interval notation (see Chapter 2):

- $k(x) = x^2 + 3$. The range of this function consists of the number 3 and any number greater than 3. You write the range as $k \geq 3$ or, in interval notation, $[3, \infty)$. The outputs can never be less than 3 because the numbers you input are squared. The result of squaring a real number is always positive (or if you input zero, you square zero). If you add a positive number or 0 to 3, you never get anything smaller than 3.

✔ $m(x) = \sqrt{x+7}$. The range of this function consists of all positive numbers and zero. You write the range as $m \geq 0$ or, in interval notation, $[0, \infty)$. The number under the radical can never be negative, and all the square roots come out positive or zero.

✔ $p(x) = \dfrac{2}{x-5}$. Some function's equations, such as this one, don't give an immediate clue to the range values. It often helps to sketch the graphs of these functions. Figure 6-1 shows the graph of the function p. See if you can figure out the range values before peeking at the following explanation.

Figure 6-1:
Try graphing
equations
that don't
give an
obvious
range.

The graph of this function never touches the X-axis, but it gets very close. For the numbers in the domain bigger than five, the graph has some really high y values and some y values that get really close to zero. But the graph never touches the X-axis, so the function value never really reaches zero. For numbers in the domain smaller than five, the curve is below the X-axis. These function values are negative — some really small. But, again, the y values never reach zero. So, if you guessed that the range of the function is every real number except zero, you're right! You write the range as $y \neq 0$, or $(-\infty, 0) \cap (0, \infty)$. Did you also notice that the function doesn't have a value when $x = 5$? This happens because 5 isn't in the domain.

When a function's range has a lowest or highest value, it presents a case of an *absolute minimum* or an *absolute maximum*. For instance, if the range is $[2, \infty)$, the range includes the number 2 and every value bigger than 2. The *absolute minimum* value that the function can output is 2. Not all functions have these absolute minimum and maximum values, but being able to identify them is important — especially if you use functions to determine your weekly pay. Do you want a cap on your pay at the *absolute maximum* of $500?

For some tips on how to graph functions, head to Chapters 7 through 10, where I discuss the graphs of the different kinds of functions.

Betting on Even or Odd Functions

You can classify numbers as even or odd (and you can use this information to your advantage; for example, you know you can divide even numbers by two and come out with a whole number). You can also classify some functions as even or odd. The even and odd integers (like 2, 4, 6, and 1, 3, 5) play a role in this classification, but they aren't the be-all and end-all. You have to put a bit more calculation work into it. If you didn't have to, you would've mastered this stuff in third grade, and I'd have nothing to write about.

Recognizing even and odd functions

An *even function* is one in which a domain value (an input) and its opposite result in the same range value (output) — for example, $f(-x) = f(x)$. An *odd* function is one in which a domain value and its opposite produce opposite results in the range — for example, $f(-x) = -f(x)$.

To determine if a function is even or odd (or neither), you replace every x in the function equation with $-x$ and simplify. If the function is even, the simplified version looks exactly like the original. If the function is odd, the simplified version looks like what you get after multiplying the original function equation by -1.

The descriptions for even and odd functions may remind you of how even and odd numbers for exponents act (see Chapter 10). If you raise -2 to an even power, you get a positive number — $(-2)^4 = 16$. If you raise -2 to an odd power, you get a negative (opposite) result — $(-2)^5 = -32$.

The following are examples of some even and odd functions, and I explain how you label them so you can master the practice on your own:

- ✔ $f(x) = x^4 - 3x^2 + 6$ is even, because whether you input 2 or -2, you get the same output:

   • $f(2) = (2)^4 - 3(2)^2 + 6 = 16 - 12 + 6 = 10$

   • $f(-2) = (-2)^4 - 3(-2)^2 + 6 = 16 - 3(4) + 6 = 10$

   So, you can say $f(2) = f(-2)$.

- ✔ $g(x) = \dfrac{12x}{x^2 + 2}$ is odd, because the inputs 2 and -2 give you opposite answers:

- $g(2) = \dfrac{12(2)}{(2)^2 + 2} = \dfrac{24}{4+2} = \dfrac{24}{6} = 4$

- $g(-2) = \dfrac{12(-2)}{(-2)^2 + 2} = \dfrac{-24}{4+2} = \dfrac{-24}{6} = -4$

So, you can say that $g(-2) = -g(2)$.

You can't say that a function is even just because it has even exponents and coefficients, and you can't say that a function is odd just because the exponents and coefficients are odd numbers. If you do make these assumptions, you classify the functions incorrectly, which messes up your graphing. You have to apply the rules to determine which label a function has.

Applying even and odd functions to graphs

The biggest distinction of even and odd functions is how their graphs look:

✔ **Even functions:** The graphs of even functions are symmetric with respect to the Y-axis (the vertical axis). You see what appears to be a mirror image to the left and right of the vertical axis. For an example of this type of symmetry, see Figure 6-2a, which is the graph of the even function $f(x) = \dfrac{5}{x^2 + 1}$.

✔ **Odd functions:** The graphs of odd functions are symmetric with respect to the origin. The symmetry is radial, or circular, so it looks the same if you rotate the graph by 180 degrees. The graph in Figure 6-2b, which is the odd function $g(x) = x^3 - 8x$, displays origin symmetry.

Figure 6-2:
An even (a) and odd (b) function.

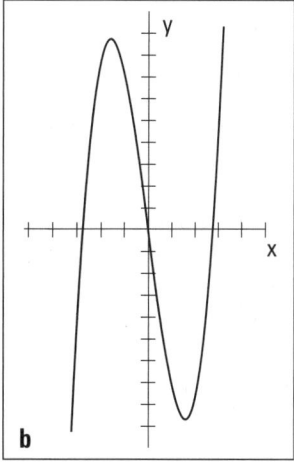

You may be wondering whether you can have symmetry with respect to the X-axis. After all, is the Y-axis all that better than the X-axis? I'll leave that up to you, but yes, X-axis symmetry does exist — just not in the world of functions. By its definition, a function can have only one *y* value for every *x* value. If you have points on either side of the X-axis, above and below an *x* value, you don't have a function. Head to Chapter 11 if you want to see some pictures of curves that are symmetric all over the place.

Facing One-to-One Confrontations

Functions can have many classifications or names, depending on the situation (maybe you want to model a business transaction or use them to figure out payments and interest) and what you want to do with them (put the formulas or equations in spreadsheets or maybe just graph them, for example). One very important classification is deciding whether a function is one-to-one.

Defining one-to-one functions

A function is *one-to-one* if you calculate exactly one output value for every input value *and* exactly one input value for every output value. Formally, you write this definition as follows:

If $f(x_1) = f(x_2)$, then $x_1 = x_2$

In simple terms, if the two output values are the same, the two input values must also be the same.

One-to-one functions are important because they're the only functions that can have inverses, and functions with inverses aren't all that easy to come by. If a function has an inverse, you can work backward and forward — find an answer if you have a question and find the original question if you know the answer (sort of like *Jeopardy*). For more on inverse functions, see the section "Singing Along with Inverse Functions" later in this chapter.

An example of a one-to-one function is $f(x) = x^3$. The rule for the function involves cubing the variable. The cube of a positive number is positive, and the cube of a negative number is negative. Therefore, every input has a unique output — no other input value gives you that output.

Some functions without the one-to-one designation may look like the previous example, which *is* one-to-one. Take $g(x) = x^3 - x$, for example. This counts as a function because only one output comes with every input. However, the function isn't one-to-one, because you can create many outputs or function values from more than one input. For instance, $g(1) = (1)^3 - (1) = 1 - 1 = 0$, and $g(-1) = (-1)^3 - (-1) = -1 + 1 = 0$. You have two inputs, 1 and -1, that result in the same output of 0.

Functions that don't qualify for the one-to-one label can be hard to spot, but you can rule out any functions with all even-numbered exponents right away. Functions with absolute value usually don't cooperate either.

Eliminating one-to-one violators

You can determine which functions are one-to-one and which are violators by *sleuthing* (guessing and trying), using algebraic techniques, and graphing. Most mathematicians prefer the graphing technique because it gives you a nice, visual answer. The basic graphing technique is the horizontal line test. But, to better understand this test, you need to meet its partner, the vertical line test. (I show you how to graph various functions in Chapters 7 through 10.)

Vertical line test

The graph of a function always passes the *vertical line test*. The test stipulates that any vertical line drawn through the graph of the function passes through that function no more than once. This is a visual illustration that only one y value (output) exists for every x value (input), a rule of functions. Figure 6-3a shows a function that passes the vertical line test, and Figure 6-3b contains a curve that isn't a function and therefore flunks the vertical line test.

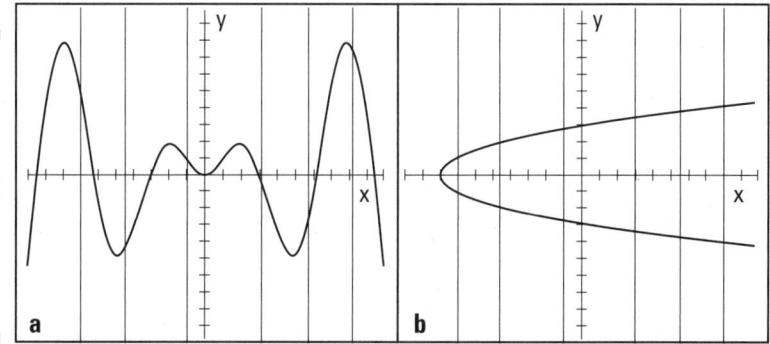

Figure 6-3: A function passes the vertical line test, but a non-function inevitably fails.

Horizontal line test

All functions pass the vertical line test, but only one-to-one functions pass the *horizontal line test*. With this test, you can see if any horizontal line drawn through the graph cuts through the function more than one time. If the line passes through the function more than once, the function fails the test and therefore isn't a one-to-one function. Figure 6-4a shows a function that passes the horizontal line test, and Figure 6-4b shows a function that flunks it.

Both graphs in Figure 6-4 are functions, however, so they both pass the vertical line test.

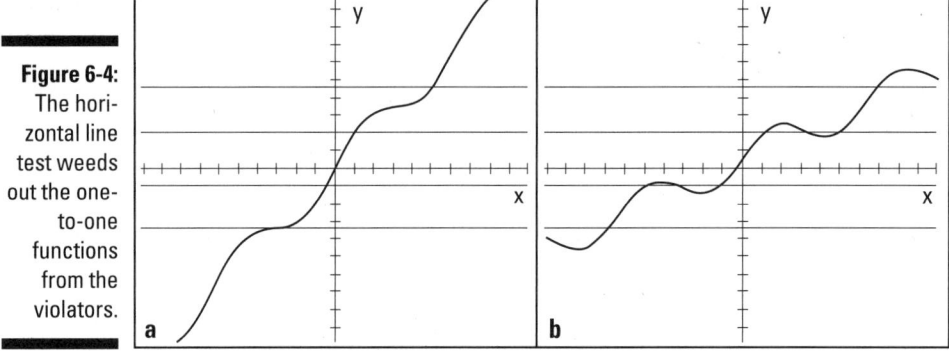

Going to Pieces with Piecewise Functions

A *piecewise function* consists of two or more function rules (function equations) pieced together (listed separately for different *x* values) to form one bigger function. A change in the function equation occurs for different values in the domain. For example, you may have one rule for all the negative numbers, another rule for numbers bigger than three, and a third rule for all the numbers between those two rules.

Piecewise functions have their place in situations where you don't want to use the same rule for everyone or everything. Should a restaurant charge a 3-year-old the same amount for a meal as it does an adult? Do you put on the same amount of clothing when the temperature is 20 degrees as you do in hotter weather? No, you place different rules on different situations. In mathematics, the piecewise function allows for different rules to apply to different numbers in the domain of a function.

Doing piecework

Piecewise functions are often rather contrived. Oh, they seem real enough when you pay your income tax or figure out your sales commission. But in an algebra discussion, it seems easier to come up with some nice equations to illustrate how piecewise functions work and then introduce the applications later. The following is an example of a piecewise function:

$$f(x) = \begin{cases} x^2 - 2 & \text{if } x \le -2 \\ 5 - x & \text{if } -2 < x \le 3 \\ \sqrt{x+1} & \text{if } x > 3 \end{cases}$$

With this function, you use one rule for all numbers smaller than or equal to –2, another rule for numbers between –2 and 3 (including the 3), and a final rule for numbers larger than 3. You still have only one output value for every input value. For instance, say you want to find the values of this function for *x* equaling –4, –2, –1, 0, 1, 3, and 5. Notice how you use the different rules depending on the input value:

$$f(-4) = (-4)^2 - 2 = 16 - 2 = 14$$
$$f(-2) = (-2)^2 - 2 = 4 - 2 = 2$$
$$f(-1) = 5 - (-1) = 5 + 1 = 6$$
$$f(0) = 5 - 0 = 5$$
$$f(1) = 5 - 1 = 4$$
$$f(3) = 5 - 3 = 2$$
$$f(5) = \sqrt{5+1} = \sqrt{6}$$

Figure 6-5 shows you the graph of the piecewise function with these function values.

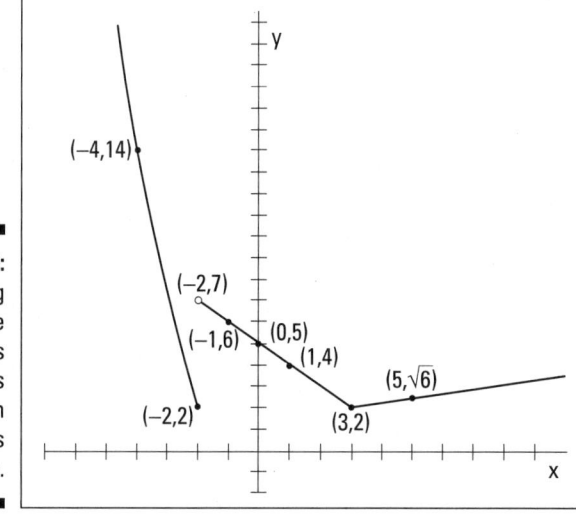

Notice the three different sections to the graph. The left curve and the middle line don't connect because a discontinuity exists when $x = -2$. A *discontinuity* occurs when a gap or hole appears in the graph. Also, notice that the left line falling toward the X-axis ends with a solid dot, and the middle section has an open circle just above it. These features preserve the definition of a function — only one output for each input. The dot tells you to use the rule on the left when $x = -2$.

The middle section connects at point (3, 2) because the rule on the right gets really, really close to the same output value as the rule in the middle when $x = 3$. Technically, you should draw both a hollow circle and a dot, but you really can't spot this feature just by looking at it.

Applying piecewise functions

Why in the world would you need to use a piecewise function? Do you have any good reason to change the rules right in the middle of things? I have two examples that aim to ease your mind — examples that you may relate to very well.

Utilizing a utility

Utility companies can use piecewise functions to charge different rates for users based on consumption levels. A big factory uses a ton of electricity and rightfully gets a different rate than a homeowner. Here's what the company Lightning Strike Utility uses to figure the charges for its customers:

$$C(h) = \begin{cases} 0.0747900h & \text{if } h < 1,000 \\ 74.79 + 0.052500(h - 1,000) & \text{if } 1,000 \leq h < 5,000 \\ 284.79 + 0.033300(h - 5,000) & \text{if } h \geq 5,000 \end{cases}$$

where h is the number of kilowatt hours and C is the cost in dollars.

So, how much does a homeowner who uses 750 kilowatt hours pay? How much does a business that uses 10,000 kilowatt hours pay?

You use the top rule for the homeowner and the bottom rule for the business, determined by what interval the input value lies in. The homeowner has the following equation:

$C(h) = 0.0747900h$

$C(750) = 56.0925$

The person using 750 kilowatt hours pays a little more than $56 per month. And the business?

$$C(h) = 284.79 + 0.033300(h - 5{,}000)$$
$$C(10{,}000) = 284.79 + 0.033300(10{,}000 - 5{,}000)$$
$$C(10{,}000) = 284.79 + 166.50 = 451.29$$

The company pays a little more than $451.

Taxing the situation

As April 15 rolls around, many people are faced with their annual struggle with income tax forms. The rate at which you pay tax is based on how much your adjusted income is — a graduated scale where (supposedly) people who make more money pay more income tax. The income values are the inputs (values in the domain), and the government determines the tax paid by putting numbers into the correct formula.

In 2005, a single taxpayer paid her income tax based on her taxable income, according to the following rules (laid out in a piecewise function):

$$X(t) = \begin{cases} 0.10t & t < 7{,}300 \\ 0.15(t - 7{,}300) + 730 & 7{,}300 \le t < 29{,}700 \\ 0.25(t - 29{,}700) + 4{,}090 & 29{,}700 \le t < 71{,}950 \\ 0.28(t - 71{,}950) + 14{,}653 & 71{,}950 \le t < 150{,}150 \\ 0.33(t - 150{,}150) + 36{,}549 & 150{,}150 \le t < 326{,}450 \\ 0.35(t - 326{,}450) + 94{,}728 & t \ge 326{,}450 \end{cases}$$

where t is the taxable income and X is the tax paid.

If her taxable income was $45,000, how much did she pay in taxes? Insert the value and follow the third rule, because 45,000 is between 29,700 and 71,950:

$$X(45{,}000) = 0.25(45{,}000 - 29{,}700) + 4{,}090$$
$$= 3{,}825 + 4{,}090 = 7{,}915$$

This person had to pay almost $8,000 in income taxes.

Composing Yourself and Functions

You can perform the basic mathematical operations of addition, subtraction, multiplication, and division on the equations used to describe functions. (You can also perform whatever simplification is possible on the different parts of the expression and write the result as a new function.) For example, you can take the two functions $f(x) = x^2 - 3x - 4$ and $g(x) = x + 1$ and perform the four operations on them:

$$f + g = (x^2 - 3x - 4) + (x + 1) = x^2 - 2x - 3$$

$$f - g = (x^2 - 3x - 4) - (x + 1) = x^2 - 3x - 4 - x - 1 = x^2 - 4x - 5$$

$$f \cdot g = (x^2 - 3x - 4)(x + 1)$$
$$= (x^2 - 3x - 4)(x) + (x^2 - 3x - 4)(1)$$
$$= x^3 - 3x^2 - 4x + x^2 - 3x - 4$$
$$= x^3 - 2x^2 - 7x - 4$$

$$f / g = \frac{x^2 - 3x - 4}{x + 1} = \frac{(x - 4)(x + 1)}{x + 1} = x - 4$$

Well done, but you have another operation at your disposal — an operation special to functions — called *composition.*

Performing compositions

No, I'm not switching the format to a writing class. The *composition* of functions is an operation in which you use one function as the input into another and perform the operations on that input function.

You indicate the composition of functions *f* and *g* with a small circle between the function names, $f \circ g$, and you define the composition as $f \circ g = f(g)$.

Here's how you perform an example composition, using the functions *f* and *g* from the preceding section:

$$f \circ g = f(g) = (g)^2 - 3(g) - 4$$
$$= (x + 1)^2 - 3(x + 1) - 4$$
$$= x^2 + 2x + 1 - 3x - 3 - 4$$
$$= x^2 - x - 6$$

The composition of functions isn't commutative (addition and multiplication are *commutative,* because you can switch the order and not change the result). The order in which you perform the composition — which function comes first — matters. The composition $f \circ g$ isn't the same as $g \circ f$, save one exception: when the two functions are inverses of one another (see the section "Singing Along with Inverse Functions" later in this chapter).

Simplifying the difference quotient

The *difference quotient* shows up in most high school Algebra II classes as an exercise you do after your instructor shows you the composition of functions. You perform this exercise because the difference quotient is the basis of the definition of the derivative. The difference quotient allows you to find the derivative, which allows you to be successful in calculus (because everyone wants to be successful in calculus, of course). So, where does the composition of functions come in? With the difference quotient, you do the composition of some designated function $f(x)$ and the function $g(x) = x + h$ or $g(x) = x + \Delta x$, depending on what calculus book you use.

The difference quotient for the function f is $\dfrac{f(x+h)-f(x)}{h}$. Yes, you have to memorize it.

Now, for an example, perform the difference quotient on the same function f from the previous section:

$$\frac{f(x+h)-f(x)}{h} = \frac{(x+h)^2 - 3(x+h) - 4 - (x^2 - 3x - 4)}{h}$$

Notice that you find the expression for $f(x + h)$ by putting $x + h$ in for every x in the function — $x + h$ is the input variable. Now, continuing on with the simplification:

$$= \frac{x^2 + 2xh + h^2 - 3x - 3h - 4 - x^2 + 3x + 4}{h}$$

$$= \frac{2xh + h^2 - 3h}{h}$$

Did you notice that x^2, $3x$, and 4 all appear in the numerator with their opposites? That's why they disappear. Now, to finish:

$$= \frac{h(2x + h - 3)}{h} = 2x + h - 3$$

Now, this may not look like much to you, but you've created a wonderful result. You're one step away from finding the derivative. Tune in next week at the same time . . . no, I lied. You need to look at *Calculus For Dummies,* by Mark Ryan (Wiley), if you can't stand the wait and really want to find the derivative. For now, you've just done some really decent algebra.

Singing Along with Inverse Functions

Some functions are *inverses* of one another, but a function can have an inverse only if it's one-to-one (see the section "Facing One-to-One Confrontations" earlier in this chapter if you need a refresher). If two functions *are* inverses of one another, each function undoes what the other does. In other words, you use them to get back where you started.

The notation for inverse functions is the exponent –1 written after the function name. The inverse of function $f(x)$, for example, is $f^{-1}(x)$. Here are two inverse functions and how they undo one another:

$$f(x) = \frac{x+3}{x-4}$$

$$f^{-1}(x) = \frac{4x+3}{x-1}$$

If you put 5 into function f, you get 8 as a result. If you put 8 into f^{-1}, you get 5 as a result — you're back where you started:

$$f(5) = \frac{5+3}{5-4} = \frac{8}{1} = 8$$

$$f^{-1}(8) = \frac{4(8)+3}{8-1} = \frac{32+3}{7} = \frac{35}{7} = 5$$

Now, what was the question? How can you tell with the blink of an eye when functions are inverses? Read on!

Determining if functions are inverses

In an example from the intro to this section, I tell you that two functions are inverses and then demonstrate how they work. You can't really *prove* that two functions are inverses by plugging in numbers, however. You may face a situation where a couple numbers work, but, in general, the two functions aren't really inverses.

The only way to be sure that two functions are inverses of one another is to use the following general definition: Functions f and f^{-1} are inverses of one another only if $f\left(f^{-1}(x)\right) = x$ *and* $f^{-1}\left(f(x)\right) = x$.

In other words, you have to do the composition in both directions (do $f \circ g$ and then do $g \circ f$ in the opposite order) and show that both result in the single value x.

For some practice, show that $f(x) = \sqrt[3]{2x - 3} + 4$ and $g(x) = \dfrac{(x-4)^3 + 3}{2}$ are inverses of one another. First, you perform the composition $f \circ g$:

$$f \circ g = f(g) = \sqrt[3]{2(g) - 3} + 4$$

$$= \sqrt[3]{2\left(\dfrac{(x-4)^3 + 3}{2}\right) - 3} + 4$$

$$= \sqrt[3]{(x-4)^3 + 3 - 3} + 4$$

$$= \sqrt[3]{(x-4)^3} + 4$$

$$= (x-4) + 4$$

$$= x$$

Now you perform the composition in the opposite order:

$$g \circ f = \dfrac{(f-4)^3 + 3}{2}$$

$$= \dfrac{\left((\sqrt[3]{2x-3} + 4) - 4\right)^3 + 3}{2}$$

$$= \dfrac{\left(\sqrt[3]{2x-3}\right)^3 + 3}{2}$$

$$= \dfrac{(2x-3) + 3}{2}$$

$$= \dfrac{2x}{2}$$

$$= x$$

Both come out with a result of x, so the functions are inverses of one another.

Solving for the inverse of a function

Up until now in this section, I've given you two functions and told you that they're inverses of one another. How did I know? Was it magic? Did I pull the functions out of a hat? No, you have a nice process to use. I can show you my secret so you can create all sorts of inverses for all sorts of functions. Lucky you! The following lists gives you step-by-step process that I use (more memorization to come).

To find the inverse of the one-to-one function $f(x)$, follow these steps:

1. **Rewrite the function, replacing $f(x)$ with y to simplify the notation.**
2. **Change each y to an x and each x to a y.**
3. **Solve for y.**
4. **Rewrite the function, replacing the y with $f^{-1}(x)$.**

Here's an example of how you work it. Find the inverse of the function $f(x) = \dfrac{x}{x-5}$:

1. **Rewrite the function, replacing $f(x)$ with y to simplify the notation.**

$$y = \dfrac{x}{x-5}$$

2. **Change each y to an x and each x to a y.**

$$x = \dfrac{y}{y-5}$$

3. **Solve for y.**

$$x = \dfrac{y}{y-5}$$
$$x(y-5) = y$$
$$xy - 5x = y$$
$$xy - y = 5x$$
$$y(x-1) = 5x$$
$$y = \dfrac{5x}{x-1}$$

4. **Rewrite the function, replacing the y with $f^{-1}(x)$.**

$$f^{-1}(x) = \dfrac{5x}{x-1}$$

This process helps you find the inverse of a function, if it has one. If you can't solve for the inverse, the function may not be one-to-one to begin with. For instance, if you try to solve for the inverse of the function $f(x) = x^2 + 3$, you get stuck when you have to take a square root and don't know if you want a positive or negative root. These are the kinds of roadblocks that alert you to the fact that the function has no inverse — and that the function isn't one-to-one.

Chapter 7

Sketching and Interpreting Quadratic Functions

A quadratic function is one of the more recognizable and useful polynomial (multi-termed) functions found in all of algebra. The function describes a graceful U-shaped curve called a *parabola* that you can quickly sketch and easily interpret. People use quadratic functions to model economic situations, physical training progress, and the paths of comets. How much more useful can math get?

The most important features to recognize in order to sketch a parabola are the opening (up or down, steep or wide), the intercepts, the vertex, and the axis of symmetry. In this chapter, I show you how to identify all these features within the standard form of the quadratic function. I also show you some equations of parabolas that model events.

Interpreting the Standard Form of Quadratics

A *parabola* is the graph of a quadratic function. The graph is a nice, gentle, U-shaped curve that has points located an equal distance on either side of a line running up through its middle — called its *axis of symmetry*. Parabolas can be turned upward, downward, left, or right, but parabolas that represent functions only turn up or down. (In Chapter 11, you find out more about the other types of parabolas in the general discussion of conics.) The standard form for the quadratic function is

$$f(x) = ax^2 + bx + c$$

The coefficients (multipliers of the variables) a, b, and c are real numbers; a can't be equal to zero because you'd no longer have a quadratic function. You have plenty to discover from the simple standard form equation. The coefficients a and b are important, and some equations may not have all three of the terms in them. As you can see, there's meaning in everything (or nothing)!

Starting with "a" in the standard form

As the lead coefficient of the standard form of the quadratic function $f(x) = ax^2 + bx + c$, a gives you two bits of information: the direction in which the graphed parabola opens, and whether the parabola is steep or flat. Here's the breakdown of how the sign and size of the lead coefficient, a, affect the parabola's appearance:

- ✔ If a is positive, the graph of the parabola opens upward (see Figures 7-1a and 7-1b).

- ✔ If a is negative, the graph of the parabola opens downward (see Figures 7-1c and 7-1d).

- ✔ If a has an absolute value greater than one, the graph of the parabola is steep (see Figures 7-1a and 7-1c). (See Chapter 2 for a refresher on absolute values.)

- ✔ If a has an absolute value less than one, the graph of the parabola flattens out (see Figures 7-1b and 7-1d).

Figure 7-1:
Parabolas
opening up
and down,
appearing
steep and
flat.

If you remember the four rules that identify the lead coefficient, you don't even have to graph the equation to describe how the parabola looks. Here's how you can describe some parabolas from their equations:

$y = 4x^2 - 3x + 2$: You say that this parabola is steep and opens upward because the lead coefficient is positive and greater than one.

$y = -\frac{1}{3}x^2 + x - 11$: You say that this parabola is flattened out and opens downward because the lead coefficient is negative, and the absolute value of the fraction is less than one.

$y = 0.002x^2 + 3$: You say that this parabola is flattened out and opens upward because the lead coefficient is positive, and the decimal value is less than one. In fact, the coefficient is so small that the flattened parabola almost looks like a horizontal line.

Following up with "b" and "c"

Much like the lead coefficient in the quadratic function (see the previous section), the terms b and c give you plenty of information. Mainly, the terms tell you a lot if they're *not* there. In the next section, you find out how to use the terms to find intercepts (or zeros). For now, you concentrate on their presence or absence.

The lead coefficient, a, can never be equal to zero. If that happens, you no longer have a quadratic function, and this discussion is finished. As for the other two terms:

✔ If the second coefficient, b, is zero, the parabola straddles the Y-axis. The parabola's *vertex* — the highest or lowest point on the curve, depending on which way it faces — is on that axis, and the parabola is symmetric about the axis (see Figure 7-2a, a graph of a quadratic function where $b = 0$). The second term is the x term, so if the coefficient b is zero, the second term disappears. The standard equation becomes $y = ax^2 + c$, which makes finding intercepts very easy (see the following section).

✔ If the last coefficient, c, is zero, the graph of the parabola goes through the origin — in other words, one of its intercepts is the origin (see Figure 7-2b, a graph of a quadratic function where $c = 0$). The standard equation becomes $y = ax^2 + bx$, which you can easily factor into $y = x(ax + b)$. (See Chapters 1 and 3 for more on factoring.)

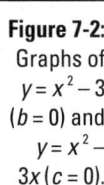

Figure 7-2:
Graphs of
$y = x^2 - 3$
($b = 0$) and
$y = x^2 - 3x$ ($c = 0$).

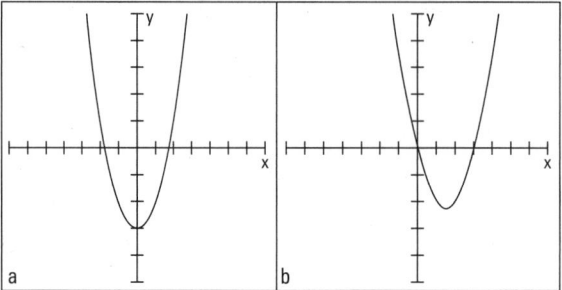

Investigating Intercepts in Quadratics

The *intercepts* of a quadratic function (or any function) are the points where the graph of the function crosses the X-axis or Y-axis. The graph of a function can cross the X-axis any number of times, but it can cross the Y-axis only once.

Why be concerned about the intercepts of a parabola? In real-life situations, the intercepts occur at points of interest — for instance, at the initial value of an investment or at the break-even point for a business.

Intercepts are also very helpful when you're graphing a parabola. The points are easy to find because one of the coordinates is always zero. If you have the intercepts, the vertex (see "Going to the Extreme: Finding the Vertex" later in the chapter), and what you know about the symmetry of the parabola (as I discuss in the section "Lining Up along the Axis of Symmetry"), you have a good idea of what the graph looks like.

Finding the one and only y-intercept

The *y*-intercept of a quadratic function is $(0, c)$. A parabola with the standard equation $y = ax^2 + bx + c$ is a function, so by definition (as I cover in Chapter 6), only one *y* value can exist for every *x* value. When $x = 0$, as it does at the *y*-intercept, the equation becomes $y = a(0)^2 + b(0) + c = 0 + 0 + c = c$, or $y = c$. The equalities $x = 0$ and $y = c$ combine to become the *y*-intercept, $(0, c)$.

To find the *y*-intercepts of the following functions, you let $x = 0$:

$y = 4x^2 - 3x + 2$: When $x = 0$, $y = 2$ (or $c = 2$). The *y*-intercept is $(0, 2)$.

$y = -x^2 - 5$: When $x = 0$, $y = -5$ (or $c = -5$); don't let the missing *x* term throw you. The *y*-intercept is $(0, -5)$.

$y = \frac{1}{2}x^2 + \frac{3}{2}x$: When $x = 0$, $y = 0$. The equation provides no constant term; you could also say the missing constant term is zero. The y-intercept is $(0, 0)$.

People can model many situations with quadratic functions, and the places where the input variables or output variables equal zero are important. For example, a candle-making company has figured out that its profit is based on the number of candles it produces and sells. The company uses the function $P(x) = -0.05x^2 + 8x - 140$ — where x represents the number of candles — to determine P, the profit. As you can see from the equation, the graph of this parabola opens downward (because a is negative; see the section "Starting with 'a' in the standard form"). Figure 7-3 is a sketch of the graph of the profit, with the Y-axis representing profit and the X-axis representing the number of candles.

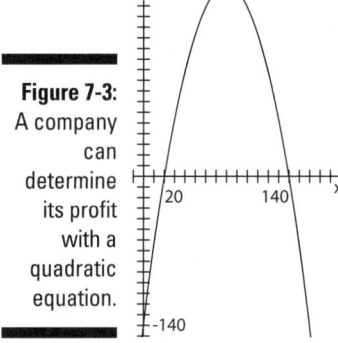

Figure 7-3:
A company can determine its profit with a quadratic equation.

Does it make sense to use a quadratic function to model profit? Why would the profit decrease after a certain point? Does that make business sense? It does if you consider that perhaps, when you make too many candles, the cost of overtime and the need for additional machinery plays a part.

What about the y-intercept? What part does it play, and what does it mean in this candle-making case? You can say that $x = 0$ represents not producing or selling any candles. According to the equation and graph, the y-intercept has a y-coordinate of -140. It makes sense to find a negative profit if the company has costs that it has to pay no matter what (even if it sells no candles): insurance, salaries, mortgage payments, and so on. With some interpretation, you can find a logical explanation for the y-intercept being negative in this case.

Finding the x-intercepts

You find the *x*-intercepts of quadratics when you solve for the *zeros,* or solutions, of the quadratic equation. The method you use to solve for the zeros is the same method you use to solve for the intercepts, because they're really just the same thing. The names change (intercept, zero, solution), depending on the application, but you find the intercepts the same way.

Parabolas with an equation of the standard form $y = ax^2 + bx + c$ open upward or downward and may or may not have *x*-intercepts. Look at Figure 7-4, for example. You see a parabola with two *x*-intercepts (Figure 7-4a), one with a single *x*-intercept (Figure 7-4b), and one with no *x*-intercept (Figure 7-4c). Notice, however, that they all have a *y*-intercept.

Figure 7-4:
Parabolas can intercept the X-axis multiple times, a single time, or not at all.

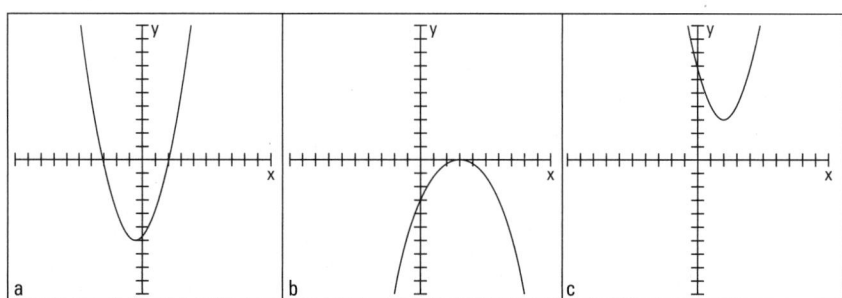

The coordinates of all *x*-intercepts have zeros in them. An intercept's *y* value is zero, and you write it in the form $(h, 0)$. How do you find the value of *h?* You let $y = 0$ in the general equation and then solve for *x*. You have two options to solve the equation $0 = ax^2 + bx + c$:

 ✔ Use the quadratic formula (refer to Chapter 3 for a refresher on the formula).

 ✔ Try to factor the expression and use the multiplication property of zero (you can find more on this in Chapter 1).

Regardless of the path you take, you have some guidelines at your disposal to help you determine the number of *x*-intercepts you should find.

When finding *x*-intercepts by solving $0 = ax^2 + bx + c$,

 ✔ You find **two** *x*-intercepts if

 • The expression factors into two different binomials.

 • The quadratic formula gives you a value greater than zero under the radical.

> ✔ You find **one** *x*-intercept (a double root) if
>
> - The expression factors into the square of a binomial.
> - The quadratic formula gives you a value of zero under the radical.
>
> ✔ You find **no** *x*-intercept if
>
> - The expression doesn't factor *and*
> - The quadratic formula gives you a value less than zero under the radical (indicating an imaginary root; Chapter 14 deals with imaginary and complex numbers).

To find the *x*-intercepts of $y = 3x^2 + 7x - 40$, for example, you can set *y* equal to zero and solve the quadratic equation by factoring:

$$0 = 3x^2 + 7x - 40$$
$$= (3x - 8)(x + 5)$$
$$3x - 8 = 0, \ x = \frac{8}{3}$$
$$x + 5 = 0, \ x = -5$$

The two intercepts are $\left(\frac{8}{3}, 0\right)$ and (–5, 0); you can see that the equation factors into two different factors. In cases where you can't figure out how to factor the quadratic, you can get the same answer by using the quadratic formula. Notice in the following calculation that the value under the radical is a number greater than zero, meaning that you have two answers:

$$x = \frac{-7 \pm \sqrt{7^2 - 4(3)(-40)}}{2(3)}$$
$$= \frac{-7 \pm \sqrt{49 - (-480)}}{6}$$
$$= \frac{-7 \pm \sqrt{529}}{6} = \frac{-7 \pm 23}{6}$$
$$x = \frac{-7 + 23}{6} = \frac{16}{6} = \frac{8}{3}$$
$$x = \frac{-7 - 23}{6} = \frac{-30}{6} = -5$$

Here's another example, with a different result. To find the *x*-intercepts of $y = -x^2 + 8x - 16$, you can set *y* equal to zero and solve the quadratic equation by factoring:

$$0 = -x^2 + 8x - 16$$
$$0 = -(x^2 - 8x + 16)$$
$$0 = -(x - 4)^2$$
$$4 = x$$

The only intercept is (4, 0). The equation factors into the square of a binomial — a *double root*. You can get the same answer by using the quadratic formula — notice that the value under the radical is equal to zero:

$$x = \frac{-8 \pm \sqrt{8^2 - 4(-1)(-16)}}{2(-1)}$$

$$= \frac{-8 \pm \sqrt{64 - (64)}}{-2}$$

$$= \frac{-8 \pm \sqrt{0}}{-2} = \frac{-8}{-2} = 4$$

This last example shows how you determine that an equation has no x-intercept. To find the x-intercepts of $y = -2x^2 + 4x - 7$, you can set y equal to zero and try to factor the quadratic equation, but you may be on the trial-and-error train for some time, because you can't do it. The equation has no factors that give you this quadratic.

When you try the quadratic formula, you see that the value under the radical is less than zero; a negative number under the radical is an imaginary number:

$$x = \frac{-4 \pm \sqrt{4^2 - 4(-2)(-7)}}{2(-2)}$$

$$= \frac{-4 \pm \sqrt{16 - (56)}}{-4}$$

$$= \frac{-4 \pm \sqrt{-40}}{-4}$$

Alas, you find no x-intercept for this parabola.

Going to the Extreme: Finding the Vertex

Quadratic functions, or parabolas, that have the standard form $y = ax^2 + bx + c$ are gentle, U-shaped curves that open either upward or downward. When the lead coefficient, a, is a positive number, the parabola opens upward, creating a *minimum value* for the function — the function values never go lower than that minimum. When a is negative, the parabola opens downward, creating a *maximum value* for the function — the function values never go higher than that maximum.

The two extreme values, the minimum and maximum, occur at the parabola's *vertex.* The *y*-coordinate of the vertex gives you the numerical value of the extreme — its highest or lowest point.

The vertex of a parabola is very useful for finding the extreme value, so certainly algebra provides an efficient way of finding it. Right? Well, sure it does! The *vertex* serves as a sort of anchor for the two parts of the curve to flare out from. The axis of symmetry (see the following section) runs through the vertex. The *y*-coordinate of the vertex is the function's maximum or minimum value — again, depending on which way the parabola opens.

The parabola $y = ax^2 + bx + c$ has its vertex where $x = \frac{-b}{2a}$. You plug in the a and b values from the equation to come up with x, and then you find the y-coordinate of the vertex by plugging this x value into the equation and solving for y.

To find the coordinates of the vertex of the equation $y = -3x^2 + 12x - 7$, for example, you substitute the coefficients a and b into the equation for x:

$$x = \frac{-12}{2(-3)} = \frac{-12}{-6} = 2$$

You solve for y by putting the x value back into the equation:

$$y = -3(2)^2 + 12(2) - 7 = -12 + 24 - 7 = 5$$

The coordinates of the vertex are (2, 5). You find a maximum value, because a is a negative number, which means the parabola opens downward from this point. The graph of the parabola never goes higher than five units up.

When an equation leaves out the b value, make sure you don't substitute the c value for the b value in the vertex equation. For example, to find the coordinates of the vertex of $y = 4x^2 - 19$, you substitute the coefficients a (4) and b (0) into the equation for x:

$$x = \frac{-0}{2(4)} = 0$$

You solve for y by putting the x value into the equation:

$$y = 4(0)^2 - 19 = -19$$

The coordinates of the vertex are (0, –19). You have a minimum value, because a is a positive number, meaning the parabola opens upward from the minimum point.

Lining Up along the Axis of Symmetry

The *axis of symmetry* of a quadratic function is a vertical line that runs through the vertex of the parabola (see the previous section) and acts as a mirror — half the parabola rests on one side of the axis, and half rests on the other. The *x*-value in the coordinates of the vertex appears in the equation for the axis of symmetry. For instance, if a vertex has the coordinates (2, 3), the axis of symmetry is $x = 2$. All vertical lines have an equation of the form $x = h$. In the case of the axis of symmetry, the *h* is always the *x*-coordinate of the vertex.

The axis of symmetry is useful because when you're sketching a parabola and finding the coordinates of a point that lies on it, you know you can find another point that exists as a partner to your point, in that

✔ It lies on the same horizontal line.

✔ It lies on the other side of the axis of symmetry.

✔ It covers the same distance from the axis of symmetry as your point.

Maybe I should just show you what I mean with a sketch! Figure 7-5 shows points on a parabola that lie on the same horizontal line and on either side of the axis of symmetry.

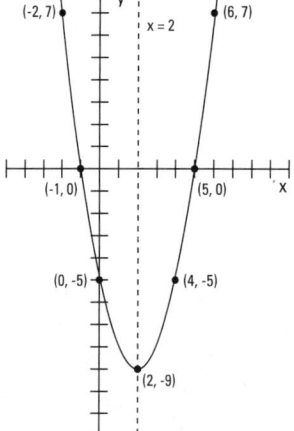

Figure 7-5:
Points
resting on
the same
horizontal
line and
equidistant
from the
axis of
symmetry.

The points (4, –5) and (0, –5) are each two units from the axis — the line $x = 2$. The points (–2, 7) and (6, 7) are each four units from the axis. And the points (–1, 0) and (5, 0), the *x*-intercepts, are each three units from the axis.

Sketching a Graph from the Available Information

You have all sorts of information available when it comes to a parabola and its graph. You can use the intercepts, the opening, the steepness, the vertex, the axis of symmetry, or just some random points to plot the parabola; you don't really need all the pieces. As you practice sketching these curves, it becomes easier to figure out which pieces you need for different situations. Sometimes the x-intercepts are hard to find, so you concentrate on the vertex, direction, and axis of symmetry. Other times you find it more convenient to use the y-intercept, a point or two on the parabola, and the axis of symmetry. This section provides you with a couple examples. Of course, you can go ahead and check off all the information that's possible. Some people are very thorough that way.

To sketch the graph of $y = x^2 - 8x + 1$, first notice that the equation represents a parabola that opens upward (see the section "Starting with 'a' in the standard form"), because the lead coefficient, a, is positive (+1). The y-intercept is (0, 1), which you get by plugging in zero for x. If you set y equal to zero to solve for the x-intercepts, you get $0 = x^2 - 8x + 1$, which doesn't factor. You could whip out the quadratic formula — but wait. You have other possibilities to consider.

The vertex is more helpful than finding the intercepts in this case because of its convenience — you don't have to work so hard to get the coordinates. Use the formula for the x-coordinate of the vertex to get $x = \dfrac{-(-8)}{2(1)} = \dfrac{8}{2} = 4$ (see the section "Going to the Extreme: Finding the Vertex"). Plug the 4 into the formula for the parabola, and you find that the vertex is at (4, –15). This coordinate is below the X-axis, and the parabola opens upward, so the parabola does have x-intercepts; you just can't find them easily because they're *irrational numbers* (square roots of numbers that aren't perfect squares).

You can try whipping out your graphing calculator to get some decimal approximations of the intercepts (see Chapter 5 for info on using a graphing calculator). Or, instead, you can find a point and its partner point on the other side of the axis of symmetry, which is $x = 4$ (see the section "Lining Up along the Axis of Symmetry"). If you let $x = 1$, for example, you find that $y = -6$. This point is three units from $x = 4$, to the left; you find the distance by subtracting $4 - 1 = 3$. Use this distance to find three units to the right, $4 + 3 = 7$. The corresponding point is (7, –6).

If you sketch all that information in a graph first — the y-intercept, vertex, axis of symmetry, and the points $(1, -6)$ and $(7, -6)$ — you can identify the shape of the parabola and sketch in the whole thing. Figure 7-6 shows the two steps: Putting in the information (Figure 7-6a), and sketching in the parabola (Figure 7-6b).

Figure 7-6: Using the various pieces of a quadratic as steps for sketching a graph ($y = x^2 - 8x + 1$).

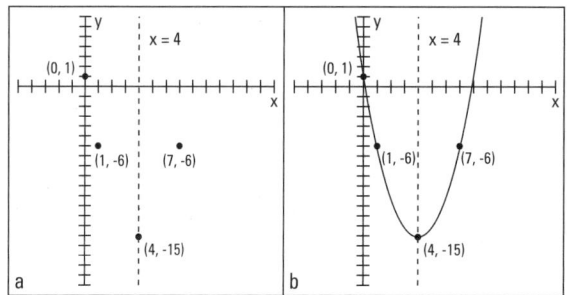

Here's another example for practice. To sketch the graph of $y = -0.01x^2 - 2x$, look for the hints. The parabola opens downward (because a is negative) and is pretty flattened out (because the absolute value of a is less than zero). The graph goes through the origin because the constant term (c) is missing. Therefore, the y-intercept and one of the x-intercepts is $(0, 0)$. The vertex lies at $(-100, 100)$. To solve for the other x-intercept, let $y = 0$ and factor:

$$0 = -0.01x(x + 200)$$

The second factor tells you that the other x-intercept occurs when $x = -200$. The intercepts and vertex are sketched in Figure 7-7a.

Figure 7-7: Using intercepts and the vertex to sketch a parabola.

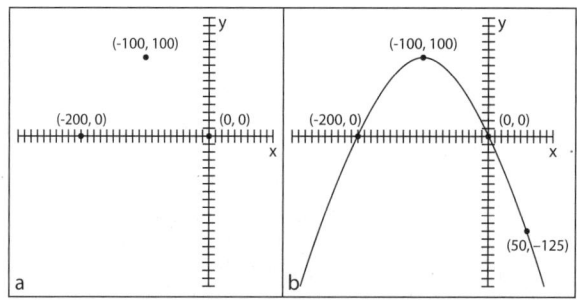

You can add the point $(50, -125)$ for a little more help with the shape of the parabola by using the axis of symmetry (see the section "Lining Up along the Axis of Symmetry"). See how you can draw the curve in? Figure 7-7b shows you the way. It really doesn't take much to do a decent sketch of a parabola.

Applying Quadratics to the Real World

Quadratic functions are wonderful models for many situations that occur in the real world. You can see them at work in financial and physical applications, just to name a couple. This section provides a few applications for you to consider.

Selling candles

A candle-making company has figured out that its profit is based on the number of candles it produces and sells. The function $P(x) = -0.05x^2 + 8x - 140$ applies to the company's situation, where x represents the number of candles, and P represents the profit. You may recognize this function from the section "Investigating Intercepts in Quadratics" earlier in the chapter. You can use the function to find out how many candles the company has to produce to garner the greatest possible profit.

You find the two x-intercepts by letting $y = 0$ and solving for x by factoring:

$$0 = -0.05x^2 + 8x - 140$$
$$= -0.05\left(x^2 - 160x + 2800\right)$$
$$= -0.05\left(x - 20\right)\left(x - 140\right)$$
$$x - 20 = 0, \; x = 20$$
$$x - 140 = 0, \; x = 140$$

The intercept (20, 0) represents where the function (the profit) changes from negative values to positive values. You know this because the graph of the profit function is a parabola that opens downward (because a is negative), so the beginning and ending of the curve appear below the X-axis. The intercept (140, 0) represents where the profit changes from positive values to negative values. So, the maximum value, the vertex, lies somewhere between and above the two intercepts (see the section "Going to the Extreme: Finding the Vertex"). The x-coordinate of the vertex lies between 20 and 140. Refer to Figure 7-3 if you want to see the graph again.

You now use the formula for the x-coordinate of the vertex, $x = \frac{-b}{2a}$, to find $x = \frac{-8}{2(-0.05)} = \frac{-8}{-0.1} = 80$. The number 80 lies between 20 and 140; in fact, it rests halfway between them. The nice, even number is due to the symmetry of the graph of the parabola and the symmetric nature of these functions. Now you can find the P value (the y-coordinate of the vertex): $P(80) = -0.05(80)^2 + 8(80) - 140 = -320 + 640 - 140 = 180$.

Your findings say that if the company produces and sells 80 candles, the maximum profit will be $180. That seems like an awful lot of work for $180, but maybe the company runs a small business. Work such as this shows you how important it is to have models for profit, revenue, and cost in business so you can make projections and adjust your plans.

Shooting basketballs

A local youth group recently raised money for charity by having a *Throw-A-Thon.* Participants prompted sponsors to donate money based on a promise to shoot baskets over a 12-hour period. This was a very successful project, both for charity and for algebra, because you can find some interesting bits of information about shooting the basketballs and the number of misses that occurred.

Participants shot baskets for 12 hours, attempting about 200 baskets each hour. The quadratic equation $M(t) = \frac{17}{6}t^2 - \frac{77}{3}t + 100$ models the number of baskets they *missed* each hour, where t is the time in hours (numbered from 0 through 12) and M is the number of misses.

This quadratic function opens upward (because a is positive), so the function has a minimum value. Figure 7-8 shows a graph of the function.

From the graph, you see that the initial value, the y-intercept, is 100. At the beginning, participants were missing about 100 baskets per hour. The good news is that they got better with practice. $M(2) = 60$, which means that at hour two into the project, the participants were missing only 60 baskets per hour. The number of misses goes down and then goes back up again. How do you interpret this? Even though the participants got better with practice, they let the fatigue factor take over.

Figure 7-8: The downs and ups of shooting baskets.

What's the fewest number of misses per hour? When did the participants shoot their best? To answer these questions, find the vertex of the parabola by using the formula for the *x*-coordinate (you can find this in the "Going to the Extreme: Finding the Vertex" section earlier in the chapter):

$$h = \frac{-\left(-\dfrac{77}{3}\right)}{2\left(\dfrac{17}{\cancel{6}_3}\right)} = \frac{77}{3} \cdot \frac{3}{17} = \frac{77}{17} = 4\frac{9}{17}$$

The best shooting happened about 4.5 hours into the project. How many misses occurred then? The number you get represents what's happening the entire hour — although that's fudging a bit:

$$M\left(\frac{77}{17}\right) = \frac{17}{6}\left(\frac{77}{17}\right)^{2} - \frac{77}{3}\left(\frac{77}{17}\right) + 100$$
$$= \frac{4{,}271}{102} \approx 41.87$$

The fraction is rounded to two decimal places. The best shooting is about 42 misses that hour.

Launching a water balloon

One of the favorite springtime activities of the engineering students at a certain university is to launch water balloons from the top of the engineering building so they hit the statue of the school's founder, which stands 25 feet from the building. The launcher sends the balloons up in an arc to clear a tree that sits next to the building. To hit the statue, the initial velocity and angle of the balloon have to be just right. Figure 7-9 shows a successful launch.

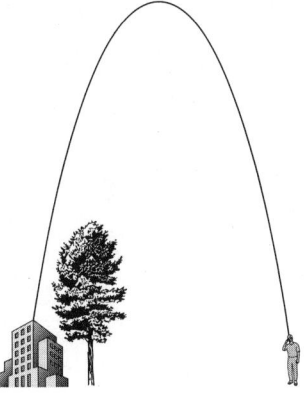

Figure 7-9:
Launching a water balloon over a tree requires more math than you think.

This year's launch was successful. The students found that by launching the water balloons 48 feet per second at a precise angle, they could hit the statue. Here's the equation they worked out to represent the path of the balloons: $H(t) = -2t^2 + 48t + 60$. The t represents the number of seconds, and H is the height of the balloon in feet. From this quadratic function, you can answer the following questions:

1. **How high is the building?**

   Solving the first question is probably easy for you. The launch occurs at time $t = 0$, the initial value of the function. When $t = 0$, $H(0) = -2(0)^2 + 48(0) + 60 = 0 + 0 + 60 = 60$. The building is 60 feet high.

2. **How high did the balloon travel?**

   You answer the second question by finding the vertex of the parabola: $t = \dfrac{-48}{2(-2)} = \dfrac{-48}{-4} = 12$. This gives you t, the number of seconds it takes the balloon to get to its highest point — 12 seconds after launch. Substitute the answer into the equation to get the height: $H(12) = -2(12)^2 + 48(12) + 60 = -288 + 576 + 60 = 348$. The balloon went 348 feet into the air.

3. **If the statue is 10 feet tall, how many seconds did it take for the balloon to reach the statue after the launch?**

   To solve the third question, use the fact that the statue is 10 feet high; you want to know when $H = 10$. Replace H with 10 in the equation and solve for t by factoring (see Chapters 1 and 3):

   $$10 = -2t^2 + 48t + 60$$
   $$0 = -2t^2 + 48t + 50$$
   $$= -2(t^2 - 24t - 25)$$
   $$= -2(t - 25)(t + 1)$$
   $$t - 25 = 0, \ t = 25$$
   $$t + 1 = 0, \ t = -1$$

According to the equation, t is either 25 or –1. The –1 doesn't really make any sense because you can't go back in time. The 25 seconds, however, tells you how long it took the balloon to reach the statue. Imagine the anticipation!

Chapter 8

Staying Ahead of the Curves: Polynomials

*T*he word *polynomial* comes from *poly-*, meaning many, and *-nomial*, meaning name or designation. *Binomial* (two terms) and *trinomial* (three terms) are two of the many names or designations used for selected polynomials. The terms in a polynomial are made up of numbers and letters that get stuck together with multiplication.

Although the name may seem to imply complexity (much like Albert Einstein, Pablo Picasso, or Mary Jane Sterling), polynomials are some of the easier functions or equations to work with in algebra. The exponents used in polynomials are all whole numbers — no fractions or negatives. Polynomials get progressively more interesting as the exponents get larger — they can have more intercepts and turning points. This chapter outlines what you can do with polynomials: factor them, graph them, analyze them to pieces — everything but make a casserole with them. The graph of a polynomial looks like a Wisconsin landscape — smooth, rolling curves. Are you ready for this ride?

Taking a Look at the Standard Polynomial Form

A *polynomial function* is a specific type of function that can be easily spotted in a crowd of other types of functions and equations. The exponents on the variable terms in a polynomial function are always whole numbers. And, by convention, you write the terms from the largest exponent to the smallest. Actually, the exponent 0 on the variable makes the variable factor equal to 1, so you don't see a variable there at all.

The traditional equation for the standard way to write the terms of a polynomial is shown below. Don't let all the subscripts and superscripts throw you. The letter *a* is repeated with numbers, rather than going *a, b, c,* and so on, because a polynomial with a degree higher than 26 would run out of letters in the English alphabet.

The general form for a polynomial function is

$$f(x) = a_n x^n + a_{n-1} x^{n-1} + a_{n-2} x^{n-2} + \ldots + a_1 x^1 + a_0$$

Here, the *a*'s are real numbers and the *n*'s are whole numbers. The last term is technically $a_0 x^0$, if you want to show the variable in every term.

Exploring Polynomial Intercepts and Turning Points

The *intercepts* of a polynomial are the points where the graph of the curve of the polynomial crosses the X- and Y-axes. A polynomial function has *exactly* one *y*-intercept, but it can have any number of *x*-intercepts, depending on the degree of the polynomial (the powers of the variable). The higher the degree, the more *x*-intercepts you might have.

The *x-intercepts* of a polynomial are also called the *roots, zeros,* or *solutions.* You may think that mathematicians can't make up their minds about what to call these values, but they do have their reasons; depending on the application, the *x*-intercept has an appropriate name for what you're working on (see Chapter 3 for more information on this algebra name game). The nice thing is that you use the same technique to solve for the intercepts, no matter what they're called. (Lest the *y*-intercept feel left out, it's frequently called the *initial value.*)

The *x*-intercepts are often where the graph of the polynomial goes from positive values (above the X-axis) to negative values (below the X-axis) or from negative values to positive values. Sometimes, though, the values on the graph don't change sign at an *x*-intercept: These graphs take on a *touch and go* appearance. The graphs approach the X-axis, seem to change their minds about crossing the axis, touch down at the intercepts, and then go back to the same side of the axis.

A *turning point* of a polynomial is where the graph of the curve changes direction. It can change from going upward to going downward, or vice versa. A turning point is where you find a relative maximum value of the polynomial, an absolute maximum value, a relative minimum value, or an absolute minimum value.

Interpreting relative value and absolute value

As I introduce in Chapter 5, any function can have an *absolute maximum* or an *absolute minimum* value — the point at which the graph of the function has no higher or lower value, respectively. For example, a parabola opening downward has an absolute maximum — you see no point on the curve that's higher than the maximum. In other words, no value of the function is greater than that number. (Check out Chapter 6 for more on quadratic functions and their parabola graphs.) Some functions, however, also have relative maximum or minimum values:

- ✔ **Relative maximum:** A point on the graph — a value of the function — that's *relatively* large; the point is bigger than anything around it, but you might be able to find a bigger point somewhere else.

- ✔ **Relative minimum:** A point on the graph — a value of the function — that's smaller than anything close to it; it's smaller or lower relative to all the points on the curve near it.

In Figure 8-1, you can see five turning points. Two are relative maximum values, which means they're higher than any points close to them. Three are minimum values, which means they're lower than any points around them. Two of the minimums are relative minimum values, and one is absolutely the lowest point on the curve. This function has no absolute maximum value because it keeps going up and up without end.

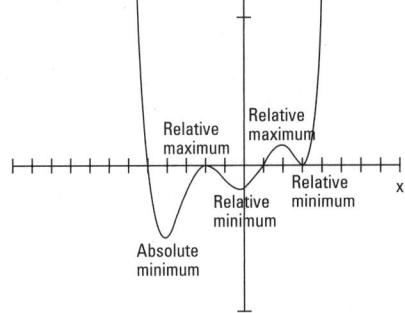

Figure 8-1:
Extreme points on a polynomial.

Counting intercepts and turning points

The number of potential turning points and *x*-intercepts of a polynomial function is good to know when you're sketching the graph of the function. You can count the number of *x*-intercepts and turning points of a polynomial if you have the graph of it in front of you, but you can also make an estimate of the number if you have the equation of the polynomial. Your estimate is actually a number that represents the most points that can occur. You can say, "There are at most *m* intercepts and at most *n* turning points." The estimate is the best you can do, but that's usually not a bad thing.

To determine the rules for the greatest number of possible intercepts and greatest number of possible turning points from the equation of a polynomial, you look at the general form of a polynomial function.

Given the polynomial $f(x) = a_n x^n + a_{n-1} x^{n-1} + a_{n-2} x^{n-2} + \ldots + a_1 x^1 + a_0$, the maximum number of *x*-intercepts is *n*, the degree or highest power of the polynomial. The maximum number of turning points is $n - 1$, or one less than the number of possible intercepts. You may find fewer intercepts than *n*, or you may find exactly that many.

If *n* is an odd number, you know right away that you have to find at least one *x*-intercept. If *n* is even, you may not find any *x*-intercepts.

Examine the following two function equations as examples of polynomials. To determine the possible number of intercepts and turning points for the functions, look for the values of *n*, the exponents that have the highest values:

$$f(x) = 2x^7 + 9x^6 - 75x^5 - 317x^4 + 705x^3 + 2700x^2$$

This graph has at most seven *x*-intercepts (7 is the highest power in the function) and six turning points (7 – 1).

$$f(x) = 6x^6 + 24x^5 - 120x^4 - 480x^3 + 384x^2 + 1536x - 2000$$

This graph has at most six *x*-intercepts and five turning points.

You can see the graphs of these two functions in Figure 8-2. According to its function, the graph of the first example (Figure 8-2a) could have at least seven *x*-intercepts, but it has only five; it does have all six turning points, though. You can also see that two of the intercepts are *touch-and-go* types, meaning that they approach the X-axis before heading away again. The graph of the second example (Figure 8-2b) can have at most six *x*-intercepts, but it has only two; it does have all five turning points.

Figure 8-2:
The
intercept
and turning-
point
behavior
of two
polynomial
functions.

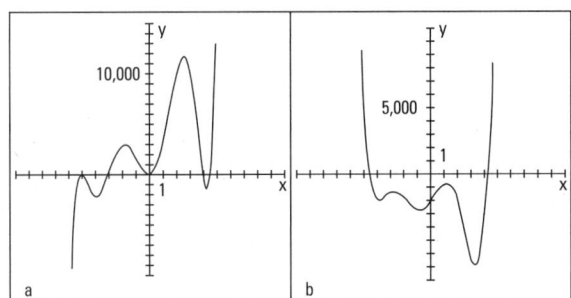

Figure 8-3 provides you with two extreme polynomial examples. The graphs of $y = x^8 + 1$ (Figure 8-3a) and $y = x^9$ (Figure 8-3b) seem to have great possibilities . . . that don't pan out. The graph of $y = x^8 + 1$, according to the rules of polynomials, could have as many as eight intercepts and seven turning points. But, as you can see from the graph, it has no intercepts and just one turning point. The graph of $y = x^9$ has only one intercept and no turning points.

The moral to the story of Figure 8-3 is that you have to use the rules of polynomials wisely, carefully, and skeptically. Also, think about the basic graphs of polynomials. The graphs are gently rolling curves that go from the left to the right across you graph. They cross the Y-axis exactly once and may or may not cross the X-axis. You get hints from the standard equation of the polynomial and from the factored form. Refer to Chapter 5 if you need to review them.

Figure 8-3:
A
polynomial's
highest
power
provides
the most
possible
turning
points and
intercepts.

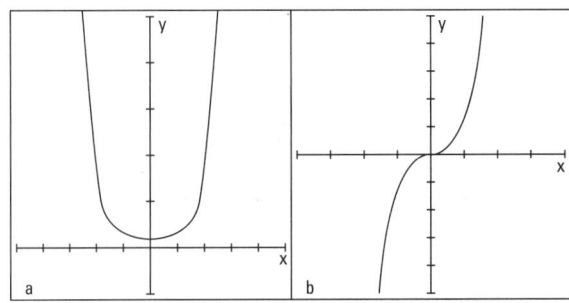

Solving for polynomial intercepts

You can easily solve for the *y*-intercept of a polynomial function, of which you can find only one. The *y*-intercept is where the curve of the graph crosses the

Y-axis, and that's when $x = 0$. So, to determine the y-intercept for any polynomial, simply replace all the x's with zeros and solve for y (that's the y part of the coordinate of that intercept):

$$y = 3x^4 - 2x^2 + 5x - 3$$

$$x = 0, y = 3(0)^4 - 2(0)^2 + 5(0) - 3 = -3$$

The y-intercept is $(0, -3)$.

$$y = 8x^5 - 2x^3 + x^2 - 3x$$

$$x = 0, y = 8(0)^5 - 2(0)^3 + (0)^2 - 3(0) = 0$$

The y-intercept is $(0, 0)$, at the origin.

After you complete the easy task of solving for the y-intercept, you find out that the x-intercepts are another matter altogether. The value of y is zero for all x-intercepts, so you let $y = 0$ and solve.

Here, however, you don't have the advantage of making everything disappear except the constant number like you do when solving for the y-intercept. When solving for the x-intercepts, you may have to factor the polynomial or perform a more elaborate process — techniques you can find later in this chapter in the section "Factoring for polynomial roots" or "Saving your sanity: The Rational Root Theorem," respectively. For now, just apply the process of factoring and setting the factored form equal to zero to some carefully selected examples. This is essentially using the *multiplication property of zero* — setting the factored form equal to zero to find the intercepts (see Chapter 1).

To determine the x-intercepts of the following three polynomials, replace the y's with zeros and solve for the x's:

$$y = x^2 - 16$$

$y = 0, 0 = x^2 - 16, x^2 = 16, x = \pm 4$ (using the square root rule from Chapter 3)

$$y = x(x - 5)(x - 2)(x + 1)$$

$y = 0, 0 = x(x - 5)(x - 2)(x + 1), x = 0, 5, 2,$ or -1 (using the multiplication property of zero from Chapter 1)

$$y = x^4 (x + 3)^8$$

$y = 0, 0 = x^4 (x + 3)^8, x = 0$ or -3 (using the multiplication property of zero)

Both of these intercepts come from *multiple roots* (when a solution appears more than once). Another way of writing the factored form is $x \cdot x \cdot x \cdot x \cdot (x + 3) \cdot (x + 3) \cdot (x + 3) \cdot (x + 3) \cdot (x + 3) \cdot (x + 3) \cdot (x + 3) \cdot (x + 3) = 0$.

You could list the answer as 0, 0, 0, 0, –3, –3, –3, –3, –3, –3, –3, –3. The number of times a root repeats is significant when you're graphing. A multiple root has a different kind of look or graph where it intersects the axis. (For more, head to the section "Changing from roots to factors" later in this chapter.)

Determining Positive and Negative Intervals

When a polynomial has positive y values for some interval — between two x-values — its graph lies above the X-axis. When a polynomial has negative values, its graph lies below the X-axis in that interval. The only way to change from positive to negative values or vice versa is to go through zero — in the case of a polynomial, at an x-intercept. Polynomials can't jump from one side of the X-axis to the other because their domains are all real numbers — nothing is skipped to allow such a jump. The fact that x-intercepts work this way is good news for you because x-intercepts play a large role in the big picture of solving polynomial equations and determining the positive and negative natures of polynomials.

The positive versus negative values of polynomials are important in various applications in the real world, especially where money is involved. If you use a polynomial function to model the profit in your business or the depth of water (above or below flood stage) near your house, you should be interested in positive versus negative values and in what intervals they occur. The technique you use to find the positive and negative intervals also plays a big role in calculus, so you get an added bonus by using it here first.

Using a sign-line

If you're a visual person like me, you'll appreciate the interval method I present in this section. Using a *sign line* and marking the intervals between x-values allows you to determine where a polynomial is positive or negative, and it appeals to your artistic bent!

The function $f(x) = x(x - 2)(x - 7)(x + 3)$, for example, changes sign at every intercept. Setting $f(x) = 0$ and solving, you find that the x-intercepts are at $x = 0, 2, 7,$ and –3. Now you can put this information about the problem to work.

To determine the positive and negative intervals for a polynomial function, follow this method:

1. **Draw a number line, and place the values of the *x*-intercepts in their correct positions on the line.**

2. **Choose random values to the right and left of the intercepts to test whether the function is positive or negative in those positions.**

   If the function equation is factored, you determine whether each factor is positive or negative, and you find the sign of the product of all the factors.

   Some possible random number choices are $x = -4, -1, 1, 3,$ and 8. These values represent numbers in each interval determined by the intercepts. (*Note:* These aren't the only possibilities; you can pick your favorites.)

   $f(-4) = (-4)(-4 - 2)(-4 - 7)(-4 + 3)$. You don't need the actual number value, just the sign of the result, so $f(-4) = (-)(-)(-)(-) = +$.

   $f(-1) = (-1)(-1 - 2)(-1 - 7)(-1 + 3)$. $f(-1) = (-)(-)(-)(+) = -$.

   $f(1) = (1)(1 - 2)(1 - 7)(1 + 3)$. $f(1) = (+)(-)(-)(+) = +$.

   $f(3) = (3)(3 - 2)(3 - 7)(3 + 3)$. $f(3) = (+)(+)(-)(+) = -$.

   $f(8) = (8)(8 - 2)(8 - 7)(8 + 3)$. $f(8) = (+)(+)(+)(+) = +$.

   You need to check only one point in each interval; the function values all have the same sign within that interval.

3. **Place a + or – symbol in each interval to show the function sign.**

   The graph of this function is positive, or above the X-axis, whenever x is smaller than –3, between 0 and 2, and bigger than 7. You write this answer as: $x < -3$ or $0 < x < 2$ or $x > 7$.

The function $f(x) = (x - 1)^2(x - 3)^5(x + 2)^4$ doesn't change at each intercept. The intercepts are where $x = 1, 3,$ and –2.

1. **Draw the number line and insert the intercepts.**

2. **Test values to the left and right of each intercept. Some possible random choices are to let** $x = -3, 0, 2,$ **and** $4.$

   When you can, you should always use 0, because it combines so nicely.

   $f(-3) = (-3 - 1)^2(-3 - 3)^5(-3 + 2)^4 = (-)^2(-)^5(-)^4 = (+)(-)(+) = -.$

   $f(0) = (0 - 1)^2(0 - 3)^5(0 + 2)^4 = (-)^2(-)^5(+)^4 = (+)(-)(+) = -.$

   $f(2) = (2 - 1)^2(2 - 3)^5(2 + 2)^4 = (+)^2(-)^5(+)^4 = (+)(-)(+) = -.$

   $f(4) = (4 - 1)^2(4 - 3)^5(4 + 2)^4 = (+)^2(+)^5(+)^4 = (+)(+)(+) = +.$

3. **Mark the signs in the appropriate places on the number line.**

You probably noticed that the factors raised to an even power were always positive. The factor raised to an odd power is only positive when the result in the parenthesis is positive.

Interpreting the rule

Look back at the two polynomial examples in the previous section. Did you notice that in the first example, the sign changed every time, and in the second, the signs were sort of stuck on a sign for a while? When the signs of functions don't change, the graphs of the polynomials don't cross the X-axis at any intercepts, and you see touch-and-go graphs. Why do you suppose this is? First, look at graphs of two functions from the previous section in Figures 8-4a and 8-4b, $y = x(x - 2)(x - 7)(x + 3)$ and $y = (x - 1)^2(x - 3)^5(x + 2)^4.$

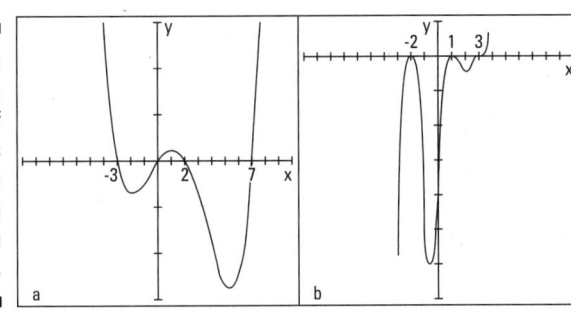

Figure 8-4:
Comparing graphs of polynomials that have differing sign behaviors.

The rule for whether a function displays sign changes or not at the intercepts is based on the exponent on the factor that provides you with a particular intercept.

If a polynomial function is factored in the form $y = (x - a_1)^{n1}(x - a_2)^{n2} \ldots$, you see a sign change whenever $n1$ is an odd number (meaning it crosses the X-axis), and you see no sign change whenever $n1$ is even (meaning the graph of the function is touch and go; see the section "Exploring Polynomial Intercepts and Turning Points").

So, for example, with the function $y = x^4(x - 3)^3(x + 2)^8(x + 5)^2$, shown in Figure 8-5a, you see a sign change at $x = 3$ and no sign change at $x = 0, -2,$ or -5. With the function $y = (2 - x)^2(4 - x)^2(6 - x)^2(2 + x)^2$, shown in Figure 8-5b, you never see a sign change.

Figure 8-5:
The powers of a polynomial determine whether the curve crosses the X-axis.

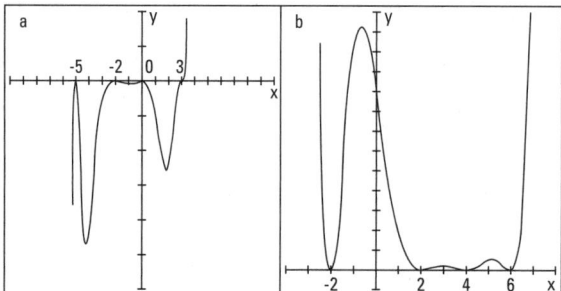

Finding the Roots of a Polynomial

Finding intercepts (or roots or zeros) of polynomials can be relatively easy or a little challenging, depending on the complexity of the function. Factored polynomials have roots that just stand up and shout at you, "Here I am!" Polynomials that factor easily are very desirable. Polynomials that don't factor at all, however, are relegated to computers or graphing calculators.

The polynomials that remain are those that factor — with a little planning and work. The planning process involves counting the number of possible positive and negative real roots and making a list of potential rational roots. The work is using synthetic division to test the list of choices to find the roots.

A polynomial function of degree n (the highest power in the polynomial equation is n) can have as many as n roots.

Factoring for polynomial roots

Finding intercepts of polynomials isn't difficult — as long as you have the polynomial in nicely factored form. You just set the y equal to zero and use the multiplication property of zero (see Chapter 1) to pick the intercepts off like flies. But what if the polynomial isn't in factored form (and it should be)? What do you do? Well, you factor it, of course. This section deals with easily recognizable factors of polynomials — types that probably account for 70 percent of any polynomials you'll have to deal with. (I cover other, more challenging types in the following sections.)

Applying factoring patterns and groupings

Half the battle is recognizing the patterns in factorable polynomial functions. You want to take advantage of the patterns. If you don't see any patterns (or if none exist), you need to investigate further. The most easily recognizable factoring patterns used on polynomials are the following (which I brush up on in Chapter 1):

$a^2 - b^2 = (a + b)(a - b)$	Difference of squares
$ab \pm ac = a(b \pm c)$	Greatest common factor
$a^3 - b^3 = (a - b)(a^2 + ab + b^2)$	Difference of cubes
$a^3 + b^3 = (a + b)(a^2 - ab + b^2)$	Sum of cubes
$a^2 \pm 2ab + b^2 = (a \pm b)^2$	Perfect square trinomial
UnFOIL	Trinomial factorization
Grouping	Common factors in groups

The following examples incorporate the different methods of factoring. They contain perfect cubes and squares and all sorts of good combinations of factorization patterns.

To factor the following polynomial, for example, you should use the greatest common factor and then the difference of squares:

$$y = x^3 - 9x$$
$$= x(x^2 - 9)$$
$$= x(x + 3)(x - 3)$$

This polynomial requires factoring, using the sum of two perfect cubes:

$$y = 27x^3 + 8$$
$$= (3x + 2)(9x^2 - 6x + 4)$$

You initially factor the next polynomial by grouping. The first two terms have a common factor of x^3, and the second two terms have a common factor of -125. The new equation has a common factor of $x^2 - 1$. After performing the factorization, you see that the first factor is the difference of squares and the second is the difference of cubes:

$$y = x^5 - x^3 - 125x^2 + 125$$
$$= x^3(x^2 - 1) - 125(x^2 - 1)$$
$$= (x^2 - 1)(x^3 - 125)$$
$$= (x + 1)(x - 1)(x - 5)(x^2 + 5x + 25)$$

This last polynomial shows how you first use the greatest common factor and then factor with a perfect square trinomial:

$$y = x^6 - 12x^5 + 36x^4$$
$$= x^4(x^2 - 12x + 36)$$
$$= x^4(x - 6)^2$$

Considering the unfactorable

Life would be wonderful if you would always find your paper on the doorstep in the morning and if all polynomials would factor easily. I won't get into the trials and tribulations of newspaper delivery, but polynomials that can't be factored do need to be discussed. You can't just give up and walk away. In some cases, polynomials can't be factored, but they do have intercepts that are decimal values that go on forever. Other polynomials both can't be factored and have no x-intercepts.

If a polynomial doesn't factor, you can attribute the roadblock to one of two things:

✔ **The polynomial doesn't have x-intercepts.** You can tell that from its graph using a graphing calculator.

✔ **The polynomial has irrational roots or zeros.** These can be estimated with a graphing calculator or computer.

Irrational roots means that the x-intercepts are written with radicals or rounded-off decimals. Irrational numbers are those that you can't write as fractions; you usually find them as square roots of numbers that aren't perfect squares or cube roots of numbers that aren't perfect cubes. You can sometimes solve for irrational roots if one of the factors of the polynomial is a quadratic. You can apply the quadratic formula to find that solution (see Chapter 3).

A quicker way to find irrational roots is to use a graphing calculator. Some will find the zeros for you one at a time; others will list the irrational roots for you all at once. You can also save some time if you recognize the patterns in polynomials that don't factor (you won't try to factor them).

For instance, polynomials in these forms don't factor:

$x^{2n} + c$. A polynomial in this form has no real roots or solutions. Refer to Chapter 14 on how to deal with imaginary numbers.

$x^2 + ax + a^2$. The quadratic formula will help you determine the imaginary roots.

$x^2 - ax + a^2$. This form requires the quadratic formula, too. You'll get imaginary roots.

The second and third examples above are part of the factorization of the difference or sum of two cubes. In Chapter 3, you see how the difference or sum of cubes are factored, and you're told, there, that the resulting trinomial doesn't factor.

Saving your sanity: The Rational Root Theorem

What do you do if the factorization of a polynomial doesn't leap out at you? You have a feeling that the polynomial factors, but the necessary numbers escape you. Never fear! Your faithful narrator has just saved your day. My help is in the form of the *Rational Root Theorem*. This theorem is really neat because it's so orderly and predictable, and it has an obvious end to it; you know when you're done so you can stop looking for solutions. But before you can put the theorem into play, you have to be able to recognize a rational root — or a rational number, for that matter.

A *rational number* is any real number that you can write as a fraction with one integer divided by another. (An *integer* is a positive or negative whole number or zero.) You often write a rational number in the form $\frac{p}{q}$, with the understanding that the denominator, q, can't equal zero. All whole numbers are rational numbers because you can write them as fractions, such as $4 = \frac{12}{3}$.

What distinguishes rational numbers from their opposites, irrational numbers, has to do with the decimal equivalences. The decimal associated with a fraction (rational number) will either terminate or repeat (have a pattern of numbers that occurs over and over again). The decimal equivalent of an irrational number never repeats and never ends; it just wanders on aimlessly.

Without further ado, here's the *Rational Root Theorem:* If the polynomial $f(x) = a_n x^n + a_{n-1}x^{n-1} + a_{n-2}x^{n-2} + \ldots + a_1 x^1 + a_0$ has any rational roots, they all meet the requirement that you can write them as a fraction equal to $\dfrac{\text{factor of } a_0}{\text{factor of } a_n}$.

In other words, according to the theorem, any rational root of a polynomial is formed by dividing a factor of the constant term by a factor of the lead coefficient.

Putting the theorem to good use

The Rational Root Theorem gets most of its workout by letting you make a list of numbers that may be roots of a particular polynomial. After using the theorem to make your list of potential roots (and check it twice), you plug the numbers into the polynomial to determine which, if any, work. You may run across an instance where none of the candidates work, which tells you that there are no rational roots. (And if a given rational number isn't on the list of possibilities that you come up with, it can't be a root of that polynomial.)

Before you start to plug and chug, however, check out the section "Letting Descartes make a ruling on signs," later in this chapter. It helps you with your guesses. Also, you can refer to "Synthesizing Root Findings" for a quicker method than plugging in.

Polynomials with constant terms

To find the roots of the polynomial $y = x^4 - 3x^3 + 2x^2 + 12$, for example, you test the following possibilities: ±1, ±2, ±3, ±4, ±6, and ±12. These values are all the factors of the number 12. Technically, you divide each of these factors of 12 by the factors of the lead coefficient, but because the lead coefficient is one (as in $1x^4$), dividing by that number won't change a thing. ***Note:*** You ignore the signs, because the factors for both 12 and 1 can be positive or negative, and you find several combinations that can give you a positive or negative root. You find out the sign when you test the root in the function equation.

To find the roots of another polynomial, $y = 6x^7 - 4x^4 - 4x^3 + 2x - 20$, you first list all the factors of 20: ±1, ±2, ±4, ±5, ±10, and ±20. Now divide each of those factors by the factors of 6. You don't need to bother dividing by 1, but you need to divide each by 2, 3, and 6:

$$\pm\frac{1}{2}, \pm\frac{2}{2}, \pm\frac{4}{2}, \pm\frac{5}{2}, \pm\frac{10}{2}, \pm\frac{20}{2},$$
$$\pm\frac{1}{3}, \pm\frac{2}{3}, \pm\frac{4}{3}, \pm\frac{5}{3}, \pm\frac{10}{3}, \pm\frac{20}{3},$$
$$\pm\frac{1}{6}, \pm\frac{2}{6}, \pm\frac{4}{6}, \pm\frac{5}{6}, \pm\frac{10}{6}, \pm\frac{20}{6}$$

You may have noticed some repeats in the previous list that occur when you reduce fractions. For instance, $\pm\frac{2}{2}$ is the same as ±1, and $\pm\frac{10}{6}$ is the same as $\pm\frac{5}{3}$. Even though this looks like a mighty long list, between the integers and fractions, it still gives you a reasonable number of candidates to try out. You can check them off in a systematic manner.

Polynomials without constant terms

When a polynomial doesn't have a constant term, you first have to factor out the greatest power of the variable that you can. If you're looking for the possible rational roots of $y = 5x^8 - 3x^4 - 4x^3 + 2x$, for example, and you want to use the Rational Root Theorem, you get nothing but zeros. You have no constant term — or you can say the constant is zero, so all the numerators of the fractions would be zero.

You can overcome the problem by factoring out the factor of x: $y = x(5x^7 - 3x^3 - 4x^2 + 2)$. This gives you the root zero. Now you apply the Rational Root Theorem to the new polynomial in the parenthesis to get the possible roots:

$$\pm1,\ \pm2,\ \pm\frac{1}{5},\ \pm\frac{2}{5}$$

Changing from roots to factors

When you have the factored form of a polynomial and set it equal to 0, you can solve for the solutions (or x-intercepts, if that's what you want). Just as important, if you have the solutions, you can go backwards and write the factored form. Factored forms are needed when you have polynomials in the numerator and denominator of fractions and you want to reduce the fraction. Factored forms are easier to compare with one another.

How can you use the Rational Root Theorem to factor a polynomial function? Why would you want to? The answer to the second question, first, is that you can reduce a factored form if it's in a fraction. Also, a factored form is more easily graphed. Now, for the first question: You use the Rational Root Theorem to find roots of a polynomial and then translate those roots into binomial factors whose product is the polynomial. (For methods on how to actually find the roots from the list of possibilities that the Rational Root Theorem gives you, see the sections "Letting Descartes make a ruling on signs" and "Using synthetic division to test for roots.")

If $x = \frac{b}{a}$ is a root of the polynomial $f(x)$, the binomial $(ax - b)$ is a factor. This works because

$$x = \frac{b}{a}$$
$$ax = b$$
$$ax - b = 0$$

To find the factors of a polynomial with the five roots $x = 1$, $x = -2$, $x = 3$, $x = \frac{3}{2}$, and $x = -\frac{1}{2}$, for example, you apply the previously stated rule: $f(x) = (x - 1)(x + 2)(x - 3)(2x - 3)(2x + 1)$. Notice that the positive roots give factors of the form $x - c$, and the negative roots give factors of the form $x + c$, which comes from $x - (-c)$.

To show *multiple roots,* or roots that occur more than once, use exponents on the factors. For instance, if the roots of a polynomial are $x = 0$, $x = 2$, $x = 2$, $x = -3$, $x = -3$, $x = -3$, $x = -3$, and $x = 4$, the corresponding polynomial is $f(x) = x(x - 2)^2(x + 3)^4(x - 4)$.

Letting Descartes make a ruling on signs

Rene Descartes was a French philosopher and mathematician. One of his contributions to algebra is the *Descartes' Rule of Signs.* This handy, dandy rule is a weapon in your arsenal for the fight to find roots of polynomial functions. If you pair this rule with the Rational Root Theorem from the previous section, you'll be well equipped to succeed.

The Rule of Signs tells you how many positive and negative *real* roots you may find in a polynomial. A *real number* is just about any number you can think of. It can be positive or negative, rational or irrational. The only thing it can't be is imaginary. (I cover imaginary numbers in Chapter 14 if you want to know more about them.)

Counting up the positive roots

The first part of the Rule of Signs helps you identify how many of the roots of a polynomial are positive.

Descartes' Rule of Signs (Part I): The polynomial $f(x) = a_n x^n + a_{n-1} x^{n-1} + a_{n-2} x^{n-2} + \ldots + a_1 x^1 + a_0$ has at most n roots. Count the number of times the sign changes in f, and call that value p. The value of p is the maximum number of *positive* real roots of f. If the number of positive roots isn't p, it is $p - 2$, $p - 4$, or some number less by a multiple of two.

To use Part I of Descartes' Rule of Signs on the polynomial $f(x) = 2x^7 - 19x^6 + 66x^5 - 95x^4 + 22x^3 + 87x^2 - 90x + 27$, for example, you count the number of sign changes. The sign of the first term starts as a positive, changes to a negative, and moves to positive; negative; positive; stays positive; negative; and then positive. Whew! In total, you count six sign changes. Therefore, you conclude that the polynomial has six positive roots, four positive roots, two positive roots, or none at all. Out of seven roots possible, it looks like at least one has to be negative. (By the way, this polynomial does have six positive roots; I built it that way! The only way you'd know that without being told is to go ahead and find the roots, with help from the Rule of Signs.)

Changing the function to count negative roots

Along with the positive roots (see the previous section), Descartes' Rule of Signs deals with the possible number of negative roots of a polynomial. After you count the possible number of positive roots, you combine that value to the number of possible negative roots to make your guesses and solve the equation.

Descartes' Rule of Signs (Part II): The polynomial $f(x) = a_n x^n + a_{n-1} x^{n-1} + a_{n-2} x^{n-2} + \ldots + a_1 x^1 + a_0$ has at most n roots. Find $f(-x)$, and then count the number of times the sign changes in $f(-x)$ and call that value q. The value of q is the maximum number of *negative* roots of f. If the number of negative roots isn't q, the number is $q - 2$, $q - 4$, and so on, for as many multiples of two as necessary.

To determine the possible number of negative roots of the polynomial $f(x) = 2x^7 - 19x^6 + 66x^5 - 95x^4 + 22x^3 + 87x^2 - 90x + 27$, for example, you first find $f(-x)$ by replacing each x with $-x$ and simplifying:

$$f(-x) = 2(-x)^7 - 19(-x)^6 + 66(-x)^5 - 95(-x)^4 + 22(-x)^3 + 87(-x)^2 - 90(-x)^1 + 27$$
$$= -2x^7 - 19x^6 - 66x^5 - 95x^4 - 22x^3 + 87x^2 + 90x + 27$$

As you can see, the function has only one sign change, from negative to positive. Therefore, the function has exactly one negative root — no more, no less.

Knowing the potential number of positive and negative roots for a polynomial is very helpful when you want to pinpoint an exact number of roots. The example polynomial I present in this section has only one negative real root. That fact tells you to concentrate your guesses on positive roots; the odds are better that you'll find a positive root first. When you're using synthetic division (see the "Synthesizing Root Findings" section) to find the roots, the steps get easier and easier as you find and eliminate roots. By picking off the roots that you have a better chance of finding, first, you can save the harder to find for the end.

Synthesizing Root Findings

You use synthetic division to test the list of possible roots for a polynomial that you come up with by using the Rational Root Theorem (see the section "Saving your sanity: The Rational Root Theorem" earlier in the chapter.) *Synthetic division* is a method of dividing a polynomial by a binomial, using only the coefficients of the terms. The method is quick, neat, and highly accurate — usually even more accurate than long division — and it uses most of the information from earlier sections in this chapter, putting it all together for the search for roots/zeros/intercepts of polynomials. (You can find more information about, and practice problems for, long division and synthetic division in one of my other spine-tingling thrillers, *Algebra Workbook For Dummies*.)

You can interpret your results in three different ways, depending on what purpose you're using synthetic division for. I explain each way in the following sections.

Using synthetic division to test for roots

When you want to use synthetic division to test for roots in a polynomial, the last number on the bottom row of your synthetic division problem is the telling result. If that number is zero, the division had no remainder, and the number is a root. The fact that there's no remainder means that the binomial represented by the number is dividing the polynomial evenly. The number is a root because the binomial is a factor.

The polynomial $y = x^5 + 5x^4 - 2x^3 - 28x^2 - 8x + 32$, for example, has zeros or roots when $y = 0$. You could find as many as five real roots, which you can tell from the exponent 5 on the first x. Using Descartes' Rule of Sign (see the previous section), you find two or zero positive real roots (indicating two sign changes). Replacing each x with $-x$, the polynomial now reads $y = -x^5 + 5x^4 + 2x^3 - 28x^2 + 8x + 32$. Again, using the Rule of Sign, you find three or one negative real roots. (Counting up the number of positive and negative roots helps when you're making a guess as to what a root may be.)

Now, using the Rational Root Theorem (see the section "Saving your sanity: The Rational Root Theorem"), your list of the potential rational roots is $\pm 1, \pm 2, \pm 4, \pm 8, \pm 16$, and ± 32. You choose one of these and apply synthetic division.

You should generally go with the smaller numbers, first, when using synthetic division, so use 1 and –1, 2 and –2, and so on.

Keeping in mind that smaller is better in this case, the following process shows a guess that $x = 1$ is a root.

The steps for performing synthetic division on a polynomial to find its roots are as follows:

1. **Write the polynomial in order of decreasing powers of the exponents. Replace any missing powers with zero to represent the coefficient.**

   In this case, you've lucked out. The polynomial is already in the correct order: $y = x^5 + 5x^4 - 2x^3 - 28x^2 - 8x + 32$.

2. **Write the coefficients in a row, including the zeros.**

   1   5   -2   -28   -8   32

3. **Put the number you want to divide by in front of the row of coefficients, separated by a half-box.**

   In this case, the guess is $x = 1$.

   $\underline{1|}$ 1   5   -2   -28   -8   32

4. **Draw a horizontal line below the row of coefficients, leaving room for numbers under the coefficients.**

   $\underline{1|}$ 1   5   -2   -28   -8   32

   _____

5. **Bring the first coefficient straight down below the line.**

   $\underline{1|}$ 1   5   -2   -28   -8   32
          ↓
   _____
       1

6. **Multiply the number you bring below the line by the number that you're dividing into everything. Put the result under the second coefficient.**

   $\underline{1|}$ 1   5   -2   -28   -8   32
        ×↘   1
   _____
       1 ↗ =

7. Add the second coefficient and the product, putting the result below the line.

```
1│  1    5    -2   -28   -8   32
           ↓ +
           1
    ─────────────────────────
    1    6
```

8. Repeat the multiplication/addition from Steps 6 and 7 with the rest of the coefficients.

```
1│  1    5    -2   -28    -8    32
         1     6     4   -24   -32
    ──────────────────────────────
    1    6     4   -24   -32     0
```

The last entry on the bottom is a zero, so you know one is a root. Now, you can do a modified synthetic division when testing for the next root; you just use the numbers across the bottom except the bottom right zero entry. (These values are actually coefficients of the quotient, if you do long division; see the following section.)

If your next guess is to see if $x = -1$ is a root, the modified synthetic division appears as follows:

```
-1│  1    6     4   -24   -32
         -1    -5     1    23
    ─────────────────────────
    1    5    -1   -23    -9
```

The last entry on the bottom row isn't zero, so -1 isn't a root.

The really good guessers amongst you decide to try $x = 2$, $x = -4$, $x = -2$, and $x = -2$ (a second time). These values represent the rest of the roots, and the synthetic division for all of the guesses looks like this:

First, trying the $x = 2$,

```
2│  1    6     4   -24   -32
         2    16    40    32
    ─────────────────────────
    1    8    20    16     0
```

The last number in the bottom row is 0. That's the remainder in the division. So now just look at all the numbers that come before the 0; they're the new coefficients to divide into. Notice that the last coefficient is now 16, so you can modify your list of possible roots to be just the factors of 16. Now, dividing by –4:

$$
\begin{array}{r|rrrr}
-4 & 1 & 8 & 20 & 16 \\
   &   & -4 & -16 & -16 \\
\hline
   & 1 & 4 & 4 & 0 \\
\end{array}
$$

In the next division, you consider only the factors of 4:

This time, dividing by –2,

$$
\begin{array}{r|rrr}
-2 & 1 & 4 & 4 \\
   &   & -2 & -4 \\
\hline
   & 1 & 2 & 0 \\
\end{array}
$$

The last number n the row is 0, so –2 is a root. Repeat the division, and you find that –2 is a double root:

$$
\begin{array}{r|rr}
-2 & 1 & 2 \\
   &   & -2 \\
\hline
   & 1 & 0 \\
\end{array}
$$

Your job is finished when you see the number one remaining in the last row, before the zero.

Now you can collect all the numbers that divided evenly — the roots of the equation — and use them to write the answer to the equation that's set equal to zero or to write the factorization of the polynomial or to sketch the graph with these numbers as *x*-intercepts.

Synthetically dividing by a binomial

Finding the roots of a polynomial isn't the only excuse you need to use synthetic division. You can also use synthetic division to replace the long, drawn-out process of dividing a polynomial by a binomial. Divisions like this are found in lots of calculus problems — where you need to make the expression more simplified.

The polynomial can be any degree; the binomial has to be either $x + c$ or $x - c$, and the coefficient on the x is one. This may seem rather restrictive, but a huge number of long divisions you'd have to perform fit in this category, so it helps to have a quick, efficient method to perform these basic division problems.

To use synthetic division to divide a polynomial by a binomial, you first write the polynomial in decreasing order of exponents, inserting a zero for any missing exponent. The number you put in front or divide by is the *opposite* of the number in the binomial. So, if you divide $x^5 + 3x^4 - 8x^2 - 5x + 2$ by the binomial $x + 2$, you use -2 in the synthetic division, as shown here:

$$
\begin{array}{r|rrrrrr}
-2 & 1 & 3 & 0 & -8 & -5 & 2 \\
   &   & -2 & -2 & 4 & 8 & -6 \\
\hline
   & 1 & 1 & -2 & -4 & 3 & -4
\end{array}
$$

As you can see, the last entry on the bottom row isn't zero. If you're looking for roots of a polynomial equation, this fact tells you that -2 isn't a root. In this case, because you're working on a long division application (which you know because you need to divide to simplify the expression), the -4 is the remainder of the division — in other words, the division doesn't come out even.

You obtain the answer (quotient) of the division problem from the coefficients across the bottom of the synthetic division. You start with a power one value lower than the original polynomial's power, and you use all the coefficients, dropping the power by one with each successive coefficient. The last coefficient is the remainder, which you write over the divisor.

Here's the division problem and its solution. The original division problem is written first. After the problem, you see the coefficients from the synthetic division written in front of variables — starting with one degree lower than the original problem. The remainder of -4 is written in a fraction on top of the divisor, $x + 2$.

$$\left(x^5 + 3x^4 - 8x^2 - 5x + 2\right) \div \left(x + 2\right) = x^4 + x^3 - 2x^2 - 4x + 3 - \frac{4}{x + 2}$$

Wringing out the Remainder (Theorem)

In the two previous sections, you use synthetic division to test for roots of a polynomial equation and then to do a long division problem. You use the same synthetic division process, but you read and use the results differently. In this section, I present yet another use of synthetic division involving the

Remainder Theorem. When you're looking for roots or solutions of a polynomial equation, you always want the remainder from the synthetic division to be zero. In this section, you get to see how to make use of all those pesky remainders that weren't zeros.

The Remainder Theorem: When the polynomial $f(x) = a_n x^n + a_{n-1} x^{n-1} + a_{n-2} x^{n-2} + \ldots + a_1 x^1 + a_0$ is divided by the binomial $x - c$, the remainder of the division is equal to $f(c)$.

For instance, in the division problem from the previous section, $(2x^5 + 3x^4 - 8x^2 - 5x + 2) \div (x + 2)$ has a remainder of -4. Therefore, according to the Remainder Theorem, for the function $f(x) = 2x^5 + 3x^4 - 8x^2 - 5x + 2$, $f(-2) = -4$.

The Remainder Theorem comes in very handy for root problems because you'll find it much easier to do synthetic division, where you multiply and add repeatedly, than to have to substitute numbers in for variables, raise the numbers to high powers, multiply by the coefficients, and then combine the terms.

Using the Remainder Theorem to find $f(3)$ when $f(x) = x^8 - 3x^7 + 2x^5 - 14x^3 + x^2 - 15x + 11$, for example, you apply synthetic division to the coefficients using 3 as the divider in the half-box.

3│	1	-3	0	2	0	-14	1	-15	11
		3	0	0	6	18	12	39	72
	1	0	0	2	6	4	13	24	83

The remainder of the division by $x - 3$ is 83, and, by the Remainder Theorem, $f(3) = 83$. Compare the process you use here with substituting the 3 into the function: $f(3) = (3)^8 - 3(3)^7 + 2(3)^5 - 14(3)^3 + (3)^2 - 15(3) + 11$. These numbers get really large. For instance, $3^8 = 6,561$. The numbers are much more manageable when you use synthetic division and the Remainder Theorem.

Chapter 9

Relying on Reason: Rational Functions

*T*he word "rational" has many uses. You say that rational people act reasonably and predictably. You can also say that *rational numbers* are reasonable and predictable — their decimals either repeat (have a distinctive pattern as they go on forever and ever) or terminate (come to an abrupt end). This chapter gives your rational repertoire another boost — it deals with rational functions.

A rational function may not appear to be reasonable, but it's definitely predictable. In this chapter, you refer to the intercepts, the asymptotes, any removable discontinuities, and the limits of rational functions to tell where the function values have been, what they're doing for particular values of the domain, and what they'll be doing for large values of *x*. You also need all this information to discuss or graph a rational function.

A nice feature of rational functions is that you can use the intercepts, asymptotes, and removable discontinuities to help you sketch the graphs of the functions. And, by the way, you can throw in a few limits to help you finish the whole thing off with a bow on top.

Whether you're graphing rational functions by hand (yes, of course, it's holding a pencil) or with a graphing calculator, you need to be able to recognize their various characteristics (intercepts, asymptotes, and so on). If you don't know what these characteristics are and how to find them, your calculator is no better than a paperweight to you.

Exploring Rational Functions

You see *rational functions* written, in general, in the form of a fraction:

$$y = \frac{f(x)}{g(x)}$$

where *f* and *g* are *polynomials* (expressions with whole-number exponents; see Chapter 8).

Rational functions (and more specifically their graphs) are distinctive because of what they do and don't have. The graphs of rational functions *do* have *asymptotes* (lines drawn in to help with the shape and direction of the curve; a new concept that I cover in the "Adding Asymptotes to the Rational Pot" section later in the chapter), and the graphs often *don't* have all the real numbers in their domains. Polynomials and exponential functions (which I cover in Chapters 8 and 10, respectively) make use of all the real numbers — their domains aren't restricted.

Sizing up domain

As I explain in Chapter 6, the *domain* of a function consists of all the real numbers that you can use in the function equation. Values in the domain have to work in the equation and avoid producing imaginary or nonexistent answers.

You write the equations of rational functions as fractions — and fractions have denominators. The denominator of a fraction can't equal zero, so you exclude anything that makes the denominator of a rational function equal to zero from the domain of the function.

The following list illustrates some examples of domains of rational functions:

- ✔ The domain of $y = \frac{x-1}{x-2}$ is all real numbers except 2. In interval notation (see Chapter 2), you write the domain as $(-\infty, 2) \cup (2, \infty)$. (The symbol ∞ signifies that the numbers increase without end; and the $-\infty$ signifies decreasing without end. The $\cup$ between the two parts of the answer means *or*.)

- ✔ The domain of $y = \frac{x+1}{x(x+4)}$ is all real numbers except 0 and –4. In interval notation, you write the domain as $(-\infty, -4) \cup (-4, 0) \cup (0, \infty)$.

- ✔ The domain of $y = \frac{x}{x^2+3}$ is all real numbers; no number makes the denominator equal to zero.

Introducing intercepts

Functions in algebra can have intercepts. A rational function may have an *x*-intercept and/or a *y*-intercept, but it doesn't have to have either. You can determine whether a given rational function has intercepts by looking at its equation.

Using zero to find y-intercepts

The coordinate (0, *b*) represents the *y*-intercept of a rational function. To find the value of *b*, you substitute a zero for *x* and solve for *y*.

For instance, if you want to find the *y*-intercept of the rational function $y = \frac{x+6}{x-3}$, you replace each *x* with zero to get $y = \frac{0+6}{0-3} = \frac{6}{-3} = -2$. The *y*-intercept is (0, –2).

If zero is in the domain of a rational function, you can be sure that the function at least has a *y*-intercept. A rational function doesn't have a *y*-intercept if its denominator equals zero when you substitute zero in the equation for *x*.

X marks the spot

The coordinate (*a*, 0) represents an *x*-intercept of a rational function. To find the value(s) of *a*, you let *y* equal zero and solve for *x*. (Basically, you just set the numerator of the fraction equal to zero — after you completely reduce the fraction.) You could also multiply each side of the equation by the denominator to get the same equation — it just depends on how you look at it.

To find the *x*-intercepts of the rational function $y = \frac{x^2 - 3x}{x^2 + 2x - 48}$, for example, you set $x^2 - 3x$ equal to zero and solve for *x*. Factoring the numerator, you get $x(x - 3) = 0$. The two solutions of the equation are $x = 0$ and $x = 3$. The two intercepts, therefore, are (0, 0) and (3, 0).

Adding Asymptotes to the Rational Pot

The graphs of rational functions take on some distinctive shapes because of asymptotes. An asymptote is a sort of ghost line. *Asymptotes* are drawn into the graph of a rational function to show the shape and direction of the function. The asymptotes aren't really part of the graphs, though, because they aren't made up of function values. Rather, they indicate where the function *isn't*. You lightly sketch in the asymptotes when you're graphing to help you with the final product. The types of asymptotes that you usually find in a rational function include the following:

✔ Vertical asymptotes

✔ Horizontal asymptotes

✔ Oblique (slant) asymptotes

In this section, I explain how you crunch the numbers of rational equations to identify asymptotes and graph them.

Determining the equations of vertical asymptotes

The equations of vertical asymptotes appear in the form $x = h$. This equation of a line has only the x variable — no y variable — and the number h. A vertical asymptote occurs in the rational function $y = \dfrac{f(x)}{g(x)}$ if $f(x)$ and $g(x)$ have no common factors, and it appears at whatever values the denominator equals zero — $g(x) = 0$. (In other words, vertical asymptotes occur at values that don't fall in the domain of the rational function.)

A *discontinuity* is a place where a rational function doesn't exist — you find a break in the flow of the numbers being used in the function equation. A discontinuity is indicated by a numerical value that tells you where the function isn't defined; this number isn't in the domain of the function. You know that a function is *discontinuous* wherever a vertical asymptote appears in the graph because vertical asymptotes indicate breaks or gaps in the domain.

To find the vertical asymptotes of the function $y = \dfrac{x}{x^2 - 4x + 3}$, for example, you first note that there's no common factor in the numerator and denominator. Then you set the denominator equal to zero. Factoring $x^2 - 4x + 3 = 0$, you get $(x - 1)(x - 3) = 0$. The solutions are $x = 1$ and $x = 3$, which are the equations of the vertical asymptotes.

Determining the equations of horizontal asymptotes

The horizontal asymptote of a rational function has an equation that appears in the form $y = k$. This linear equation has only the variable y — no x — and the k is some number. A rational function $y = \dfrac{f(x)}{g(x)}$ has only one horizontal asymptote — if it has one at all (some rational functions have no horizontal asymptotes, others have one, and none of them have more than one). A rational function has a horizontal asymptote when the degree (highest power) of $f(x)$, the polynomial in the numerator, is less than or equal to the degree of $g(x)$, the polynomial in the denominator.

Here's a rule for determining the equation of a horizontal asymptote. The horizontal asymptote of $y = \dfrac{f(x)}{g(x)} = \dfrac{a_n x^n + a_{n-1} x^{n-1} + \ldots + a_0}{b_m x^m + b_{m-1} x^{m-1} + \ldots + b_0}$ is $y = \dfrac{a_n}{b_m}$ when $n = m$, meaning that the highest degrees of the polynomials are the same. The fraction here is made up of the lead coefficients of the two polynomials. When $n < m$, meaning that the degree in the numerator is less than the degree in the denominator, $y = 0$.

If you want to find the horizontal asymptote for $y = \dfrac{3x^4 - 2x^3 + 7}{x^5 - 3x^2 - 5}$, for example, you use the previously stated rules. Because $4 < 5$, the horizontal asymptote is $y = 0$. Now look at what happens when the degree of the denominator is the same as the degree of the numerator. The horizontal asymptote of $y = \dfrac{3x^4 - 2x^3 + 7}{x^4 - 3x^2 - 5}$ is $y = 3$ (a_n over b_m). The fraction formed by the lead coefficients is $y = \dfrac{3}{1} = 3$.

Graphing vertical and horizontal asymptotes

When a rational function has one vertical asymptote and one horizontal asymptote, its graph usually looks like two flattened-out, C-shaped curves that appear diagonally opposite one another from the intersection of the asymptotes. Occasionally, the curves appear side by side, but that's the exception rather than the rule. Figure 9-1 shows you two examples of the more frequently found graphs in the one horizontal and one vertical classification.

Figure 9-1: Rational functions approaching vertical and horizontal asymptotes.

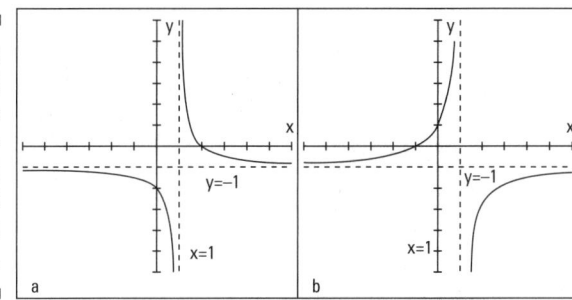

In both graphs, the vertical asymptotes are at $x = 1$, and the horizontal asymptotes are at $y = -1$. In Figure 9-1a, the intercepts are $(0, -2)$ and $(2, 0)$. Figure 9-1b has intercepts of $(-1, 0)$ and $(0, 1)$.

You can have only one horizontal asymptote in a rational function, but you can have more than one vertical asymptote. Typically, the curve to the right of the right-most vertical asymptote and to the left of the left-most vertical asymptote are like flattened-out or slowly turning C's. They nestle in the corner and follow along the asymptotes. Between the vertical asymptotes is where some graphs get more interesting. Some graphs between vertical asymptotes can be U-shaped, going upward or downward (see Figure 9-2a), or they can cross in the middle, clinging to the vertical asymptotes on one side or the other (see Figure 9-2b). You find out which case you have by calculating a few points — intercepts and a couple more — to give you clues as to the shape. The graphs in Figure 9-2 show you some of the possibilities.

Figure 9-2:
Rational
functions
curving
between
vertical
asymptotes.

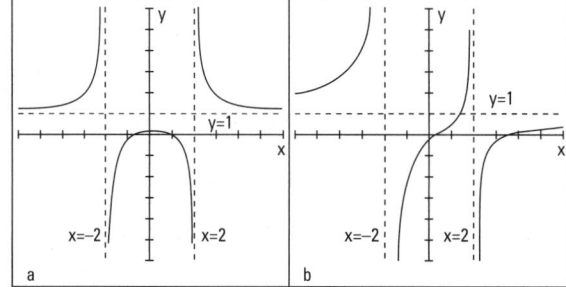

In Figure 9-2, the vertical asymptotes are at $x = 2$ and $x = -2$. The horizontal asymptotes are at $y = 1$. Figure 9-2a has two x-intercepts lying between the two vertical asymptotes; the y-intercept is there, too. In Figure 9-2b, the y-intercept and one x-intercept lie between the two vertical asymptotes; another x-intercept is to the right of the right-most vertical asymptote.

The graph of a rational function can cross a horizontal asymptote, but it never crosses a vertical asymptote. Horizontal asymptotes show what happens for very large or very small values of x.

Crunching the numbers and graphing oblique asymptotes

An *oblique* or *slant asymptote* takes the form $y = ax + b$. You may recognize this form as the slope-intercept form for the equation of a line (as seen in Chapter 5). A rational function has a slant asymptote when the degree of the polynomial in the numerator is exactly one value greater than the degree in the denominator (x^4 over x^3, for example).

You can find the equation of the slant asymptote by using long division. You divide the denominator of the rational function into the numerator and use the first two terms in the answer. Those two terms are the $ax + b$ part of the equation of the slant asymptote.

To find the slant asymptote of $y = \dfrac{x^4 - 3x^3 + 2x - 7}{x^3 + 3x - 1}$, for example, do the long division:

$$
\begin{array}{r}
x - 3 \\
x^3 + 3x - 1 \overline{\smash{\big)}\ x^4 - 3x^3 \qquad\quad + 2x - 7} \\
\underline{-\left(x^4 \qquad\ + 3x^2 - x\right)} \\
-3x^3 - 3x^2 + 3x - 7 \\
\underline{-\left(-3x^3 \qquad\ - 9x + 3\right)} \\
-3x^2 + 12x - 10
\end{array}
$$

You can ignore the remainder at the bottom. The slant asymptote for this example is $y = x - 3$. (For more on long division of polynomials, see *Algebra Workbook For Dummies,* by yours truly and published by Wiley.)

An oblique (or slant) asymptote creates two new possibilities for the graph of a rational function. If a function has an oblique asymptote, its curve tends to be a very-flat C on opposite sides of the intersection of the slant asymptote and a vertical asymptote (see Figure 9-3a), or the curve has U-shapes between the asymptotes (see Figure 9-3b).

Figure 9-3:
Rational
graphs
between
vertical and
oblique
asymptotes.

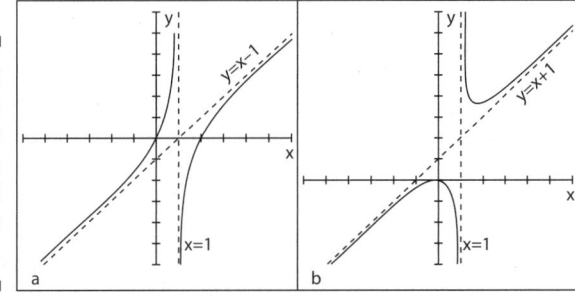

Figure 9-3a has a vertical asymptote at $x = 1$ and a slant asymptote at $y = x - 1$; its intercepts are at $(0, 0)$ and $(2, 0)$. Figure 9-3b has a vertical asymptote at $x = 1$ and a slant asymptote at $y = x + 1$; its only intercept is at $(0, 0)$.

Accounting for Removable Discontinuities

Discontinuities at vertical asymptotes (see the "Determining the equations of vertical asymptotes" section earlier in the chapter for a definition) can't be removed. But rational functions sometimes have *removable discontinuities* in other places. The removable designation is, however, a bit misleading. The gap in the domain still exists at that "removable" spot, but the function values and graph of the curve tend to behave a little better than at x values where there's a nonremovable discontinuity. The function values stay close together — they don't spread far apart — and the graphs just have tiny holes, not vertical asymptotes where the graphs don't behave very well (they go infinitely high or infinitely low). Refer to "Going to Infinity," later in this chapter, for more on this.

You have the option of removing discontinuities by factoring the original function statement — if it does factor. If the numerator and denominator don't have a common factor, then there isn't a removable discontinuity.

You can recognize removable discontinuities when you see them graphed on a rational function; they appear as holes in the graph — big dots with spaces in the middle rather than all shaded in. Removable discontinuities aren't big, obvious discontinuities like vertical asymptotes; you have to look carefully for them. If you just can't wait to see what these things look like, skip ahead to the "Showing removable discontinuities on a graph" section a bit later in the chapter.

Removal by factoring

Discontinuities are *removed* when they no longer have an effect on the rational function equation. You know this is the case when you find a factor that's common to both the numerator and the denominator. You accomplish the removal process by factoring the polynomials in the numerator and denominator of the rational function and then reducing the fraction.

To remove the discontinuity in the rational function $y = \dfrac{x^2 - 4}{x^2 - 5x - 14}$, for example, you first factor the fraction into this form (see Chapter 3):

$$y = \frac{(x-2)(x+2)}{(x-7)(x+2)} = \frac{(x-2)\cancel{(x+2)}}{(x-7)\cancel{(x+2)}}$$

Now you reduce the fraction to the new function statement:

$$y = \frac{x-2}{x-7}$$

By getting rid of the removable discontinuity, you simplify the equation that you're graphing. It's easier to graph a line with a little hole in it than deal with an equation that has a fraction — and all the computations involved.

Evaluating the removal restrictions

The function $y = \frac{x^2 - 4}{x^2 - 5x - 14}$, which you work with in the previous section, starts out with a quadratic in the denominator (see Chapters 3 and 7). You factor the denominator, and when you set it equal to zero, you find that the solutions $x = -2$ and $x = 7$ don't appear in the domain of the function. Now what?

Numbers excluded from the domain stay excluded even after you remove the discontinuity. The function still isn't defined for the two values you find. Therefore, you can conclude that the function behaves differently at each of the discontinuities. When $x = -2$, the graph of the function has a hole; the curve approaches the value, skips it, and goes on. It behaves in a reasonable fashion: The function values skip over the discontinuity, but the values get really close to it. When $x = 7$, however, a vertical asymptote appears; the discontinuity doesn't go away. The function values go haywire at that x value and don't settle down at all.

Showing removable discontinuities on a graph

A vertical asymptote of a rational function indicates a *discontinuity,* or a place on its graph where the function isn't defined. On either side of a vertical asymptote, the graph rises toward positive infinity or falls toward negative infinity. The rational function has no limit wherever you see a vertical asymptote (see the "Pushing the Limits with Reational Functions" section later in the chapter). Some rational functions may have discontinuities at which limits exist. When a function has a *removable discontinuity,* a limit exists, and its graph shows this by putting a hollow circle in place of a piece of the graph.

Figure 9-4 shows a rational function with a vertical asymptote at $x = -2$ and a removable discontinuity at $x = 3$. The horizontal asymptote is the X-axis (written $y = 0$). Unfortunately, graphing calculators don't show the little hollow circles indicating removable discontinuities. Oh, sure, they leave a gap there, but the gap is only one pixel wide, so you can't see it with the naked eye. You just have to know that the discontinuity is there. We're still better than the calculators!

Vulgar uncles versus vulgar fractions

You should try to avoid vulgar people because they can't speak without introducing some unpleasant words or thoughts — like an overbearing uncle at Christmas trying to induce some laughs. A *vulgar fraction,* however, should elicit the opposite reaction on your part; you should in no way avoid it. Unlike the uncle, vulgar fractions are perfectly proper and G-rated.

A fraction is *vulgar* if it has one integer divided by another integer, as long as the integer that's doing the dividing isn't equal to zero. For instance, the fraction "two divided by three," or

⅔, is considered vulgar. The value 2 is in the numerator of the fraction, and the value 3 is in the denominator. The word *numerator* comes from *enumerate,* which means to announce an amount. So, the numerator announces how many of the fractional parts you're talking about at that time. In the case of ⅔, you have 2 of the 3 pieces of the whole picture. The word *denominator* means to name something or to tell what kind of item you're talking about. The 3 in ⅔ tells what kind of divisions of the whole have been made.

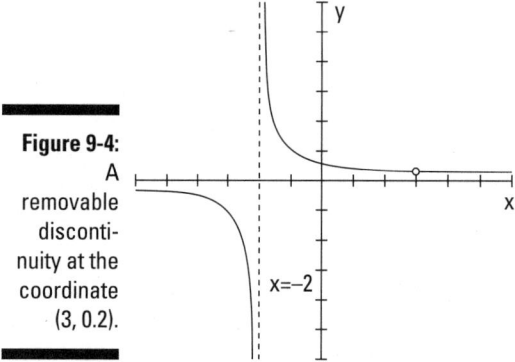

Figure 9-4:
A removable discontinuity at the coordinate (3, 0.2).

Pushing the Limits of Rational Functions

The limit of a rational function is something like the speed limit on a road. The speed limit tells you how fast you can go (legally). As you approach the speed limit, you adjust the pressure you put on the gas pedal accordingly, trying to keep close to the limit. Most drivers want to stay at least slightly above or slightly below the limit.

The *limit* of a rational function acts this way, too — homing in on a specific number, either slightly above or slightly below it. If a function has a limit at a particular number, as you approach the designated number from the left or from the right (from below or above the value, respectively), you approach the same place or function value. The function doesn't have to be defined at the number you're approaching (sometimes they are and sometimes not) — there could be a discontinuity. But, if a limit rests at the number, the values of the function have to be really close together — but not touching.

The special notation for limits is as follows:

$$\lim_{x \to a} f(x) = L$$

You read the notation as, "The limit of the function, $f(x)$, as x approaches the number a, is equal to L." The number a doesn't have to be in the domain of the function. You can talk about a limit of a function whether a is in the domain or not. And you can approach a, as long as you don't actually reach it.

Allow me to relate the notation back to following the speed limit. The value a is the exact pressure you need to put on the pedal to achieve the exact speed limit — often impossible to attain.

Look at the function $f(x) = x^2 + 2$. Suppose that you want to see what happens on either side of the value $x = 1$. In other words, you want to see what's happening to the function values as you get close to 1 coming from the left and then coming from the right. Table 9-1 shows you some selected values.

Table 9-1	Approaching $x = 1$ from Both Sides in $f(x) = x^2 + 2$		
x Approaching 1 from the Left	*Corresponding Behavior in $x^2 + 2$*	*x Approaching 1 from the Right*	*Corresponding Behavior in $x^2 + 2$*
0.0	2.0	2.0	6.0
0.5	2.25	1.5	4.25
0.9	2.81	1.1	3.21
0.999	2.998001	1.001	3.002001
0.99999	2.9999800001	1.00001	3.0000200001

As you approach $x = 1$ from the left or the right, the value of the function approaches the number 3. The number 3 is the limit. You may wonder why

I didn't just plug the number 1 into the function equation: $f(1) = 1^2 + 2 = 3$. My answer is, in this case, you can. I just used the table to illustrate how the concept of a limit works.

Evaluating limits at discontinuities

The beauty of a limit is that it can also work when a rational function isn't defined at a particular number. The function $y = \dfrac{x-2}{x^2 - 2x}$, for example, is discontinuous at $x = 0$ and at $x = 2$. You find these numbers by factoring the denominator, setting it equal to zero — $x(x-2) = 0$ — and solving for x. This function has no limit when x approaches zero, but it has a limit when x approaches two. Sometimes it's helpful to actually see the numbers — see what you get from evaluating a function at different values — so I've included Table 9-2. It shows what happens as x approaches zero from the left and right, and it illustrates that the function has no limit at that value.

Table 9-2	Approaching $x = 0$ from Both Sides in $y = \dfrac{x-2}{x^2-2x}$		
x Approaching 0 from the Left	Corresponding Behavior of $y = \dfrac{x-2}{x^2-2x}$	x Approaching 0 from the Right	Corresponding Behavior of $y = \dfrac{x-2}{x^2-2x}$
−1.0	−1	1.0	1
−0.5	−2	0.5	2
−0.1	−10	0.1	10
−0.001	−1,000	0.001	1,000
−0.00001	−100,000	0.00001	100,000

Table 9-2 shows you that $\lim\limits_{x \to 0} \dfrac{x-2}{x^2-2x}$ doesn't exist. As x approaches from below the value of zero, the values of the function drop down lower and lower toward negative infinity. Coming from above the value of zero, the values of the function raise higher and higher toward positive infinity. The sides will never come to an agreement; no limit exists.

Table 9-3 shows you how a function can have a limit even when the function isn't defined at a particular number. Sticking with the previous example function, you find a limit as x approaches 2.

Table 9-3	Approaching $x = 2$ from Both Sides in $y = \dfrac{x-2}{x^2-2x}$		
x Approaching 2 from the Left	*Corresponding Behavior of* $y = \dfrac{x-2}{x^2-2x}$	*x Approaching 2 from the Right*	*Corresponding Behavior of* $y = \dfrac{x-2}{x^2-2x}$
1.0	1.0	3.0	0.3333 . . .
1.5	0.6666 . . .	2.5	0.4
1.9	0.526316 . . .	2.1	0.476190 . . .
1.99	0.502512 . . .	2.001	0.499750 . . .
1.999	0.500250 . . .	2.00001	0.4999975 . . .

Table 9-3 shows $\lim\limits_{x \to 2} \dfrac{x-2}{x^2-2x} = 0.5$. The numbers get closer and closer to 0.5 as x gets closer and closer to 2 from both directions. You find a limit at $x = 2$, even though the function isn't defined there.

Determining an existent limit without tables

If you've examined the two tables from the previous section, you may think that the process of finding limits is exhausting. Allow me to tell you that algebra offers a much easier way to find limits — if they exist.

Functions with removable discontinuities have limits at the values where the discontinuities exist. To determine the values of these limits, you follow these steps:

1. **Factor the rational function equation.**

2. **Reduce the function equation.**

3. **Evaluate the new, revised function equation at the value of x in question.**

To solve for the limit when $x = 2$ in the rational function $y = \dfrac{x-2}{x^2-2x}$, an example from the previous section, you first factor and then reduce the fraction:

$$y = \frac{x-2}{x^2-2x} = \frac{\cancel{x-2}}{x\,(\cancel{x-2})} = \frac{1}{x}$$

Now you replace the x with 2 and get $y = 0.5$, the limit when $x = 2$. Wow! How simple! In general, if a rational function factors, then you'll find a limit at the number excluded from the domain if the factoring makes that exclusion seem to disappear.

Determining which functions have limits

Some rational functions have limits at discontinuities and some don't. You can determine whether to look for a removable discontinuity in a particular function by first trying the x value in the function. Replace all the x's in the function with the number in the limit (what x is approaching). The result of that substitution tells you if you have a limit or not. You use the following rules of thumb:

- If $\lim\limits_{x \to a} \dfrac{f(x)}{g(x)} = \dfrac{\text{some number}}{0}$, the function has no limit at a.

- If $\lim\limits_{x \to a} \dfrac{f(x)}{g(x)} = \dfrac{0}{0}$, the function has a limit at a. You reduce the fraction and evaluate the newly formed function equation at a (as I explain how to do in the preceding section).

A fraction has no value when a zero sits in the denominator, but a zero divided by a zero does have a form — called an *indeterminate form*. Take this form as a signal that you can look for a value for the limit.

For example, here's a function that has no limit when $x = 1$: You're looking at what x is approaching in the limit statement, so you're only concerned about the 1.

$$\lim_{x \to 1} \frac{x^2 - 4x - 5}{x^2 - 1} = \frac{1 - 4 - 5}{1 - 1} = \frac{-8}{0}$$

The function has no limit at 1 because it has a number over zero.

If you try this function at $x = -1$, you see that the function has a removable discontinuity:

$$\lim_{x \to -1} \frac{x^2 - 4x - 5}{x^2 - 1} = \frac{1 - 4(-1) - 5}{1 - 1} = \frac{0}{0}$$

The function has a limit at -1 because it has zero over zero.

Going to infinity

When a rational function doesn't have a limit at a particular value, the function values and graph have to go somewhere. A particular function may not have the number 3 in its domain, and its graph may have a vertical asymptote when $x = 3$. Even though the function has no limit, you can still say something about what's happening to the function as it approaches 3 from the left and the right. The graph has no numerical limit at that point, but you can still tell something about the behavior of the function. The behavior is attributed to *one-sided limits*.

A one-sided limit tells you what a function is doing at an *x*-value as the function approaches from one side or the other. One-sided limits are more restrictive; they work only from the left or from the right.

The notation for indicating one-sided limits from the left or right is shown here:

- The limit as *x* approaches the value *a* from the left is $\lim\limits_{x \to a^-} f(x)$.
- The limit as *x* approaches the value *a* from the right is $\lim\limits_{x \to a^+} f(x)$.

Do you see the little positive or negative sign after the *a?* You can think of *from the left* as coming from the same direction as all the negative numbers on the number line and *from the right* as coming from the same direction as all the positive numbers.

Table 9-4 shows some values of the function $y = \dfrac{1}{x-3}$, which has a vertical asymptote at *x* = 3.

Table 9-4	Approaching *x* = 3 from Both Sides in $y = \dfrac{1}{x-3}$		
x Approaching 3 from the Left	*Corresponding Behavior of* $\dfrac{1}{x-3}$	*x* Approaching 3 from the Right	*Corresponding Behavior of* $\dfrac{1}{x-3}$
2.0	−1	4.0	1
2.5	−2	3.5	2
2.9	−10	3.1	10
2.999	−1000	3.001	1,000
2.99999	−100,000	3.00001	100,000

You express the one-sided limits for the function from Table 9-4 as follows:

$$\lim_{x \to 3^-} \frac{1}{x-3} = -\infty, \ \lim_{x \to 3^+} \frac{1}{x-3} = +\infty$$

The function goes down to negative infinity as it approaches 3 from below the value and up to positive infinity as it approaches 3 from above the value. "And nary the twain shall meet."

Catching rational limits at infinity

The previous section describes how function values can go to positive or negative infinity as x approaches some specific number. This section also talks about infinity, but it focuses on what rational functions do as their x values become very large or very small (approaching infinity themselves).

A function such as the parabola $y = x^2 + 1$ opens upward. If you let x be some really big number, y gets very big, too. Also, when x is very small (a "big" negative number), you square the value, making it positive, so y is very big for the small x. In function notation, you describe what's happening to this function as the x values approach infinity with $\lim_{x \to \infty} (x^2 + 1) = +\infty$.

You can indicate that a function approaches positive infinity going in one direction and negative infinity going in another direction with the same kind of notation you use for one-sided limits (see the previous section).

For example, the function $y = -x^3 + 6$ approaches negative infinity as x gets very large — think about what $x = 1,000$ does to the y value (you'd get $-1,000,000,000+6$). On the other hand, when $x = -1,000$, the y value gets very large, because you have $y = -(-1,000,000,000) + 6$, so the function approaches positive infinity.

In the case of rational functions, the limits at infinity — as x gets very large or very small — may be specific, finite, describable numbers. In fact, when a rational function has a horizontal asymptote, its limit at infinity is the same value as the number in the equation of the asymptote.

If you're looking for the horizontal asymptote of the function $y = \dfrac{4x^2 + 3}{2x^2 - 3x - 7}$, for example, you can use the rules in the section "Determining the equations of horizontal asymptotes" to determine that the horizontal asymptote of the function is $y = 2$. Using limit notation, you can write the solution as $\lim_{x \to \infty} \dfrac{4x^2 + 3}{2x^2 - 3x - 7} = 2$.

The proper algebraic method for evaluating limits at infinity is to divide every term in the rational function by the highest power of x in the fraction and then look at each term. Here's an important property to use: As x approaches infinity, any term with $\frac{1}{x}$ in it approaches zero — in other words, gets very small — so you can replace those terms with zero and simplify.

Here's how the property works when evaluating the limit of the previous example function, $\dfrac{4x^2 + 3}{2x^2 - 3x - 7}$. The highest power of the variable in the fraction is x^2, so every term is divided by x^2:

$$\lim_{x \to \infty} \frac{4x^2 + 3}{2x^2 - 3x - 7} =$$

$$\lim_{x \to \infty} \frac{\frac{4x^2}{x^2} + \frac{3}{x^2}}{\frac{2x^2}{x^2} - \frac{3x}{x^2} - \frac{7}{x^2}} =$$

$$\lim_{x \to \infty} \frac{4 + \frac{3}{x^2}}{2 - \frac{3}{x} - \frac{7}{x^2}} = \frac{4 + 0}{2 - 0 - 0} = \frac{4}{2} = 2$$

The limit as x approaches infinity is 2. As predicted, the number 2 is the number in the equation of the horizontal asymptote. The quick method for determining horizontal asymptotes is an easier way to find limits at infinity, and this procedure is also the correct *mathematical* way of doing it — and it shows why the other rule (the quick method) works. You also need this quicker method for more involved limit problems found in calculus and other higher mathematics.

Putting It All Together: Sketching Rational Graphs from Clues

The graphs of rational functions can include intercepts, asymptotes, and removable discontinuities (topics I cover earlier in this chapter). In fact, some graphs include all three. Sketching the graph of a rational function is fairly simple if you prepare carefully. Make use of any and all information you can glean from the function equation, sketch in any intercepts and asymptotes, and then plot a few points to determine the general shape of the curve.

To sketch the graph of the function $y = \frac{x^2 - 2x - 3}{x^2 - x - 2}$, for example, you should first look at the powers of the numerator and denominator. The degrees, or highest powers, are the same, so you find the horizontal asymptote by making a fraction of the lead coefficients. Both coefficients are one, and one divided by one is still one, so the equation of the horizontal asymptote is $y = 1$.

The rest of the necessary information for graphing is more forthcoming if you factor the numerator and denominator:

$$y = \frac{(x-3)(x+1)}{(x-2)(x+1)} = \frac{(x-3)\cancel{(x+1)}}{(x-2)\cancel{(x+1)}} = \frac{x-3}{x-2}$$

You factor out a common factor of $x + 1$ from the numerator and denominator. This action tells you two things. First, because $x = -1$ makes the denominator equal to zero, you know that -1 isn't in the domain of the function. Furthermore, the fact that -1 is *removed* by the factoring signals a removable discontinuity when $x = -1$. You can plug -1 into the new equation to find out where to graph the hole or open circle:

$$y = \frac{x-3}{x-2} = \frac{-1-3}{-1-2} = \frac{-4}{-3} = \frac{4}{3}$$

The hole is at $\left(-1, \frac{4}{3}\right)$. The remaining terms in the denominator tell you that the function has a vertical asymptote at $x = 2$. You find the y-intercept by letting $x = 0$: $\left(0, \frac{3}{2}\right)$. You find the x-intercepts by setting the new numerator equal to zero and solving for x. When $x - 3 = 0$, $x = 3$, so the x-intercept is $(3, 0)$. You place all this information on a graph, which is shown in Figure 9-5. Figure 9-5a shows how you sketch in the asymptotes, intercepts, and removable discontinuity.

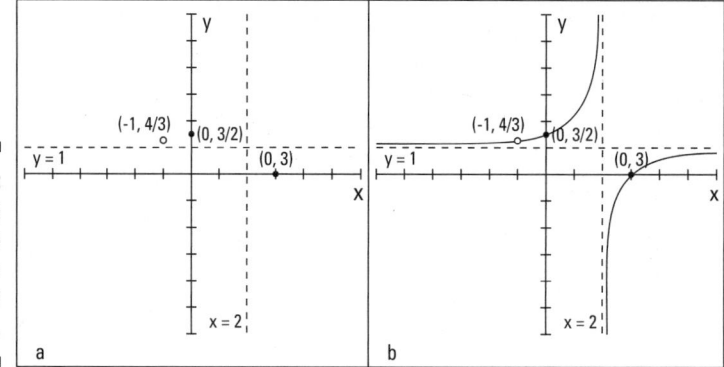

Figure 9-5: Following the steps to graph a rational function.

Figure 9-5a seems to indicate that the curve will have soft-C shapes in the upper left and lower right parts of the graph, opposite one another through the asymptotes. If you plot a few points to confirm this, you see that the graph approaches positive infinity as it approaches $x = 2$ from the left and goes to negative infinity from the right. You can see the completed graph in Figure 9-5b.

The graphs of other rational functions, such as $y = \frac{x-1}{x^2 - 5x - 6}$, don't give you quite as many clues before you have to do the actual graphing. For this example, by factoring the denominator and setting it equal to zero, you get $(x + 1)(x - 6) = 0$; the vertical asymptotes are at $x = -1$ and $x = 6$. The horizontal asymptote is at $y = 0$, which is the X-axis. The only x-intercept is $(1, 0)$, and a y-intercept is located at $\left(0, \frac{1}{6}\right)$. Figure 9-6a shows the asymptotes and intercepts on a graph.

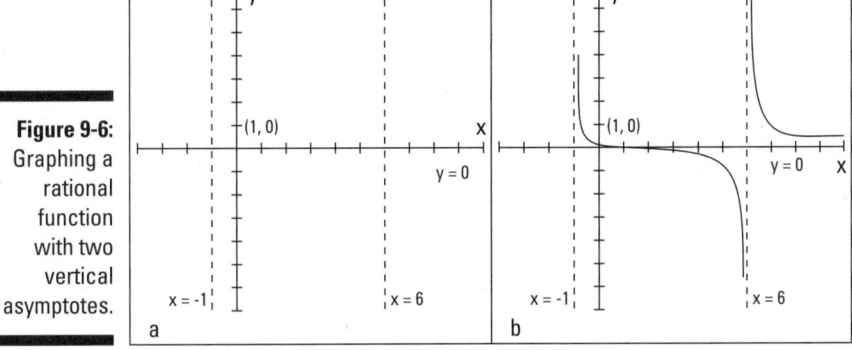

Figure 9-6: Graphing a rational function with two vertical asymptotes.

a   b

The only place the graph of the function crosses the X-axis is at $(1, 0)$, so the curve must come from the left side of that middle section separated by the vertical asymptotes and continue to the right side. If you try a couple of points — for instance, $x = 0$ and $x = 4$ — you get the points $\left(0, \frac{1}{6}\right)$ and $\left(4, -\frac{3}{10}\right)$. These points tell you that the curve is above the X-axis to the left of the x-intercept and below the X-axis to the right of the intercept. You can use this information to sketch a curve that drops down through the whole middle section.

Two other random points you may choose are $x = -2$ and $x = 7$. You choose points such as these to test the extreme left and right sections of the graph. From these values, you get the points $\left(-2, -\frac{3}{8}\right)$ and $\left(7, \frac{3}{4}\right)$. Plot these points and sketch in the rest of the graph, which is shown in Figure 9-6b.

Chapter 10

Exposing Exponential and Logarithmic Functions

*E*xponential growth and decay are natural phenomena. They happen all around us. And, being the thorough, worldly people they are, mathematicians have come up with ways of describing, formulating, and graphing these phenomena. You express the patterns observed when exponential growth and decay occur mathematically with exponential and logarithmic functions.

Why else are these functions important? Certain algebraic functions, such as polynomials and rational functions, share certain characteristics that exponential functions lack. For example, algebraic functions all show their variables as bases raised to some power, such as x^2 or x^8. Exponential functions, on the other hand, show their variables up in the powers of the expressions and lower some numbers as the bases, such as 2^x or e^x. In this chapter, I discuss the properties, uses, and graphs of exponential and logarithmic functions in detail.

Evaluating Exponential Expressions

An exponential function is unique because its variable appears in the exponential position and its constant appears in the base position. You write an *exponent*, or power, as a superscript just after the *base*. In the expression 3^x, for example, the variable x is the exponent, and the constant 3 is the base. The general form for an exponential function is $f(x) = a \cdot b^x$, where

✔ The base b is any positive number.

✔ The coefficient a is any real number.

✔ The exponent x is any real number.

The base of an exponential function can't be zero or negative; the domain is all real numbers; and the range is all positive numbers when a is positive (see Chapter 6 for more on these topics).

When you enter a number into an exponential function, you evaluate it by using the *order of operations* (along with other rules for working with exponents; I discuss these topics in Chapters 1 and 4). The order of operations dictates that you evaluate the function in the following order:

1. Powers and roots

2. Multiplication and division

3. Addition and subtraction

If you want to evaluate $f(x) = 3^x + 1$ for $x = 2$, for example, you replace the x with the number 2. So, $f(2) = 3^2 + 1 = 9 + 1 = 10$. To evaluate the exponential function $g(x) = 4\left(\frac{1}{2}\right)^x - 3$ for $x = -2$, you write the steps as follows:

$$g(-2) = 4\left(\frac{1}{2}\right)^{-2} - 3$$
$$= 4\left(\frac{2}{1}\right)^{2} - 3$$
$$= 4(4) - 3$$
$$= 16 - 3$$
$$= 13$$

You raise to the power first, multiply by 4, and then subtract 3.

Exponential Functions: It's All About the Base, Baby

The base of an exponential function can be any positive number. The bigger the number, the bigger it becomes as you raise it to higher and higher powers. (Sort of like the more money you have, the more money you make.) The bases can get downright small, too. In fact, when the base is some number between zero and one, you don't have a function that grows; instead, you have a function that falls.

Observing the trends in bases

The base of an exponential function tells you so much about the nature and character of the function, making it one of the first things you should look at and classify. The main way to classify the bases of exponential functions is to determine whether they're larger or smaller than one. After you make that designation, you look at how much larger or how much smaller. The exponents themselves affect the expressions that contain them in somewhat predictable ways, making them prime places to look when grouping.

Because the domain (or *input;* see Chapter 6) of exponential functions is all real numbers, and the base is always positive, the result of b^x is always a positive number. Even when you raise a positive base to a negative power, you get a positive answer. Exponential functions can end up with negative values if you subtract from the power or multiply by a negative number, but the power itself is always positive.

Grouping exponential functions according to their bases

Algebra offers three classifications for the base of an exponential function, due to the fact that the numbers used as bases appear to react in distinctive ways when raised to positive powers:

- When $b > 1$, the values of b^x grow larger as x gets bigger — for instance, $2^2 = 4$, $2^5 = 32$, $2^7 = 128$, and so on.

- When $b = 1$, the values of b^x show no movement. Raising the number 1 to higher powers always results in the number 1: $1^2 = 1$, $1^5 = 1$, $1^7 = 1$, and so on. You see no exponential growth or decay.

  In fact, some mathematicians leave the number 1 out of the listing of possible bases for exponentials. Others leave it in as a bridge between functions that increase in value and those that decrease in value. It's just a matter of personal taste.

- When $0 < b < 1$, the value of b^x grows smaller as x gets bigger. The base b has to be positive, and the numbers $0 < b < 1$ are all proper fractions (fractions with the numerators smaller than the denominators). Look at what happens to a fractional base when you raise it to the second, fifth, and eighth degrees: $\left(\frac{1}{3}\right)^2 = \frac{1}{9}$, $\left(\frac{1}{3}\right)^5 = \frac{1}{243}$, $\left(\frac{1}{3}\right)^8 = \frac{1}{6,561}$. The numbers get smaller and smaller as the powers get bigger.

Grouping exponential functions according to their exponents

An exponent placed with a number can affect the expression that contains the number in somewhat predictable ways. The exponent makes the result take on different qualities, depending on whether the exponent is greater than zero, equal to zero, or smaller than zero:

- ✔ When the base $b > 1$ and the exponent $x > 0$, the values of b^x get bigger and bigger as x gets larger — for instance, $4^3 = 64$ and $4^6 = 4,096$. You say that the values grow *exponentially*.

- ✔ When the base $b > 1$ and the exponent $x = 0$, the only value of b^x you get is 1. The rule is that $b^0 = 1$ for any number except $b = 0$. So, an exponent of zero really flattens things out.

- ✔ When the base $b > 1$ and the exponent $x < 0$ — a negative number — the values of b^x get smaller and smaller as the exponents get further and further from zero. Take these expressions, for example: $6^{-1} = \frac{1}{6}$ and $6^{-4} = \frac{1}{6^4} = \frac{1}{1,296}$. These numbers can get very small very quickly.

Meeting the most frequently used bases: 10 and e

Exponential functions feature bases represented by numbers greater than zero. The two most frequently used bases are 10 and *e,* where $b = 10$ and $b = e$.

It isn't too hard to understand why mathematicians like to use base 10 — in fact, just hold all your fingers in front of your face! All the powers of 10 are made up of ones and zeros — for instance, $10^2 = 100$, $10^9 = 1,000,000,000$, and $10^{-5} = 0.00001$. How much more simple can it get? Our number system, the decimal system, is based on tens.

Like the value 10, base *e* occurs naturally. Members of the scientific world prefer base *e* because powers and multiples of *e* keep creeping up in models of natural occurrences. Including *e*'s in computations also simplifies things for financial professionals, mathematicians, and engineers.

If you use a scientific calculator to get the value of *e,* you see only some of *e.* The numbers you see estimate only what *e* is; most calculators give you seven or eight decimal places. Here are the first nine decimal places in the value of *e,* in rounded form:

$e \approx 2.718281828$

The decimal value of e actually goes on forever without repeating in a pattern. The hint of a pattern you see in the nine digits doesn't hold for long. The equation $\lim_{x \to \infty} \left(1 + \frac{1}{x}\right)^x$ represents the exact value of e. The larger and larger x gets, the more correct decimal places you get.

Most algebra courses call for you to memorize only the first four digits of e — 2.718.

Now for the bad news. As wonderful as e is for taming function equations and scientific formulas, its powers aren't particularly easy to deal with. The base e is approximately 2.718, and when you take a number with a decimal value that never ends and raise it to a power, the decimal values get even more unwieldy. The common practice in mathematics is to leave your answers as multiples and powers of e instead of switching to decimals, unless you need an approximation in some application. Scientific calculators change your final answer, in terms of e, to an answer correct to as many decimal places as needed.

You follow the same rules when simplifying an expression that has a factor of e that you use on a variable base (see Chapter 4). The following list presents some examples of how you simplify expressions with a base of e:

When you multiply expressions with the same base, you add the exponents:

$$e^{2x} \cdot e^{3x} = e^{5x}$$

When you divide expressions with the same base, you subtract the exponents:

$$e^{15x^2} \Big/ e^{3x} = e^{15x^2 - 3x}$$

The two exponents don't combine any further.

Use the order of operations — powers first, followed by multiplication, addition, or subtraction:

$$e^2 \cdot e^4 + 2\left(3e^3\right)^2 = e^6 + 2\left(9e^6\right)$$
$$= e^6 + 18e^6 = 19e^6$$

Change radicals to fractional exponents:

$$\frac{e^{-2}\sqrt{e^8}}{\left(e \cdot e^2\right)^2} = \frac{e^{-2}\left(e^8\right)^{1/2}}{\left(e^3\right)^2}$$
$$= \frac{e^{-2} \cdot e^4}{e^6} = \frac{e^2}{e^6} = \frac{1}{e^4} = e^{-4}$$

Solving Exponential Equations

To solve an algebraic equation, you work toward finding the numbers that replace any variables and make a true statement. The process of solving exponential equations incorporates many of the same techniques you use in algebraic equations — adding to or subtracting from each side, multiplying or dividing each side by the same number, factoring, squaring both sides, and so on.

Solving exponential equations requires some additional techniques, however. That's what makes them so much fun! Some techniques you use when solving exponential equations involve changing the original exponential equations into new equations that have matching bases. Other techniques involve putting the exponential equations into more recognizable forms — such as quadratic equations — and then using the appropriate formulas. (If you can't change to matching bases or put the equations into quadratic or linear forms, you have to switch to logarithms or use a change-of-base formula — neither of which is within the scope of this book.)

Making bases match

If you see an equation written in the form $b^x = b^y$, where the same number represents the bases b, the following rule holds:

$$b^x = b^y \longleftrightarrow x = y$$

You read the rule as follows: "If b raised to the xth power is equal to b raised to the yth power, that implies that $x = y$." The double-pointed arrow indicates that the rule is true in the opposite direction, too.

Using the base rule to solve the equation $2^{3+x} = 2^{4x-9}$, you see that the bases (the 2s) are the same, so the exponents must also be the same. You just solve the linear equation $3 + x = 4x - 9$ for the value of x: $12 = 3x$, or $x = 4$. You then put the 4 back into the original equation to check your answer: $2^{3+4} = 2^{4(4)-9}$, which simplifies to $2^7 = 2^7$, or $128 = 128$.

Seems simple enough. But what do you do if the bases aren't the same? Unfortunately, if you can't change the problem to make the bases the same, you can't solve the problem with this rule. (In this case, you either use logarithms — change to a logarithmic equation — or resort to technology.)

When bases are related

Many times, bases are related to one another by being powers of the same number.

For example, to solve the equation $4^{x+3} = 8^{x-1}$, you need to write both the bases as powers of 2 and then apply the rules of exponents (see Chapter 4). Here are the steps of the solution:

1. **Change the 4 and 8 to powers of 2.**

   $$4^{x+3} = 8^{x-1}$$
   $$\left(2^2\right)^{x+3} = \left(2^3\right)^{x-1}$$

2. **Raise a power to a power.**

   $$2^{2(x+3)} = 2^{3(x-1)}$$
   $$2^{2x+6} = 2^{3x-3}$$

3. **Equate the two exponents, because the bases are now the same, and then solve for x.**

   $$2x + 6 = 3x - 3$$
   $$9 = x$$

4. **Check your answer in the original equation.**

   $$4^{9+3} = 8^{9-1}$$
   $$4^{12} = 8^8$$
   $$16,777,216 = 16,777,216$$

When other operations are involved

Changing all the bases in an equation to a single base is especially helpful when you have operations such as roots, multiplication, or division involved. If you want to solve the equation $\dfrac{27^{x+1}}{\sqrt{3}} = 9^{2x-3}$ for x, for example, your best approach is to change each of the bases to a power of 3 and then apply the rules of exponents (see Chapter 4).

Here's how you change the bases and powers in the example equation (Steps 1 and 2 from the list in the previous section):

$$\frac{27^{x+1}}{\sqrt{3}} = 9^{2x-3}$$

$$\frac{\left(3^3\right)^{x+1}}{3^{1/2}} = \left(3^2\right)^{2x-3}$$

$$\frac{3^{3x+3}}{3^{1/2}} = 3^{4x-6}$$

You change the bases 9 and 27 to powers of 3, replace the radical with a fractional exponent, and then raise the powers to other powers by multiplying the exponents.

When you divide two numbers with the same base, you subtract the exponents. After you have a single power of 3 on each side, you can equate the exponents and solve for x:

$$3^{3x+3-(1/2)} = 3^{4x-6}$$
$$3^{3x+(5/2)} = 3^{4x-6}$$
$$3x + \frac{5}{2} = 4x - 6$$
$$6 + \frac{5}{2} = x$$
$$\frac{17}{2} = x$$

Recognizing and using quadratic patterns

When exponential terms appear in equations with two or three terms, you may be able to treat the equations as you do quadratic equations (see Chapter 3) to solve them with familiar methods. Using the methods for solving quadratic equations is a big advantage because you can factor the exponential equations, or you can resort to the quadratic formula.

You factor quadratics by dividing every term by a common factor or, with trinomials, by using *unFOIL* to determine the two binomials whose product is the trinomial. (Refer to Chapters 1 and 3 if you need a refresher on these types of factoring.)

You can make use of just about any equation pattern that you see when solving exponential functions. If you can simplify the exponential to the form of a quadratic or cubic and then factor, find perfect squares, find sums and differences of squares, and so on, you've made life easier by changing the equation into something recognizable and doable. In the sections that follow, I provide examples of the two most common types of problems you're likely to run up against: those involving common factors and unFOIL.

Taking out a greatest common factor

When you solve a quadratic equation by factoring out a greatest common factor (GCF), you write that greatest common factor outside the parenthesis and show all the results of dividing by it inside the parenthesis.

In the equation $3^{2x} - 9 \cdot 3^x = 0$, for example, you factor 3^x from each term and get $3^x(3^x - 9) = 0$. After factoring, you use the *multiplication property of zero* by setting each of the separate factors equal to zero. (If the product of two numbers is zero, at least one of the numbers must be zero; see Chapter 1.) You set the factors equal to zero to find out what x satisfies the equation:

$3^x = 0$ has no solution; 3 raised to a power can't be equal to 0.

$$3^x - 9 = 0$$
$$3^x = 9$$
$$3^x = 3^2$$
$$x = 2$$

The factor is equal to 0 when $x = 2$; you find only one solution to the entire equation.

Factoring like a quadratic trinomial

A *quadratic trinomial* has a term with the variable squared, a term with the variable raised to the first power, and a constant term. This is the pattern you're looking for if you want to solve an exponential equation by treating it like a quadratic. The trinomial $5^{2x} - 26 \cdot 5^x + 25 = 0$, for example, resembles a quadratic trinomial that you can factor. One option is the quadratic $y^2 - 26y + 25 = 0$, which would look something like the exponential equation if you replace each 5^x with a y. The quadratic in y's factors into $(y - 1)(y - 25) = 0$. Using the same pattern on the exponential version, you get the factorization $(5^x - 1)(5^x - 25) = 0$. Setting each factor equal to zero, when $5^x - 1 = 0$, $5^x = 1$. This equation holds true when $x = 0$, making that one of the solutions. Now, when $5^x - 25 = 0$, you say that $5^x = 25$, or $5^x = 5^2$. In other words, $x = 2$. You find two solutions to this equation: $x = 0$ and $x = 2$.

Showing an "Interest" in Exponential Functions

Professionals (and you, although you may not know it) use exponential functions in many financial applications. If you have a mortgage on your home, an annuity for your retirement, or a credit-card balance, you should be interested in interest — and in the exponential functions that drive it.

Applying the compound interest formula

When you deposit your money in a savings account, an individual retirement account (IRA), or other investment vehicle, you get paid for the money you invest; this payment comes from the proceeds of *compound interest* — interest that earns interest. For instance, if you invest $100 and earn $2.00 in interest, the two amounts come together, and you now earn interest on $102. The interest *compounds*. This, no doubt, is a wonderful thing.

Here's the formula you can use to determine the total amount of money you have *(A)* after you deposit the principal *(P)* that earns interest at the rate of *r* percent (written as a decimal), compounding *n* times each year for *t* years:

$$A = P\left(1 + \frac{r}{n}\right)^{nt}$$

For example, say you receive a windfall of $20,000 for an unexpected inheritance, and you want to sock it away for 10 years. You invest the cash in a fund at 4.5-percent interest, compounded monthly. How much money will you have at the end of 10 years if you can manage to keep your hands off it? Apply the formula as follows:

$$A = 20,000\left(1 + \frac{0.045}{12}\right)^{(12)(10)}$$
$$= 20,000(1.00375)^{120}$$
$$= 20,000(1.566993)$$
$$= 31,339.86$$

You'd have over $31,300. This growth in your money shows you the power of compounding and exponents.

Planning on the future: Target sums

You can figure out what amount of money you'll have in the future if you make a certain deposit now by applying the compound interest formula. But how about going in the opposite direction? Can you figure out how much you need to deposit in an account in order to have a target sum in a certain number of years? You sure can.

If you want to have $100,000 available 18 years from now, when your baby will be starting college, how much do you have to deposit in an account that earns 5-percent interest, compounded monthly? To find out, you take the compound interest formula and work backward:

$$100,000 = P\left(1 + \frac{0.05}{12}\right)^{(12)(18)}$$
$$100,000 = P(1.0041667)^{216}$$

You solve for *P* in the equation by dividing each side by the value in the parenthesis raised to the power 216. According to the order of operations (see Chapter 1), you raise to a power before you multiply or divide. So, after you set up the division to solve for *P*, you raise what's in the parenthesis to the 216th power and divide the result into 100,000:

$$100,000 = P(1.0041667)^{216}$$
$$\frac{100,000}{(1.0041667)^{216}} = P$$
$$\frac{100,000}{2.455026} = P$$
$$40,732.77 = P$$

A deposit of almost $41,000 will result in enough money to pay for college in 18 years (not taking into account the escalation of tuition fees). You may want to start talking to your baby about scholarships now!

If you have a target amount of money in mind and want to know how many years it will take to get to that level, you can use the formula for compound interest and work backward. Put in all the specifics — principal, rate, compounding time, the amount you want — and solve for *t*. You may need a scientific calculator and some logarithms to finish it out, but you'll do whatever it takes to plan ahead, right?

Measuring the actual compound: Effective rates

When you go into a bank or credit union, you see all sorts of interest rates posted. You may have noticed the words "nominal rate" and "effective rate" in previous visits. The *nominal rate* is the named rate, or the value entered into the compounding formula. The named rate may be 4 percent or 7.5 percent, but that value isn't indicative of what's really happening because of the compounding. The *effective rate* represents what's really happening to your money after it compounds. A nominal rate of 4 percent translates into an effective rate of 4.074 percent when compounded monthly. This may not seem like much of a difference — the effective rate is about 0.07 higher — but it makes a big difference if you're talking about fairly large sums of money or long time periods.

You compute the effective rate by using the middle portion of the compound interest formula: $(1 + r/n)^n$. To determine the effective rate of 4 percent compounded monthly, for example, you use $\left(1 + \frac{0.04}{12}\right)^{12} = 1.040741543$.

The 1 before the decimal point in the answer indicates the original amount. You subtract that 1, and the rest of the decimals are the percentage values used for the effective rate.

Table 10-1 shows what happens to a nominal rate of 4 percent when you compound it different numbers of times per year.

Table 10-1	Compounding a Nominal 4-Percent Interest Rate	
Times Compounded	*Computation*	*Effective Rate*
Annually	$(1 + 0.04/1)^1 = 1.04$	4.00%
Biannually	$(1 + 0.04/2)^2 = 1.0404$	4.04%
Quarterly	$(1 + 0.04/4)^4 = 1.04060401$	4.06%
Monthly	$(1 + 0.04/12)^{12} = 1.04074154292$	4.07%
Daily	$(1 + 0.04/365)^{365} = 1.04080849313$	4.08%
Hourly	$(1 + 0.04/8,760)^{8,760} = 1.04081067873$	4.08%
Every second	$(1 + 0.04/31,536,000)^{31,536,000} = 1.04081104727$	4.08%

Looking at continuous compounding

Typical compounding of interest occurs annually, quarterly, monthly, or perhaps even daily. *Continuous compounding* occurs immeasurably quickly or often. To accomplish continuous compounding, you use a different formula than you use for other compounding problems.

Here's the formula you use to determine a total amount *(A)* when the initial value or principal is *P* and when the amount grows continuously at the rate of *r* percent (written as a decimal) for *t* years:

$$A = Pe^{rt}$$

The *e* represents a constant number (the *e* base; see the section "Meeting the most frequently used bases: 10 and *e*" earlier in this chapter) — approximately 2.71828. You can use this formula to determine how much money you'd have after 10 years of investment, for example, when the interest rate is 4.5 percent and you deposit $20,000:

$$A = 20,000e^{(0.045)(10)}$$
$$= 20,000(1.568312)$$
$$= 31,366.24$$

You should use the continuous compounding formula as an approximation in appropriate situations — when you're not actually paying out the money. The compound interest formula is much easier to deal with and gives a good estimate of total value.

Using the continuous compounding formula to approximate the effective rate of 4 percent compounded continuously, you get $e^{0.04} = 1.0408$ (an effective rate of 4.08 percent). Compare this with the value in Table 10-1 in the previous section.

Logging On to Logarithmic Functions

A *logarithm* is the exponent of a number. Logarithmic (log) functions are the inverses of exponential functions. They answer the question, "What power gave me that answer?" The log function associated with the exponential function $f(x) = 2^x$, for example, is $f^{-1}(x) = \log_2 x$. The superscript -1 after the function name f indicates that you're looking at the inverse of the function f. So, $\log_2 8$, for example, asks, "What power of 2 gave me 8?"

A logarithmic function has a *base* and an *argument*. The logarithmic function $f(x) = \log_b x$ has a base b and an argument x. The base must always be a positive number and not equal to one. The argument must always be positive.

You can see how a function and its inverse work as exponential and log functions by evaluating the exponential function for a particular value and then seeing how you get that value back after applying the inverse function to the answer. For example, first let $x = 3$ in $f(x) = 2^x$; you get $f(3) = 2^3 = 8$. You put the answer, 8, into the inverse function $f^{-1}(x) = \log_2 x$, and you get $f^{-1}(8) = \log_2 8 = 3$. The answer comes from the definition of how logarithms work; the 2 raised to the power of 3 equals 8. You have the answer to the fundamental logarithmic question, "What power of 2 gave me 8?"

Meeting the properties of logarithms

Logarithmic functions share similar properties with their exponential counterparts. When necessary, the properties of logarithms allow you to manipulate log expressions so you can solve equations or simplify terms. As with exponential functions, the base b of a log function has to be positive. I show the properties of logarithms in Table 10-2.

Table 10-2	Properties of Logarithms	
Property Name	*Property Rule*	*Example*
Equivalence	$y = \log_b x \leftrightarrow b^y = x$	$y = \log_9 3 \leftrightarrow 9^y = 3$
Log of a product	$\log_b xy = \log_b x + \log_b y$	$\log_2 8z = \log_2 8 + \log_2 z$

(continued)

Table 10-2 *(continued)*

Property Name	Property Rule	Example
Log of a quotient	$\log_b x/y = \log_b x - \log_b y$	$\log_2 8/5 = \log_2 8 - \log_2 5$
Log of a power	$\log_b{}^{xn} = n\log_b x$	$\log_3 8^{10} = 10\log_3 8$
Log of 1	$\log_b 1 = 0$	$\log_4 1 = 0$
Log of the base	$\log_b b = 1$	$\log_4 4 = 1$

Exponential terms that have a base e (see the section "Meeting the most frequently used bases: 10 and e") have special logarithms just for the e's (the ease?). Instead of writing the log base e as $\log_e x$, you insert a special symbol, *ln*, for the log. The symbol *ln* is called the *natural logarithm*, and it designates that the base is e. The equivalences for base e and the properties of natural logarithms are the same, but they look just a bit different. Table 10-3 shows them.

Table 10-3   **Properties of Natural Logarithms**

Property Name	Property Rule	Example
Equivalence	$y = \ln x \leftrightarrow e^y = x$	$6 = \ln x \leftrightarrow e^6 = x$
Natural log of a product	$\ln xy = \ln x + \ln y$	$\ln 4z = \ln 4 + \ln z$
Natural log of a quotient	$\ln x/y = \ln x - \ln y$	$\ln 4/z = \ln 4 - \ln z$
Natural log of a power	$\ln x^n = n\ln x$	$\ln x^5 = 5\ln x$
Natural log of 1	$\ln 1 = 0$	$\ln 1 = 0$
Natural log of e	$\ln e = 1$	$\ln e = 1$

As you can see in Table 10-3, the natural logs are much easier to write — you have no subscripts. Professionals use natural logs extensively in mathematical, scientific, and engineering applications.

Putting your logs to work

You can use the basic exponential/logarithmic equivalence $\log_b x = y \leftrightarrow b^y = x$ to simplify equations that involve logarithms. Applying the equivalence makes the equation much nicer. If you're asked to evaluate $\log_9 3$, for example (or if you have to change it into another form), you can write it as an equation, $\log_9 3 = x$, and use the equivalence: $9^x = 3$. Now you have it in a form that you can solve for x (the x that you get is the answer or value of the original expression). You solve by changing the 9 to a power of 3 and then finding x in the new, more familiar form:

$$\left(3^2\right)^x = 3$$
$$3^{2x} = 3^1$$
$$2x = 1$$
$$x = \frac{1}{2}$$

The result tells you that $\log_9 3 = \frac{1}{2}$ — much simpler than the original log expression. (If you need to review solving exponential equations, refer to the section "Making bases match" earlier in this chapter.)

Now look at the process of determining that $10\log_3 27$ is equal to 30. You have to admit that the number 30 is much easier to understand and deal with than $10\log_3 27$, so here are the steps:

$$10\log_3 27 = 10\left(\log_3 27\right) = 10\left(x\right)$$
$$\text{If } x = \log_3 27,$$
$$3^x = 27$$
$$3^x = 3^3$$
$$x = 3$$
$$10\left(x\right) = 10\left(3\right) = 30$$
$$10\log_3 27 = 30$$

As you can see from the preceding equivalence example, the properties of log functions allow you to do simplifications that you just can't do with other types of functions. For example, because $\log_b b = 1$, you can replace $\log_3 3$ with the number 1.

Using the rules for the log of 1, the log of the base, the log of a power, and the log of a quotient (see Table 10-2), you can change a complicated log expression into something equal to –2, for example:

$$\log_5 \left(\frac{1}{25}\right) = \log_5 1 - \log_5 25 \quad \left[\text{Log of a quotient}\right]$$
$$= \log_5 1 - \log_5 5^2 \quad \left[\text{Rewriting 25 as a power of 5}\right]$$
$$= \log_5 1 - 2\log_5 5 \quad \left[\text{Log of a power}\right]$$
$$= 0 - 2\left(1\right) = -2 \quad \left[\text{Log of 1 and log of the base}\right]$$

Expanding expressions with log notation

You write logarithmic expressions and create logarithmic functions by combining all the usual algebraic operations of addition, subtraction, multiplication, division, powers, and roots. Expressions with two or more of these operations can get pretty complicated. A big advantage of logs, though, is their properties. Because of the log properties, you can change multiplication to addition and powers to products. Put all the log properties together, and you can change a single complicated expression into several simpler terms.

If you want to simplify $\log_3 \dfrac{x^3 \sqrt{x^2+1}}{(x-2)^7}$ by using the properties of logarithms, for example, you first use the property for the log of a quotient and then use the property for the log of a product on the first term you get (see Table 10-2 to review these properties):

$$\log_3 \frac{x^3 \sqrt{x^2+1}}{(x-2)^7} = \log_3 x^3 \sqrt{x^2+1} - \log_3 (x-2)^7$$

$$= \log_3 x^3 + \log_3 \sqrt{x^2+1} - \log_3 (x-2)^7$$

The last step is to use the log of a power on each term, changing the radical to a fractional exponent first:

$$\log_3 x^3 + \log_3 (x^2+1)^{1/2} - \log_3 (x-2)^7$$

$$= 3\log_3 x + \frac{1}{2}\log_3 (x^2+1) - 7\log_3 (x-2)$$

The three new terms you create are each much simpler than the whole expression.

Rewriting for compactness

Results of computations in science and mathematics can involve sums and differences of logarithms. When this happens, experts usually prefer to have the answers written all in one term, which is where the properties of logarithms come in. You apply the properties in just the opposite way that you break down expressions for greater simplicity (see the previous section). Instead of spreading the work out, you want to create one compact, complicated expression.

To simplify $4ln(x + 2) - 8ln(x^2 - 7) - \frac{1}{2}ln(x + 1)$, for example, you first apply the property involving the natural log *(ln)* of a power to all three terms (see the section "Meeting the properties of logarithms"). You then factor out -1 from the last two terms and write them in a bracket:

$$ln(x+2)^4 - ln(x^2-7)^8 - ln(x+1)^{1/2}$$

$$= ln(x+2)^4 - \left[ln(x^2-7)^8 + ln(x+1)^{1/2} \right]$$

You now use the property involving the *ln* of a product on the terms in the bracket, change the ½ exponent to a radical, and use the property for the *ln* of a quotient to write everything as the *ln* of one big fraction:

$$ln(x+2)^4 - \left[ln(x^2-7)^8 + ln(x+1)^{1/2} \right]$$

$$= ln(x+2)^4 - \left[ln(x^2-7)^8 (x+1)^{1/2} \right]$$

$$= ln(x+2)^4 - ln(x^2-7)^8 \sqrt{x+1}$$

$$= ln \frac{(x+2)^4}{(x^2-7)^8 \sqrt{x+1}}$$

The expression is messy and complicated, but it sure is compact.

Solving Logarithmic Equations

Logarithmic equations can have one or more solutions, just like other types of algebraic equations. What makes solving log equations a bit different is that you get rid of the log part as quickly as possible, leaving you to solve either a polynomial or an exponential equation in its place. Polynomial and exponential equations are easier and more familiar, and you may already know how to solve them (if not, see Chapter 8 and the section "Solving Exponential Equations" earlier in this chapter).

The only caution I present before you begin solving logarithmic equations is that you need to check the answers you get from the new, revised forms. You may get answers to the polynomial or exponential equations, but they may not work in the logarithmic equation. Switching to another type of equation introduces the possibility of *extraneous roots* — answers that fit the new, revised equation that you choose but sometimes don't fit in with the original equation.

Setting log equal to log

One type of log equation features each term carrying a logarithm in it (all the logarithms have to have the same base). You need to have exactly one log term on each side, so if an equation has more, you apply any properties of logarithms that form the equation to fit the rule (see Table 10-2 for these properties). After you do, you can apply the following rule:

If $\log_b x = \log_b y$, $x = y$

When presented with the equation $\log_4 x^2 = \log_4(x+6)$, for example, you apply the rule so that you can write and solve the equation $x^2 = x + 6$:

$$x^2 = x + 6$$
$$x^2 - x - 6 = 0$$
$$(x-3)(x+2) = 0$$
$$x = 3 \text{ or } x = -2$$

The $x = 3$ and $x = -2$ you find are solutions of the quadratic equation, and both work in the original logarithmic equation:

If $x = 3$:
$$\log_4 3^2 = \log_4(3+6)$$
$$\log_4 9 = \log_4 9$$

So, 3 is a solution.

If $x = -2$:
$$\log_4(-2)^2 = \log_4(-2+6)$$
$$\log_4 4 = \log_4 4$$

You have another winner.

When no log base is shown you assume that the log's base is 10. These are *common logarithms.*

The following equation shows you how you may get an extraneous solution.

When solving $\log(x-8) + \log x = \log 9$, you first apply the property involving the log of a product to get just one log term on the left: $\log(x-8)x = \log 9$. Next, you use the property that allows you to drop the logs and get the equation $(x-8)x = 9$. This is a quadratic equation that you can solve with factoring (see Chapters 1 and 3):

$$(x-8)x = 9$$
$$x^2 - 8x - 9 = 0$$
$$(x-9)(x+1) = 0$$
$$x = 9 \text{ or } x = -1$$

Checking the answers, you see that the solution 9 works just fine:

$$\log(9-8) + \log 9 = \log 9$$
$$\log 1 + \log 9 = \log 9$$
$$0 + \log 9 = \log 9$$

However, the solution –1 doesn't work:

$$\log(-1-8) + \log(-1) = \log 9$$

You can stop right there. Both of the logs on the left have negative arguments. The argument in a logarithm has to be positive, so the –1 doesn't work in the log equation (even though it was just fine in the quadratic equation). You determine that –1 is an extraneous solution.

Rewriting log equations as exponentials

When a log equation has log terms and a term that doesn't have a logarithm in it, you need to use algebra techniques and log properties (see Table 10-2) to put the equation in the form $y = \log_b x$. After you create the right form, you can apply the equivalence to change it to a purely exponential equation.

For instance, to solve $\log_3(x + 8) - 2 = \log_3 x$, you first subtract $\log_3 x$ from each side and add 2 to each side to get $\log_3(x + 8) - \log_3 x = 2$. Now you apply the property involving the log of a quotient, rewrite the equation by using the equivalence, and solve for x:

$$\log_3 \frac{x+8}{x} = 2$$
$$3^2 = \frac{x+8}{x}$$
$$9x = x + 8$$
$$8x = 8$$
$$x = 1$$

The only solution is $x = 1$, which works in the original logarithmic equation:

$$\log_3(x+8) - 2 = \log_3 x$$
$$\log_3(1+8) - 2 = \log_3 1$$
$$\log_3 9 - 2 = 0$$
$$\log_3 9 = 2$$
$$3^2 = 9$$

Graphing Exponential and Logarithmic Functions

Exponential and logarithmic functions have rather distinctive graphs because they're so plain and simple. The graphs are lazy C's that can slope upward or downward. The main trick when graphing them is to determine any intercepts, which way the graphs move as you go from left to right, and how steep the curves are.

Expounding on the exponential

Exponential functions have curves that usually look like the graphs you see in Figures 10-1a and 10-1b.

Figure 10-1:
Exponential graphs rise away from the X-axis or fall toward the X-axis.

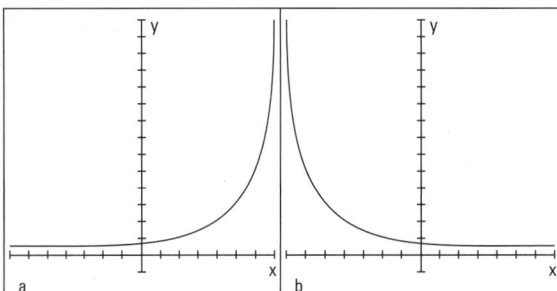

The graph in Figure 10-1a depicts *exponential growth* — when the values of the function are increasing. Figure 10-1b depicts *exponential decay* — when the values of the function are decreasing. Both graphs intersect the Y-axis but not the X-axis, and they both have a horizontal asymptote: the X-axis.

Identifying a rise or fall

You can tell whether a graph will feature exponential growth or decay by looking at the equation of the function.

In order to determine whether a function represents exponential growth or decay, you look at its base:

✓ If the exponential function $y = b^x$ has a base $b > 1$, the graph of the function rises as you read from left to right, meaning you observe exponential growth.

✔ If the exponential function $y = b^x$ has a base $0 < b < 1$, the graph falls as you read from left to right, meaning you observe exponential decay.

The values of the functions $f(x) = 3(2)^x$ and $g(x) = 4e^{3x}$, for example, both rise as you read from left to right because their bases are greater than 1. The graphs of $h(x) = 3(0.2)^x$ and $g(x) = 4(0.9)^{3x}$ both fall as you look at increasing values of x, because 0.2 and 0.9 are both between 0 and 1.

Sketching exponential graphs

In general, exponential functions have no x-intercepts, but they do have single y-intercepts. The exception to this rule is when you change the function equation by subtracting a number from the exponential term; this action drops the curve down below the X-axis.

To find the y-intercept of an exponential function, you set $x = 0$ and solve for y. If you want to find the y-intercept of $y = 3(2)^{0.4x}$, for example, you replace x with 0 to get $y = 3(2)^{0.4(0)} = 3(2)^0 = 3(1) = 3$. So, the y-intercept is $(0, 3)$. This function rises from left to right because the base is greater than one. The multiplier 0.4 on the x in the exponent acts like the slope of a line — in this case, making the graph rise more slowly or gently (see Chapter 2).

Before you try to graph an equation, you should find another point or two for help with the shape. For instance, if $x = 5$ in the previous example, $y = 3(2)^{0.4(5)} = 3(2)^2 = 3(4) = 12$. So, the point $(5, 12)$ falls on the curve. Also, if $x = -5$, $y = 3(2)^{0.4(-5)} = 3(2)^{-2} = 3(0.25) = 0.75$ (see Chapter 4 for info on dealing with negative exponents). The point $(-5, 0.75)$ is also on the curve. Figure 10-2 shows the graph of this example curve with the intercept and points drawn in.

Figure 10-2:
The graph
of the
exponential
function
$y = 3(2)^{0.4x}$.

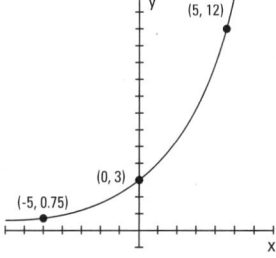

To graph the function $y = 10(0.9)^x$, you first find the y-intercept. When $x = 0$, $y = 10(0.9)^0 = 10(1) = 10$. So, the y-intercept is $(0, 10)$. Two other points you may decide to use are $(1, 9)$ and $(-4, 15.24)$. The graph of this function falls as you read from left to right because the base is smaller than one. Figure 10-3 shows the graph of this example curve and the random points of reference.

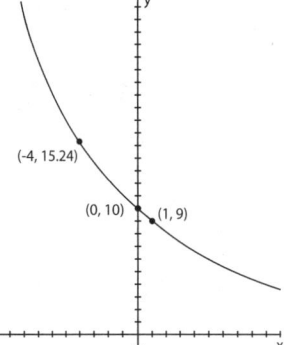

(-4, 15.24)

(0, 10) (1, 9)

Figure 10-3:
The graph
of the
exponential
function
$y = 10(0.9)^x$.

Not seeing the logs for the trees

The graphs of logarithmic functions either rise or fall, and they usually look like one of the sketches in Figure 10-4. The graphs have a single vertical asymptote: the Y-axis. Having the Y-axis as an asymptote is the opposite of an exponential function, whose asymptote is the X-axis (see the previous section). Log functions are also different from exponential functions in that they have an x-intercept but not (usually) a y-intercept.

Figure 10-4:
Logarithmic
functions
rise or fall,
breaking
away
from the
asymptote:
the Y-axis.

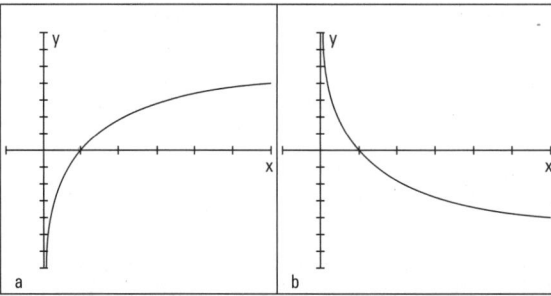

Graphing log functions with the use of intercepts

When you graph a log function, you look at its x-intercept and its base:

- If the base is a number greater than one, the graph rises from left to right.

- If the base is between zero and one, the graph falls as you go from left to right.

For instance, the graph of $y = \log_2 x$ has an x-intercept of $(1, 0)$ and rises from left to right. You get the intercept by letting $y = 0$ and solving the equation $0 = \log_2 x$ for x. You should choose a couple other points on the curve to help with shaping the graph. The graph of $y = \log_2 x$ contains the points $(2, 1)$ and $\left(\frac{1}{8}, -3\right)$. You compute those points by substituting the chosen x value into the function equation and solving for y. (If you need a refresher on how to solve those equations, refer to the section "Solving Logarithmic Equations.") Figure 10-5 shows you the graph of the example function (the points are indicated).

Reflecting on inverses of exponential functions

Exponential and logarithmic functions are inverses of one another. You may have noticed in previous sections that the flattened, C-shaped curves of log functions look vaguely familiar. In fact, they're mirror images of the graphs of exponential functions.

Figure 10-5: With a log base of 2, the curve of the function rises.

An exponential function, $y_1 = b^x$, and its logarithmic inverse, $y_2 = \log_b x$, have graphs that mirror one another over the line $y = x$.

For example, the exponential function $y = 3^x$ has a y-intercept of $(0, 1)$, the X-axis as its horizontal asymptote, and the plotted points $(1, 3)$ and $\left(-2, \frac{1}{9}\right)$. You can compare that function to its inverse function, $y = \log_3 x$, which has an x-intercept at $(1, 0)$, the Y-axis as its vertical asymptote, and the plotted points $(3, 1)$ and $\left(\frac{1}{9}, -2\right)$. Figure 10-6 shows you both of the graphs and some of the points.

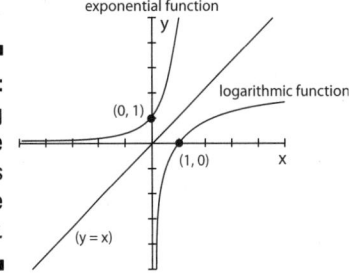

Figure 10-6:
Graphing
inverse
curves
over the
line $y = x$.

The symmetry of exponential and log functions about the diagonal line $y = x$ is very helpful when graphing the functions. For instance, if you want to graph $y = \log_{1/4} x$ and don't want to mess with its fractional base, you can graph $y = \left(\frac{1}{4}\right)^x$ instead and flip the graph over the diagonal line to get the graph of the log function. The graph of $y = \left(\frac{1}{4}\right)^x$ contains the points $(0, 1)$, $\left(1, \frac{1}{4}\right)$, and $(-2, 16)$. These points are easier to compute than the log values. You just reverse the coordinates of those points to $(1, 0)$, $\left(\frac{1}{4}, 1\right)$, and $(16, -2)$; you now have points on the graph of the log function. Figure 10-7 illustrates this process.

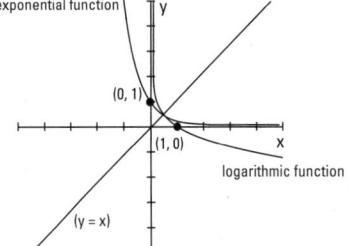

Figure 10-7:
Using an
exponential
function as
an inverse
to graph a
log function.

Notice that the exponential function and its inverse log function cross the line $y = x$ at the same point. This is true of all functions and their inverses — they cross the line $y = x$ in the same place or places. Keep this in mind when you're graphing.

Part III
Conquering Conics and Systems of Equations

The 5th Wave   By Rich Tennant

"I love this time of year when the Algebra II students start finding new ways to visualize conic sections."

In this part . . .

Conic sections have information-packed equations and beautifully symmetric graphs. In this part, you find out how to take advantage of the information in the equations and translate it into the graphs. You also solve systems of equations and inequalities by determining what the equations and inequalities share or have in common. By graphing the systems, you can see how the curves can share the same space and how you determine exactly where they cross or touch one another. Don't worry; if you're not into graphing, you have algebraic methods to employ.

Chapter 11

Cutting Up Conic Sections

*C*onic is the name given to a special group of curves. What they have in common is how they're constructed — points lying relative to an anchored point or points with respect to a line. But that sounds a bit stuffy, doesn't it? Maybe it works better to think of conic sections in terms of how you can best describe the curves visually. Picture the curved cables hanging between the uprights of a suspension bridge. Imagine the earth's path swinging around the sun. Trace the progress of a comet coming toward earth and veering off again. All these pictures circling your mind are related to curves called conics.

If you take a cone — imagine one of those yummy sugar cones that you put ice cream in — and slice it through in a particular fashion, the resulting edge you create will trace one of the four conic sections: a parabola, circle, ellipse, or hyperbola. (You can see a sketch of a cone in the following section. Hope it doesn't make you too hungry!)

Each conic section has a specific equation, and I cover each thoroughly in this chapter. You can glean a good deal of valuable information from a conic section's equation, such as where it's centered in a graph, how wide it opens, and its general shape. I also discuss the techniques that work best for you when you're called on to graph conics. Grab a pizza and an ice-cream cone for visual motivation and read on!

Cutting Up a Cone

A *conic section* is a curve formed by the intersection of a cone and a plane (a *cone* is a shape whose base is a circle and whose sides taper up to a point). The curve formed depends on where the cone is sliced:

✔ If you slice the cone straight across, you make the edge a *circle,* just like the top.

✔ If you slice a side piece off at an angle, you form a U-shaped *parabola* along the edge.

✔ If you slice the cone at a slant, you make the edge an *ellipse,* or an oval shape.

✔ If you picture two cones, tip to tip, and a slice going straight down through both, you picture a *hyperbola.* You have two wide, U-shaped edges that sort of come nose to nose. A hyperbola takes some special gymnastics!

Figure 11-1 shows each of the four conic sections sliced and diced. Each conic section has a specific equation that you use to graph the conic or to use it in an application (such as when the conic equation represents the curvature of a tunnel). You can head to the sections in this chapter that deal with the different conics to cover the topics in detail.

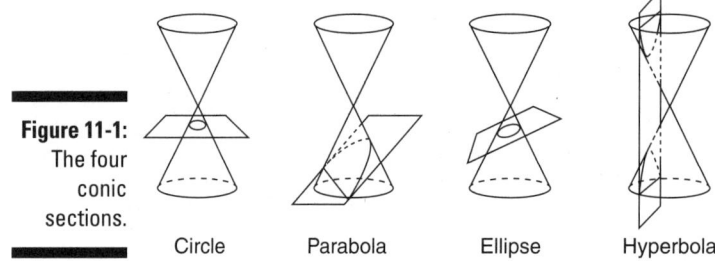

Figure 11-1: The four conic sections.

Circle   Parabola   Ellipse   Hyperbola

Opening Every Which Way with Parabolas

A *parabola,* a U-shaped conic that I first introduce in Chapter 7 (the parabola is the only conic section that fits the definition of a polynomial), is defined as all the points that fall the same distance from some fixed point, called its *focus,* and a fixed line, called its *directrix.* The focus is denoted by F, and the directrix by $y = d$. Figure 11-2 shows you some of the points on a parabola and how they each appear the same distance from the parabola's focus and directrix.

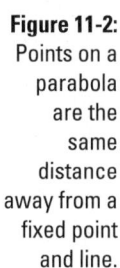

Figure 11-2:
Points on a
parabola
are the
same
distance
away from a
fixed point
and line.

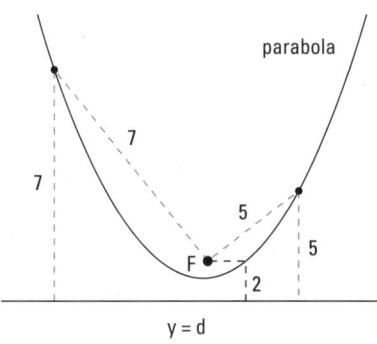

A parabola has a couple other defining features. The *axis of symmetry* of a parabola is a line that runs through the focus and is perpendicular to the directrix (imagine a line running through F in Figure 11-2). The axis of symmetry does just what its name suggests: It shows off how symmetric a parabola is. A parabola is a mirror image on either side of its axis. Another feature is the parabola's *vertex*. The vertex is the curve's extreme point — the lowest or highest point, or the point on the curve farthest right or farthest left. The vertex is also the point where the axis of symmetry crosses the curve (you can create this point by putting a pencil on the curve in Figure 11-2 where the imaginary axis crosses the curve after spearing through F).

Looking at parabolas with vertices at the origin

Parabolas can have graphs that go every which way and have vertices at any point in the coordinate system. When possible, though, you want to deal with parabolas that have their vertices at the origin. The equations are easier to deal with, and the applications are easier to solve. Therefore, I put the cart before the horse with this section to cover these specialized parabolas. (In the section "Observing the general form of parabola equations" later in the chapter, I deal with all the other parabolas you may run across.)

Opening to the right or left

Parabolas with their vertices (the plural form of *vertex* is *vertices* — just a little Latin for you) opening to the right or left have a standard equation $y^2 = 4ax$ and are known as *relations* — you see a relationship between the variables. The standard form comes packed with information about the focus, directrix, vertex, axis of symmetry, and direction of a parabola. The equation also gives you a hint as to whether the parabola is narrow or opens wide.

The general form of a parabola with the equation $y^2 = 4ax$ gives you the following info:

Focus: $(a, 0)$

Directrix: $x = -a$

Vertex: $(0, 0)$

Axis of symmetry: $y = 0$

Opening: To the right if a is positive; to the left if a is negative

Shape: Narrow if $|4a|$ is less than 1; wide if $|4a|$ is greater than 1

I use the absolute value operation, $|\ |$, instead of saying that $4a$ has to be between 0 and 1 or between -1 and 0. I think it's just a neater way of dealing with the rule.

For example, if you want to extract information about the parabola $y^2 = 8x$, you can put it in the form $y^2 = 4(2)x$ by substituting known information. In this case, you extract the following info:

The value of a is 2 (from $4 \cdot 2 = 8$).

The focus is at $(2, 0)$.

The directrix is the line $x = -2$.

The vertex is at $(0, 0)$.

The axis of symmetry is $y = 0$.

The parabola opens to the right.

The parabola is wide $|4(2)|$ is greater than 1.

Figure 11-3 shows the graph of the parabola $y^2 = 8x$ with all the key information listed in the sketch.

Figure 11-3:
The parabola $y^2 = 8x$ with all its properties on display.

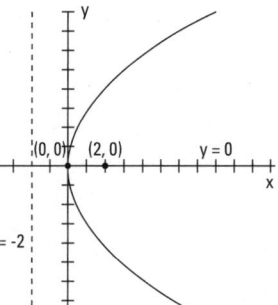

Of course, extracting information is always more convenient when the equation of the parabola has an even-number coefficient, such as $y^2 = 8x$, but the process works for odd numbers, too. You just have to deal with fractions. For instance, you can write the parabola $y^2 = 7x$ as $y^2 = 4(\frac{7}{4})x$, so the value of a is $\frac{7}{4}$, the focus is $(\frac{7}{4}, 0)$, and so on. The value of a is just what it takes to multiply 4 by to get the coefficient.

Opening upward or downward

Parabolas that open left or right are relations, but parabolas that have vertices opening upward or downward are a bit more special. The parabolas that open upward or downward are functions — they have only one y value for every x value. (For more on functions, refer to Chapter 6.) Parabolas of this variety have the following standard equation: $x^2 = 4ay$.

You can tell that you have a relation (opening left or right) rather than a function (opening up or down) if the y variable is squared. The quadratic polynomials (from Chapter 7) and the functions all have the x variable squared.

The information you glean from the standard equation tells much the same story as the equation of parabolas that open sideways; however, many of the rules are reversed.

From the general form for the parabola $x^2 = 4ay$, you extract the following information:

Focus: $(0, a)$

Directrix: $y = -a$

Vertex: $(0, 0)$

Axis of symmetry: $x = 0$

Opening: Upward if a is positive; downward if a is negative

Shape: Narrow if $|4a|$ is less than 1; wide if $|4a|$ is greater than 1

You can convert the parabola $x^2 = -\frac{1}{2}y$ to the form $x^2 = 4\left(-\frac{1}{8}\right)y$ because the general form gives you quick, easily accessible information. Just divide the coefficient by 4, and then write the coefficient as 4 times the result of the division. You haven't changed the value of the coefficient; you've just changed how it looks.

In this case, you extract the following info:

The value of a is $-\frac{1}{8}$.

The focus falls on the point $\left(0, -\frac{1}{8}\right)$.

The directrix forms a line at $y = \frac{1}{8}$.

The vertex is at $(0, 0)$.

The axis of symmetry is $x = 0$.

The graph opens downward.

The parabola is narrow because $\left| 4\left(a\right) \right| = \left| 4\left(-\frac{1}{8}\right) \right| = \left| -\frac{1}{2} \right| < 1$.

Figure 11-4 shows the graph of the parabola $x^2 = -\frac{1}{2}\,y$ with all the elements illustrated in the sketch.

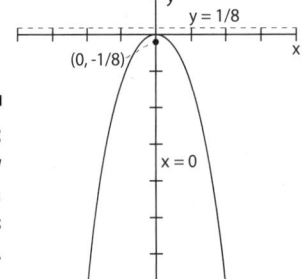

Figure 11-4:
A narrow
parabola
that opens
downward.

Observing the general form of parabola equations

The curves of parabolas can open upward, downward, to the left, or to the right, but the curves don't always have to have their vertices at the origin. Parabolas can wander all around a graph. So, how do you track the curves down to pin them on a graph? You look to their equations, which give you all the information you need to find out where they've wandered to.

When the vertex of a parabola is at the point (h, k), the general form for the equation is one of the following:

✔ **Opening left or right: $(y - k)^2 = 4a(x - h)$.** When the y variable is squared, the parabola opens left or right. From this equation, just like with parabolas that have their vertices at the origin (see the previous section), you can extract information about the elements:

   • If $4a$ is positive, the curve opens right; if $4a$ is negative, the curve opens left.

   • If $|4a| > 1$, the parabola is relatively wide; if $|4a| < 1$, the parabola is relatively narrow.

✔ **Opening up or down: $(x - h)^2 = 4a(y - k)$.** When the x variable is squared, the parabola opens up or down. Here's the info you can extract from this equation:

- If $4a$ is positive, the parabola opens upward; if $4a$ is negative, the curve opens downward.
- If $|4a| > 1$, the parabola is wide; if $|4a| < 1$, the parabola is narrow.

The standard forms used for parabolas with their vertices at the origin (which I discuss in the previous sections) are special cases of these more general parabolas. If you replace the h and k coordinates with zeros, you have the special parabolas anchored at the origin.

A move in the position of the vertex (away from the origin, for example) changes the focus, directrix, and axis of symmetry of a parabola. In general, a move of the vertex just adds the value of h or k to the basic form. For instance, when the vertex is at (h, k), the focus is at $(h + a, k)$ for parabolas opening to the right and at $(h, k + a)$ for parabolas opening upward. The directrix is also affected by h or k in its equation; it becomes $x = h - a$ for parabolas opening sideways and $y = k - a$ for those opening up or down. The whole graph shifts its position, but the shift doesn't affect which direction it opens or how wide it opens. The shape and direction stay the same.

Sketching the graphs of parabolas

Parabolas have distinctive U-shaped graphs, and with just a little information, you can make a relatively accurate sketch of the graph of a particular parabola. The first step is to think of all parabolas as being in one of the general forms I list in the previous two sections. (Refer to the rules in the previous two sections when you graph a parabola.)

Taking the necessary graphing steps

Here's the full list of steps to follow when sketching the graph of a parabola — either $(x - h)^2 = 4a(y - k)$ or $(y - k)^2 = 4a(x - h)$:

1. **Determine the coordinates of the vertex, (h, k), and plot it.**

   If the equation contains $(x + h)$ or $(y + k)$, change the forms to $(x - [-h])$ or $(y - [-k])$, respectively, to determine the correct signs. Actually, you're just reversing the sign that's already there.

2. **Determine the direction the parabola opens, and decide if it's wide or narrow, by looking at the $4a$ portion of the general parabola equation.**

3. **Lightly sketch in the axis of symmetry that goes through the vertex ($x = h$ when the parabola opens upward or downward and $y = k$ when it opens sideways).**

4. **Choose a couple other points on the parabola and find each of their partners on the other side of the axis of symmetry to help you with the sketch.**

For example, if you want to graph the parabola $(y + 2)^2 = 8(x - 1)$, you first note that this parabola has its vertex at the point $(1, -2)$ and opens to the right, because the y is squared (if the x had been squared, it would open up or down) and a (2) is positive. The graph is relatively wide about the axis of symmetry, $y = -2$, because $a = 2$, which makes $|4a|$ greater than 1. To find a random point on the parabola, try letting $y = 6$ and solve for x:

$$(6 + 2)^2 = 8(x - 1)$$
$$8^2 = 8(x - 1)$$
$$64 = 8(x - 1)$$
$$8 = x - 1$$
$$9 = x$$

Another example point, which you find by using the same process that gave you (9, 6), is (5.5, 4). Figure 11-5a shows the vertex, axis of symmetry, and the two points placed in a sketch.

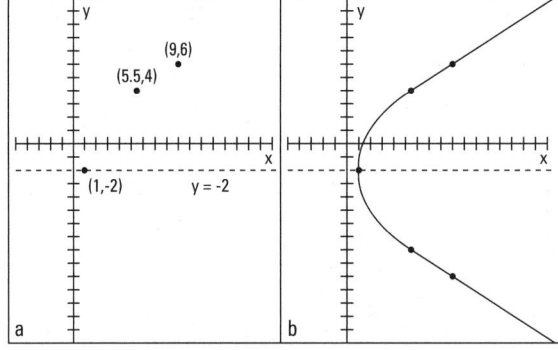

The two randomly chosen points have counterparts on the opposite side of the axis of symmetry. The point (9, 6) is 8 units above the axis of symmetry, so 8 units below the axis puts you at (9, -10). The point (5.5, 4) is 6 units above the axis of symmetry, so its partner is the point (5.5, -8). Figure 11-5b shows the two new points and the parabola sketched in.

Applying suspense to the parabola

Sketching parabolas helps you visualize how they're used in an application, so you want to be able to sketch quickly and accurately if need be. A rather natural occurrence of the parabola involves the cables that hang between the towers of a suspension bridge. These cables form a parabolic curve. Consider the following situation: An electrician wants to put a decorative light on the cable of a suspension bridge at a point 100 feet (horizontally) from where the cable touches the roadway at the middle of the bridge. You can see the electrician's blueprint in Figure 11-6.

Figure 11-6:
The suspended cable on this bridge forms a parabola.

80 feet tall

100 feet

400 feet wide

The electrician needs to know how high the cable is at a point 100 feet from the center of the bridge so he can plan his lighting experiment. The towers holding the cable are 80 feet high, and the total length of the bridge is 400 feet.

You can help the electrician solve this problem by writing the equation of the parabola that fits all these parameters. The easiest way to handle the problem is to let the bridge roadbed be the X-axis and the center of the bridge be the origin (0, 0). The origin, therefore, is the vertex of the parabola. The parabola opens upward, so you use the equation of a parabola that opens upward with its vertex at the origin, which is $x^2 = 4ay$ (see the section "Looking at parabolas with vertices at the origin").

To solve for a, you put the coordinates of the point (200, 80) in the equation. Where do you get these seemingly random numbers? Half the total of 400 feet of bridge is 200 feet. You move 200 feet to the right of the middle of the bridge and 80 feet up to get to the top of the right tower. Replacing the x in the equation with 200 and the y with 80, you get 40,000 = 4a(80). Dividing each side of the equation by 80, you find that $4a$ is 500, so the equation of the parabola that represents the cable is $x^2 = 500y$.

So, how high is the cable at a point 100 feet from the center? In Figure 11-6, the point 100 feet from the center is to the left, so –100 represents x. A parabola is symmetric about its vertex, so it doesn't matter whether you use positive or negative 100 to solve this problem. But, sticking to the figure, let $x = -100$ in the equation; you get $(-100)^2 = 500y$, which becomes 10,000 = 500y. Dividing each side by 500, you get $y = 20$. The cable is 20 feet high at the point where the electrician wants to put the light. He needs a ladder!

Converting parabolic equations to the standard form

When the equation of a parabola appears in standard form, you have all the information you need to graph it or to determine some of its characteristics, such as direction or size. Not all equations come packaged that way, though. You may have to do some work on the equation first to be able to identify anything about the parabola.

The standard form of a parabola is $(x - h)^2 = a(y - k)$ or $(y - k)^2 = a(x - h)$, where (h, k) is the vertex.

The methods used here to rewrite the equation of a parabola into its standard form also apply when rewriting equations of circles, ellipses, and hyperbolas. (See the last section in this chapter, "Identifying Conics from Their Equations, Standard or Not," for a more generalized view of changing the forms of conic equations.) The standard forms for conic sections are factored forms that allow you to immediately identify needed information. Different algebra situations call for different standard forms — the form just depends on what you need from the equation.

For instance, if you want to convert the equation $x^2 + 10x - 2y + 23 = 0$ into the standard form, you act out the following steps, which contain a method called *completing the square* (a method you use to solve quadratic equations; refer to Chapter 3 for a review of completing the square):

1. **Rewrite the equation with the x^2 and x terms (or the y^2 and y terms) on one side of the equation and the rest of the terms on the other side.**

   $$x^2 + 10x - 2y + 23 = 0$$
   $$x^2 + 10x = 2y - 23$$

2. **Add a number to each side to make the side with the squared term into a perfect square trinomial (thus completing the square; see Chapter 3 for more on trinomials).**

   $$x^2 + 10x + 25 = 2y - 23 + 25$$
   $$x^2 + 10x + 25 = 2y + 2$$

3. **Rewrite the perfect square trinomial in factored form, and factor the terms on the other side by the coefficient of the variable.**

   $$(x + 5)^2 = 2(y + 1)$$

You now have the equation in standard form. The vertex is at $(-5, -1)$; it opens upward and is fairly wide (see the section "Sketching the graphs of parabolas" to see how to make these determinations).

Going Round and Round in Conic Circles

A *circle,* probably the most recognizable of the conic sections, is defined as all the points plotted at the same distance from a fixed point — the circle's center, C. The fixed distance is the radius, *r,* of the circle.

The standard form for the equation of a circle with radius *r* and with its center at the point (h, k) is $(x - h)^2 + (y - k)^2 = r^2$.

Figure 11-7 shows the sketch of a circle.

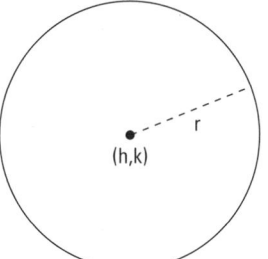

Standardizing the circle

When the equation of a circle appears in the standard form, it provides you with all you need to know about the circle: its center and radius. With these two bits of information, you can sketch the graph of the circle. The equation $x^2 + y^2 + 6x - 4y - 3 = 0$, for example, is the equation of a circle. You can change this equation to the standard form by *completing the square* for each of the variables (refer to Chapter 3 if you need a review of this process). Just follow these steps:

1. **Change the order of the terms so that the *x*'s and *y*'s are grouped together and the constant appears on the other side of the equal sign.**

   Leave a space after the groupings for the numbers that you need to add:

   $$x^2 + y^2 + 6x - 4y - 3 = 0$$
   $$x^2 + 6x \quad + y^2 - 4y \quad = 3$$

2. **Complete the square for each variable, adding the number that creates perfect square trinomials.**

   $$x^2 + 6x + 9 + y^2 - 4y + 4 = 3 + 9 + 4$$
   $$x^2 + 6x + 9 + y^2 - 4y + 4 = 16$$

3. **Factor each perfect square trinomial.**

The standard form for the equation of this circle is $(x + 3)^2 + (y - 2)^2 = 16$.

The example circle has its center at the point (–3, 2) and has a radius of 4 (the square root of 16). To sketch this circle, you locate the point (–3, 2) and then count 4 units up, down, left, and right; sketch in a circle that includes those points. Figure 11-8 shows you the way.

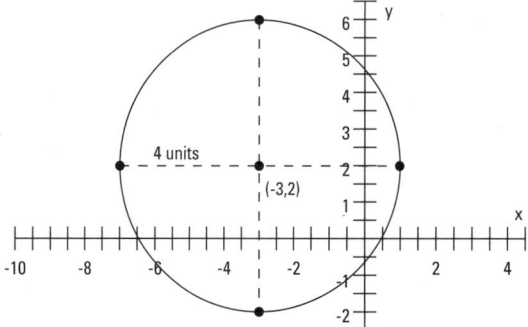

Figure 11-8: With a center, radius, and compass, you too can sketch this circle.

Specializing in circles

Two circles you should consider special are the circle with its center at the origin and the unit circle. A circle with its center at the origin has a center at (0, 0), so its standard equation becomes $x^2 + y^2 = r^2$.

The equation of the center origin is simple and easy to work with, so you should take advantage of its simplicity and try to manipulate any application you're working with that uses a circle into one with its center at the origin.

The *unit circle* also has its center at the origin, but it always has a radius of one. The equation of the unit circle is $x^2 + y^2 = 1$. This circle is also convenient and nice to work with. You use it to define trigonometric functions, and you find it in analytic geometry and calculus applications.

For example, a circle with its center at the origin and a radius of 5 units is created from the equation $(x - h)^2 + (y - k)^2 = r^2$, where (h, k) is (0, 0) and $r^2 = 5^2 = 25$; therefore, it has the equation $x^2 + y^2 = 25$. Any circle has an infinite number of points on it, but this clever choice for the radius gives you plenty of integers for coordinates. Points lying on the circle include (3, 4), (–3, 4), (3, –4), (–3, –4), (4, 3), (–4, 3), (4, –3), (–4, –3), (5, 0), (–5, 0), (0, 5), and (0, –5). Not all circles offer this many integral coordinates, which is why this one is one of the favorites. (The rest of the infinite number of points on the circle have coordinates that involve fractions and radicals.)

What goes around, comes around

Imagine that the earth is actually a smooth surface, like the globes that model it, and that you have a metal band stretched tightly around the earth at the equator. The circumference (distance around the earth at the equator) you find is almost 25,000 miles. Now you add *1 yard* of material to the band. Do you think you can slip your finger under the band? How high do you think this extra yard of material would raise the metal band? Believe it or not, this extra yard would raise the band by 6 inches *all the way around* the world. You could roll a baseball under it. Don't believe me?

Here's the explanation. The circumference of a circle — or a globe at its equator — is equal to *pi* times the diameter, $C = \pi \cdot d$. The value of π is a little more than 3. If you add 36 inches to the circumference, you add 36 to each side of the equation: $C + 36 = (\pi \cdot d) + 36$. Now you factor π from each term on the right (let π be about 3, for simplicity). $C + 36 \approx \pi(d + 12)$. The diameter increases by about 12 inches, so the radius increases by about 6 inches. The metal band is now 6 inches above the surface of the earth.

Preparing Your Eyes for Solar Ellipses

The ellipse is considered the most aesthetically pleasing of all the conic sections. It has a nice oval shape often used for mirrors, windows, and art forms. Our solar system seems to agree: All the planets take an elliptical path around the sun.

The definition of an *ellipse* is all the points where the sum of the distances from the points to two fixed points is a constant. The two fixed points are the *foci* (plural of *focus*), denoted by F. Figure 11-9 illustrates this definition. You can pick a point on the ellipse, and the two distances from that point to the two foci sum to a number equal to any other distance sum from other points on the ellipse. In Figure 11-9, the distances from point A to the two foci are 3.2 and 6.8, which add to 10. The distances from point B to the two foci are 5 and 5, which also add to 10.

Figure 11-9: The summed distances to the foci are equal for all points on an ellipse.

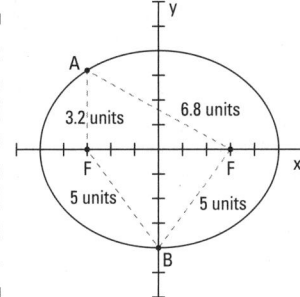

Raising the standards of an ellipse

You can think of the ellipse as a sort of squished circle. Of course, there's much more to ellipses than that, but the label sticks because the standard equation of an ellipse has a vague resemblance to the equation for a circle (see the previous section).

The standard equation for an ellipse with its center at the point (h, k) is

$$\frac{(x-h)^2}{a^2} + \frac{(y+k)^2}{b^2} = 1, \text{ where}$$

- ✔ x and y are points on the ellipse.
- ✔ a is half the length of the ellipse from left to right at its widest point.
- ✔ b is half the distance up or down the ellipse at its tallest point.

To be successful in elliptical problems, you need to find out more from the standard equation than just the center. You want to know if the ellipse is long and narrow or tall and slim. How long is it across, and how far up and down does it go? You may even want to know the coordinates of the foci. You can determine all these elements from the equation.

Determining the shape

An ellipse is criss-crossed by a *major axis* and a *minor axis*. Each axis divides the ellipse into two equal halves, with the *major axis* being the longer of the segments (such as the X-axis in Figure 11-9; if no axis is longer, you have a circle). The two axes intersect at the center of the ellipse. At the ends of the major axis, you find the *vertices* of the ellipse. Figure 11-10 shows two ellipses with their axes and vertices identified.

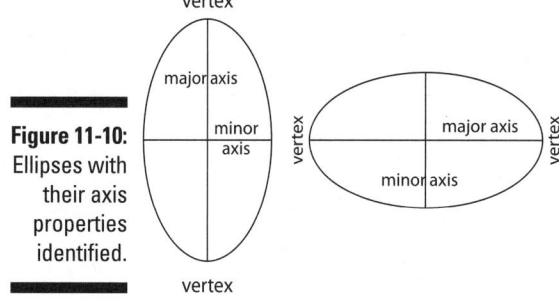

Figure 11-10:
Ellipses with
their axis
properties
identified.

To determine the shape of an ellipse, you need to pinpoint two characteristics:

- ✔ **Lengths of the axes:** You can determine the lengths of the two axes from the standard equation of the ellipse. You take the square roots of the numbers in the denominators of the fractions. Whichever value is larger, a^2 or b^2, tells you which one is the major axis. The square roots of these numbers represent the distances from the center to the points on the ellipse along their respective axes. In other words, a is half the length of one axis, and b is half the length of the other. Therefore, $2a$ and $2b$ are the lengths of the axes.

  To find the lengths of the major and minor axes of the ellipse $\dfrac{(x-4)^2}{25} + \dfrac{(y+1)^2}{49} = 1$, for example, you take the square roots of 25 and 49. The square root of the larger number, the 49, is 7. Twice 7 is 14, so the major axis is 14 units long. The square root of 25 is 5, and twice 5 is 10. The minor axis is 10 units long.

- ✔ **Assignment of the axes:** The positioning of the axes is significant. The denominator that falls under the x signifies the axis that runs parallel to the X-axis. In the previous example, the 25 falls under the x, so the minor axis runs horizontal. The denominator that falls under the y factor is the axis that runs parallel to the Y-axis. In the previous example, 49 falls under the y, so the major axis runs up and down parallel to the Y-axis. This is a tall, narrow ellipse.

Finding the foci

You can find the two foci of an ellipse by using information from the standard equation. The foci, for starters, always lie on the major axis. They lie c units from the center. To find the value of c, you use parts of the ellipse equation to form the equation $c^2 = a^2 - b^2$ or $c^2 = b^2 - a^2$, depending on which is larger, a^2 or b^2. The value of c^2 has to be positive.

In the ellipse $\dfrac{x^2}{25} + \dfrac{y^2}{9} = 1$, for example, the major axis runs across the ellipse, parallel to the X-axis (see the previous section to find out why). Actually, the major axis *is* the X-axis, because the center of this ellipse is the origin. You know this because the h and k are missing from the equation (actually, they're both equal to zero). You find the foci of this ellipse by solving the foci equation:

$$c^2 = a^2 - b^2$$
$$c^2 = 25 - 9$$
$$c^2 = 16$$
$$c = \pm\sqrt{16} = \pm 4$$

So, the foci are 4 units on either side of the center of the ellipse. In this case, the coordinates of the foci are (–4, 0) and (4, 0). Figure 11-11 shows the graph of the ellipse with the foci identified.

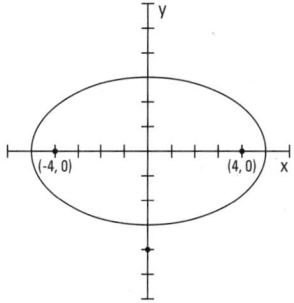

Figure 11-11:
The foci
always lie
on the major
axis (in this
case, the
X-axis).

Also, for this example ellipse, the major axis is 10 units long, running from (–5, 0) to (5, 0). These two points are the vertices. The minor axis is 6 units long, running from (0, 3) down to (0, –3).

When the center of the ellipse isn't at the origin, you find the foci the same way and adjust for the center. The ellipse $\frac{(x+1)^2}{625} + \frac{(y-3)^2}{49} = 1$ has its center at (–1, 3). You find the foci by solving $c^2 = a^2 - b^2$, which, in this case, is $c^2 = 25^2 - 7^2 = 625 - 49 = 576$. The value of c is either 24 (the root of 576) or –24 from the center, so the foci are (23, 3) and (–25, 3). The major axis is 2(25) = 50 units, and the minor axis is 2(7) = 14 units. The 25 and 7 come from the square roots of 625 and 49, respectively. And the vertices, the endpoints of the major axis, are at (24, 3) and (–26, 3).

Sketching an elliptical path

Have you ever been in a whispering gallery? I'm talking about a room or auditorium where you can stand at a spot and whisper a message, and a person standing a great distance away from you can hear your message. This phenomenon was much more impressive before the days of hidden microphones, which tend to make us more skeptical of how this works. Anyway, here's the algebraic principle behind a whispering gallery. You're standing at one focus of an ellipse, and the other person is standing at the other focus. The sound waves from one focus reflect off the surface or ceiling of the gallery and move over to the other focus.

Suppose you run across a problem on a test that asks you to sketch the ellipse associated with a whispering gallery that has foci 240 feet apart and a major axis (length of the room) of 260 feet. Your first task is to construct the equation of the ellipse.

The foci are 240 feet apart, so they each stand 120 feet from the center of the ellipse. The major axis is 260 feet long, so the vertices are each 130 feet from the center. Using the equation $c^2 = a^2 - b^2$ — which gives the relationship between c, the distance of a focus from the center; a, the distance from the center to the end of the major axis; and *b*, the distance from the center to the end of the minor axis — you get $120^2 = 130^2 - b^2$, or $b^2 = 50^2$. Armed with the values of a^2 and b^2, you can write the equation of the ellipse representing the curvature of the ceiling: $\frac{x^2}{130^2} + \frac{y^2}{50^2} = 1$.

To sketch the graph of this ellipse, you first locate the center at (0, 0). You count 130 units to the right and left of the center and mark the vertices, and then you count 50 units up and down from the center for the endpoints of the minor axis. You can sketch the ellipse by using these endpoints. The sketch in Figure 11-12 shows the points described and the ellipse. It also shows the foci — where the two people would stand in the whispering gallery.

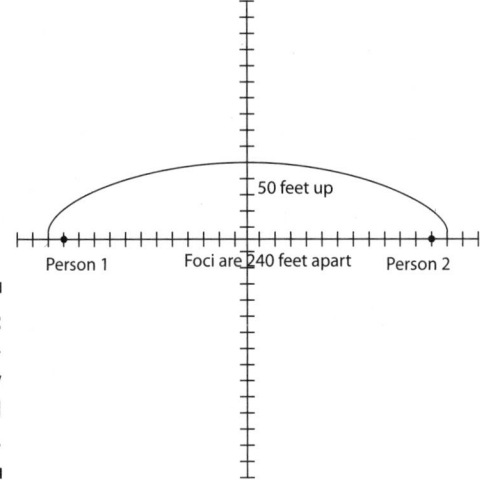

50 feet up

Person 1   Foci are 240 feet apart   Person 2

Figure 11-12:
A whisper-
ing gallery
is long and
narrow.

Feeling Hyper about Hyperbolas

The hyperbola is a conic section that seems to be in combat with itself. It features two completely disjoint curves, or *branches,* that face away from one another but are mirror images across a line that runs halfway between them.

A *hyperbola* is defined as all the points such that the difference of the distances between two fixed points (called *foci*) is a constant value. In other words, you pick a value, such as the number 6; you find two distances whose difference is 6, such as 10 and 4; and then you find a point that rests 10 units from the one point and 4 units from the other point. The hyperbola has two axes, just as the ellipse has two axes (see the previous section). The axis of the hyperbola that goes through its two foci is called the *transverse axis.* The other axis, the *conjugate axis,* is perpendicular to the transverse axis, goes through the center of the hyperbola, and acts as the mirror line for the two branches.

Figure 11-13 shows two hyperbolas with their axes and foci identified. Figure 11-13a shows the distances P and q. The difference between these two distances to the foci is a constant number (which is true no matter what points you pick on the hyperbola).

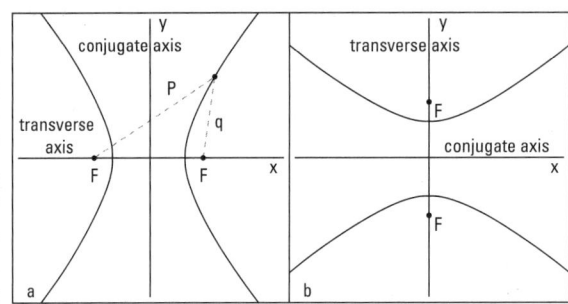

Figure 11-13: The branches of hyperbolas face away from one another.

There are two basic equations for hyperbolas. You use one when the hyperbola opens to the left and right: $\dfrac{(x-h)^2}{a^2} - \dfrac{(y-k)^2}{b^2} = 1$. You use the other when the hyperbola opens upward and downward: $\dfrac{(y-k)^2}{b^2} - \dfrac{(x-h)^2}{a^2} = 1$.

In both cases, the center of the hyperbola is at (h, k), and the foci are c units away from the center, where the relationship $b^2 = c^2 - a^2$ describes the relationship between the different parts of the equation. For example, $\dfrac{(x-4)^2}{25} - \dfrac{(y+5)^2}{144} = 1$ is the equation of a hyperbola with its center at $(4, -5)$.

Including the asymptotes

A very helpful tool you can use to sketch hyperbolas is to first lightly sketch in the two diagonal asymptotes of the hyperbola. *Asymptotes* aren't actual parts of the graph; they just help you determine the shape and direction of the curves. The asymptotes of a hyperbola intersect at the center of the hyperbola. You find the equations of the asymptotes by replacing the one in the equation of the hyperbola with a zero and simplifying the resulting equation into the equations of two lines.

If you want to find the equations of the asymptotes of the hyperbola $\dfrac{(x-3)^2}{9} - \dfrac{(y+4)^2}{16} = 1$, for example, you change the one to zero, set the two fractions equal to one another, and take the square root of each side:

$$\frac{(x-3)^2}{9} - \frac{(y+4)^2}{16} = 0$$

$$\frac{(x-3)^2}{9} = \frac{(y+4)^2}{16}$$

$$\sqrt{\frac{(x-3)^2}{9}} = \pm\sqrt{\frac{(y+4)^2}{16}}$$

$$\frac{x-3}{3} = \pm\frac{y+4}{4}$$

Now you multiply each side by 4 to get the equations of the asymptotes in better form:

$$4\left(\frac{x-3}{3}\right) = \pm\left(\frac{y+4}{4}\right)4$$

$$\frac{4}{3}(x-3) = \pm(y+4)$$

Consider the two cases — one using the positive sign, and the other using the negative sign:

$$\frac{4}{3}(x-3) = +(y+4) \quad \text{or} \quad \frac{4}{3}(x-3) = -(y+4)$$

$$\frac{4}{3}x - 4 = y + 4 \qquad \text{or} \qquad \frac{4}{3}x - 4 = -y - 4$$

$$\frac{4}{3}x - 8 = y \qquad \text{or} \qquad \frac{4}{3}x = -y$$

$$-\frac{4}{3}x = y$$

The two asymptotes you find are $y = \frac{4}{3}x - 8$ and $y = -\frac{4}{3}x$, and you can see them in Figure 11-14. Notice that the slopes of the lines are the opposites of one another. (For a refresher on graphing lines, see Chapter 2.)

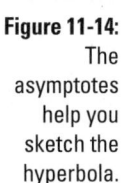

Figure 11-14:
The asymptotes help you sketch the hyperbola.

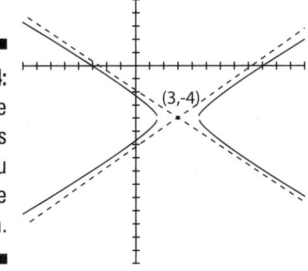

(3,-4)

Graphing hyperbolas

Hyperbolas are relatively easy to sketch, *if* you pick up the necessary information from the equations (see the intro to this section). To graph a hyperbola, use the following steps as guidelines:

1. **Determine if the hyperbola opens to the sides or up and down by noting whether the *x* term is first or second.**

   The *x* term first means it opens to the sides.

2. **Find the center of the hyperbola by looking at the values of *h* and *k*.**

3. **Lightly sketch in a rectangle twice as wide as the square root of the denominator under the *x* value and twice as high as the square root of the denominator under the *y* value.**

   The rectangle's center is the center of the hyperbola.

4. **Lightly sketch in the asymptotes through the vertices of the rectangle (see the previous section to find out how).**

5. **Draw in the hyperbola, making sure it touches the midpoints of the sides of the rectangle.**

You can use these steps to graph the hyperbola $\dfrac{(x+2)^2}{9} - \dfrac{(y-3)^2}{16} = 1$. First, note that this equation opens to the left and right because the *x* value comes first in the equation. The center of the hyperbola is at (–2, 3).

Now comes the mysterious rectangle. In Figure 11-15a, you see the center placed on the graph at (–2, 3). You count 3 units to the right and left of center (totaling 6), because twice the square root of 9 is 6. Now you count 4 units up and down from center, because twice the square root of 16 is 8. A rectangle 6 units wide and 8 units high is shown in Figure 11-15b.

Figure 11-15: Drawing a rectangle before drawing the hyperbola is a sketching help.

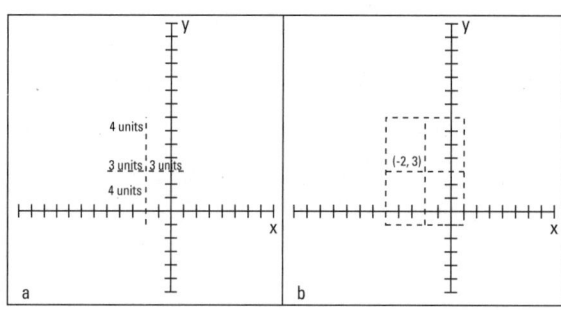

When the rectangle is in place, you draw in the asymptotes of the hyperbola diagonally through the vertices (corners) of the rectangle. Figure 11-16a shows the asymptotes drawn in. The equations of those asymptotes are $y = \frac{4}{3}x + \frac{17}{3}$ and $y = -\frac{4}{3}x + \frac{1}{3}$ (see the previous section to calculate these equations). ***Note:*** When you're just sketching the hyperbola, you usually don't need the equations of the asymptotes.

Lastly, with the asymptotes in place, you draw in the hyperbola, making sure it touches the sides of the rectangle at its midpoints and slowly gets closer and closer to the asymptotes as the curves get farther from the center. You can see the full hyperbola in Figure 11-16b.

Figure 11-16:
The
hyperbola
takes
its shape
with the
asymptotes
in place.

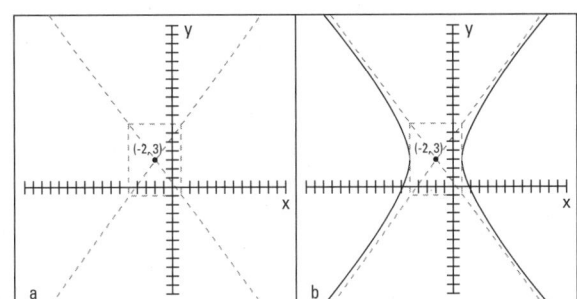

Identifying Conics from Their Equations, Standard or Not

When you write the equations of the four conic sections in their standard forms, you can easily tell which type of conic you have most of the time.

- A parabola has only one squared variable:

  $y^2 = 4ax$ or $x^2 = 4ay$

- A circle doesn't have fractions:

  $(x - h)^2 + (y - k)^2 = r^2$

- An ellipse has the *sum* of the two variable terms:

  $$\frac{(x - h)^2}{a^2} + \frac{(y - k)^2}{b^2} = 1$$

✔ A hyperbola has the *difference* of the two variable terms:

$$\frac{(x-h)^2}{a^2} - \frac{(y-k)^2}{b^2} = 1 \text{ or } \frac{(y-k)^2}{b^2} - \frac{(x-h)^2}{a^2} = 1$$

Sometimes, however, the equation you're given isn't in standard form. It has yet to be changed, using completing the square (see Chapter 3) or whatever method it takes. So, in these situations, how do you tell which type of conic you have by just looking at the equation?

Consider all the equations of the form $Ax^2 + By^2 + Cx + Dy + E = 0$. You have squared terms and first-degree terms of the two variables x and y. By observing what A, B, C, and D are, you can determine which type of conic you have.

When you have an equation in the form $Ax^2 + By^2 + Cx + Dy + E = 0$, follow these rules:

✔ If A = B, you have a circle (as long as A and B don't equal zero).

✔ If A ≠ B, and A and B are the same sign, you have an ellipse.

✔ If A and B are different signs, you have a hyperbola.

✔ If A or B is zero, you have a parabola (they can't both be zero).

Use these rules to determine which conic you have from some equations:

$9x^2 + 4y^2 - 72x - 24y + 144 = 0$ is an ellipse, because A ≠ B, and A and B are both positive. In fact, the standard form for this equation (which you find by completing the square) is $\frac{(x-4)^2}{4} + \frac{(y-3)^2}{9} = 1$.

$2x^2 + 2y^2 + 12x - 20y + 24 = 0$ is a circle, because A = B. Its standard form is $(x+3)^2 + (y-5)^2 = 22$.

$9y^2 - 8x^2 - 18y - 16x - 71 = 0$ is a hyperbola, because A and B have opposite signs. This hyperbola's standard form is $\frac{(y-1)^2}{8} - \frac{(x+1)^2}{9} = 1$.

$x^2 + 8x - 6y + 10 = 0$ is a parabola. You see only an x^2 term — the only variable raised to the second power. The standard form for this parabola is $(x+4)^2 = 6(y+1)$.

Chapter 12

Solving Systems of Linear Equations

A *system of equations* consists of a number of equations with an equal (or sometimes different) amount of variables — variables that are linked in a specific way. The solution of a system of equations uncovers these links in one of two ways: with a list of numbers that makes each equation in the system a true statement or a list of relationships between numbers that makes each equation in the system a true statement.

In this chapter, I cover systems of linear equations. As I explain in Chapter 2, *linear equations* feature variables that reach only the first degree, meaning that the highest power of any variable you solve for is one. You have a number of techniques at your disposal to solve systems of linear equations, including graphing lines, adding multiples of one equation to another, substituting one equation into another, and by taking advantage of a neat rule that Gabriel Cramer (an 18th century Swiss mathematician) developed. I cover each of the many different ways to solve a linear system of equations in this chapter.

Looking at the Standard Linear-Systems Form and Its Possible Solutions

The standard form for a system of linear equations is as follows:

$$\begin{cases} a_1 x_1 + a_2 x_2 + a_3 x_3 + \ldots = k_1 \\ b_1 x_1 + b_2 x_2 + b_3 x_3 + \ldots = k_2 \\ c_1 x_1 + c_2 x_2 + c_3 x_3 + \ldots = k_3 \\ \vdots \end{cases}$$

If a system has only two equations, the equations appear in the $Ax + By = C$ form I introduce in Chapter 2, and a brace groups them together. But don't let me fool you — a system of equations can contain any number of equations (I show you how to work through larger systems toward the end of the chapter).

Linear equations with two variables, like $Ax + By = C$, have lines as graphs. In order to solve a system of linear equations, you need to determine what values for x and y are true at the same time. Your job is to account for three possible solutions (if you count "no solution" as a solution) that can make this happen:

✔ **One solution:** The solution appears at the point where the lines intersect — the same x and the same y work at the same time in all the equations.

✔ **An infinite number of solutions:** The equations are describing the same line.

✔ **No solution:** Occurs when the lines are parallel — no value for (x, y) works in all the equations.

Graphing Solutions of Linear Systems

To solve a system of two linear equations (with integers as solutions), you can graph both equations on the same axes (X and Y). (Check out Chapter 5 for instructions on graphing lines.) With the graphs on paper, you see one of three things — intersecting lines (one solution), identical lines (infinite solutions), or parallel lines (no solution).

Solving linear systems by graphing the lines created by the equations is very satisfying to your visual senses, but beware: Using this method to find a solution requires careful plotting of the lines. Therefore, it works best for lines that have integers for solutions. If the lines don't intersect at a place where the grid on the graph crosses, you have a problem. The task of determining rational (fractions) or irrational (square roots) solutions from graphs on graph paper is too difficult, if not impossible. (In noninteger cases, you have to use substitution, elimination, or Cramer's Rule, which I show you in this chapter.) You can estimate or approximate fractional or irrational values, but you don't get an exact answer.

Pinpointing the intersection

Lines are made up of many, many points. When two lines cross one another, they share just one of those points. Sketching the graphs of two intersecting lines allows you to determine that one special point by observing where, on the graph, the two lines cross. You need to graph very carefully, using a sharpened pencil and ruler with no bumps or holes. The resulting graph is very gratifying.

Take the following straightforward linear system, for example:

$$\begin{cases} 2x + 3y = 12 \\ x - y = 11 \end{cases}$$

A quick way to sketch these lines is to find their *intercepts* — where they cross the axes. For the first equation, if you let $x = 0$ and solve for y, you get $y = 4$, so the y-intercept is $(0, 4)$; in the same equation, when you let $y = 0$, you get $x = 6$, so the x-intercept is $(6, 0)$. Plot those two points on a graph and draw a line through them. You do the same for the other equation, $x - y = 11$; you find the intercepts $(0, -11)$ and $(11, 0)$. Figure 12-1 shows the two lines graphed, using their intercepts.

Figure 12-1:
Two lines
from a linear
system
crossing at
a single
point.

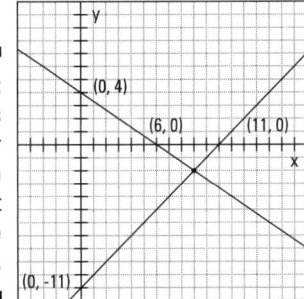

The two lines intersect at the point $(9, -2)$. You mark the point by counting the grid marks in the figure. This method shows you how important it is to graph the lines very carefully!

What exactly does the point $(9, -2)$ mean as far as a *solution* of the system? It means that if you let $x = 9$ and $y = -2$, both equations in the system are true statements. Try putting these values to work. Insert them into the first equation: $2(9) + 3(-2) = 12$; $18 - 6 = 12$; $12 = 12$. In the second equation, $9 - (-2) = 11$; $9 + 2 = 11$; $11 = 11$. The solution $x = 9$ and $y = -2$ is the only one that works for both equations.

Toeing the same line twice

A unique situation that occurs with systems of linear equations happens when everything seems to work. Every point you find that works for one equation works for the other, too. This match-made-in-heaven scenario plays out when the equations are just two different ways of describing the same line. Kind of like finding out you're dating a set of twins, no?

 When two equations in a system of linear equations represent the same line, the equations are multiples of one another. For instance, consider the following system of equations:

$$\begin{cases} x + 3y = 7 \\ 2x + 6y = 14 \end{cases}$$

You can tell that the second equation is twice the first. Sometimes, however, this sameness is disguised when the equations appear in different forms. Here's the same system as before, only with the second equation written in slope-intercept form:

$$\begin{cases} x + 3y = 7 \\ y = -\dfrac{1}{3}x + \dfrac{7}{3} \end{cases}$$

The sameness isn't as obvious here, but when you graph the two equations, you can't tell one graph from the other because they're the same line (for more on graphing lines, see Chapter 5).

Dealing with parallel lines

Parallel lines never intersect and never have anything in common except the direction they move in (their *slope;* refer to Chapter 5 for more on a line's slope). So, when you solve systems of equations that have no solutions at all, you should know right away that the lines represented by the equations are parallel.

The system $\begin{cases} x + 2y = 8 \\ 3x + 6y = 7 \end{cases}$, for example, has no solution. When you graph the two lines [with the *x*-intercepts at (⅞, 0) and (8, 0) and the *y*-intercepts at (0, ⅙) and (0, 4)], you see that they never touch — even if you extend the graph forever and ever. The lines are parallel. Figure 12-2 shows what the lines look like.

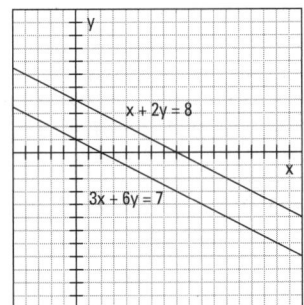

Figure 12-2:
Parallel
lines in a
linear
system of
equations
never
intersect.

One way you can predict that two lines are parallel — and that no solution exists for the system of equations — is by checking the slopes of the lines. You can write each equation in *slope-intercept form* (refer to Chapter 5 for a refresher on this form). The slope-intercept form for the line $x + 2y = 8$, for example, is $y = -\frac{1}{2}x + 4$, and the slope-intercept form for $3x + 6y = 7$ is $y = -\frac{1}{2}x + \frac{7}{6}$. The lines both have the slope $-\frac{1}{2}$, and their y-intercepts are different, so you know the lines are parallel.

Eliminating Systems of Two Linear Equations with Addition

Even though graphing lines to solve systems of equations is heaps of fun (see the previous section if you don't believe me), the graphing method has one big drawback: You'll find it almost impossible to find noninteger answers. Graphing is also time-consuming and requires careful plotting of points. The methods mathematicians prefer for solving systems of linear equations involve using algebra. The two most preferred (and common) methods for solving systems of two linear equations are *elimination,* which I cover in this section, and *substitution,* which I cover in the section "Solving Systems of Two Linear Equations with Substitution" later in the chapter. Determining which method you should use depends on what form the equations start out in and, often, personal preference.

The elimination method may also be called *linear combinations* or just *add/ subtract.* The word "elimination" describes exactly *what* you accomplish with this method, but add/subtract tells you *how* you accomplish the elimination.

Getting to an elimination point

To carry out the elimination method, you want to add the two equations together, or subtract one from the other, and eliminate (get rid of) one of the variables. Sometimes you have to multiply one or both of the equations by a carefully selected number before you add them together (or subtract them).

When you solve the system of equations $\begin{cases} 3x - 5y = 2 \\ 2x + 5y = 18 \end{cases}$, for example, you see that by adding the two equations together, you eliminate the y variable. The two y terms are opposites in the two different equations. The resulting expression is $5x = 20$. Dividing each side by 5, you get $x = 4$. When you put $x = 4$ into the first equation, you get $3(4) - 5y = 2$. Solving that equation for y, you get $y = 2$. The solution, therefore, is $x = 4$, $y = 2$. If you graph the two lines corresponding to the equations, you see them intersecting at the point $(4, 2)$.

You should always check to be sure you have the right answer. One way to check the previous example is to replace the x and y in the second equation (the one you didn't use to solve for the second variable) with the 4 and the 2 to see if the statement you get is true. Substituting, you get $2(4) + 5(2) = 18$; $8 + 10 = 18$; $18 = 18$. It works!

The system of equations $\begin{cases} 3x - 2y = 17 \\ 2x - 5y = 26 \end{cases}$ requires some adjustments before you add or subtract the two equations. If you add the two equations in their current forms, you get just another equation: $5x - 7y = 43$. You create a perfectly nice equation, and it has the same solution as the other two, but it doesn't help you find the solution of the system. Before you add or subtract, you need to make sure one of the variables in the two equations has the same or opposite coefficient as its counterpart; this way, when you add or subtract the variable, you eliminate it.

You have several different options to choose from to make the equations in this example system ready for elimination:

✔ You can multiply the first equation by 2 and the second by 3 and then subtract to eliminate the x's.

✔ You can multiply the first equation by 2 and the second by –3 and then add to eliminate the x's.

✔ You can multiply the first equation by 5 and the second by 2 and then subtract to eliminate the y's.

✔ You can multiply the first equation by 5 and the second by –2 and then add to eliminate the y's.

When faced with multiplying an equation through by a negative number, be sure that you multiply every term in the equation by the negative sign; the sign of every term will change.

If you choose to solve the previous system by multiplying the first equation by 2 and the second by –3, you get a new version of the system:

$$\begin{cases} 6x - 4y = 34 \\ -6x + 15y = -78 \end{cases}$$

Adding the two equations together, you get $11y = -44$, eliminating the x's. Dividing each side of the new equation by 11, you get $y = -4$. Substitute this value into the first *original* equation.

Always go back to the original equations to solve for the other variable or to check your work. You have a better chance of catching errors that way.

Substituting –4 for the y value, you get $3x - 2(-4) = 17$. Solving for x, you get $x = 3$. Now check your work by putting the 3 and –4 into the second original equation. You get $2(3) - 5(-4) = 26$; $6 + 20 = 26$; $26 = 26$. Check! The solution is $(3, -4)$.

Recognizing solutions for parallel and coexisting lines

When you graph systems of linear equations, it becomes pretty apparent when the systems produce parallel lines or have equations that represent the same lines. You don't have to graph the lines to recognize these situations algebraically, however; you just have to know what to look for.

For example, when solving the following system, you multiply the second equation through by –1 and add the equations together. The result that you get is $0 = 5$. That just isn't so. The false statement is your signal that the system doesn't have a solution and that the lines are parallel.

$$\begin{cases} 3x + 5y = 12 \\ 3x + 5y = 7 \end{cases}$$

$$\begin{array}{r} 3x + 5y = 12 \\ -3x - 5y = -7 \\ \hline 0 + 0 = 5 \end{array}$$

By contrast, when you have two equations for the same line and add the equations with the elimination method, you get an equation that's *always* true, such as $0 = 0$ or $5 = 5$.

Solving Systems of Two Linear Equations with Substitution

Another method used to solve systems of linear equations is called *substitution*. Some people prefer to use this method most of the time because you need to use it for equations with higher exponents; that way, you only have to master one method. Substitution in algebra works something like substitution in a basketball game — you replace a player with another player who can play that position and hope for better results. Isn't that what you hope for in algebra, too? One drawback of substitution is that you may have to work with fractions (oh, the horror), which you can avoid by using elimination (see the previous section). The method used is often a matter of personal choice.

Variable substituting made easy

Executing substitution in systems of two linear equations is a two-step process:

1. **Solve one of the equations for one of the variables, *x* or *y*.**

2. **Substitute the value of the variable into the other equation.**

To solve the system $\begin{cases} 2x - y = 1 \\ 3x - 2y = 8 \end{cases}$ by substitution, for example, you first look for a variable you can tab as a likely candidate for the first step. In other words, you want to solve for it.

Before substituting, look for a variable with a coefficient of 1 or –1 to use to solve one of the equations. By sticking with terms that have coefficients of 1 or –1, you avoid having to substitute fractions into the other equation. Sometimes fractions are unavoidable; in those cases, you should choose a term with a small coefficient so the fractions don't get too unwieldy.

In the previous example equation, the *y* term in the first equation has a coefficient of –1, so you should solve this equation for *y* (rewrite it so *y* is alone on one side of the equation). You get $y = 2x - 1$. Now you can substitute the $2x - 1$ for the *y* in the other equation:

$$3x - 2y = 8$$
$$3x - 2(2x - 1) = 8$$
$$3x - 4x + 2 = 8$$
$$-x = 6$$
$$x = -6$$

You've already created the equation $y = 2x - 1$, so you can put the value $x = -6$ into the equation to get y: $y = 2(-6) - 1 = -12 - 1 = -13$. To check your work, put both values, $x = -6$ and $y = -13$, into the equation that you didn't change (the second equation, in this case): $3(-6) - 2(-13) = 8$; $-18 + 26 = 8$; $8 = 8$. Your work checks out.

Identifying parallel and coexisting lines

As I mention in the section "Recognizing solutions for parallel and coexisting lines" earlier in the chapter, your job is well and good when you come up with a simple point of intersection for your solution. But you also have to identify the impossible (parallel lines) and always possible (coexisting lines) when using the substitution method to find solutions.

Here are some hints for recognizing these two special cases:

- ✔ When lines are parallel, the algebraic result is an impossible statement. You get an equation that can't be true, such as $2 = 6$.

- ✔ When lines are coexistent (the same), the algebraic result is a statement that's always true (the equation is always correct). An example is the equation $7 = 7$.

Facing the impossible: Parallel lines

The system of equations $\begin{cases} 3x - 2y = 4 \\ y = \dfrac{3}{2}x + 2 \end{cases}$ doesn't have a solution. If you graph the lines, you see that the graphs the equations represent are parallel. You also get an impossible statement when you try to solve the system with algebra. Using substitution to solve the system, you insert the equivalence of y from the second equation into the first:

$$3x - 2y = 4$$
$$3x - 2\left(\frac{3}{2}x + 2\right) = 4$$
$$3x - 3x - 4 = 4$$
$$-4 = 4$$

Substitution produces an incorrect statement. This equation is always wrong, so you can never find a solution.

Discerning the always possible: Coexisting lines

The system of equations $\begin{cases} 3x - 2y = 4 \\ y = \dfrac{3}{2}x - 2 \end{cases}$ represents two ways of saying the same equation — two equations that represent the same line. When you graph the equations, you produce one identical line. When you solve the system by using substitution, you substitute the equivalent of y into the first equation:

$$3x - 2y = 4$$
$$3x - 2\left(\frac{3}{2}x - 2\right) = 4$$
$$3x - 3x + 4 = 4$$
$$4 = 4$$

Substitution creates an equation that's always true. So, any pair of values that works for one equation will work for the other.

You can write the answer in the (x, y) form for the coordinates of a point by using one variable — in this case the x — and writing the other variable in terms of x. The solutions for the previous example are $\left(x, \dfrac{3}{2}x - 2\right)$. The y value is always two less than three-halves the x value. If you pick a number for the x value, you can put it into the second coordinate to get the y. For instance, if you choose $x = 6$, plug it in the solution to get y:

$$\left(6, \frac{3}{2}(6) - 2\right)$$
$$= (6, 9 - 2)$$
$$= (6, 7)$$

The point $(6, 7)$ works for both equations, as do infinitely more pairs of values.

Using Cramer's Rule to Defeat Unwieldy Fractions

Solving systems of linear equations with graphing, elimination, or substitution is usually quite doable and simple. You have another alternative to keep in mind, though, called *Cramer's Rule*. For *Seinfeld* fans, this Cramer is unlike

Jerry's Kramer, a wild and crazy man. Cramer's Rule remains very nice and calm when the solutions involve messy fractions with denominators like 47, 319, or some value equally nasty. Cramer's Rule gives you the exact fractional value of the solutions you find — not some rounded-off decimal value you're likely to get with a calculator or computer solution — and you can use the rule for any system of linear equations. It isn't the method of choice, however, because it's more complicated and takes longer than the other methods. The extra effort is worth it, though, when the answers are huge fractions.

Setting up the linear system for Cramer

To use *Cramer's Rule,* you first have to write the two linear equations in the following form:

$$a_1 x + b_1 y = c_1$$
$$a_2 x + b_2 y = c_2$$

The two equations have the variables on one side — in x, y order — and the constant, c, on the other side. The coefficients bear the subscripts 1 or 2 to identify which equation they come from.

Your next step is to assign d to represent the difference of the two products of the coefficients of x and y: $d = a_1 b_2 - b_1 a_2$. The order of the subtraction is very important here.

An easy way to remember the order of the products and difference is to imagine the coefficients in a square where you multiply and subtract diagonally.

You cross the top left times the bottom right and subtract the top right times the bottom left:

$$\begin{vmatrix} a_1 & b_1 \\ a_2 & b_2 \end{vmatrix}$$

You find the solution of the system, the values of x and y, by dividing two other cleverly created differences by the value of d. You can memorize these differences, or you can picture the squares you form by replacing the a's and b's with the c's and doing the criss-cross multiplication and subtraction:

$$\begin{vmatrix} c_1 & b_1 \\ c_2 & b_2 \end{vmatrix} \qquad \begin{vmatrix} a_1 & c_1 \\ a_2 & c_2 \end{vmatrix}$$

To solve for x and y by using Cramer's Rule when you have two linear equations written in the correct form, use the following equations:

$$x = \frac{c_1 b_2 - b_1 c_2}{d} = \frac{c_1 b_2 - b_1 c_2}{a_1 b_2 - b_1 a_2}$$

$$y = \frac{a_1 c_2 - c_1 a_2}{d} = \frac{a_1 c_2 - c_1 a_2}{a_1 b_2 - b_1 a_2}$$

Applying Cramer's Rule to a linear system

You can solve most systems of equations with elimination or substitution, but some systems can get pretty gruesome because of the fractions that pop up. If you use Cramer's Rule, the work is much easier (see the previous section for the necessary setup).

To solve $\begin{cases} 13x + 7y = 25 \\ 10x - 9y = 13 \end{cases}$ with Cramer's Rule, for example, you first find the value of the denominator you'll create, d. Using the formula $d = a_1 b_2 - b_1 a_2$ (the subscripts point to which equations the a and b values come from), $d = 13(-9) - 7(10) = -117 - 70 = -187$.

With your d value in tow, try solving for x first:

$$x = \frac{c_1 b_2 - b_1 c_2}{d} = \frac{25(-9) - 7(13)}{-187} = \frac{-225 - 91}{-187} = \frac{-316}{-187} = \frac{316}{187}$$

Cramer: A man of many talents

Gabriel Cramer (1704-1752) was a Swiss mathematician/astronomer/philosopher who modern algebra students recognize for his slick method for solving systems of linear equations. Cramer was a bit of a child prodigy, earning his doctorate when he was 18. His thesis was on the theory of sound. He aspired to teach philosophy in Geneva but decided to share teaching responsibilities in mathematics with Calandrini instead. Cramer taught geometry and mechanics in French rather than the traditional Latin, which was a bit of a departure for the members of the Academy.

Cramer spent a good deal of time with mathematicians throughout Europe, sometimes editing and other times distilling their work. He made contributions to the general knowledge of planets — their shapes and positions in their orbits. He also published articles on the date of Easter, the history of mathematics, and the aurora borealis and even dabbled in the probability applied to witnesses in a court case. His mathematical expertise aided him in his roles in local government, where he advised on fortification, building construction, and artillery. If there was ever a case for the practicality of mathematics, Cramer made it.

You don't want to substitute such a scary number into one of the equations to solve for y, so move on to the formula for y:

$$y = \frac{a_1 c_2 - c_1 a_2}{d} = \frac{13(13) - 25(10)}{-187} = \frac{169 - 250}{-187} = \frac{-81}{-187} = \frac{81}{187}$$

Checking your solution isn't much fun because of the fractions, but substituting the values of x and y into both equations shows you that you do, indeed, have the solution.

If you get $d = 0$, you have to stop. You can't divide by zero. The zero value for d indicates that you have either no solution or an infinite number of solutions — the lines are parallel, or you have two equations for the same line. In any case, Cramer's Rule has failed you, and you have to go back and use a different method.

Raising Linear Systems to Three Linear Equations

Systems of three linear equations also have solutions: sets of numbers (all the same for each equation) that make each of the equations true. When a system has three variables rather than two, you no longer graph the equations as lines. To graph these equations, you have to do a three-dimensional graph of the planes represented by the equations containing the three variables. In other words, you really can't find the solution by graphing. The best method for solving systems with three linear equations involves using your algebra skills.

Solving three-equation systems with algebra

When you have a system of three linear equations and three unknown variables, you solve the system by reducing the three equations with three variables to a system of two equations with two variables. At that point, you're back to familiar territory and have all sorts of methods at your disposal to solve the system (see the previous sections in this chapter). After you determine the values of the two variables in the new system, you *back-substitute* into one of the original equations to solve for the value of the third variable.

To solve the following system, for example, you choose a variable to eliminate:

$$\begin{cases} 3x - 2y + z = 17 \\ 2x + y + 2z = 12 \\ 4x - 3y - 3z = 6 \end{cases}$$

The prime two candidates for elimination are the y and z because of the coefficients of 1 or –1 that occur in their equations. It makes your job easier if you can avoid larger coefficients on the variables when you have to multiply an equation through by a number to create sums of zero. Assume that you choose to eliminate the z variable.

To eliminate the z's from the equations, you add two of the equations together — after multiplying by an appropriate number — to get a new equation. You then repeat the process with a different combination of two equations. Your result is two equations that contain only the x and y variables.

For this example problem, you start by multiplying the terms in the top equation by –2 and adding them to the terms in the middle equation:

$$-2\left(3x - 2y + z = 17\right) \rightarrow -6x + 4y - 2z = -34$$
$$\underline{2x + y + 2z \ = 12}$$
$$-4x + 5y \qquad = -22$$

Next, you multiply the terms in the top equation (the original top equation — not the one that you multiplied earlier) by 3 and add them to the terms in the bottom equation (again, the original one):

$$3\left(3x - 2y + z = 17\right) \rightarrow 9x - 6y + 3z = 51$$
$$\underline{4x - 3y - 3z = 6}$$
$$13x - 9y \qquad = 57$$

The two equations you create by adding comprise a new system of equations with just two variables:

$$\begin{cases} -4x + 5y = -22 \\ 13x - 9y = 57 \end{cases}$$

To solve this new system, you can multiply the terms in the first equation by 9 and the terms in the second equation by 5 to create coefficients of 45 and –45 on the y terms. You add the two equations together, getting rid of the y terms, and solve for x:

$$-36x + 45y = -198$$
$$\underline{65x - 45y = 285}$$
$$29x \qquad = 87$$
$$x = 3$$

Now you substitute $x = 3$ into the equation $-4x + 5y = -22$. Choosing this equation is just an arbitrary choice — either equation will do. When you substitute $x = 3$, you get $-4(3) + 5y = -22$. Adding 12 to each side, you get $5y = -10$, or $y = -2$.

You can check your work by taking the $x = 3$ and $y = -2$ and substituting them into one of the original equations. A good habit is to substitute the values into the first equation and then check by substituting all three answers into the other two.

Putting $x = 3$ and $y = -2$ into the first equation, you get $3(3) - 2(-2) + z = 17$, giving you $9 + 4 + z = 17$. You subtract 13 from each side for a result of $z = 4$. Now check these three values in the other two equations:

$$2(3) + (-2) + 2(4) = 6 - 2 + 8 = 12$$
$$4(3) - 3(-2) - 3(4) = 12 + 6 - 12 = 6$$

They both check — of course!

You can write the solution of the system as $x = 3$, $y = -2$, $z = 4$, or you can write it as an *ordered triple.* An ordered triple consists of three numbers in a parenthesis, separated by commas. The order of the numbers matters. The first value represents x, the second y, and the third z. You write the previous example solution as $(3, -2, 4)$. The ordered triple is a simpler and neater method — as long as everyone agrees what the numbers stand for.

Settling for a generalized solution for linear combinations

When dealing with three linear equations and three variables, you may come across a situation where one of the equations is a linear combination of the other two. This means you won't find one single solution for the system, such as $(3, -2, 4)$. A more generalized solution may be $(-z, 2z, z)$, where you pick a number for z that determines what the x and y values are. In this case, where the solution is $(-z, 2z, z)$, if you let $z = 7$, the ordered triple becomes $(-7, 14, 7)$. You can find an infinite number of solutions for this particular system of equations, but the solutions are very specific in form — the variables all have a relationship.

You first get an inkling that a system has a generalized answer when you find out that one of the reduced equations you create is a multiple of another. Take the following system, for example:

$$\begin{cases} 2x + 3y - z = 12 \\ x - 3y + 4z = -12 \\ 5x - 6y + 11z = -24 \end{cases}$$

To solve this system, you can eliminate the z's by multiplying the terms in the first equation by 4 and adding them to the second equation. You then multiply the terms in the first equation by 11 and add them to the third equation:

$$4(2x + 3y - z = 12) \rightarrow 8x + 12y - 4z = 48$$
$$\underline{x - 3y + 4z = -12}$$
$$9x + 9y = 36$$
$$11(2x + 3y - z = 12) \rightarrow 22x + 33y - 11z = 132$$
$$\underline{5x - 6y + 11z = -24}$$
$$27x + 27y = 108$$

The second equation, $27x + 27y = 108$, is three times the first equation. Because these equations are multiples of one another, you know that the system has no single solution; it has an infinite number of solutions.

To find those solutions, you take one of the equations and solve for a variable. You may choose to solve for y in $9x + 9y = 36$. Dividing through by 9, you get $x + y = 4$. Solving for y, you get $y = 4 - x$. You substitute that equation into one of the original equations in the system to solve for z in terms of x. After you solve for z this way, you have the three variables all written as some version of x.

Substituting $y = 4 - x$ into $2x + 3y - z = 12$, for example, you get

$$2x + 3(4 - x) - z = 12$$
$$2x + 12 - 3x - z = 12$$
$$-x - z = 0$$
$$-x = z$$

The ordered triple giving the solutions of the system is $(x, 4 - x, -x)$. You can find an infinite number of solutions, all determined by this pattern. Just pick an x, such as $x = 3$. The solution is $(3, 1, -3)$. These values of x, y, and z all work in the equations of the original system.

Upping the Ante with Increased Equations

Systems of linear equations can be any size. You can have two, three, four, or even 100 linear equations. (After you get past three or four, you resort to technology.) Some of these systems have solutions and others don't. You have to dive in to determine whether you can find a solution or not. You can try to solve a system of just about any number of linear equations, but you find a single, unique solution (one set of numbers for the answer) only when the number of variables in the system has at least that many equations. If a system has three different variables, you need at least three different equations. Having enough equations for the variables doesn't guarantee a unique solution, but you have to at least start out that way.

The general process for solving n equations with n variables is to keep eliminating variables. A systematic way is to start with the first variable, eliminate it, move to the second variable, eliminate it, and so on until you create a reduced system with two equations and two variables. You solve for the solutions of that system and then start substituting values into the original equations. This process can be long and tedious, and errors are easy to come by, but if you have to do it by hand, this is a very effective method. Technology, however, is most helpful when systems get unmanageable.

The following system has five equations and five variables:

$$\begin{cases} x+y+z+w+t=3 \\ 2x-y+z-w+3t=28 \\ 3x+y-2z+w+t=-8 \\ x-4y+z-w+2t=28 \\ 2x+3y+z-w+t=6 \end{cases}$$

You begin the process by eliminating the x's:

1. **Multiply the terms in the first equation by –2 and add them to the second equation.**

2. **Multiply the first equation through by –3 and add the terms to the third equation.**

3. **Multiply the first equation through by –1 and add the terms to the fourth equation.**

4. **Multiply the first equation through by –2 and add the terms to the last equation.**

After you finish (whew!), you get a system with the x's eliminated:

$$\begin{cases} -3y - z - 3w + t = 22 \\ -2y - 5z - 2w - 2t = -17 \\ -5y - 2w + t = 25 \\ y - z - 3w - t = 0 \end{cases}$$

Now you eliminate the y's in the new system by multiplying the last equation by 3, 2, and 5 and adding the results to the first, second, and third equations, respectively:

$$\begin{cases} -4z - 12w - 2t = 22 \\ -7z - 8w - 4t = -17 \\ -5z - 17w - 4t = 25 \end{cases}$$

You eliminate the z's in the latest system by multiplying the terms in the first equation by 7 and the second by –4 and adding them together. You then multiply the terms in the second equation by 5 and the third by –7 and add them together. The new system you create has only two variables and two equations:

$$\begin{cases} -52w + 2t = 222 \\ 79w + 8t = -260 \end{cases}$$

To solve the two-variable system in the most convenient way, you multiply the first equation through by –4 and add the terms to the second:

$$\begin{aligned} 208w - 8t &= -888 \\ \underline{79w + 8t} &= \underline{-260} \\ 287w &= -1{,}148 \\ w &= -4 \end{aligned}$$

You find $w = -4$. Now back-substitute w into the equation $-52w + 2t = 222$ to get $-52(-4) + 2t = 222$, which simplifies to $208 + 2t = 222 \rightarrow 2t = 14 \rightarrow t = 7$.

Take these two values and plug them into $-4z - 12w - 2t = 22$. Substituting, you get $-4z - 12(-4) -2(7) = 22$, which simplifies to $-4z + 34 = 22 \rightarrow -4z = -12 \rightarrow z = 3$.

Put the three values into $y - z - 3w - t = 0$: $y - (3) - 3(-4) - 7 = 0$, or $y + 2 = 0 \rightarrow y = -2$. Only one more to go!

Move back to the equation $x + y + z + w + t = 3$, and plug in values: $x + (-2) + 3 + (-4) + 7 = 3$, which simplifies to $x + 4 = 3 \rightarrow x = -1$.

The solution reads: $x = -1$, $y = -2$, $z = 3$, $w = -4$, and $t = 7$.

You can put this into an *ordered quintuple* (five numbers in parenthesis), as long as everyone knows the proper order: (x, y, z, w, t). The order isn't alphabetical, in this case, but this is pretty typical — to have the x, y, and z first and then list other variables. You know what order to use from the way the problem is stated. The ordered quintuple is $(-1, -2, 3, -4, 7)$.

Applying Linear Systems to Our 3-D World

Being able to solve systems of two, three, and even more linear equations is just grand, but what's the point? The point is that the real-world applications are many. And, outside of the algebra classroom, technology steps in when the number of equations gets large and unwieldy. Check out the following situation that you can solve with three equations; it may be something you're interested in.

You're trying to figure out the cost of a hamburger, fries, and soft drink. You know that when a friend purchased four hamburgers, two fries, and three soft drinks, he paid $14; another friend purchased six hamburgers and six soft drinks for $18; and a third friend purchased five hamburgers, six fries, and eight soft drinks for $27.

You can figure out the cost of a hamburger, fries, and drink by setting up three equations. Let h represent the cost of a hamburger, f represent the cost of fries, and d represent the cost of a soft drink. Writing the three friends' purchases in terms of these variables, you get the following:

$$\begin{cases} 4h + 2f + 3d = 14 \\ 6h \quad\quad + 6d = 18 \\ 5h + 6f + 8d = 27 \end{cases}$$

Because an equation is missing the variable f, you choose that variable for elimination (see the section "Eliminating Systems of Two Linear Equations with Addition"). Multiply the terms in the first equation by -3 and add them to the terms in the last equation:

$$-3\,(4h + 2f + 3d = 14) \rightarrow -12h - 6f - 9d = -42$$
$$\underline{5h + 6f + 8d = 27}$$
$$-7h \qquad -d = -15$$

Multiply the new equation through by 6 and add the terms to the original middle equation. Divide and solve for h:

$$6\,(-7h - d = -15) \rightarrow -42h - 6d = -90$$
$$\underline{6h + 6d = 18}$$
$$-36h \qquad = -72$$
$$h = 2$$

You find that a hamburger costs \$2. Substitute this value into the original middle equation to get $6(2) + 6d = 18$. Subtracting 12 from each side, you get $6d = 6$, or $d = 1$. Soft drinks are \$1. Take these two values and substitute them into the first (original) equation to get $4(2) + 2f + 3(1) = 14$, which simplifies to $2f + 11 = 14$. Subtracting 11 from each side, you get $2f = 3$, or $f = 1.5$. Fries cost \$1.50. So, if you want a hamburger, fries, and a soft drink, you have to spend $2 + \$1.50 + \$1 = \$4.50$. Would you like to do a problem on the calories in that meal? Okay. I won't spoil your appetite.

Using Systems to Decompose Fractions

If you have an algebraic fraction such as $\dfrac{7x - 1}{x^2 - x - 6}$ — a rational function with polynomials in the numerator and denominator — you can determine what two fractions came together to create the equation. This process is called *decomposing fractions*. The reason you want to decompose fractions is that you have an advantage when denominators are linear expressions (the powers on the variables are first degree). You can use this technique with two, three, four, and more fractions. Of course, the systems of equations grow larger with respect to the number of fractions.

The name suggests some sort of autopsy is taking place. No, this isn't an episode of *CSI, Crime Scene Investigation*. But, wait, maybe it is. You can call this *Carefully Selecting Integers*. Do you suppose this will catch on? Okay, I'll drop the CSI stuff. Decomposing fractions is also a very handy method to know if you're in calculus and want to determine an antiderivative. (You can find out more about antiderivatives in *Calculus For Dummies*, by Mark Ryan [Wiley].)

The fraction $\dfrac{7x - 1}{x^2 - x - 6}$ has a denominator, $x^2 - x - 6$, that factors into $(x + 2)$ $(x - 3)$. This means you have two fractions, one with a denominator of $x + 2$

and the other with a denominator of $x - 3$, that you can add to find the fraction with a numerator of $7x - 1$. (See Chapter 3 for more about factoring.)

When adding fractions, you need a common denominator. If the two denominators of the fractions you're adding have nothing in common, the common denominator is the product of the two fractions' denominators. For instance, to add $\frac{3}{4} + \frac{1}{3}$, the common denominator is 12, the product of the 4 and 3.

Back to the original fraction. To find the numerators of the two fractions needed to give you the resulting fraction, write out the following equation:

$$\frac{7x - 1}{(x + 2)(x - 3)} = \frac{A}{x + 2} + \frac{B}{x - 3}$$

The numerators are A and B. Now you combine the two fractions by creating the common denominator and multiplying by the appropriate terms:

$$\frac{7x - 1}{(x + 2)(x - 3)} = \left(\frac{A}{x + 2} \cdot \frac{x - 3}{x - 3} \right) + \left(\frac{B}{x - 3} \cdot \frac{x + 2}{x + 2} \right)$$
$$= \frac{A(x - 3) + B(x + 2)}{(x + 2)(x - 3)}$$

The example fraction and your newly created fraction are equal if their numerators are equal. So, you can write the equation expressing this relationship as $7x - 1 = A(x - 3) + B(x + 2)$. You solve the equation by first distributing the letters A and B over the respective terms on the right and then rearranging the terms so that the two terms with x are together and the two terms without x are together. Now factor the x out of the first two terms and group the last two terms:

$$\begin{aligned} 7x - 1 &= A(x - 3) + B(x + 2) \\ &= Ax - 3A + Bx + 2B \\ &= (A + B)x + (-3A + 2B) \end{aligned}$$

In the original numerator, you see that the coefficient of x is 7. In your latest numerator creation, the coefficient of x is $A + B$. You set these equal to one another to get $7 = A + B$. The constant term in the original numerator is -1. In the newest numerator, the constant term is $-3A + 2B$. Set these equal to one another to get $-1 = -3A + 2B$. You now have your system of linear equations:

$$\begin{cases} 7 = A + B \\ -1 = -3A + 2B \end{cases}$$

Multiply the terms in the top equation by 3 and add them to the second equation:

$$
\begin{aligned}
3A + 3B &= 21 \\
-3A + 2B &= -1 \\
\hline
5B &= 20 \\
B &= 4
\end{aligned}
$$

You find that $B = 4$. Substitute $B = 4$ into $A + B = 7$ to get $A = 3$. You now have the numerators of the fractions:

$$\frac{7x - 1}{(x + 2)(x - 3)} = \frac{3}{x + 2} + \frac{4}{x - 3}$$

Chapter 13

Solving Systems of Nonlinear Equations and Inequalities

*I*n systems of linear equations, the variables have exponents of one, and you typically find only one solution (see Chapter 12). The possibilities for multiple solutions seem to grow as the exponents get larger, creating systems of nonlinear equations. For example, a line and parabola may intersect at two points, one point, or at no point at all. A circle and ellipse can intersect in four different points. And consider inequalities. The graphs of inequalities involve many solutions. When you put two inequalities together, the possibilities are exciting. (What can I say? I'm an algebra professor.)

One of the most important parts of solving nonlinear systems is planning. If you have an inkling as to what's coming, you'll have an easy time planning for the solution, and you'll be more convinced when your predictions come true. In this chapter, you find out at how many points a line and a parabola can cross and how many ways a parabola and a circle can cross. I also help you visualize a circle and an ellipse — when you put one on top of the other, you can plan on how many points of intersection you expect to find. Finally, you see how graphing inequalities takes on a whole new picture — catching curvy areas between circles and the like.

Crossing Parabolas with Lines

A *parabola* is a predictable, smooth, U-shaped curve (which I first introduce in detail in Chapter 7). A line is also very predictable; it goes up or down and left or right at the same rate forever and ever. If you put these two characteristics together, you can predict with a fair amount of accuracy what will happen when a line and a parabola share the same space.

When you combine the equations of a line and a parabola, you get one of three results (which you can check out in Figure 13-1):

- ✔ Two common solutions (Figure 13-1a)
- ✔ One common solution (Figure 13-1b)
- ✔ No solution at all (Figure 13-1c)

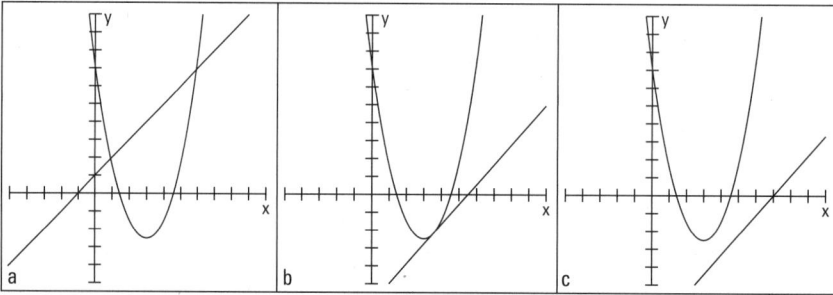

Figure 13-1: A line and a parabola sharing space on a graph.

The easiest way to find the common solutions, or sets of values, for a line and a parabola is to solve their system of equations algebraically. A graph is helpful for confirming your work and putting the problem into perspective. When solving a system of equations involving a line and a parabola, most mathematicians use the *substitution* method. For a complete look at how to use substitution, refer to Chapter 12. You can also pick up on the method by following the work in the coming pages.

You almost always substitute x's for the y in an equation, because you often see functions written with the y's equal to so many x's. You may have to replace x's with y's, but that's the exception. Just be flexible. (If you want to see an exception, check out the "Sorting out the solutions" section later in this chapter.)

Determining the point(s) where a line and parabola cross paths

The graphs of a line and a parabola can cross in two places, one place, or no place at all (see Figure 13-1). In terms of equations, these assertions translate to two common solutions, one solution, or no solution at all. Doesn't that fit together nicely?

Finding two solutions

The parabola $y = 3x^2 - 4x - 1$ and the line $x + y = 5$ have two points in common. To solve for the two solutions by using the substitution method, you first solve for y in the equation of the line: $y = -x + 5$. Now you substitute this equivalence of y into the first equation, set the new equation equal to zero, and factor as you do any quadratic equation (see Chapter 3):

$$y = 3x^2 - 4x - 1$$
$$-x + 5 = 3x^2 - 4x - 1$$
$$0 = 3x^2 - 3x - 6$$
$$0 = 3\left(x^2 - x - 2\right)$$
$$0 = 3\left(x - 2\right)\left(x + 1\right)$$

Setting each of the binomial factors equal to zero, you get $x = 2$ and $x = -1$. When you substitute those values into the equation $y = -x + 5$, you find that when $x = 2$, $y = 3$, and when $x = -1$, $y = 6$. The two points of intersection, therefore, are $(2, 3)$ and $(-1, 6)$. Figure 13-2 shows the graphs of the parabola $(y = 3x^2 - 4x - 1)$, the line $(y = -x + 5)$, and the two points of intersection.

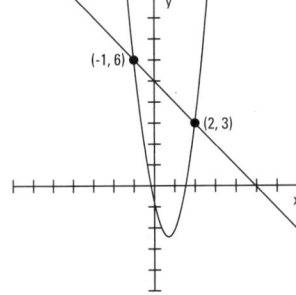

Figure 13-2:
You find
the two
points of
intersection
with
substitution.

Settling for one solution

When a line and a parabola have one point of intersection, and therefore share one common solution, the line is *tangent* to the parabola. A line and a curve can be tangent to one another if they touch or share exactly one point and if the line appears to follow the curvature at that point. (Two curves can also be tangent to one another — they touch at a point and then go their own merry ways.) The parabola $y = -x^2 + 5x + 6$ and the line $y = 3x + 7$, for example, have only one point in common — at their point of *tangency.* Figure 13-3 shows how a line and a parabola can be tangent.

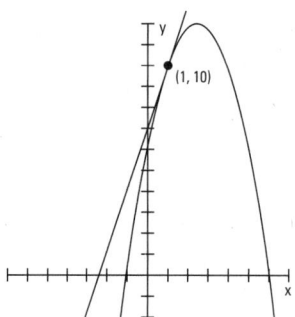

Figure 13-3:
The line touches the parabola in just one point — their point of tangency.

You find the coordinate of the point that the parabola and line share by solving the system of equations formed by the parabola and line. You substitute the equivalence of y in the line equation into the parabolic equation and solve for x (see Chapter 12):

$$y = -x^2 + 5x + 6$$
$$3x + 7 = -x^2 + 5x + 6$$
$$0 = -x^2 + 2x - 1$$
$$0 = -1(x^2 - 2x + 1)$$
$$0 = -1(x - 1)^2$$
$$x = 1$$

TIP

The dead giveaway that the parabola and line are tangent is the quadratic equation that results from the substitution. It has a *double root* — the same solution appears twice — when the binomial factor is squared.

Substituting $x = 1$ into the equation of the line, you get $y = 3(1) + 7 = 10$. The coordinates of the point of tangency are (1, 10).

Dealing with a solution that's no solution

You can see when no solution exists in a system of equations involving a parabola and line if you graph the two figures and find that their paths never cross. You also discover that a parabola and line don't intersect when you get a no-answer answer to the algebra problem — no need to even graph the figures. For instance, if you solve the system of equations containing the parabola $x = y^2 - 4y + 3$ and the line $y = 2x + 5$ by using substitution (see Chapter 12), you get the following:

$$x = y^2 - 4y + 3$$
$$x = (2x + 5)^2 - 4(2x + 5) + 3$$
$$x = 4x^2 + 20x + 25 - 8x - 20 + 3$$
$$0 = 4x^2 + 11x + 8$$

The equation looks perfectly good so far, even though the quadratic doesn't factor. You have to resort to the quadratic formula. (You can find details on using the quadratic formula in Chapter 3, if you need a refresher.) Substituting the numbers from the quadratic equation into the formula, you get the following:

$$x = \frac{-11 \pm \sqrt{121 - 4(4)(8)}}{2(4)} = \frac{-11 \pm \sqrt{121 - 128}}{8} = \frac{-11 \pm \sqrt{-7}}{8}$$

Whoa! You can stop right there. You see that a negative value sits under the radical. The square root of –7 isn't real, so no answer exists for x (for more on nonreal numbers, see Chapter 14). The nonexistent answer is your big clue that the system of equations doesn't have a common solution, meaning that the parabola and line never intersect (hey, even Sherlock Holmes had to dig around a bit before finding his clue). Figure 13-4 shows the graphs of the parabola and line. You can see why you found no solution. I wish I could give you an easy way to tell that a system has no solution before you go to all that work. Think of it this way: An answer of *no solution* is a perfectly good answer.

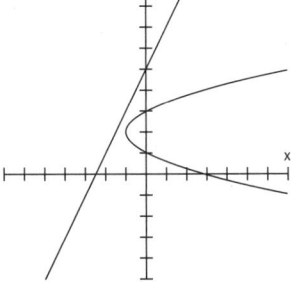

Intertwining Parabolas and Circles

The graph of a parabola is a U-shaped curve, and a circle — well, you could go round and round about a circle. When a parabola and circle share the same gridded plot, they can interact in several different ways (much like you and your neighbors, I suppose). The figures can intersect at four different points, three points, two points, one point, or no points at all. The possibilities may seem endless, but that's wishful thinking. The five possibilities I list here are what you have to work with. Your challenge is to determine which situation you have and to find the solutions of the system of equations (not as challenging as getting your neighbor to return borrowed items, you'll be glad to know).

Managing multiple intersections

A parabola and a circle can intersect at up to four different points, meaning that their equations can have up to four common solutions. Figure 13-5 shows just such a situation.

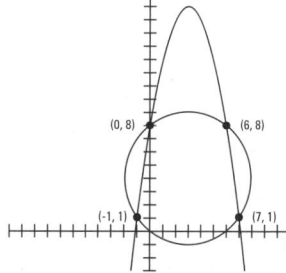

Figure 13-5: A parabola and circle intersecting at four points.

To solve for the common solutions you see in Figure 13-5, you have to solve the system of equations that includes $y = -x^2 + 6x + 8$, the equation of a parabola, and $x^2 + y^2 - 6x - 8y = 0$, the equation of a circle. Here are the steps for solving this system:

1. **If the parabola and the circle don't appear in their standard forms, you have to get them there if you want to graph them.**

   In this case, you need to change the circle to its standard form — $(x - h)^2 + (y - k)^2 = r^2$ — by completing the square (see Chapter 11). The circle's equation, written in standard form, is $(x - 3)^2 + (y - 4)^2 = 25$. (For more on the parabola's standard form, check out Chapters 3 and 7.)

2. **To solve for common points, replace each y in the equation of the circle with the equivalence of y in the parabola.**

$$x^2 + y^2 - 6x - 8y = 0$$
$$x^2 + \left(-x^2 + 6x + 8\right)^2 - 6x - 8\left(-x^2 + 6x + 8\right) = 0$$
$$x^2 + x^4 - 12x^3 + 20x^2 + 96x + 64 - 6x + 8x^2 - 48x - 64 = 0$$
$$x^4 - 12x^3 + 29x^2 + 42x = 0$$

TIP

Solving a system with a parabola and circle by substitution (see Chapter 12) involves squaring a trinomial and factoring a third-degree polynomial. When squaring a trinomial, you may find it easier to distribute the terms instead of stacking the terms like in a multiplication problem. To find $(-x^2 + 6x + 8)^2$, for example, think of the product $(-x^2 + 6x + 8)(-x^2 + 6x + 8)$. You multiply each term by $-x^2$, and then by $6x$, and lastly by 8. Finish by combining the like terms (see Chapter 8 for more on polynomials):

$$-x^2(-x^2 + 6x + 8) + 6x(-x^2 + 6x + 8) + 8(-x^2 + 6x + 8)$$
$$= x^4 - 6x^3 - 8x^2 - 6x^3 + 36x^2 + 48x - 8x^2 + 48x + 64$$
$$= x^4 - 12x^3 + 20x^2 + 96x + 64$$

3. **Set the resulting terms equal to zero and solve for x (this usually requires factoring or the use of the quadratic formula; see Chapter 3).**

   The terms in the equation have a common factor of x. Factoring out the x, you get $x(x^3 - 12x^2 + 29x + 42) = 0$. The expression in the parenthesis factors into the product of three binomials. You can do this factorization and find these binomials by using the Rational Root Theorem, which leads you to try the factors of 42 — 1, 6, and 7 — and synthetic division. (Chapter 8 has a full explanation of the Rational Root Theorem and factoring.) The final factorization of the equation is $x(x + 1)(x - 6)(x - 7) = 0$. The solutions are $x = 0, -1, 6$, and 7.

4. **Substitute the solutions you find into the equation of the curve with the smaller exponents to find the coordinates of the points of intersection.**

   In this case, you substitute into the equation of the parabola. You find that when $x = 0$, $y = 8$; when $x = -1$, $y = 1$; when $x = 6$, $y = 8$; and when $x = 7$, $y = 1$. The points of intersection are, therefore, (0, 8), (–1, 1), (6, 8), and (7, 1).

A circle and a parabola can also intersect at three points, two points, one point, or no points. Figure 13-6 shows what the three-point and two-point situations look like. In Figure 13-6a, the parabola's vertex is tangent to a point on the circle, and the parabola cuts the circle at two other points. In Figure 13-6b, the parabola cuts the circle at only two points.

Figure 13-6:
Parabolas and circles tangling, offering up different solutions.

a   b

You use the same substitution method to solve systems of equations with fewer than four intersections. The algebra leads you to the solutions — but beware the false promises. You have to watch out for extraneous solutions by checking your answers.

After substituting one equation into another, take a look at the resulting equation. The highest power of the equation tells you what to expect as far as the number of common solutions. When the power is three or four, you can have as many as three or four solutions, respectively (see Chapter 8 on solving polynomials). When the power is two, you can have up to two common solutions (Chapter 3 covers these equations). A power of one indicates only one possible solution (see Chapter 2). If you end up with an equation that has no solutions, you know the system has no points of intersection — the graphs just pass by like ships in the night.

Sorting out the solutions

In the section "Crossing Parabolas with Lines" earlier in the chapter, the examples I provide use substitution where the x's replace the y variable. Most of the time, this is the method of choice, but I suggest you remain flexible and open for other opportunities. The next example is just such an opportunity — taking advantage of a situation where it makes more sense to replace the x term with the y term.

To find the common solutions of the parabola $y = x^2$, which has its vertex at the origin (see Chapter 7), and the circle $x^2 + (y-1)^2 = 9$, which has its center at $(0, 1)$ and has a radius of 3 (see Chapter 11), you take advantage of the simplicity of $y = x^2$ by replacing the x^2 in the circle equation with y. That sets you up with an equation of y's to solve:

$$x^2 + (y-1)^2 = 9$$
$$y + (y-1)^2 = 9$$
$$y + y^2 - 2y + 1 = 9$$
$$y^2 - y - 8 = 0$$

This quadratic equation doesn't factor, so you have to use the quadratic formula (see Chapter 3) to solve for y:

$$y = \frac{1 \pm \sqrt{1 - 4(1)(-8)}}{2(1)} = \frac{1 \pm \sqrt{33}}{2}$$

You find two different values for y, according to this solution. When you use the positive part of the $\pm$, you find that y is close to 3.37. When you use the negative part, you find that y is about -2.37. Something doesn't seem right. What is it that's bothering you? It has to be the negative value for y. The common solutions of a system should work in both equations, and $y = -2.37$ doesn't work in $y = x^2$, because when you square x, you don't get a negative number. So, only the positive part of the solution, where $y \approx 3.37$, works.

Substitute $\dfrac{1 + \sqrt{33}}{2}$ into the equation $y = x^2$ to get x:

$$\frac{1 + \sqrt{33}}{2} = x^2$$

$$\pm\sqrt{\frac{1 + \sqrt{33}}{2}} = x$$

The value of x comes out to about ± 1.84. The graph in Figure 13-7 shows you the parabola, the circle, and the points of intersection at about $(1.84, 3.37)$ and about $(-1.84, 3.37)$.

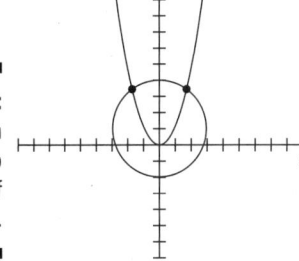

When $y = -2.37$, you get points that lie on the circle, but these points don't fall on the parabola. The algebra shows that, and the picture agrees.

When substituting into one of the original equations to solve for the other variable, always substitute into the *simpler* equation — the one with smaller exponents. This helps you catch any extraneous solutions.

Planning Your Attack on Other Systems of Equations

I deal with intersections of lines and different conic sections first in this chapter because the curves of the conics are easy to visualize and the results of the intersections are somewhat predictable. However, you can also find the intersections (common solutions) of other functions and curves; you may just have to start out with no clue as to what's going to happen. Not to worry, though; using the correct processes — substitution and equation solving (see Chapter 12) — assures you of honest answers and results.

You can deal with a system of linear equations in many ways. (I cover the methods for solving systems of linear equations in great detail in Chapter 12.) A system of equations that contains one or more polynomial (nonlinear) functions, however, presents fewer options for finding the solutions. Throw in a rational function or exponential function, and the plot thickens. But as long as the equations cooperate, the different algebraic methods — elimination and substitution — will work. Lucky you!

Cooperative equations are the ones I discuss in this book. When systems of equations defy algebraic methods, you have to turn to calculators, computers, and high-powered college mathematics courses. In the meantime, you can concentrate on the nicely defined, manageable systems I present in this section for nonlinear entities.

Mixing polynomials and lines

A *polynomial* is a continuous, smooth curve. (Chapter 8 gives you plenty of information on the behavior of polynomial curves and how to graph them.) The more the curve of a polynomial changes direction and moves up and down across a graph, the more opportunities a line has to cross it. For instance, the line $y = 3x + 21$ intersects the polynomial $y = -x^3 + 5x^2 + 20x$ three times. Figure 13-8 shows you the graphs of the line, the polynomial, and the points of intersection.

Figure 13-8:
A line crossing the curves of a polynomial.

Substitution that replaces the y in the polynomial with the equivalence of y in the line is usually your most effective route for solving for the common solutions of the intersections. (You apply this method when the solutions are cooperative integers. If the common points involve fractions and radicals, technology is the only way to go.)

To solve for the solutions of this system of equations by using substitution, you start by replacing the y in the polynomial with the equivalence of y in the line and setting the equation equal to zero:

$$y = -x^3 + 5x^2 + 20x$$
$$3x + 21 = -x^3 + 5x^2 + 20x$$
$$0 = -x^3 + 5x^2 + 17x - 21$$

The simplified equation factors into $0 = -(x + 3)(x - 1)(x - 7)$. (If you need help with this factorization, refer to the Rational Root Theorem and synthetic division information in Chapter 8.) The zeros, or solutions, of the equation are $x = -3$, 1, and 7. You solve for the y values of the intersection points by plugging these x values into the equation of the line. The points of intersection that you find are $(-3, 12)$, $(1, 24)$, and $(7, 42)$.

Always use the equation with the smaller exponents when you solve for the full solution. The substitution is easier with smaller exponents, and, more importantly, you won't end up with extraneous solutions — nonexistent answers for the problem (for more on imaginary numbers, see Chapter 14).

Crossing polynomials

"Crossing polynomials" almost sounds like you're doing a genetics experiment and creating a new, hybrid curve — a nonlinear monster of sorts. But before I start my sinister laugh, I must admit that crossing polynomials is bloomin' wonderful stuff. (Hey, beauty is in the eye of the beholder.)

Just to give you an example of how intersecting polynomials can provide several solutions, I've chosen a quartic and a cubic to cross-pollinate (oops, I mean intersect). A fourth-degree (quartic) polynomial (the power 4 is the highest power) and a third-degree (cubic) polynomial can share as many as four common solutions. The graphs of $y = x^4 + 2x^3 - 13x^2 - 14x + 24$ and $y = x^3 + 8x^2 - 13x + 4$, for example, are shown in Figure 13-9. The W-shaped curve is the fourth-degree polynomial, and the sideways S is the third-degree polynomial.

Crossing music and chemistry

If you took a poll for the most unlikely couple when it comes to professions or interests, music and chemistry would be high on the list. It may surprise you, then, that these topics have an amazing similarity. Music is made up of octaves of tones — eight tones, running from *do, re, mi* up to *do* (are you humming the song from *The Sound of Music?*) and then starting all over again. In 1869, Dimitri Ivanovich Mendeléev found that if he arranged the elements in chemistry in order of increasing atomic weight, every eighth element (starting from a given element) had chemical properties similar to the first one. He used his discovery to predict that elements existed that scientists hadn't yet discovered. What he predicted was the existence of *blanks,* or missing elements in the pattern. Mathematics — counting to eight and starting over again — guided Pythagoras with the musical scale and Mendeléev with the atomic chart.

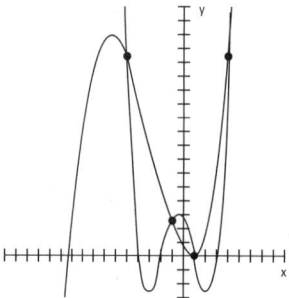

Figure 13-9: Counting the intersections of quartic and cubic polynomials.

To solve systems of equations containing two polynomials, you use the substitution method (see Chapter 12). Set *y* equal to *y,* move all the terms to the left, and simplify:

$$x^4 + 2x^3 - 13x^2 - 14x + 24 = x^3 + 8x^2 - 13x + 4$$
$$x^4 + x^3 - 21x^2 - x + 20 = 0$$

This equation factors into $(x + 5)(x + 1)(x - 1)(x - 4) = 0$ (see Chapter 8), which gives you the solutions $x = -5, -1, 1,$ and 4. Substituting these values into the cubic (third-degree) equation (you should always substitute into the equation with the smallest exponential values), you get $y = 144$ when $x = -5$, $y = 24$ when $x = -1$, $y = 0$ when $x = 1$, and $y = 144$ when $x = 4$. You now have all the points of intersection.

Navigating exponential intersections

Exponential functions are flattened, C-shaped curves when graphed on a grid (I cover exponentials in Chapter 10). When exponentials intersect with one another, they usually do so in only one place, creating one common solution. Mixing exponential curves with other types of curves produces results similar to those you see when mixing lines and parabolas — you may get more than one solution.

Visualizing exponential solutions

The exponential functions $y = 5^x$ and $y = 3^x$ have one common solution: They both cross the Y-axis at the point $(0, 1)$. If you let $x = 0$ in $y = 5^x$, you get $y = 5^0 = 1$. Anything to the zero power equals one. So, that means substituting 0 for x in $y = 3^x$ gives you $y = 3^0 = 1$. You get the same number for both equations. You know $(0, 1)$ is the only possible solution for the two exponential functions because any other power of 5 and 3 won't give you the same number. The numbers 3 and 5 are both prime numbers, and raising them to powers won't create any common solutions. You can discover the solution through algebra, through reasoning it out, or by looking at the graphs of the equations. The graphs of the exponential functions $y = 5^x$ and $y = 3^x$ are shown in Figure 13-10a. The steeper of the two exponential curves is $y = 5^x$.

Figure 13-10:
Graphing
systems
containing
exponential
equations.

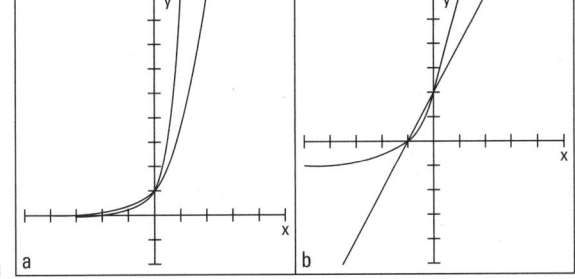

Figure 13-10b shows the intersections of the line $y = 2x + 2$ and the exponential function $y = 2^{2x+1} + 2^x - 1$. Due to the complexities of the exponential function, it isn't practical to solve this system of equations algebraically. It really takes some help from technology. But, if you can determine the solutions by looking at the graphs, take advantage of the situation. The two solutions that the line and exponential function appear to have in common — where they intersect — are at $(-1, 0)$ and $(0, 2)$, assuming that each notch on each axis moves one unit at a time (you expect only two solutions because the line keeps going on in one direction, and the exponential function just keeps growing, like exponential functions do, and doesn't double back on itself). You can check to be sure that these answers are correct by substituting the x values into the equations to see if you get the correct y values:

When $x = -1$ in the equation of the line, $y = 2(-1) + 2 = -2 + 2 = 0$.

When $x = -1$ in the equation of the exponential, you get the following:

$$y = 2^{2(-1)+1} + 2^{-1} - 1$$
$$= 2^{-2+1} + 2^{-1} - 1$$
$$= 2^{-1} + 2^{-1} - 1$$
$$= \frac{1}{2} + \frac{1}{2} - 1$$
$$= 0$$

The line and the exponential have the common solution $(-1, 0)$.

When $x = 0$ in the equation of the line, $y = 2(0) + 2 = 0 + 2 = 2$.

When $x = 0$ in the equation of the exponential, you get the following:

$$y = 2^{2(0)+1} + 2^0 - 1$$
$$= 2^1 + 1 - 1$$
$$= 2$$

The line and the exponential have the common solution $(0, 2)$.

Solving for exponential solutions

You can solve some systems that contain exponential functions by using algebraic techniques. What should you look for? When the bases of the exponential functions are the same number or are powers of the same number, an algebraically found solution is possible.

For instance, you can solve the system $y = 4^x$ and $y = 2^{x+1}$ algebraically because the base 4 in the first equation is a power of 2, the base in the second equation. You can write the number 4 as 2^2. (If you need a refresher on dealing with exponential equations, head to Chapter 10.)

To solve a system of exponential functions when the bases of the functions are the same number or are powers of the same number, you set the two y values equal to one another, set the exponents equal to one another, and solve for x. You change it from the exponential form by setting those exponents equal to one another and discarding the bases.

Setting the two y values equal to one another for the previous example, you have

$$4^x = 2^{x+1}$$
$$\left(2^2\right)^x = 2^{x+1}$$
$$2^{2x} = 2^{x+1}$$

Now you can set the exponents equal to one another. The solution of $2x = x + 1$ is $x = 1$. When $x = 1$, $y = 4$ in both equations. Figure 13-11 shows the graphs of the intersecting equations.

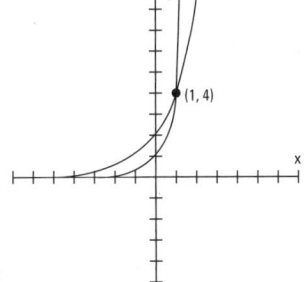

Figure 13-11:
Two exponential functions intersecting at (1, 4).

Rounding up rational functions

A *rational function* is a fraction that contains a polynomial expression in both its numerator and denominator. A polynomial has one or more terms that have whole-number exponents, so a rational function has all whole-number exponents — just in fractional form. The graph of a rational function typically has vertical and/or horizontal asymptotes that reveal its shape. Also, rational functions usually have pieces of hyperbolas in their graphs. (You can find plenty of information on rational functions in Chapter 9.)

Solving and graphing systems of equations that include rational functions means dealing with fractions — every student and teacher's favorite task. No worries, though. I prepare you in the following sections.

Intersections of a rational function and a line

The rational function $y = \dfrac{x-1}{x+2}$ and the line $3x + 4y = 7$ intersect at two points, meaning they have two common solutions. You can see that from their graphs in Figure 13-12. But don't confuse the intersections of the line with the asymptotes of the rational function as parts of the solution. You consider only the intersections with the curves of the rational function. The algebraic solution you find also confirms that you use only the points on the curve.

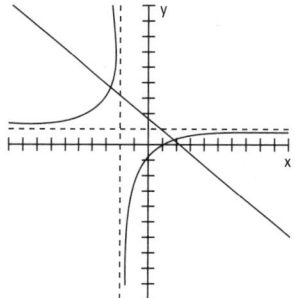

Figure 13-12:
A line
crossing a
rational
function,
forming two
solutions.

To solve this system of equations, you solve the equation of the line for y, and then you substitute this equivalence for y into the equation of the rational function:

$$3x + 4y = 7$$
$$4y = 7 - 3x$$
$$y = \frac{7}{4} - \frac{3}{4}x$$
$$\frac{7}{4} - \frac{3}{4}x = \frac{x-1}{x+2}$$

The equation that remains looks like a bit of a mess, doesn't it? You can make the equation look much nicer by multiplying each side by $4(x + 2)$, the common denominator of the fractions in the equation:

$$4(x+2)\left(\frac{7}{4} - \frac{3}{4}x\right) = \left(\frac{x-1}{x+2}\right)4(x+2)$$
$$\cancel{4}(x+2)\frac{7}{\cancel{4}} - \cancel{4}(x+2)\frac{3}{\cancel{4}}x = \left(\frac{x-1}{\cancel{x+2}}\right)4\,\cancel{(x+2)}$$
$$7(x+2) - 3x(x+2) = 4(x-1)$$

Now you can simplify the resulting equation by distributing and combining like terms (see Chapter 1 to find out how) and setting the whole equation equal to zero. The result is a quadratic equation that you can factor (see Chapter 3):

$$7(x+2) - 3x(x+2) = 4(x-1)$$
$$7x + 14 - 3x^2 - 6x = 4x - 4$$
$$x + 14 - 3x^2 = 4x - 4$$
$$0 = 3x^2 + 3x - 18$$
$$0 = 3(x^2 + x - 6)$$
$$0 = 3(x+3)(x-2)$$

When you change the form of an equation that contains fractions, radicals, or exponentials, you have to be cautious about *extraneous solutions* — answers that satisfy the new form but not the original. Always check your work by substituting your answers into the original equation.

The solutions of the quadratic equation are $x = -3$ and $x = 2$. You now substitute these values into the rational function to check your work. When $x = -3$, you get $y = 4$. When $x = 2$, you get $y = \frac{1}{4}$. These values represent the common solutions (coordinates of intersection) of the rational function and the line (you can see them in Figure 13-12).

Discovering that functions are inverses when solving a system

With a little algebra and graphing, you'll notice something special about the two rational functions in the following system:

$$\begin{cases} y = \dfrac{3x-4}{x-2} \\ y = \dfrac{2x-4}{x-3} \end{cases}$$

When you solve the system by identifying the solutions, you notice something that may or may not be a coincidence. A graph helps confirm that your discovery isn't a coincidence. Okay, enough with the suspense. To solve the system, you set the two fractions equal to one another — the same as substituting the y equivalences for one another:

$$\frac{3x-4}{x-2} = \frac{2x-4}{x-3}$$

The equation you create is a *proportion,* meaning that you have two different ratios (fractions) set equal to one another. Cross-multiply and simplify the products, and then move every term over to the left to set the equation equal to zero:

$$(3x-4)(x-3) = (2x-4)(x-2)$$
$$3x^2 - 13x + 12 = 2x^2 - 8x + 8$$
$$x^2 - 5x + 4 = 0$$

The quadratic equation factors into $(x-4)(x-1) = 0$, so the two solutions are $x = 4$ and $x = 1$. You substitute those values into either of the original equations. You find that when $x = 4$, $y = 4$, and when $x = 1$, $y = 1$. So, back to the big mystery. Do you notice anything special about these values? Yep, the x and y values are the same, because they both lie on the line $y = x$. This phenomenon happens because the two rational functions are *inverses* of one another. (You can find more on function inverses in Chapter 6.)

The special characteristic about the graphs of functions and their inverses is that they're always symmetric with respect to one another over the line $y = x$. Also, if they intersect that line of symmetry, they intersect in the same place, making for a rather artsy picture. Figure 13-13 shows the two rational functions I previously introduce and how they're symmetric about the line $y = x$. (I leave out the lines showing the asymptotes so the picture doesn't get too cluttered.)

Figure 13-13:
Two inverse rational functions reflected over the line $y = x$.

Playing Fair with Inequalities

Systems of inequalities appear in applications used for business ventures *and* calculus problems. A system of inequalities, for instance, can represent a set of constraints in a problem that involves production of some item — the *constraints* put limits on the resources being used or the time available. In calculus, systems of inequalities represent areas between curves that you need to compute. Graphically, the solutions of systems of inequalities appear as shaded areas between curves. This gives you a visual solution and helps you determine values of x and y that work.

You find so many answers for systems of inequalities — infinitely many solutions — that you can't list them all; you just give rules in terms of the inequality statements. Algebraically, the solutions are statements that involve inequalities — telling what x or y is bigger than or smaller than. Often, the graph of a system gives you more information than the listing of the inequalities shown in an algebraic solution. You can *see* that all the points in the solution lie above a certain line, so you pick numbers that work in the system based on what you see.

Drawing and quartering inequalities

The simplest inequality to graph and solve is one that falls above or below a horizontal line or to the right or left of a vertical line. A system of inequalities that involves two such lines (one vertical and one horizontal) has a graph that appears as one quarter of the plane. For instance, the graph of the system of inequalities $\begin{cases} x \geq 2 \\ y \leq 3 \end{cases}$ is shown in Figure 13-14.

Figure 13-14: Two inequalities intersecting to share one quarter of the plane (the heavy shading).

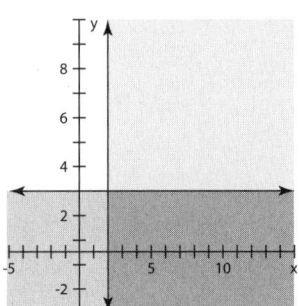

Everything to the right of the line $x = 2$ represents the graph of $x \geq 2$, and everything below the line $y = 3$ represents the graph of $y \leq 3$. Their intersection, the heavily shaded area in the lower-right quadrant formed by the intersecting lines, consists of all the points in that shaded area. You can find an infinite number of solutions. Some examples are the points $(3, 1)$, $(4, 2)$, and $(2, -1)$. You could never list all the answers.

Graphing areas with curves and lines

You find the solution of a system of inequalities involving a line and a curve (such as a parabola), two curves, or any other such combination by graphing the individual equations, determining which side to shade for each curve, and identifying where the equations share the shading.

To solve the system $\begin{cases} x \geq y^2 + 2y - 3 \\ y \geq x - 3 \end{cases}$, for example, make your way through the following steps:

1. **Graph the line $y = x - 3$ and determine which side of the line to shade by checking a *test point* (a random point that's clearly on one side or the other) to see if it satisfies the inequality. (For info on graphing lines, see Chapters 2 and 5.)**

   If the point satisfies the inequality, you shade that side. In this case, you can use the test point $(0, 0)$, which is clearly above and to the left of the line. When you put the coordinate $(0, 0)$ in the inequality $y \geq x - 3$, you get $0 \geq 0 - 3$. Yes, 0 is greater than -3, so the test point is in the area that you need to shade — above the line.

2. **Graph the parabola $x = y^2 + 2y - 3$ and use a test point to see whether you need to shade inside the parabola or outside the parabola. (For info on graphing parabolas, see Chapters 5 and 7.)**

   Again, the point $(0, 0)$ is handy. Test the point in the inequality for the parabola, $x \geq y^2 + 2y - 3$. You find $0 \geq 0 + 0 - 3$, which is true. So, the point $(0, 0)$ falls in the area you need to shade — inside the parabola.

3. **Determine where the two shaded areas overlap to find your solution to the system of inequalities.**

   The two shaded areas overlap where the inside of the parabola and area above the line intersect.

Figure 13-15 shows the line and parabola corresponding to the inequalities. The shaded area indicates the solution — where the two inequalities overlap.

Figure 13-15:
A parabola
and line
outline a
solution
wedge
for the
inequalities.

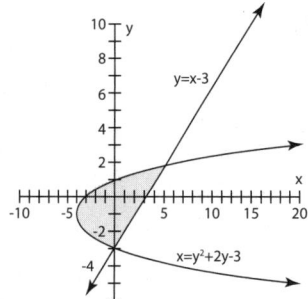

Part IV
Shifting into High Gear with Advanced Concepts

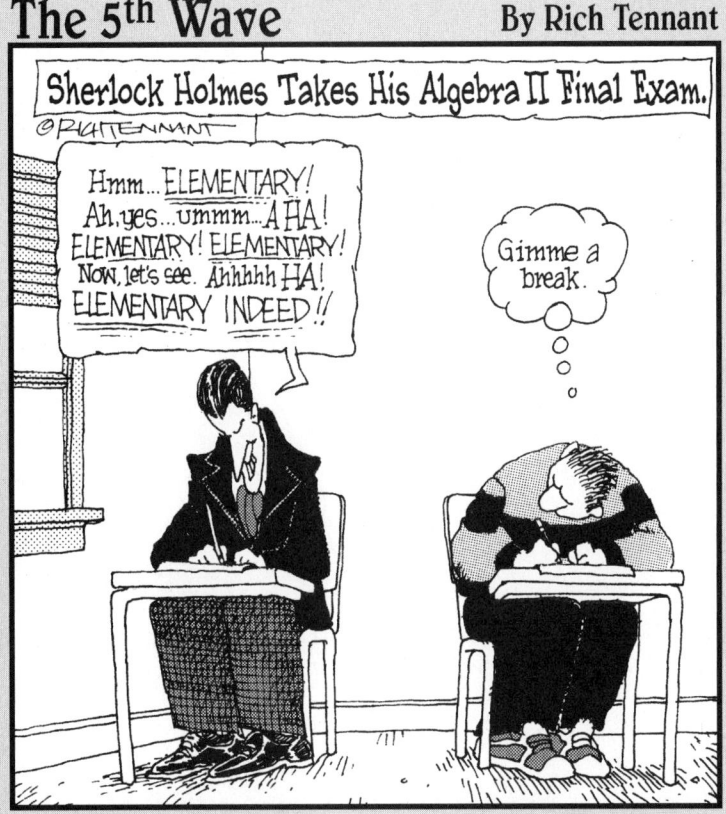

Sherlock Holmes Takes His Algebra II Final Exam.

Hmm... ELEMENTARY!
Ah, yes... ummm... A HA!
ELEMENTARY! ELEMENTARY!
Now, let's see. Ahhhh HA!
ELEMENTARY INDEED!!

Gimme a break.

In this part . . .

The chapters in Part IV appear very different, and they are on the surface, but what they share are the same basic algebraic operations and ideas. Matrices look a lot different from single x variables, but the arithmetic you apply to them is familiar, and you use algebra to work on their properties. Sequences and series are lists of numbers and sums of lists of numbers, and you use some great algebra when discussing both lists and sums. Sets and set notation introduce some more interesting notation, and sets are a basic premise of the study of probability and statistics.

So, as you can see, Part IV is rather eclectic. Think of it as sort of a blended family — you see ties from one member to another, but the members have their own roots, too. As an added bonus, the material in this chapter leaves you well prepared for more advanced math studies such as calculus, finite math, and statistics.

Chapter 14

Simplifying Complex Numbers in a Complex World

*I*maginary numbers are the results of mathematicians' imaginations. No, imaginary numbers aren't real — although, is any number real? Can you touch a number (if you don't count playing with a toddler's educational toy set)? Can you feel it? Who decided a 9 should be shaped the way it is, and what makes that person right? Does your brain hurt yet?

Mathematicians define *real numbers* as all the whole numbers, negative and positive numbers, fractions and decimals, radicals — anything you can think of to use in counting, graphing, and comparing amounts. Mathematicians introduced imaginary numbers when they couldn't finish problems without them. For example, when solving for roots of quadratic equations such as $x^2 + x + 4 = 0$, you quickly discover that you can find no real answers. Using the quadratic formula (see Chapter 3), the solutions come out to be

$$x = \frac{-1 \pm \sqrt{1^2 - 4(1)(4)}}{2(1)} = \frac{-1 \pm \sqrt{-15}}{2}$$

The answers seem sort of final — like you have nowhere to go. But you can't find the square root of a negative number, because no real number multiplies itself and ends up with a negative result. So, rather than staying stuck there, with no final solution, mathematicians came up with a new rule: They let $i^2 = -1$. They made up a number to replace $\sqrt{-1}$, and they called it i (didn't take much *imagination* to come up with that). When you take the square root of each side, you get $\sqrt{i^2} = \sqrt{-1}$, $i = \sqrt{-1}$. In this format, you can complete the problems and write the answers with i's in them.

In this chapter, you find out how to create, work with, and analyze imaginary numbers and the complex expressions they appear in. Just remember to use your imagination!

Using Your Imagination to Simplify Powers of i

The powers of x (representing real numbers) — x^2, x^3, x^4, and so on — follow the rules of exponents, such as adding exponents when multiplying the powers together or subtracting exponents when dividing them (see Chapter 1). The powers of i (representing imaginary numbers) are rule-followers, too. The powers of i, however, have some neat features that set them apart from other numbers.

You can write all the powers of i as one of four different numbers: $i, -i, 1$, and -1; all it takes is some simplifying of products, using the properties of exponents, to rewrite the powers of i:

- ✔ $i = i$: Just plain old i

- ✔ $i^2 = -1$: From the definition of an imaginary number (see the introduction to this chapter)

- ✔ $i^3 = -i$: Use the rule for exponents — $i^3 = i^2 \cdot i$ — and then replace i^2 with -1; so, $i^3 = (-1) \cdot i = -i$

- ✔ $i^4 = 1$: Because $i^4 = i^2 \cdot i^2 = (-1)(-1) = 1$

- ✔ $i^5 = i$: Because $i^5 = i^4 \cdot i = (1)(i) = i$

- ✔ $i^6 = -1$: Because $i^6 = i^4 \cdot i^2 = (1)(-1) = -1$

- ✔ $i^7 = -i$: Because $i^7 = i^4 \cdot i^2 \cdot i = (1)(-1)(i) = -i$

- ✔ $i^8 = 1$: Because $i^8 = i^4 \cdot i^4 = (1)(1) = 1$

Consider the two powers of i presented in the following list and how you determine the rewritten values; if you want to find a power of i, you make use of the rules for exponents and the first four powers of i:

- ✔ $i^{41} = i$: Because $i^{41} = i^{40} \cdot i = (i^4)^{10}(i) = (1)^{10} \cdot i = 1 \cdot i = i$

- ✔ $i^{935} = -i$: Because $i^{935} = i^{932} \cdot i^3 = (i^4)^{233}(i^3) = (1)^{233}(-i) = 1(-i) = -i$

The process of changing the powers of i seems like a lot of work — plus you need to figure out what to multiply four by to get a high power (you want to find a multiple of four — the biggest possible value that's smaller than the exponent). But you really don't need to go through all the raising of powers if you recognize a particular pattern in the powers of i.

Every power of i that's a multiple of four is equal to one. If the power is one value greater than a multiple of four, the power of i is equal to i. And so the process goes. Here's the list in full:

$$i^{4n} = 1$$

$$i^{4n+1} = i$$

$$i^{4n+2} = -1$$

$$i^{4n+3} = -i$$

So, all you need to do to change the powers of i is figure out where a power of i is in relation to some multiple of four. If you need the value of $i^{5,001}$, for example, you know that 5,000 is a multiple of 4 (because it ends in 00), and 5,001 is 1 value greater than 5,000, so $i^{5,001} = i$. It really simplifies expressions when you have a rule that reduces the powers of numbers.

Understanding the Complexity of Complex Numbers

An imaginary number, i, is a part of the numbers called *complex numbers,* which arose after mathematicians established imaginary numbers. The standard form of complex numbers is $a + bi$, where a and b are real numbers, and i^2 is –1. The fact that i^2 is equal to –1 and i is equal to $\sqrt{-1}$ is the foundation of the complex numbers. If you didn't have the imaginary numbers with the i's, you'd have no need for complex numbers with imaginaries in them.

Some examples of complex numbers include $3 + 2i$, $-6 + 4.45i$, and $7i$. In the last number, $7i$, the value of a is zero. If the value of b is zero as well, you no longer have a complex number — you have a real number without the imaginary part.

So, a is the *real* part of a complex number and the bi is the *complex* part (even though b is a real number). Is that complex enough for you?

Complex numbers have many applications, and mathematicians study them extensively. In fact, whole math courses and fields of study are devoted to complex numbers. And, imagine this, you get a glimpse of this ethereal world right here and now in this section.

Operating on complex numbers

You can add, subtract, multiply, and divide complex numbers — in a very careful manner. The rules used to perform operations on complex numbers look very much like the rules used for any algebraic expression, with two big exceptions:

✔ You simplify the powers of i, change them to their equivalent in the first four powers of i (see the section "Using Your Imagination to Simplify Powers of i" earlier in the chapter), and then combine like terms.

✔ You don't really divide complex numbers; you multiply by the conjugate (I tell you all about this in the section "Multiplying by the conjugate to perform division" later in the chapter).

Adding complex numbers

When you add two complex numbers $a + bi$ and $c + di$, you get the sum of the real parts and the sum of the imaginary parts:

$$(a + bi) + (c + di) = (a + c) + (b + d)i$$

The result of the addition is now in the form of a complex number, where $a + c$ is the real part and $(b + d)i$ is the imaginary part.

When you add $(-4 + 5i) + (3 + 2i)$, for example, you get $(-4 + 3) + (5 + 2)i$ $= -1 + 7i$.

Subtracting complex numbers

When you subtract the complex numbers $a + bi$ and $c + di$, you get the difference of the real parts and the difference of the imaginary parts:

$$(a + bi) - (c + di) = (a - c) + (b - d)i$$

The result of the subtraction is now in the form of a complex number, where $a - c$ is the real part and $(b - d)i$ is the imaginary part.

When you subtract $(-4 + 5i) - (3 + 2i)$, for example, you get $(-4 - 3) + (5 - 2)i$ $= -7 + 3i$.

Multiplying complex numbers

When you multiply complex numbers, the operations get a little more exciting.

To multiply complex numbers, you can't just multiply the real parts together and the complex parts together; you have to treat the numbers like binomials and distribute both the terms of one complex number over the other. Another way to look at it is that you have to FOIL the terms (for the details on FOIL, refer to Chapter 1):

$$(a + bi)(c + di) = (ac - bd) + (ad + bc)i$$

The result of the multiplication shown here is in the form of a complex number, with $ac - bd$ as the real part and $(ad + bc)i$ as the imaginary part. You can see from the following distribution where the values in this rule come from:

$$(a + bi)(c + di) = ac + adi + bci + bdi^2$$
$$= ac + (ad + bc)i + bd(-1)$$
$$= ac - bd + (ad + bc)i$$

In the previous equations, the product of the two first terms is $ac;$ the product of the two outer terms is $adi;$ the product of the two inner terms is $bci;$ and the product of the two last terms is bdi^2. You factor out the i in the second and third terms, replace the i^2 with -1 (see the first section of this chapter), and then combine that term with the other real term.

To find the product of $(-4 + 5i)(3 + 2i)$, for example, you FOIL to get $-12 - 8i + 15i + 10i^2$. You simplify the last term to -10 and combine it with the first term. Your result is $-22 + 7i$, a complex number.

You don't really have to memorize the form of the rule. You can just as easily FOIL the complex binomials and simplify the terms.

Multiplying by the conjugate to perform division

The complex thing about dividing complex numbers is that you don't really divide. Do you remember when you first found out how to multiply and divide fractions? You never really *divided* fractions; you changed the second fraction to its reciprocal, and then you changed the problem to multiplication. You found that the answer to the multiplication problem is the same as the answer to the original division problem. You avoid division in much the same way with complex numbers. You do a multiplication problem — one that has the same answer as the division problem. But before you tackle the "division," you should know more about the *conjugate* of a complex number.

Defining the conjugate

A complex number and its *conjugate* have opposite signs between the two terms. The conjugate of the complex number $a + bi$ is $a - bi$, for instance.

Here's a couple numerical examples: The conjugate of $-3 + 2i$ is $-3 - 2i$, and the conjugate of $5 - 3i$ is $5 + 3i$. Seems simple enough, because you don't see the special trait attributed to the conjugate of a complex number until you multiply the complex number and its conjugate together.

The product of an imaginary number and its conjugate is a real number (no imaginary part) and takes the following form:

$$(a + bi)(a - bi) = a^2 + b^2$$

Here's the product of the complex number and its conjugate, all worked out by using FOIL (see Chapter 1): $(a + bi)(a - bi) = a^2 - abi + abi - b^2i^2 = a^2 - b^2(-1) = a^2 + b^2$. The middle terms are opposites of one another, and $i^2 = -1$ (the definition of an imaginary number) gets rid of the i^2 factor.

Using conjugates to divide complex numbers

When a problem calls for you to divide one complex number by another, you write the problem as a fraction and then multiply by the number one. You don't actually multiply by one; you multiply by a fraction that has the conjugate of the denominator in both numerator and denominator (because the same value appears in the numerator and denominator, the fraction is equal to one):

$$(a + bi) \div (c + di) = \frac{a + bi}{c + di} \cdot \frac{c - di}{c - di}$$
$$= \frac{(ac + bd) + (bc - ad)i}{c^2 + d^2}$$

To write the result of the division of complex numbers in a strictly real-and-imaginary-parts format, you break up the fraction:

$$\frac{(ac + bd)}{c^2 + d^2} + \frac{(bc - ad)}{c^2 + d^2}i$$

The form of the fraction that results looks awfully complicated, and you don't really want to have to memorize it. When doing a complex division problem, you can use the same process I use to get the previous form for dividing complex numbers. To divide $(-4 + 5i)$ by $(3 + 2i)$, for example, perform the following steps:

$$\frac{-4 + 5i}{3 + 2i} \cdot \frac{3 - 2i}{3 - 2i} = \frac{-12 + 8i + 15i - 10i^2}{3^2 + 2^2}$$
$$= \frac{-12 + (8 + 15)i - 10(-1)}{9 + 4}$$
$$= \frac{-12 + 10 + (8 + 15)i}{13}$$
$$= \frac{-2 + 23i}{13} = -\frac{2}{13} + \frac{23}{13}i$$

Divide, average, and conquer square roots

You can use one of several methods to find the square root of a number to a specified amount of decimal places. The main methods mathematicians used before hand-held calculators involved a pencil and a piece of paper (or, before that, feather quill and ink on birch bark). One of the easiest methods to remember is the *divide/average* method. For instance, if you want to find the square root of 56 to two decimal places, you make a guess and then divide 56 by your guess. For example, I'll guess 7 as the square root: $56 \div 7 = 8$.

I now *average* my guess, the 7, and the answer to the division problem, 8. The average of 7 and 8 (add together and divide by two) is 7.5. I now use 7.5 as my next guess: $56 \div 7.5 = 7.467$, rounded to three places (one more place than my target). Averaging 7.5 and 7.467, I get 7.484. I divide 56 by 7.484 and get 7.483. The numbers 7.484 and 7.483 round to the same number, 7.48, so you conclude that 7.48 is the square root of 56, correct to two decimal places (no other number with two decimal places has a square closer to 56 than this one). The simple divide/average method is very accurate and easy to remember!

Simplifying radicals

Until mathematicians defined imaginary numbers, many problems had no answers because the answers involved square roots of negative numbers, or *radicals*. After the definition of an imaginary number, $i^2 = -1$, came into being, doors opened; windows flung wide; parades were held; children danced in the streets; and problems were solved. Eureka!

To simplify the square root of a negative number, you write the square root as the product of square roots and simplify: $\sqrt{-a} = \sqrt{-1}\sqrt{a} = i\sqrt{a}$.

If you want to simplify $\sqrt{-24}$, for example, you first split up the radical into the square root of −1 and the square root of the rest of the number, and then you do any simplifying by factoring out perfect squares:

$$
\begin{aligned}
\sqrt{-24} &= \sqrt{-1}\sqrt{24} \\
&= \sqrt{-1}\sqrt{4}\sqrt{6} \\
&= i \cdot 2\sqrt{6}
\end{aligned}
$$

By convention, you write the previous solution as $2i\sqrt{6}$. This form isn't at all different in value from the other form. Most mathematicians just want to put the numerical part of the coefficient first, the variables or other letters next (in alphabetical order), and the radicals last. They think it looks better that way.

Technically, in complex form, you write a number with the i at the end, after all the other numbers — even after the radical. If you write the number this way, be sure not to tuck the i under the radical — keep it clearly to the right. For instance, $\sqrt{6} \cdot i \neq \sqrt{6i}$. To avoid confusion, write it as $i\sqrt{6}$.

Solving Quadratic Equations with Complex Solutions

You can always solve quadratic equations with the quadratic formula. It may be easier to solve quadratic equations by factoring, but when you can't factor, the formula comes in handy. (Refer to Chapter 3 if you need a refresher on the quadratic formula.)

Until mathematicians began recognizing imaginary numbers, however, they couldn't complete many results of the quadratic formula. Whenever a negative value appeared under a radical, the equation stumped the mathematicians.

The modern world of imaginary numbers, to the rescue! To solve the quadratic equation $2x^2 + x + 8 = 0$, for example, you can use the quadratic formula to get the following:

$$x = \frac{-1 \pm \sqrt{1^2 - 4(2)(8)}}{2(2)}$$
$$= \frac{-1 \pm \sqrt{1 - 64}}{4}$$
$$= \frac{-1 \pm \sqrt{-63}}{4}$$
$$= \frac{-1 \pm \sqrt{-1}\sqrt{9}\sqrt{7}}{4}$$
$$= \frac{-1 \pm 3i\sqrt{7}}{4}$$

Along with the quadratic formula that produces a complex solution, it helps to look at the curve that corresponds to the equation on a graph (see Chapter 7).

A parabola that opens upward or downward always has a y-intercept; these parabolas are functions and have domains that contain all real numbers. However, a parabola that opens upward or downward doesn't necessarily have any x-intercepts.

To solve a quadratic function for its x-intercepts, you set the equation equal to zero. When no real solution for the equation exists, you find no x-intercepts.

You still have a y-intercept, because all quadratic functions cross the Y-axis somewhere, but its graph may stay above or below the X-axis without crossing it.

The quadratic equation $2x^2 + x + 8 = 0$ corresponds to the parabola $2x^2 + x + 8 = y$. Substituting zero for the y variable allows you to solve for the x-intercepts of the parabola. The fact that you find no real solutions for the equation tells you that the parabola has no x-intercepts. (Figure 14-1 shows you a graph of this situation.)

Figure 14-1: A parabola with no real solutions for the x-intercepts never crosses the X-axis.

Parabolas that open to the left or right aren't functions. A function has only one y-value for every x-value, so these curves violate that requirement. A parabola opening to the left or right may never cross the Y-axis if its vertex is to the left or right of that axis, so you find no real solution for the y-intercept. A parabola that opens to the left or right always has an x-intercept, however.

To solve for any y-intercept, you set x equal to zero and then solve the quadratic equation. If the equation has no solution, it has no y-intercepts. Solving the equation $x = -y^2 + 6y - 12$ for its y-intercept, for example, you let $x = 0$. The equation doesn't factor, so you use the quadratic formula:

$$
\begin{aligned}
y &= \frac{-6 \pm \sqrt{36 - 4\,(-1)(-12)}}{2\,(-1)} \\
&= \frac{-6 \pm \sqrt{36 - 48}}{-2} \\
&= \frac{-6 \pm \sqrt{-12}}{-2} \\
&= \frac{-6 \pm 2i\sqrt{3}}{-2} = \frac{-3 \pm i\sqrt{3}}{-1} = 3 \pm i\sqrt{3}
\end{aligned}
$$

The parabola $x = -y^2 + 6y - 12$ opens to the left and never crosses the Y-axis. The graph of this parabola is shown in Figure 14-2.

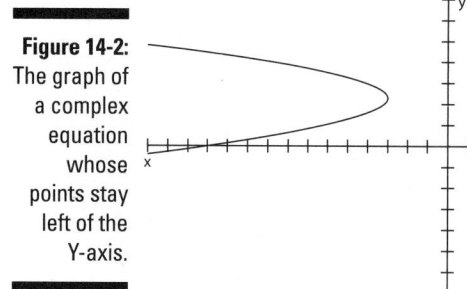

Figure 14-2:
The graph of a complex equation whose points stay left of the Y-axis.

Working Polynomials with Complex Solutions

Polynomials are functions whose graphs are nice, smooth curves that may or may not cross the X-axis. If the degree (or highest power) of a polynomial is an odd number, its graph must cross the X-axis, and it must have a real root or solution. (In Chapter 8, you discover how to find the real solutions or roots of polynomial equations.) When solving equations formed by setting polynomials equal to zero, you plan ahead as to how many solutions you can expect to find. The highest power tells you the maximum number of solutions you can find. In the case of equations where the highest powers are even numbers, you may find that the equations have absolutely no solutions, whereas if the highest powers are odd, you're guaranteed at least one solution in each equation.

Identifying conjugate pairs

A polynomial of degree (or power) n can have as many as n real zeros (also known as solutions or x-intercepts). If the polynomial doesn't have n real zeros, it has $n - 2$ zeros, $n - 4$ zeros, or some number of zeros decreased two at a time. (Refer to Chapter 8 for how to count the number of zeros by using Descartes' Rule of Sign.) The reason that the number of zeros decreases by two is that complex zeros always come in *conjugate pairs* — a complex number and its conjugate.

Complex zeros, or solutions of polynomials, come in *conjugate pairs* — $a + bi$ and $a - bi$. The product of a complex number and its conjugate, $(a + bi)(a - bi)$, is $a^2 + b^2$, so you don't see any i's in the equation of the polynomial.

The equation $0 = x^5 - x^4 + 14x^3 - 16x^2 - 32x$, for example, has three real roots and two complex roots, which you know because you apply the Rational Root Theorem and Descartes' Rule of Sign (from Chapter 8) and ferret out those real and complex solutions. The equation factors into $0 = x(x-2)(x+1)(x^2+16)$. The three real zeros are 0, 2, and –1 for the solutions of $x = 0$, $x - 2 = 0$, and $x + 1 = 0$. The two complex zeros are $4i$ and $-4i$. You say that the two complex zeros are a *conjugate pair,* and you get the roots by solving the equation $x^2 + 16 = 0$.

A more typical conjugate pair has both real and imaginary parts to the numbers. For instance, the equation $0 = x^4 + 6x^3 + 9x^2 - 6x - 10$ factors into $0 = (x-1)(x+1)(x^2+6x+10)$. The roots are $x = 1$, $x = -1$, $x = -3 + i$, and $x = -3 - i$. You obtain the last two roots by using the quadratic formula on the quadratic factor in the equation (see Chapter 3). The two roots are a conjugate pair.

Interpreting complex zeros

The polynomial function $y = x^4 + 7x^3 + 9x^2 - 28x - 52$ has two real roots and two complex roots. According to Descartes' Rule of Sign (see Chapter 8), the function could've contained as many as four real roots (you can tell by trying the roots suggested by the Rational Root Theorem; see Chapter 8). You can determine the number of complex roots in two different ways: by factoring the polynomial or by looking at the graph of the function.

The example function factors into $y = (x-2)(x+2)(x^2+7x+13)$. The first two factors give you real roots, or x-intercepts. When you set $x - 2$ equal to 0, you get the intercept $(2, 0)$. When you set $x + 2$ equal to 0, you get the intercept $(-2, 0)$. Setting the last factor, $x^2 + 7x + 13$, equal to 0 doesn't give you a real root, as you see here:

$$x^2 + 7x + 13 = 0$$
$$x = \frac{-7 \pm \sqrt{49 - 4(1)(13)}}{2(1)}$$
$$x = \frac{-7 \pm \sqrt{49 - 52}}{2}$$
$$x = \frac{-7 \pm i\sqrt{3}}{2}$$

You can also tell that a polynomial function has complex roots by looking at its graph. You can't tell what the roots are, but you *can* see that the graph has some. If you need the values of the roots, you can resort to using algebra to solve for them. Figure 14-3 shows the graph of the example function, $y = x^4 + 7x^3 + 9x^2 - 28x - 52$. You can see the two x-intercepts, which represent the two real zeros. You also see the graph flattening on the left.

A flattening-out behavior can represent a change of direction, or a *point of inflection,* where the curvature of the graph changes. Areas like this on a graph indicate that complex zeros exist for the polynomial.

Figure 14-3:
A flattening curve indicates a complex root.

Figure 14-4 can tell you plenty about the number of real zeros and complex zeros the graph of the polynomial has . . . before you ever see the equation it represents.

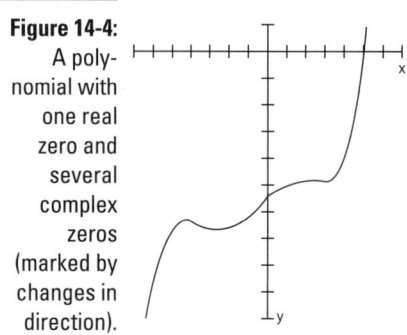

Figure 14-4:
A polynomial with one real zero and several complex zeros (marked by changes in direction).

The polynomial in Figure 14-4 appears to have one real zero and several complex zeros. Do you see how it changes direction all over the place under the X-axis? These changes indicate the presence of complex zeros. The graph represents the polynomial function $y = 12x^5 + 15x^4 - 320x^3 - 120x^2 + 2880x - 18,275$. The function has four complex zeros — two complex (conjugate) pairs — and one real zero (when $x = 5$).

You need the efforts and capabilities of good algebra; the Rational Root Theorem and Descartes' Rule of Sign combined with a graphing calculator or computer; and some luck and common sense to solve these high-order equations. When the zeros are nice integers, life is good. When the zeros are irrational or complex, life is still good, but it's a bit more complicated. Assume the best, and work through the challenges if they arise.

Chapter 15

Making Moves with Matrices

In This Chapter

▶ Perusing the roster of varying matrices

▶ Operating on matrices

▶ Singling out single rows for operations

▶ Identifying the inverses of matrices

▶ Employing matrices to solve systems of equations

A *matrix* is a rectangular array of numbers. It has the same number of elements in each of its rows, and each column shares the same number of elements. If you've seen the *Matrix* films, you may remember the lines and columns of green code scrolling down the characters' computer screens. This matrix of codes represented the abstract "matrix" of the films. Putting numbers or elements in an orderly array allows you to organize information, access information quickly, do computations involving some of the entries in the matrix, and communicate your results efficiently.

The word matrix is singular — you have just one of them. When you have more than one matrix, you have *matrices,* the plural form. Bet you weren't expecting an English (or Latin) lesson in this book.

In this chapter, you discover how to add and subtract matrices, how to multiply matrices, and how to solve systems of equations by using matrices. The processes in this chapter are easily adaptable for use in technology (as evidenced by the machines in *The Matrix!*) — you can transfer the information to a spreadsheet or graphing calculator for help with computing large amounts of data.

Describing the Different Types of Matrices

A matrix has a size, or *dimension,* that you have to recognize before you can proceed with any matrix operations.

You give the dimensions of a matrix in a particular order. You identify the number of rows in the matrix first, and then you mention the number of columns. Usually, you put the numbers for the rows and columns on either side of a × sign: *rows × columns.*

In the following figure, you see four matrices. You read their dimensions, in order from left to right, as follows: 2×4, 3×3, 1×5, and 3×1. The brackets around the array of numbers serve as clear indicators that you're dealing with matrices.

$$\begin{bmatrix} 1 & 3 & -2 & 4 \\ 0 & -3 & 5 & 9 \end{bmatrix} \quad \begin{bmatrix} 4 & -2 & 1 \\ -1 & 0 & 0 \\ -3 & 3 & 1 \end{bmatrix} \quad \begin{bmatrix} 1 & 2 & 3 & 2 & 4 \end{bmatrix} \quad \begin{bmatrix} 2 \\ 0 \\ -4 \end{bmatrix}$$

When you have to deal with more than one matrix, you can keep track of them by labeling the matrices with different names. You don't call the matrices Bill or Ted; that wouldn't be very mathematical of you! Matrices are traditionally labeled with capital letters, such as matrix A or matrix B, to avoid confusion.

The numbers that appear in the rectangular array of a matrix are called its *elements.* You refer to each element in a matrix by listing the same letter as the matrix's name, in lowercase form, and following it with subscript numbers for the row and then the column. For example, the item in the first row and third column of matrix B is b_{13}. If the number of rows or columns grows to a number bigger than nine, you put a comma between the two numbers in the subscript.

Row and column matrices

Matrices come in many sizes (or dimensions), just like rectangles, but instead of measuring width and length, you count a matrix's rows and columns. You call matrices that have only one row or one column *row matrices* or *column matrices,* respectively. A row matrix has the dimension $1 \times n$, where n is the number of columns. The following matrix, matrix A, is a row matrix with dimension 1×5:

A = [1 –3 4 5 0]

A column matrix has the dimension $m \times 1$. Matrix B, shown in the following figure, is a column matrix with dimension 4×1.

$$B = \begin{bmatrix} 6 \\ -3 \\ 0 \\ 1 \end{bmatrix}$$

Square matrices

A *square matrix* has the same number of rows and columns. Square matrices have dimensions such as 2×2, 3×3, 8×8, and so on. The elements in square matrices can take on any number — although some special square matrices have the label *identity matrices* (coming up later in this section). All matrices are rectangular arrays, and a square is a special type of rectangle.

Zero matrices

Zero matrices can have any dimension — they can have any number of rows or columns. The matrices in the following figure are zero matrices because zeros make up all their elements.

Zero matrices may not look very impressive — after all, they don't have much in them — but they're necessary for matrix arithmetic. Just like you need a zero to add and subtract numbers, you need zero matrices to do matrix addition and matrix subtraction.

$$C = \begin{bmatrix} 0 & 0 & 0 \\ 0 & 0 & 0 \\ 0 & 0 & 0 \end{bmatrix} \quad D = \begin{bmatrix} 0 & 0 & 0 & 0 \\ 0 & 0 & 0 & 0 \end{bmatrix}$$

Identity matrices

Identity matrices add a couple characteristics to the zero-matrix format (see the previous section) in terms of their dimensions and elements. An identity matrix has to

- ✔ Be a square matrix.
- ✔ Have a diagonal strip of 1s that goes from the top left to the bottom right of the matrix.
- ✔ Consist of zeros outside of the diagonal strip of 1s.

The following are three identity matrices, although you come across many, many more styles.

$$E = \begin{bmatrix} 1 & 0 \\ 0 & 1 \end{bmatrix} \quad F = \begin{bmatrix} 1 & 0 & 0 \\ 0 & 1 & 0 \\ 0 & 0 & 1 \end{bmatrix} \quad G = \begin{bmatrix} 1 & 0 & 0 & 0 \\ 0 & 1 & 0 & 0 \\ 0 & 0 & 1 & 0 \\ 0 & 0 & 0 & 1 \end{bmatrix}$$

Identity matrices are instrumental to matrix multiplication and matrix inverses. Identity matrices act pretty much like the number one in the multiplication of numbers. What happens when you multiply a number times one? The number keeps its identity. You see the same behavior when you multiply a matrix by an identity matrix — the matrix stays the same.

Performing Operations on Matrices

You can add matrices, subtract one from another, multiply them by numbers, multiply them times each other, and divide them. Well, actually, you don't divide matrices; you change the division problem to a multiplication problem. You can't add, subtract, or multiply just any matrices, though. Each operation has its own set of rules. I cover the rules for addition, subtraction, and multiplication in this section. (Division comes later in the chapter, after I discuss matrix inverses.)

Adding and subtracting matrices

To add or subtract matrices, you have to make sure the matrices are same size. In other words, they need to have identical dimensions. You find the resulting matrix by combining or subtracting the corresponding elements in the matrices. If two matrices don't have the same dimension, you can't add or subtract them, and you can do nothing to fix the situation.

Figure 15-1 illustrates the rules of adding and subtracting matrices.

$$A = \begin{bmatrix} a_{11} & a_{12} & a_{13} \\ a_{21} & a_{22} & a_{23} \end{bmatrix}, \quad B = \begin{bmatrix} b_{11} & b_{12} & b_{13} \\ b_{21} & b_{22} & b_{23} \end{bmatrix}$$

Figure 15-1:
Match up the corresponding elements and add or subtract.

$$A+B = \begin{bmatrix} a_{11}+ b_{11} & a_{12}+ b_{12} & a_{13}+ b_{13} \\ a_{21}+ b_{21} & a_{22}+ b_{22} & a_{23}+ b_{23} \end{bmatrix}$$

$$A-B = \begin{bmatrix} a_{11}- b_{11} & a_{12}- b_{12} & a_{13}- b_{13} \\ a_{21}- b_{21} & a_{22}- b_{22} & a_{23}- b_{23} \end{bmatrix}$$

You can see why matrices have to have the same dimensions before you can add or subtract. Differing matrices would have some elements without partners.

Here's an example with actual numbers for the elements. If you want to add or subtract the following matrices, just combine or subtract the elements:

$$C = \begin{bmatrix} 2 & 5 & -3 & 8 \\ -1 & 0 & 7 & -4 \end{bmatrix}, \quad D = \begin{bmatrix} -2 & 4 & 3 & -6 \\ 0 & 2 & 7 & 3 \end{bmatrix}$$

$$C+D = \begin{bmatrix} 2+(-2) & 5+4 & -3+3 & 8+(-6) \\ -1+0 & 0+2 & 7+7 & -4+3 \end{bmatrix} = \begin{bmatrix} 0 & 9 & 0 & 2 \\ -1 & 2 & 14 & -1 \end{bmatrix}$$

$$C-D = \begin{bmatrix} 2-(-2) & 5-4 & -3-3 & 8-(-6) \\ -1-0 & 0-2 & 7-7 & -4-3 \end{bmatrix} = \begin{bmatrix} 4 & 1 & -6 & 14 \\ -1 & -2 & 0 & -7 \end{bmatrix}$$

Multiplying matrices by scalars

Scalar is just a fancy word for number. Algebra uses the word scalar with regard to matrix multiplication to contrast a number with a matrix, which has a dimension. A scalar has no dimension, so you can use it uniformly throughout the matrix.

Scalar multiplication of a matrix signifies that you multiply every element in the matrix by a number.

To multiply matrix A by the number k, for example, you multiply each element in A by k. Figure 15-2 illustrates how this scalar multiplication works.

Figure 15-2: In scalar multiplication, each element is a multiple of k.

$$A = \begin{bmatrix} a_{11} & a_{12} \\ a_{21} & a_{22} \\ a_{31} & a_{32} \\ a_{41} & a_{42} \end{bmatrix}, \quad kA = k\begin{bmatrix} a_{11} & a_{12} \\ a_{21} & a_{22} \\ a_{31} & a_{32} \\ a_{41} & a_{42} \end{bmatrix} = \begin{bmatrix} ka_{11} & ka_{12} \\ ka_{21} & ka_{22} \\ ka_{31} & ka_{32} \\ ka_{41} & ka_{42} \end{bmatrix}$$

Here's how the process looks with actual numbers. Multiplying the matrix F by the scalar 3, you create a matrix where every element is a multiple of 3:

$$F = \begin{bmatrix} 4 & -3 & 1 \\ 2 & 0 & -1 \\ 5 & 10 & -4 \end{bmatrix}, \quad 3F = 3\begin{bmatrix} 4 & -3 & 1 \\ 2 & 0 & -1 \\ 5 & 10 & -4 \end{bmatrix} = \begin{bmatrix} 12 & -9 & 3 \\ 6 & 0 & -3 \\ 15 & 30 & -12 \end{bmatrix}$$

Multiplying two matrices

Matrix multiplication requires that the number of columns in the first matrix is equal to the number of rows in the second matrix. This means, for instance, that a matrix with 3 rows and 11 columns can multiply a matrix with 11 rows and 4 columns — but it has to be in that order. The 11 columns in the first matrix must match up with the 11 rows in the second matrix.

Matrix multiplication requires some strict rules about the dimensions and the order in which the matrices are multiplied. Even when the matrices are square and can be multiplied in either order, you don't get the same answer when multiplying them in the different orders. In general, the product AB does not equal the product BA.

Determining dimensions

If you want to multiply matrices, the number of columns in the first matrix has to equal the number of rows in the second matrix. After you multiply matrices, you get a whole new matrix that features the number of rows the first matrix had and the number of columns from the original second matrix. The process is sort of like cross-pollinating white and red petunias and getting pink.

In algebraic terms, if matrix A has the dimension $m \times n$ and matrix B has the dimension $p \times q$, to multiply A · B, n must equal p. The dimension of the resulting matrix will be $m \times q$, the number of rows in the first matrix and the number of columns in the second.

For example, if you multiply a 2×3 matrix times a 3×7 matrix, you get a 2×7 matrix. However, you can't multiply a 2×2 matrix by a 7×2 matrix. To show that the order in which you multiply the matrices does indeed matter, you can multiply a 7×2 matrix by a 2×2 matrix and get a 7×2 matrix.

Defining the process

Multiplying matrices is no easy matter, but it isn't all that difficult — if you can multiply and add correctly. When you multiply two matrices, you compute the elements with the following rule:

If you find the element c_{ij} after multiplying matrix A times matrix B, c_{ij} is the sum of the products of the elements in the *ith* row of matrix A and the *jth* column of matrix B.

Okay, so that rule may sound like a lot of hocus pocus. How about I give you something a bit more concrete? In Figure 15-3, matrix A has the dimension 3×2, and matrix B has the dimension 2×4. According to the rules that govern multiplying matrices, you can multiply A times B because the number of columns in matrix A is two, and the number of rows in matrix B is two. The matrix that you create when multiplying A and B has the dimension 3×4.

$$A = \begin{bmatrix} a_{11} & a_{12} \\ a_{21} & a_{22} \\ a_{31} & a_{32} \end{bmatrix}, \quad B = \begin{bmatrix} b_{11} & b_{12} & b_{13} & b_{14} \\ b_{21} & b_{22} & b_{23} & b_{24} \end{bmatrix}$$

Figure 15-3:
Multiplying
matrices
3×2 and
2×4 gives
you 3×4.

$$A * B = C = \begin{bmatrix} c_{11} & c_{12} & c_{13} & c_{14} \\ c_{21} & c_{22} & c_{23} & c_{24} \\ c_{31} & c_{32} & c_{33} & c_{34} \end{bmatrix}$$

You find the element c_{11} by multiplying the elements in the first row of A times the elements in the first column of B and then adding the products: $c_{11} = a_{11}b_{11} + a_{12}b_{21}$. You find c_{23} by multiplying the second row of matrix A times the third column of matrix B and adding: $c_{23} = a_{21}b_{13} + a_{22}b_{23}$.

Although the last example was more concrete, it didn't provide you with numbers. With the following matrices, you can multiply matrix J times matrix K because the number of columns in matrix J matches the number of rows in matrix K. You can also see the computations needed to find the resulting matrix.

$$J = \begin{bmatrix} 1 & 2 & -3 \\ 0 & 4 & 2 \end{bmatrix}, \quad K = \begin{bmatrix} 4 & 5 \\ 1 & -1 \\ 2 & 3 \end{bmatrix}$$

$$J * K = \begin{bmatrix} 1 \cdot 4 + 2 \cdot 1 + (-3) \cdot 2 & 1 \cdot 5 + 2 \cdot (-1) + (-3) \cdot 3 \\ 0 \cdot 4 + 4 \cdot 1 + 2 \cdot 2 & 0 \cdot 5 + 4 \cdot (-1) + 2 \cdot 3 \end{bmatrix}$$

$$= \begin{bmatrix} 4 + 2 - 6 & 5 - 2 - 9 \\ 0 + 4 + 4 & 0 - 4 + 6 \end{bmatrix} = \begin{bmatrix} 0 & -6 \\ 8 & 2 \end{bmatrix}$$

You get a 2×2 matrix when you multiply because multiplying a 2×3 matrix times a 3×2 matrix leaves you with two rows and two columns. The multiplication process may seem a bit complicated, but after you get the hang of it, you can do the multiplication and addition in your head.

Applying matrices and operations

One of the grand features of matrices is their capacity to organize information and make it more useful. If you're running a small business, you can keep track of sales and payroll without having to resort to matrices. But big companies and factories have hundreds, if not thousands, of items to keep track of. Matrices help with the organization, and, because you can enter them in computers, the accuracy and ease of using them increases even more.

Consider the following sales situation that occurred in an electronics store where Ariel, Ben, Carlie, and Don work. In January, Ariel sold 12 televisions, 9 compact disc players, and 4 computers; Ben sold 21 CD players and 3 computers; Carlie sold 4 TVs, 10 CD players, and 1 computer; and Don sold 13

TVs, 12 CD players, and 5 computers. In Figure 15-4, you see the sales results for Ariel, Ben, Carlie, and Don during the month of January (you also see the sales for February and March). See how nicely the matrices organize the information!

When you organize the information into matrices, you can see at a glance who's selling the most, who's had some bad months, who's had some good months, and what electronics seem to be moving the best. With this information in hand, several questions may come to mind.

Figure 15-4:
The rows represent the sales-persons, and the columns represent the items sold.

$$J = \begin{array}{c} \\ A \\ B \\ C \\ D \end{array} \begin{array}{ccc} TV & CD & C \\ \left[\begin{array}{ccc} 12 & 9 & 4 \\ 0 & 21 & 3 \\ 4 & 10 & 1 \\ 13 & 12 & 5 \end{array}\right] \end{array} \quad F = \begin{array}{c} \\ A \\ B \\ C \\ D \end{array} \begin{array}{ccc} TV & CD & C \\ \left[\begin{array}{ccc} 10 & 3 & 3 \\ 5 & 15 & 0 \\ 0 & 1 & 6 \\ 10 & 10 & 10 \end{array}\right] \end{array} \quad M = \begin{array}{c} \\ A \\ B \\ C \\ D \end{array} \begin{array}{ccc} TV & CD & C \\ \left[\begin{array}{ccc} 4 & 5 & 2 \\ 2 & 15 & 4 \\ 3 & 6 & 4 \\ 9 & 9 & 8 \end{array}\right] \end{array}$$

Determining how many of each item was sold

The first question is: How many TVs, CDs, and computers did the store sell during those three months? Because all the matrices in Figure 15-4 have the same dimensions, you can add them together (see the section "Adding and subtracting matrices"). Figure 15-5 shows how you find the sales for the first three months of the year by adding the matrices (the totals of the individual salespeople).

Figure 15-5:
Add the matrices to find the total sales of electronics.

$$J + F + M = Sum = \begin{array}{c} \\ A \\ B \\ C \\ D \end{array} \begin{array}{ccc} TV & CD & C \\ \left[\begin{array}{ccc} 12+10+4 & 9+3+5 & 4+3+2 \\ 0+5+2 & 21+15+15 & 3+0+4 \\ 4+0+3 & 10+1+6 & 1+6+4 \\ 13+10+9 & 12+10+9 & 5+10+8 \end{array}\right] \end{array} = \begin{array}{c} \\ A \\ B \\ C \\ D \end{array} \begin{array}{ccc} TV & CD & C \\ \left[\begin{array}{ccc} 26 & 17 & 9 \\ 7 & 51 & 7 \\ 7 & 17 & 11 \\ 32 & 31 & 23 \end{array}\right] \end{array}$$

To find the total sales for each type of electronic, you multiply the *sum matrix* (the matrix you just found) by a *row matrix,* T = [1 1 1 1]. You're multiplying a matrix with the dimension 1×4 times a matrix with the dimension 4×3, so your result is a matrix with the dimension 1×3; the totals of each electronic appear in order from left to right. Think of the row matrix T with the sales-persons' names across the top. When you multiply this matrix by the sum matrix, you match each of the columns in T with each of the rows in the sum. The columns in T and the rows in the sum are the salespersons, so they line up. By multiplying all the numbers in the second matrix by 1s, you essentially multiply by 100 percent — you add everything up. The resulting matrix is a row matrix with the numbers for the electronics in the respective columns. Figure 15-6 shows the computation and result.

$$\begin{array}{cccc} & A & B & C & D \\ T * Sum = [1 & 1 & 1 & 1], \end{array}$$

Figure 15-6:
Totaling the
first-quarter
electronics
sales.

$$T * Sum = \begin{array}{cccc} A & B & C & D \\ [1 & 1 & 1 & 1] \end{array} * \begin{array}{c} A \\ B \\ C \\ D \end{array}\begin{bmatrix} TV & CD & C \\ 26 & 17 & 9 \\ 7 & 51 & 7 \\ 7 & 17 & 11 \\ 32 & 31 & 23 \end{bmatrix} = \begin{array}{ccc} TV & CD & C \\ [72 & 116 & 50] \end{array}$$

You're wondering why you can't just add the numbers in each column, aren't you? Go right ahead! That works, too. I show you this process with a small, reasonable number of items so you can see how you can instruct a computer to do it with hundreds or thousands of entries. You may also want to "weight" some entries — make some items worth more than others — if you're keeping track of points. The row matrix can have different entries that align with the different items.

Determining sales by salesperson

Here's another question you can answer with the matrices from Figure 15-4: How much money did each salesperson bring in? For instance, assume that the average costs of a TV, CD player, and computer are $1,500, $400, and $2,000, respectively. You can construct a column matrix containing these dollar amounts, call it the *dollar matrix,* and multiply it by the sum matrix (see the section "Multiplying two matrices" for instructions and Figure 15-5 for the sum matrix).

Figure 15-7 shows the multiplication of the sum matrix times the dollar matrix, with the resulting amount of money brought in by each salesperson. Here are the results of multiplying the entries in the row of the first matrix by the columns of the second:

A: 26 · $1,500 + 17 · $400 + 9 · $2,000 = $63,800

B: 7 · $1,500 + 51 · $400 + 7 · $2,000 = $44,900

C: 7 · $1,500 + 17 · $400 + 11 · $2,000 = $39,300

D: 32 · $1,500 + 31 · $400 + 23 · $2,000 = $106,400

Figure 15-7: Determining each salesperson's total.

$$\text{Sum} = \begin{array}{c} A \\ B \\ C \\ D \end{array}\begin{array}{ccc} TV & CD & C \\ \left[\begin{array}{ccc} 26 & 17 & 9 \\ 7 & 51 & 7 \\ 7 & 17 & 11 \\ 32 & 31 & 23 \end{array}\right.\end{array} * \begin{array}{c} TV \\ CD \\ C \end{array}\left[\begin{array}{c} \$1,500 \\ \$400 \\ \$2,000 \end{array}\right] = \begin{array}{c} A \\ B \\ C \\ D \end{array}\left[\begin{array}{c} \$63,800 \\ \$44,900 \\ \$39,300 \\ \$106,400 \end{array}\right]$$

Determining how to increase sales

Here's one last question that deals with a percentage: How many electronic devices in each category must each salesperson sell if they have to increase sales to 125 percent during the next quarter?

You identify this problem as a scalar multiplication problem (see the section "Multiplying matrices by scalars") because you multiply each entry in the sum matrix (see Figure 15-5) by 125 percent to get the sales target. The value 1.25 represents 125 percent — 25 percent more than the last quarter. Figure 15-8 shows the result of the scalar multiplication and a second matrix with the numbers rounded up. (You can't sell half a computer, and rounding up gives the salesperson the number he or she needs to make or exceed the goal [not come in slightly below].)

Figure 15-8:
Setting a
sales
challenge
with scalar
multipli-
cation.

$$125\% * Sum = 1.25 \begin{array}{ccc} \text{TV} & \text{CD} & \text{C} \\ \begin{bmatrix} 26 & 17 & 9 \\ 7 & 51 & 7 \\ 7 & 17 & 11 \\ 32 & 31 & 23 \end{bmatrix} \end{array} = \begin{array}{c} A \\ B \\ C \\ D \end{array} \begin{array}{ccc} \text{TV} & \text{CD} & \text{C} \\ \begin{bmatrix} 32.5 & 21.25 & 11.25 \\ 8.75 & 63.75 & 8.75 \\ 8.75 & 21.25 & 13.75 \\ 40 & 38.75 & 28.75 \end{bmatrix} \end{array} \approx \begin{array}{c} A \\ B \\ C \\ D \end{array} \begin{array}{ccc} \text{TV} & \text{CD} & \text{C} \\ \begin{bmatrix} 33 & 22 & 12 \\ 9 & 64 & 9 \\ 9 & 22 & 14 \\ 40 & 39 & 29 \end{bmatrix} \end{array}$$

Defining Row Operations

Along with the matrix operations I list in the previous sections of this chapter, you can perform *row operations* on the individual rows of a matrix. You perform a row operation one matrix at a time; you don't combine one matrix with another matrix. A row operation changes the look of a matrix by altering some of the elements, but the operation allows the matrix to retain the properties that enable you to use it in other applications, such as solving systems of equations. (See the section "Using Matrices to Find Solutions for Systems of Equations" later in this chapter.)

The business of changing matrices to equivalent matrices is sort of like changing fractions to equivalent fractions so they have a common denominator — the change makes the fractions more useful. The same goes for matrices.

You have several different row operations at your disposal:

✔ You can exchange two rows.

✔ You can multiply the elements in a row by a constant (not zero).

✔ You can add the elements in one row to the elements in another row.

✔ You can add a row that you multiply by some number to another row.

Figure 15-9 shows a matrix that has experienced the following row operations, in order, performed by yours truly:

I exchange the first and third rows (a).

I multiply the second row through by –1 (b).

I add the first and third rows and put the result in the third row (c).

I add twice the first row to the second row and put the result in the second row (d).

Performing row operations correctly results in a matrix that's equivalent to the original matrix. The rows themselves aren't equivalent to one another; the whole matrix and the relationships between its rows are preserved with the operations.

Row operations may seem pointless and aimless. For the illustrations of the row operations in Figure 15-9, I have no particular goal other than to illustrate the possibilities. But you choose more wisely when you do row operations to perform a task, such as solving for an inverse matrix (see the following section).

Figure 15-9:
Row operations performed on a matrix.

Finding Inverse Matrices

Inverses of matrices act somewhat like inverses of numbers. The *additive inverse* of a number is what you have to add to the number to get zero. For example, the additive inverse of the number 2 is –2, and the additive inverse of –3.14159 is 3.14159. Simple enough.

Algebra also provides you with the *multiplicative inverse*. Multiplicative inverses give you the number one. For example, the multiplicative inverse of 2 is ½.

Adding or multiplying inverses always gives you the *identity* element for a particular operation. Before you can find inverses of matrices, you need to understand identities, so I cover those in detail here.

Determining additive inverses

You can label the number zero as the *additive identity,* because adding zero to a number allows that number to keep its identity. The additive identity for matrices is a *zero matrix.* When you add a zero matrix to any matrix with the same dimension, the original matrix doesn't change.

Inverse matrices associated with addition are easy to spot and easy to create. The additive inverse of a matrix is another matrix with the same dimension, but every element has the opposite sign. When you add the matrix and its additive inverse, the sums of the corresponding elements are all zero, and you have a zero matrix — the additive identity for matrices (see the section "Zero matrices" earlier in the chapter). Figure 15-10 shows you how two matrices can add to nothing.

All matrices have additive inverses — no matter what the dimension of the matrix. This isn't the case with multiplicative inverses of matrices. Some matrices have multiplicative inverses and some do not. Read on if you're intrigued by this situation (or if the topic will be on your next test).

Figure 15-10:
Adding elements of opposite signs results in a zero matrix.

$$A = \begin{bmatrix} 4 & 1 \\ -3 & 7 \\ -2 & 0 \end{bmatrix}, \quad B = \begin{bmatrix} -4 & -1 \\ 3 & -7 \\ 2 & 0 \end{bmatrix},$$

$$A + B = \begin{bmatrix} 4 + (-4) & 1 + (-1) \\ -3 + 3 & 7 + (-7) \\ -2 + 2 & 0 + 0 \end{bmatrix} = \begin{bmatrix} 0 & 0 \\ 0 & 0 \\ 0 & 0 \end{bmatrix}$$

Determining multiplicative inverses

You can label the number one as the multiplicative identity, because multiplying a number by one doesn't change the number — it keeps its identity.

Matrices have additive identities that consist of nothing but zeros, but multiplicative identities for matrices are a bit more picky. A *multiplicative identity* for a matrix has to be a square matrix, and this square matrix has to have a diagonal of ones; the rest of the elements are zeros. This arrangement assures that when you multiply any matrix times a multiplicative identity, you don't change the original matrix. Figure 15-11 shows the process of multiplying two matrices by identities. The multiplication rule for matrices still holds here (see the section "Multiplying two matrices"): The number of columns in the first matrix must match the number of rows in the second matrix.

Like the multiplicative identity of a matrix, the multiplicative inverse of a matrix isn't quite as accommodating as its additive cousin. When you multiply two matrices, you perform plenty of multiplication and addition, and you see plenty of change in dimension. Because of this, matrices and their inverses are always square matrices; non-square matrices don't have multiplicative inverses.

$$\begin{bmatrix} 3 & 0 & -2 \\ 1 & 5 & 9 \end{bmatrix} * \begin{bmatrix} 1 & 0 & 0 \\ 0 & 1 & 0 \\ 0 & 0 & 1 \end{bmatrix} = \begin{bmatrix} 3 \cdot 1 + 0 \cdot 0 + (-2) \cdot 0 & 3 \cdot 0 + 0 \cdot 1 + (-2) \cdot 0 & 3 \cdot 0 + 0 \cdot 0 + (-2) \cdot 1 \\ 1 \cdot 1 + 5 \cdot 0 + 9 \cdot 0 & 1 \cdot 0 + 5 \cdot 1 + 9 \cdot 0 & 1 \cdot 0 + 5 \cdot 0 + 9 \cdot 1 \end{bmatrix}$$

Figure 15-11:
Multiplying
a matrix by
an identity
preserves

$$= \begin{bmatrix} 3 + 0 + 0 & 0 + 0 + 0 & 0 + 0 + (-2) \\ 1 + 0 + 0 & 0 + 5 + 0 & 0 + 0 + 9 \end{bmatrix} = \begin{bmatrix} 3 & 0 & -2 \\ 1 & 5 & 9 \end{bmatrix}$$

the original
matrix.
$$\begin{bmatrix} 1 & 0 \\ 0 & 1 \end{bmatrix} * \begin{bmatrix} -3 \\ 6 \end{bmatrix} = \begin{bmatrix} 1 \cdot (-3) + 0 \cdot 6 \\ 0 \cdot (-3) + 1 \cdot 6 \end{bmatrix} = \begin{bmatrix} -3 + 0 \\ 0 + 6 \end{bmatrix} = \begin{bmatrix} -3 \\ 6 \end{bmatrix}$$

If matrix A and matrix A^{-1} are multiplicative inverses, $A \cdot A^{-1} = I$, and $A^{-1} \cdot A = I$. The superscript –1 on matrix A identifies it as the inverse matrix of A — it isn't a reciprocal, which the exponent –1 usually represents. Also, the capital letter I identifies the identity matrix associated with these matrices — with dimensions 2×2, 3×3, 4×4, and so on.

In Figure 15-12, I multiply matrix B and its inverse B^{-1} in one order and then in the other order; the process results in the identity matrix both times.

$$B = \begin{bmatrix} 6 & 2 \\ 8 & 3 \end{bmatrix}, \quad B^{-1} = \begin{bmatrix} 1.5 & -1 \\ -4 & 3 \end{bmatrix}$$

Figure 15-12:
Order
doesn't
matter
when
$$B * B^{-1} = \begin{bmatrix} 6 \cdot 1.5 + 2(-4) & 6(-1) + 2 \cdot 3 \\ 8 \cdot 1.5 + 3(-4) & 8(-1) + 3 \cdot 3 \end{bmatrix} = \begin{bmatrix} 9 - 8 & -6 + 6 \\ 12 - 12 & -8 + 9 \end{bmatrix} = \begin{bmatrix} 1 & 0 \\ 0 & 1 \end{bmatrix}$$

multiplying
inverse
matrices.
$$B^{-1} * B = \begin{bmatrix} 1.5 \cdot 6 + (-1) \cdot 8 & 1.5 \cdot 2 + (-1) \cdot 3 \\ -4 \cdot 6 + 3 \cdot 8 & -4 \cdot 2 + 3 \cdot 3 \end{bmatrix} = \begin{bmatrix} 9 - 8 & 3 - 3 \\ -24 + 24 & -8 + 9 \end{bmatrix} = \begin{bmatrix} 1 & 0 \\ 0 & 1 \end{bmatrix}$$

Not all square matrices have inverses. But, for those that do, you have a way to find the inverse matrices. You don't always know ahead of time which matrices will fail you, but it becomes apparent as you go through the process. The first process, or *algorithm* (a process or routine that produces a result), you can use works for any size square matrix. You can also use a neat, quick method for 2×2 matrices that works only for matrices with that dimension.

Identifying an inverse for any size square matrix

The general method you use to find a matrix's inverse involves writing down the matrix, inserting an identity matrix, and then changing the original matrix into an identity matrix.

To solve for the inverse of a matrix, follow these steps:

1. **Create a large matrix — consisting of the target matrix and the identity matrix of the same size — with the identity matrix to the right of the original.**

2. **Perform row operations until the elements on the left become an identity matrix (see the "Defining Row Operations" section earlier in the chapter).**

   The elements need to feature a diagonal of ones, with zeros above and below the ones. Upon completion of this step, the elements on the right become the elements of the inverse matrix.

For example, say you want to solve for the inverse of matrix M in the following figure. You first put the 3×3 identity matrix to the right of the elements in matrix M. The goal is to make the elements on the left look like an identity matrix by using matrix row operations.

$$M = \begin{bmatrix} 1 & 2 & 4 \\ -3 & -5 & -6 \\ 2 & -3 & -36 \end{bmatrix}$$

$$\begin{bmatrix} 1 & 2 & 4 & 1 & 0 & 0 \\ -3 & -5 & -6 & 0 & 1 & 0 \\ 2 & -3 & -36 & 0 & 0 & 1 \end{bmatrix}$$

The identity matrix has a diagonal slash of ones, and zeros lie above and below the ones. The first thing you do to create your identity matrix is to get the zeros below the one in the upper left-hand corner of the original identity. Here are the row operations you use for this example:

1. Multiply Row 1 times 3 and add the result to Row 2, making the result the new Row 2.

2. Multiply Row 1 by –2 and add the result to Row 3, putting the resulting answer in Row 3.

The following shows what the matrix looks like after you perform the row operations. The notation on the arrow between the two matrices describes the row operations you use.

$$\begin{bmatrix} 1 & 2 & 4 & 1 & 0 & 0 \\ -3 & -5 & -6 & 0 & 1 & 0 \\ 2 & -3 & -36 & 0 & 0 & 1 \end{bmatrix} \xrightarrow[(-2)R_1 + R_3 = R_3]{(3)R_1 + R_2 = R_2} \begin{bmatrix} 1 & 2 & 4 & 1 & 0 & 0 \\ 0 & 1 & 6 & 3 & 1 & 0 \\ 0 & -7 & -44 & -2 & 0 & 1 \end{bmatrix}$$

You see a one in the second column and second row of the resulting matrix — along the diagonal of ones that you want. Getting the one in this position is a nice coincidence; it doesn't always work out as well. If the one hadn't fallen in that position, you'd have to divide the whole row by whatever number makes that element a one.

Now, you want zeros above and below the one middle, so you follow these steps:

1. Multiply Row 2 by –2 and add the result to Row 1.

2. Multiply Row 2 by 7 and add the result to Row 3.

$$\begin{bmatrix} 1 & 2 & 4 & 1 & 0 & 0 \\ 0 & 1 & 6 & 3 & 1 & 0 \\ 0 & -7 & -44 & -2 & 0 & 1 \end{bmatrix} \xrightarrow[(7)R_2 + R_3 = R_3]{(-2)R_2 + R_1 = R_1} \begin{bmatrix} 1 & 0 & -8 & -5 & -2 & 0 \\ 0 & 1 & 6 & 3 & 1 & 0 \\ 0 & 0 & -2 & 19 & 7 & 1 \end{bmatrix}$$

You now need to turn the element in the third row, third column of the matrix into a one, so you multiply the row through by –0.5; this multiplication is the same as dividing through by –2.

$$\begin{bmatrix} 1 & 0 & -8 & -5 & -2 & 0 \\ 0 & 1 & 6 & 3 & 1 & 0 \\ 0 & 0 & -2 & 19 & 7 & 1 \end{bmatrix} \xrightarrow{(-.5)R_3 = R_3} \begin{bmatrix} 1 & 0 & -8 & -5 & -2 & 0 \\ 0 & 1 & 6 & 3 & 1 & 0 \\ 0 & 0 & 1 & -9.5 & -3.5 & -0.5 \end{bmatrix}$$

You have one last set of row operations to perform to get zeros above the last one on the diagonal:

1. Multiply Row 3 by 8 and add the elements in the row to Row 1.

2. Multiply Row 3 by –6 and add the result to Row 2.

You now have the 3×3 identity matrix on the left and a new 3×3 matrix on the right. The elements in the matrix on the right form the inverse matrix, M^{-1}. The product of M and its inverse, M^{-1}, is the identity matrix. The following figure shows the final two steps and the result.

$$\begin{bmatrix} 1 & 0 & -8 & -5 & -2 & 0 \\ 0 & 1 & 6 & 3 & 1 & 0 \\ 0 & 0 & 1 & -9.5 & -3.5 & -0.5 \end{bmatrix} \xrightarrow[(-6)R_3 + R_2 = R_2]{(8)R_3 + R_1 = R_1} \begin{bmatrix} 1 & 0 & 0 & -81 & -30 & -4 \\ 0 & 1 & 0 & 60 & 22 & 3 \\ 0 & 0 & 1 & -9.5 & -3.5 & -0.5 \end{bmatrix}$$

$$M^{-1} = \begin{bmatrix} -81 & -30 & -4 \\ 60 & 22 & 3 \\ -9.5 & -3.5 & -0.5 \end{bmatrix}$$

The following figure checks your work — showing that $M \cdot M^{-1}$ is the identity matrix:

$$M * M^{-1} = \begin{bmatrix} 1 & 2 & 4 \\ -3 & -5 & -6 \\ 2 & -3 & -36 \end{bmatrix} * \begin{bmatrix} -81 & -30 & -4 \\ 60 & 22 & 3 \\ -9.5 & -3.5 & -0.5 \end{bmatrix} = \begin{bmatrix} 1 & 0 & 0 \\ 0 & 1 & 0 \\ 0 & 0 & 1 \end{bmatrix}$$

Here's what happens when a matrix doesn't have an inverse: You can't get ones along the diagonal by using the row operations. You usually get a whole row of zeros as a result of your row operations. A row of zeros is the warning that no inverse is possible.

Using a quick-and-slick rule for 2 × 2 matrices

You have a special rule at your disposal to find inverses of 2×2 matrices. To implement the rule for a 2×2 matrix, you have to switch two elements, negate two elements, and divide all the elements by the difference of the cross products of the elements. This may sound complicated, but the math is really neat and sweet, and the process is much quicker than the general method (see the previous section).

Figure 15-13 shows the general formula for the 2×2 matrix rule.

As you can see in Figure 15-13, the upper-left and lower-right corner elements are switched; the upper-right corner and lower-left corner elements are negated (changed to the opposite sign); and all the elements are divided by the result of doing two cross products and subtracting.

Figure 15-13:
The quick way to compute 2×2 inverses.

$$K = \begin{bmatrix} a & b \\ c & d \end{bmatrix}, \; K^{-1} = \begin{bmatrix} \dfrac{d}{ad-bc} & \dfrac{-b}{ad-bc} \\ \dfrac{-c}{ad-bc} & \dfrac{a}{ad-bc} \end{bmatrix}$$

To find the inverse of matrix Z from Figure 15-14 by using the 2×2 method, you switch the 5 and the 11, you change the 6 to –6 and the 9 to –9, and you divide by the difference of the cross products — $(5 \cdot 11) - (6 \cdot 9) = 55 - 54 = 1$. Watch the order in which you do the subtraction — the order does matter. Dividing each element by 1 doesn't change the elements, as you can see.

Figure 15-14:
Determining the inverse of matrix Z.

$$Z = \begin{bmatrix} 5 & 6 \\ 9 & 11 \end{bmatrix}, \; Z^{-1} = \begin{bmatrix} 11 & -6 \\ -9 & 5 \end{bmatrix}$$

Dividing Matrices by Using Inverses

Until this point in the chapter, I've avoided the topic of division of matrices. I haven't spent much time on the topic because you don't *really* divide matrices — you multiply one matrix by the inverse of the other (see the previous section for info on inverses). The division process resembles what you can do with real numbers. Instead of dividing 27 by 2, for example, you can multiply 27 by 2's inverse, ½.

To divide the matrices shown in Figure 15-15, you first find the inverse of the matrix in the denominator. You then multiply the matrix in the numerator by the inverse of the matrix in the denominator (see the section "Multiplying two matrices" earlier in this chapter).

$$\frac{\begin{bmatrix} 3 & -2 \\ 4 & -3 \end{bmatrix}}{\begin{bmatrix} 6 & -10 \\ 1 & -2 \end{bmatrix}} = \begin{bmatrix} 3 & -2 \\ 4 & -3 \end{bmatrix} * \begin{bmatrix} 1 & -5 \\ 0.5 & -3 \end{bmatrix} = \begin{bmatrix} 3-1 & -15+6 \\ 4-1.5 & -20+9 \end{bmatrix} = \begin{bmatrix} 2 & -9 \\ 2.5 & -11 \end{bmatrix}$$

Figure 15-15:
Avoiding
dreaded
division by
using
inverses.

$$\begin{bmatrix} 6 & -10 \\ 1 & -2 \end{bmatrix}^{-1} = \begin{bmatrix} \dfrac{-2}{-12-(-10)} & \dfrac{10}{-12-(-10)} \\ \dfrac{-1}{-12-(-10)} & \dfrac{6}{-12-(-10)} \end{bmatrix} = \begin{bmatrix} 1 & -5 \\ 0.5 & -3 \end{bmatrix}$$

Using Matrices to Find Solutions for Systems of Equations

One of the nicest applications of matrices is that you can use them to solve systems of linear equations. In Chapter 12, you find out how to solve systems of two, three, four, and more linear equations. The methods you use in that chapter involve elimination of variables and substitution. When you use matrices, you deal only with the coefficients of the variables in the problem. This way is less messy, and you can enter the matrices into graphing calculators or computer programs for an even more relaxing process.

This method is the most desirable when you have some technology available to you. Finding inverse matrices can be nasty work if fractions and decimals crop up. Simple graphing calculators can make the process quite nice.

You can use matrices to solve systems of linear equations as long as the number of equations and variables is the same. To solve a system, adhere to the following steps:

1. **Make sure all the variables in the equations appear in the same order.**

   Replace any missing variables with zeros, and write all constants on the other side of the equal sign from the variables.

2. **Create a square *coefficient matrix*, A, by using the coefficients of the variables.**

3. **Create a column *constant matrix*, B, by using the constants in the equations.**

4. **Find the inverse of the coefficient matrix, A^{-1}.**

   You can perform this step by using the procedures shown in the section "Finding Inverse Matrices" or by using a graphing calculator.

5. **Multiply the inverse of the coefficient matrix times the constant matrix — $A^{-1} \cdot B$.**

   The resulting column matrix has the solutions, or the values of the variables, in order from top to bottom.

Say, for example, that you want to solve the following system of equations:

$$\begin{cases} x - 2y + 8z = 5 \\ 2x + 15z = 3y + 6 \\ 8y + 22 = 4x + 30z \end{cases}$$

You first rewrite the equations so that the variables appear in order and the constants appear on the right side of the equations:

$$\begin{cases} x - 2y + 8z = 5 \\ 2x - 3y + 15z = 6 \\ -4x + 8y - 30z = -22 \end{cases}$$

Now you follow Steps 2 and 3 by writing in the coefficient matrix, A, and the constant matrix, B (shown in the following figure):

$$A = \begin{bmatrix} 1 & -2 & 8 \\ 2 & -3 & 15 \\ -4 & 8 & -30 \end{bmatrix}, \quad B = \begin{bmatrix} 5 \\ 6 \\ -22 \end{bmatrix}$$

Step 4 is finding the inverse of matrix A. Follow these steps:

   Write the identity matrix next to the original matrix.

   Add Row 2 to (–2 · Row 1) for a new Row 2.

   Add Row 3 to (4 · Row 1) for a new Row 3.

   Add Row 1 to (2 · Row 2) for a new Row 1.

   Multiply Row 3 by 0.5.

   Add Row 1 to (–6 · Row 3) for a new Row 1.

   Add Row 2 to Row 3 for a new Row 2.

Voila! The inverse matrix. See the following for the finished product:

$$A^{-1} = \begin{bmatrix} -15 & 2 & -3 \\ 0 & 1 & 0.5 \\ 2 & 0 & 0.5 \end{bmatrix}$$

Now you multiply the inverse of matrix A times the constant matrix, B (see the section "Multiplying two matrices"); you get a column matrix with all the solutions of x, y, and z listed in order from top to bottom:

$$A^{-1} * B = \begin{bmatrix} -15 & 2 & -3 \\ 0 & 1 & 0.5 \\ 2 & 0 & 0.5 \end{bmatrix} * \begin{bmatrix} 5 \\ 6 \\ -22 \end{bmatrix} = \begin{bmatrix} 3 \\ -5 \\ -1 \end{bmatrix}$$

The column matrix tells you that $x = 3$, $y = -5$, and $z = -1$.

Chapter 16

Making a List: Sequences and Series

A *sequence* is a list of items or individuals — and because this is an algebra book, the sequences you see feature lists of numbers. A *series* is the sum of the numbers in a list. These concepts pop up in many areas of life (outside of top-ten lists). For example, you can make a list of the number of seats in each row at a movie theater. With this list, you can add the numbers to find the total number of seats. You want to see situations where the number of items in a list isn't random; you prefer when the number follows a pattern or rule. You can describe the patterns formed by elements in a sequence with mathematical expressions containing mathematical symbols and operations. In this chapter, you discover how to describe the terms in sequences and, when you get lucky, how to add as many of the terms as you want without too much fuss or bother.

Understanding Sequence Terminology

A *sequence of events* consists of two or more happenings in which one item or event follows another, which follows another, and so on. In mathematics, a *sequence* is a list of *terms,* or numbers, created with some sort of mathematical rule. For instance, $\{3 + 4n\}$ says that the numbers in a sequence should start with the number 7 and increase by four with each additional term. The numbers in the sequence are 7, 11, 15, ...

You call the three dots following a short list of terms an *ellipsis* (not to be confused with the ellipse from Chapter 11). You use them in place of "et cetera" (etc.) or "and so on."

To get even more specific, here's the formal definition of a *sequence:* A function whose domain consists of positive integers (1, 2, 3, ...). This is really a nice feature — having to deal with only the positive integers. The rest of this section has many more useful tidbits about sequences, guaranteed to leave your brain satisfied.

Using sequence notation

One big clue that you're dealing with a sequence is when you see something like {7, 10, 13, 16, 19, ...} or $\{a_n\}$. The braces, { }, indicate that you have a list of items, called *terms;* commas usually separate the terms in a list from one another. The term a_n is the notation for the rule that represents a particular sequence. When identifying a sequence, you can list the terms in the sequence, showing enough terms to establish a pattern, or you can give the rule that creates the terms.

For instance, if you see the notation $\{2n + 1\}$, you know that the sequence consists of the terms {3, 5, 7, 9, 11, 13, ...}. $2n + 1$ is the rule that creates the sequence when you insert all the positive integers in place of the n. When $n = 1, 2(1) + 1 = 3$; when $n = 2, 2(2) + 1 = 5$; and so on. The domain of a sequence is all positive integers, so the process is easy as 1, 2, 3.

Because the terms in a sequence are linked to positive integers, you can refer to them by their positions in the listing of the integers. If the rule for a sequence is $\{a_n\}$, for example, the terms in the sequence are named a_1, a_2, a_3, a_4, and so on. This nice, orderly arrangement allows you to ask for the tenth term in the sequence $\{a_n\} = \{n^2 - 1\}$ by writing $a_{10} = 10^2 - 1 = 99$. You don't have to write the first nine terms to get to the tenth one. Sequence notation is a time saver!

No-fear factorials in sequences

A mathematical operation you see in many sequences is the *factorial.* The symbol for factorial is an exclamation mark.

Here's the formula for the factorial in a sequence: $n! = n(n-1)(n-2)(n-3) \ldots 3 \cdot 2 \cdot 1$.

When you perform a factorial, you multiply the number you're operating on times every positive integer smaller than that number.

For example, $6! = 6 \cdot 5 \cdot 4 \cdot 3 \cdot 2 \cdot 1 = 720$, and $9! = 9 \cdot 8 \cdot 7 \cdot 6 \cdot 5 \cdot 4 \cdot 3 \cdot 2 \cdot 1 = 362{,}880$.

You can apply a special rule for $0!$ (zero factorial). The rule is that $0! = 1$. "What? How can that possibly be?" Well, it is. Mathematicians discovered that assigning the number 1 to $0!$ makes everything involving factorials work out better. (Why you need this rule for $0!$ becomes more apparent in Chapter 17. For now, just be content to use the rule when writing the terms in a sequence.)

So, if $\{c_n\} = \{n! - n\}$, you write that $c_1 = 1! - 1 = 1 - 1 = 0$, $c_2 = 2! - 2 = 2 \cdot 1 - 2 = 0$, $c_3 = 3! - 3 = 3 \cdot 2 \cdot 1 - 3 = 6 - 3 = 3$, and so on. You write the terms in the sequence as $\{0, 0, 3, 20, \ldots\}$.

Alternating sequential patterns

One special type of sequence is an alternating sequence. An *alternating sequence* has terms that forever alternate back and forth from positive to negative to positive. An alternating sequence has a multiplier of -1, which is raised to some power such as n, $n - 1$, or $n + 1$. Adding the power, which is related to the number of the term, to the -1 causes the terms to alternate because the positive integers alternate between even and odd. Even powers of -1 are equal to $+1$, and odd powers of -1 are equal to -1.

For instance, the alternating sequence $\{(-1)^n 2(n + 3)\} = \{-8, 10, -12, 14, \ldots\}$, because

$$a_1 = (-1)^1 2(1 + 3) = -1 \cdot 2(4) = -8$$

$$a_2 = (-1)^2 2(2 + 3) = +1 \cdot 2(5) = 10$$

$$a_3 = (-1)^3 2(3 + 3) = -1 \cdot 2(6) = -12$$

and so on.

Here's an example of an alternating sequence that has a factorial (see the previous section) and fraction — a little bit of everything. The first four terms of the sequence $\left\{(-1)^n \dfrac{(n+1)!}{n}\right\}$ are as follows:

$$a_1 = (-1)^1 \frac{(1+1)!}{1} = -1\left[\frac{2!}{1}\right] = -1\left[\frac{2}{1}\right] = -2$$

$$a_2 = (-1)^2 \frac{(2+1)!}{2} = +1\left[\frac{3!}{2}\right] = 1\left[\frac{6}{2}\right] = 3$$

$$a_3 = (-1)^3 \frac{(3+1)!}{3} = -1\left[\frac{4!}{3}\right] = -1\left[\frac{24}{3}\right] = -8$$

$$a_4 = (-1)^4 \frac{(4+1)!}{4} = +1\left[\frac{5!}{4}\right] = 1\left[\frac{120}{4}\right] = 30$$

So, the sequence is {–2, 3, –8, 30, ...}.

You can see how the absolute value of the terms keeps getting larger while the terms alternate between positive and negative (see Chapter 2 for more on absolute value).

Looking for sequential patterns

The list of terms in a sequence may or may not display an apparent pattern. Of course, if you see the function rule — the rule that tells you how to create all the terms in the sequence — you have a huge hint about the pattern of the terms. You can always list the terms of a sequence if you have the rule, and you can often list the rule when you have enough terms in the sequence to figure out the pattern.

The patterns you can look for range from simple to a bit tricky:

- ✔ A single-number difference between each term, such as 4, 9, 14, 19, ..., where the difference between each term is 5.

- ✔ A multiplier separating the terms, such as multiplying by 5 to get 2, 10, 50, 250, ...

- ✔ A pattern within a pattern, such as with the numbers 2, 5, 9, 14, 20, ..., where the differences between the numbers get bigger by one each time.

When you have to figure out a pattern and write a rule for a sequence of numbers, you can refer to your list of possibilities — the ones I mention and others — and see which type of rule applies.

Difference between terms

The quickest, easiest pattern to find features a common difference between the terms. A *difference* between two numbers is the result of subtracting. You can usually tell when you have a sequence of this type by inspecting it — looking at how far apart the numbers sit on the number line.

When looking for differences in a list of numbers, be careful that you always subtract in the same order — the number minus the number just to the left of it.

The following three sequences have something in common: the terms in the sequences have a common *first difference,* a common *second difference,* or a common *third difference.*

First difference

When the *first difference* of the terms in a sequence is a constant number, the rule determining the terms is usually a linear expression (a linear expression has an exponent of 1 on n; see Chapter 2). For example, the sequence of numbers $\{2, 7, 12, 17, 22, 27, ...\}$ consists of terms that have a common difference of 5. The rule for this example sequence is $\{5n - 3\}$. You use the multiplier 5 to make the terms in the sequence each 5 more than the previous term. You subtract the 3 because when you replace n with 1, you get a number too big; you want to start with the number 2, so you subtract 3 from the first multiple. Sequences with a common first difference are called *arithmetic sequences* (I cover these sequences thoroughly in the section "Taking Note of Arithmetic and Geometric Sequences").

Second difference

When the *second difference* of the terms in a sequence is a constant, like 2, the rule for that sequence is usually quadratic (see Chapter 3) — it contains the term n^2. The sequence of numbers $\{-2, 1, 6, 13, 22, 33, 46, ...\}$, for example, consists of terms that have a common second difference of 2. The first differences between the terms increase by two for each interval:

$$-2 \quad 1 \quad 6 \quad 13 \quad 22 \quad 33 \quad 46$$
$$3 \quad 5 \quad 7 \quad 9 \quad 11 \quad 13$$
$$2 \quad 2 \quad 2 \quad 2 \quad 2$$

The rule used to create this example sequence is $\{n^2 - 3\}$.

I can't give you a quick, easy way to find the specific rules, but if you know that a rule should be quadratic, you have a place to start. You can try squaring the numbers 1, 2, 3, and so on, and then see how you have to "adjust" the squares by subtracting or adding so that the numbers in the sequence appear according to the rule.

Third difference

The sequence $\{0, 6, 24, 60, 120, 210, 336, ...\}$ features a common *third difference* of 6. In the following diagram, you see that the row below the sequences shows the first differences; under the first differences are the second differences; and, finally, the third row contains the third differences:

$$0 \quad 6 \quad 24 \quad 60 \quad 120 \quad 210 \quad 336$$
$$6 \quad 18 \quad 36 \quad 60 \quad 90 \quad 126$$

$$12 \quad 18 \quad 24 \quad 30 \quad 36$$

$$6 \quad 6 \quad 6 \quad 6$$

The rule for this example sequence is $\{n^3 - n\}$, which contains a cubed term.

The rule for this sequence doesn't necessarily leap off the page when you look at the terms. You have to play around with the terms a bit to determine the rule. Start with a cubed term and then try subtracting or adding constant numbers. If that doesn't help, try adding or subtracting squares of numbers or just multiples of numbers. Sounds sort of haphazard, but you have limited options without the use of some calculus. Graphing calculators have curve-fitting features that take data and figure out the rules for you, but you still have to choose which kind of rules (what powers) make the data work.

Multiples and powers

Some sequences have fairly apparent rules that generate their terms, because each term is a multiple or a power of some constant number. For instance, the sequence {3, 6, 9, 12, 15, 18, ...} consists of multiples of 3, and its rule is $\{3n\}$.

But what if the sequence starts with 21? What's the rule for {21, 24, 27, 30, 33, 36, ...}? The terms are all multiples of 3, but $\{3n\}$ doesn't work because you have to start with $n = 1$. Remember, the domain or input of a sequence is made up of positive integers — 1, 2, 3, ... — so you can't use anything smaller than 1.

The way to get around starting sequences with smaller numbers is to add a constant to n (which is like a counter). The number 21 is $3 \cdot 7$, so add 6 to n $(1 + 6 = 7)$ to form the rule $\{3(n + 6)\}$.

The sequence $\left\{1, -\frac{1}{2}, \frac{1}{3}, -\frac{1}{4}, \frac{1}{5}, -\frac{1}{6}, ...\right\}$ has two interesting features: The terms alternate signs (see the section "Alternating sequential patterns"), and the fractions have the positive integers in their denominators. To write the rule for this sequence, consider the two features. The alternating terms suggest a power of –1. The first, third, fifth, and all other odd terms are positive, so you can raise the –1 factor to $n + 1$ to make those exponents even and the terms positive. For the fractions, you can put n, the term's number, in the denominator. The rule for this sequence, therefore, is as follows:

$$\left\{(-1)^{n+1} \frac{1}{n}\right\} = \left\{\frac{(-1)^{n+1}}{n}\right\}$$

Other sequences can have terms that are all powers of the same number. These sequences are called *geometric sequences* (I discuss them at great length in the section "Taking Note of Arithmetic and Geometric Sequences"). An example of a geometric sequence is {2, 4, 8, 16, 32, 64, 128, ...}. You can see that these terms are powers of the number 2.

Taking Note of Arithmetic and Geometric Sequences

Arithmetic and geometric sequences are special types of sequences that have many applications in mathematics. Because you can usually recognize arithmetic or geometric sequences and write their general rules with ease, these sequences have become a mathematician's best friends. Arithmetic and geometric sequences also have very nice formulas for the sums of their terms, which opens up a whole new branch of mathematical activity.

Finding common ground: Arithmetic sequences

Arithmetic sequences (pronounced *air-ith-mat-ick,* with the emphasis on *mat*) are sequences whose terms have the same differences between them, no matter how far down the lists you go (in other words, how many terms the lists include).

One way to describe the general formula for arithmetic sequences is with the following:

$$a_n = a_{n-1} + d$$

The formula says that the *nth* term of the sequence is equal to the term directly before it (the $n - 1$st term) plus the common difference, d.

Another equation you can use with arithmetic sequences is the following:

$$a_n = a_1 + (n - 1)d$$

This formula says that the *nth* term of the sequence is equal to the first term, a_1, plus $n - 1$ times the common difference, d.

The equation you use depends on what you want to accomplish. You use the first formula if you single out a term in the sequence and want to find the next one in line. For example, if you want the next term after 201 in the sequence $a_n = a_{n-1} + 3$, you just add 3 to the 201 and get that the next term is 204. You use the second formula if you want to find a specific term in a sequence. For example, if you want the 50th term in the sequence where $a_n = 5 + (n-1)7$, you replace the n with 50, subtract the 1, multiply by 7, add 5, and you get 348. All that may sound like a lot of work, but it's easier than listing the 50 terms. And you can solve for one of these rules for a sequence if you have just the right information.

For instance, if you know that the common difference between the terms of an arithmetic sequence is 4 and that the sixth term is 37, you can substitute this information into the equation $a_n = a_1 + (n-1)d$, letting $a_n = 37$, $n = 6$, and $d = 4$. You get the following:

$$a_n = a_1 + (n-1)d$$
$$a_6 = a_1 + (6-1)d$$
$$37 = a_1 + (6-1) \cdot 4$$
$$37 = a_1 + 20$$
$$17 = a_1$$

You find that the first term is 17. Now you can solve for the general rule by replacing a_1 and d and simplifying:

$$a_n = a_1 + (n-1)d$$
$$a_n = 17 + (n-1) \cdot 4$$
$$a_n = 17 + 4n - 4$$
$$a_n = 13 + 4n$$

An arithmetic sequence that has a common difference of 4 and whose sixth term is 37 has the general rule $\{13 + 4n\}$.

You use this procedure when you're given the number values of the terms flat out and when you have to figure them out from some application or story problem. Here's an example of a story problem for which you need to use an arithmetic sequence. You and a group of friends have been hired to be ushers at a local theatre performance, and your payment includes free tickets to the show. The theatre allocates the whole last row in the middle section for your group. The first row in the middle section of the theatre has 26 seats, and one more seat appears in each row as you move backward for a total of 25 rows. How many seats are in the last row?

You can solve this problem quickly with an arithmetic sequence. Using the formula $a_n = a_1 + (n-1)d$, you replace a_1 with 26, n with 25, and d with 1:

$$a_n = a_1 + (n-1)d$$
$$a_{25} = 26 + (25-1) \cdot 1$$
$$a_{25} = 26 + (24) \cdot 1 = 26 + 24$$
$$a_{25} = 50$$

It looks like your friends can bring their friends, too!

Taking the multiplicative approach: Geometric sequences

A *geometric sequence* is a sequence in which each term is different from the one that follows it by a common ratio. In other words, the sequence has a constant number that multiplies with each term to create the next one. With arithmetic sequences, you add a constant; with geometric sequences, you multiply.

A general formula or rule for a geometric sequence is as follows:

$$g_n = rg_{n-1}$$

In this equation, r is the constant ratio that multiplies each term. The rule says that, to get the *nth* term, you multiply the term before it — the $(n-1)$st term — by the ratio, r.

Another way you can write the general rule for a geometric sequence is as follows:

$$g_n = g_1 r^{n-1}$$

The second form of the rule involves the first term, g_1, and applies the ratio as many times as needed. The *nth* term is equal to the first term multiplied by the ratio $n-1$ times.

You use the first rule, $g_n = rg_{n-1}$, when you're given the ratio, r, and a particular term in the sequence, and you want the next term. If the ninth term in a sequence whose ratio is 3 is 65,610, for example, and you want the tenth term, you multiply 65,610 times 3 to get 196,830. You use the second rule, $g_n = g_1 r^{n-1}$, when you're given the first term in the sequence and you want a particular term. For instance, if you know that the first term is 3, the ratio is 2, and you want the 10th term, you find that term by multiplying 3 times 2^9, which is 1,536.

You can always find the ratio or multiplier, r, if you have two consecutive terms in a geometric sequence. Just divide the second term by the term immediately preceding it — the quotient is the ratio. For example, if you have a geometric sequence where the sixth term is 1,288,408 and the fifth term is 117,128, you can find r by dividing 1,288,408 by 117,128 to get 11. The ratio, r, is 11.

Say that the rule for a particular geometric sequence is $\left\{360\left(\frac{1}{3}\right)^{n-1}\right\}$. When $n = 1$, the power on the fraction is 0, and you have 360 multiplying the number 1. So, $g_1 = 360$. When $n = 2$, the exponent is equal to 1, so the fraction multiplies the 360, and you get 120. Here are the first few terms in this sequence:

$$\left\{360, 120, 40, \frac{40}{3}, \frac{40}{9}, \frac{40}{27}, \ldots\right\}$$

You can find each term by multiplying the previous term by $\frac{1}{3}$.

Here's another example to familiarize you with geometric formulas . . . feel free to break it out at parties in the future. An unwise gambler bets a dollar on the flip of a coin and loses. Instead of paying up, he says, "Double or nothing," meaning that he wants to flip the coin again; he'll pay two bucks if he loses, and his opponent gets nothing if the gambler wins. Oops! He loses again, and again he says, "Double or nothing!" If he repeats this doubling-and-losing process 20 times, how much will he owe on the 21st try?

Using the formula $g_n = g_1 r^{n-1}$ (you know the first term — the one dollar — and the multiplier), you replace the first term, g_1, with the number 1, r with 2 for the doubling, and n with 21:

$$g_{21} = 1(2)^{21-1} = 1(2)^{20} = 1,048,576$$

The gambler will owe over one million dollars if he keeps going to 21. If he's unwilling to part with his first dollar, how's he going to deal with this number?

Maybe the gambler shouldn't start with such a big bet at the beginning. What if he starts with a quarter rather than a dollar? Using the same formula, $g_n = g_1 r^{n-1}$, the first term is 0.25, and the ratio is still 2. So, $g_{21} = (0.25)2^{20} = \$262,144$. Looks like he's still in big trouble. And, in two more flips, letting $n = 23$, he reaches the starting point for the dollar bill.

Recursively Defining Functions

An alternate way to describe the terms of a sequence, in place of giving the general rule for the sequence, is to define the sequence *recursively.* To do so, you identify the first term, or maybe a few of the initial terms, and describe how to find the rest of the terms by using the terms that come before them.

The recursive rule for arithmetic sequences is $a_n = a_{n-1} + d$, and the recursive rule for geometric sequences is $g_n = rg_{n-1}$.

Here's an example of a recursively defined sequence. Let $a_1 = 6$ and $a_n = 2a_{n-1} + 3$. The formula says that, to find a term in the sequence, you look at the previous term (a_{n-1}), double it $(2a_{n-1})$, and add 3. The first term is 6, so the second term is 3 more than the double of 6, or 15. The next term is 3 more than the double of 15, or 33. Here are some of the terms of this sequence listed in order: {6, 15, 33, 69, 141, ...}.

Sometimes, you have to list the numbers in a recursive sequence when given a rule, and, if you're really lucky, you get to create the rule yourself. You'll see the "create your own rule" option if you get into discrete mathematics or computer programming — just not here!

You can also define sequences recursively by referring to more than one previous term. For instance, assume that you let $a_n = 3a_{n-2} + a_{n-1}$. This rule says that to find the *n*th term in the sequence (you choose the *n* you want — the 5th term, the 50th term, and so on), you have to look at the two previous terms [the $(n-2)$nd and the $(n-1)$st terms], multiply the term two positions back by 3, $3(a_{n-2})$, and then add the term one position back to the product, a_{n-1}.

To start writing the terms of this sequence, you need to identify two consecutive terms. For this particular sequence, you decide to let $b_1 = 4$ and $b_2 = -1$. (Okay, you didn't get to decide, I did. I just picked random numbers, but, after they're chosen, they determine what's going to happen to the rest of the numbers in the sequence — using the given rule.) Here's how the terms go (**Note:** If you're looking for the sixth term, you need the $[n-1]$st term, which is the fifth term, and the $[n-2]$nd term, which is the fourth term):

$$b_1 = 4, \ b_2 = -1$$
$$b_n = 3b_{n-2} + b_{n-1}$$
$$b_3 = 3b_1 + b_2 = 3(4) + (-1) = 12 - 1 = 11$$
$$b_4 = 3b_2 + b_3 = 3(-1) + 11 = -3 + 11 = 8$$
$$b_5 = 3b_3 + b_4 = 3(11) + 8 = 33 + 8 = 41$$
$$b_6 = 3b_4 + b_5 = 3(8) + 41 = 24 + 41 = 65$$

Recursively formed sequences use earlier terms in the sequences to form later terms. The rules used to write these sequences aren't as handy as the rules that allow you to find the 50th or 100th term (like arithmetic and geometric sequences) without finding all the terms that come before. But the recursive rule is sometimes easier to write.

For example, if you find out that your salary at a new job would be $20,000 this year, $25,000 next year, and that every year after it will be 80 percent of the salary from 2 years ago plus 40 percent of the salary from the previous year, would you be willing to sign the contract? The rule reads: $b_n = 0.8b_{n-2} + 0.4b_{n-1}$. Using the first two terms and this rule, your salary for the first five years would be {20,000, 25,000, 26,000, 30,400, 32,960, ...}.

Making a Series of Moves

A *series* is the sum of a certain number of terms of a sequence. How many terms? That's part of the problem — either given to you or what you determine in order to answer some question.

Being able to list all the terms in a sequence is a handy tool to have in your algebra tool belt, but you can do much more with sequences. For instance, adding a certain number of terms in a sequence is helpful when the sequence is a list of how much money you're getting for an allowance during the month or how many seats sit in a theatre.

Finding the sum of the sequence means adding as many terms as you need to figure out your total allowance for the month or to find out how many people are in the first 20 rows of the theatre. This process doesn't sound like a chore, especially with hand-held calculators, but if the numbers get really big and you want the sum of many terms, the task can be daunting.

For this reason, many sequences used in business and financial applications have formulas for the sum of their terms. These formulas are a big help. In fact, for some geometric series, you can add *all* the terms — forever — and be able to predict the sum of all the terms.

Introducing summation notation

Mathematicians like to keep formulas and rules neat and concise, so they created a special symbol to indicate that you're adding the terms of a sequence. The special notation is *sigma,* Σ, or *sum notation.*

The notation $\sum_{k=1}^{n} a_k$ indicates that you want to add all the terms in the sequence with the general rule a_k, all the way from $k = 1$ until $k = n$:

$$\sum_{k=1}^{n} a_k = a_1 + a_2 + a_3 + a_4 + \ldots + a_{n-1} + a_n$$

For instance, if you want the sum $\sum_{k=1}^{5} (k^2 - 2)$, you need to find the first five terms, letting $k = 1, 2,$ and so on; you finish by adding all those terms:

$$\sum_{k=1}^{5} (k^2 - 2) = (1^2 - 2) + (2^2 - 2) + (3^2 - 2) + (4^2 - 2) + (5^2 - 2)$$
$$= -1 + 2 + 7 + 14 + 23 = 45$$

You find that the sum is 45.

If you want to add *all* the terms in a sequence, forever and ever, you use the symbol ∞ in the sigma notation, which looks like $\sum_{k=1}^{\infty} a_k$.

Summing arithmetically

An arithmetic sequence has a general rule (see the section "Finding common ground: Arithmetic sequences") that involves the first term and the common difference between consecutive terms: $a_n = a_1 + (n-1)d$. An arithmetic series is the sum of the terms that come from an arithmetic sequence. Consider the arithmetic sequence $a_n = 4 + (n-1)5 = 5n-1$. The first ten terms in this sequence are 4, 9, 14, 19, 24, 29, 34, 39, 44, and 49. The sum of these ten terms is 265. How did I get that value? Pencil and paper, my friend. Simple addition works just fine for a small list of numbers. However, your algebra teacher may not always bless you with small lists. Consider, now, a formula for the sum of the first n terms of an arithmetic sequence.

The sum of the first n terms of an arithmetic sequence, S_n, is

$$S_n = \frac{n}{2}\big[2a_1 + (n-1)d\big] = \frac{n}{2}(a_1 + a_n)$$

Here, a_1 and d are the first term and difference, respectively, of the arithmetic sequence $a_n = a_1 + (n-1)d$. The n indicates which term in the sequence you get when you put the value of n in the formula.

The first part of the formula, on the left, allows you to insert the first term, the difference, and the number of terms you want to add. The second part, on the right, is quicker and easier; you use it when you know both the first and last terms and how many terms you want to add. In the next example, n is 10, because that's how many terms you're adding. Inserting the 10 for n in the formula for the general term, you get 49.

To use the formula for the sum of the ten numbers 4, 9, 14, 19, 24, 29, 34, 39, 44, and 49 (previously added to get 265), you insert the known data:
$$S_{10} = \frac{10}{2}(4 + 49) = 5(53) = 265.$$

Now, say you want to add the first 100 numbers in a sequence that starts with 13 and features a common difference of 2 between each of the terms: 13 + 15 + 17 + 19 + ..., all the way to the 100th number. You find the sum of these 100 numbers by using the first part of the sum formula:

$$S_n = \frac{n}{2}\big[2a_1 + (n-1)d\big]$$
$$S_{100} = \frac{100}{2}\big[2(13) + (100-1)2\big] = 50(26 + 198) = 50(224) = 11{,}200$$

Summing geometrically

A geometric sequence consists of terms that differ from one another by a common ratio. You multiply a term in the sequence by a constant number or ratio to find the next term. You can use two different formulas to find the sum of the terms in a geometric sequence. You use the first formula to find the sum of a certain, finite number of terms of a geometric sequence — any geometric sequence at all. The second formula applies only to geometric sequences that have a ratio that lies between zero and one (a proper fraction); you use it when you want to add all the terms in the sequence — forever and ever (for more on geometric sequences, see the section "Taking the multiplicative approach: Geometric sequences").

Adding the first n terms

The formula you use to add a specific, finite number of terms from a geometric sequence involves a fraction where you subtract the ratio — or a power of the ratio — from one. You can't reduce the formula, so don't try. Just use it as it is.

You find the sum of the first n terms of the geometric sequence $g_n = g_1 r^{n-1}$ with the following formula:

$$S_n = \frac{g_1(1 - r^n)}{1 - r}$$

The term g_1 is the first term of the sequence, and r represents the common ratio.

For example, say you want to add the first ten terms of the geometric sequence {1, 3, 9, 27, 81, ...} You identify the first term, the 1, and then the ratio by which you multiply, 3. Substitute this information into the formula:

$$S_n = \frac{g_1(1 - r^n)}{1 - r}$$

$$S_{10} = \frac{1(1 - 3^{10})}{1 - 3} = \frac{1 - 59,049}{-2} = \frac{-59,048}{-2} = 29,524$$

Quite a big number! Isn't using the formula easier than adding 1 + 3 + 9 + 27 + 81 + 243 + 729 + 2,187 + 6,561 + 19,683?

Adding all the terms to infinity

Geometric sequences have a ratio, or multiplier, that changes one term into the next one in line. If you multiply a number by 4 and the result by 4 and keep going, you create huge numbers in a short amount of time. So, it may sound impossible to add numbers that seem to get infinitely large.

But algebra has a really wonderful property for geometric sequences with ratios between negative one and one. The numbers in these sequences get smaller and smaller, and the sums of the terms in these sequences never exceed set, constant values.

If the ratio is bigger than one, the sum just grows and grows, and you get no final answer. If the ratio is negative and between 0 and –1, the sum is a single, constant value. For ratios smaller than –1, you have chaos again.

In Figure 16-1, you see the terms in a sequence that starts with 1 and has a ratio of ½. You also see the sum of the terms up to each successive point.

The general rule for this sequence is as follows:

$$g_n = g_1 r^{n-1} = 1 \cdot \left(\frac{1}{2}\right)^{n-1} = \left(\frac{1}{2}\right)^{n-1}$$

So, you're really finding powers of ½. Here are the first few terms:

$$\left\{1, \frac{1}{2}, \frac{1}{4}, \frac{1}{8}, \frac{1}{16}, \frac{1}{32}, \frac{1}{64}, \frac{1}{128}, \frac{1}{256}, \frac{1}{512}, \cdots\right\}$$

As you see in Figure 16-1, the sum of the terms is equal to *almost* two as the number of terms increases. The number in the numerator of the fraction of the sum is always one less than twice the denominator. The sum in Figure 16-1 approaches two but never exactly hits it. The sum gets so very, very close, though, that you can round up to two. This business of *approaching* a particular value is true of any geometric sequence with a proper fraction (between zero and one) for a ratio.

Algebra also offers a formula for finding the sum of *all* the terms in a geometric sequence with a ratio between zero and one. You may think this formula is much more complicated than the formula for finding only a few terms, but that isn't the case. This formula is actually much simpler.

The sum of all the terms from a geometric sequence whose ratio, r, is between zero and one ($0 < r < 1$) is $S_n \to \frac{g_1}{1-r}$, where g_1 is the first term in the sequence.

You can apply this rule to the sum of the sequence whose first term is 1 and whose common ratio is ½:

$$S_n \to \frac{g_1}{1-r} = \frac{1}{1-\frac{1}{2}} = \frac{1}{\frac{1}{2}} = 2$$

n	g_n	S_n	Decimal
1	1	1	1
2	$\dfrac{1}{2}$	$1 + \dfrac{1}{2} = \dfrac{3}{2}$	1.5
3	$\dfrac{1}{4}$	$\dfrac{3}{2} + \dfrac{1}{4} = \dfrac{7}{4}$	2.75
4	$\dfrac{1}{8}$	$\dfrac{7}{4} + \dfrac{1}{8} = \dfrac{15}{8}$	1.875
5	$\dfrac{1}{16}$	$\dfrac{15}{8} + \dfrac{1}{16} = \dfrac{31}{16}$	1.9375
6	$\dfrac{1}{32}$	$\dfrac{31}{16} + \dfrac{1}{32} = \dfrac{63}{32}$	1.96875
$\vdots$	$\vdots$	$\vdots$	
12	$\dfrac{1}{2{,}048}$	$= \dfrac{4{,}095}{2{,}048}$	1.999511719
$\vdots$	$\vdots$	$\vdots$	
n	$\left(\dfrac{1}{2}\right)^{n-1}$	$= \dfrac{2\left(2^{n-1}\right)-1}{2^{n-1}}$	1.999999999......

Figure 16-1: Adding terms in a geometric sequence.

Applying Sums of Sequences to the Real World

Having the tools to add all the terms in a mathematical sequence is peachy keen from a homework standpoint, but what's the point outside the classroom? Why would anyone need to be able to add sequences in the real world? You may be surprised at the possible applications in many walks of life. Hopefully, the three examples I include in this section will give you a hint of how helpful figuring sums can be.

Cleaning up an amphitheater

You've been hired for cleanup duty in a huge theater after a performance by a popular band. One of your tasks is to steam-clean all the seats on the main floor. Because it takes two minutes per seat, you need to allocate enough time to do the entire job. The head of maintenance tells you that the first row has 36 seats, and each subsequent row has one more seat per row. The theater has a total of 25 rows. How many seats are in the theater? How much time will it take?

This problem calls for the sum of an arithmetic sequence (see the section "Summing arithmetically"). The first term is 36, the common difference is 1, and you have 25 terms. Using the formula for the sum of the terms of an arithmetic sequence, you calculate the following:

$$s_n = \frac{n}{2}\left[2a_1 + (n-1)d\right]$$
$$= \frac{25}{2}\left[2(36) + (25-1)\cdot 1\right]$$
$$= \frac{25}{2}\left[72 + 24\right]$$
$$= \frac{25}{2}\left[96\right] = 1{,}200$$

You have 1,200 seats to steam-clean. At 2 minutes per seat, you need to set aside 2,400 minutes, or 40 hours for the job. Better hire some help!

Negotiating your allowance

You approach your dad about increasing your allowance because $10 per week just doesn't cut it anymore. He replies, "Absolutely not. Not until you improve your math grade." You then negotiate a deal with the following proposition: You'll take 1 cent on the first day of the month, 2 cents on the second day of the month, 4 cents on the third, 8 cents on the fourth, and so on, doubling the amount each day until the end of the month. At that point, your dad can check on your math grade and see if he wants to change the system back and raise your allowance. He knows how crafty you are, so he asks you to explain what you're up to before he agrees.

How much will you get with your system for the month of January? That month has 31 days of penny-pinching allowance. Using the formula for the sum of a geometric sequence whose first term is 1, common ratio is 2, and number of terms is 31 (see the section "Summing geometrically"), you calculate the following:

$$S_n = \frac{a_1(1-r^n)}{1-r}$$
$$= \frac{1(1-2^{31})}{1-2}$$
$$= \frac{1-2{,}147{,}483{,}648}{-1}$$
$$= 2{,}147{,}483{,}647$$

Of course, your answer is in pennies, so you move the decimal point over. Your allowance comes to a total of $21,474,836.47. What does your dad think of your math ability now? Unfortunately, he wants you to care about your grades as much as your money. No deal.

Bouncing a ball

The bouncing-ball example comes from my book *Algebra For Dummies* (Wiley). I finally get a chance to explain where I got this example! In *Algebra For Dummies,* I use the bouncing ball to show you how to use exponents; however, I didn't expect to have so many questions about where I got the formula. So many people wrote to ask about this problem that I put a solution on my Web site so they could see that I didn't just make up the formula.

You want to figure the total distance (up and down and up and down . . .) that a superball travels in *n* bounces, plus the first drop, if it always bounces back 75 percent of the distance it falls. You drop the superball onto a smooth sidewalk from a window sitting 40 feet off the ground. Look at Figure 16-2 to see what the problem entails.

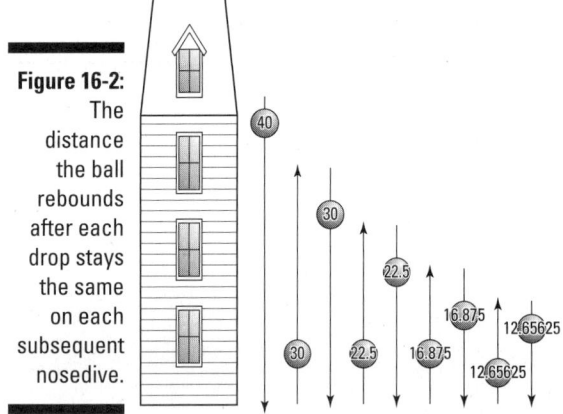

Figure 16-2: The distance the ball rebounds after each drop stays the same on each subsequent nosedive.

As you can see, the first distance is the 40-foot drop. The ball bounces up 75 percent of the distance it fell, or 30 feet, and drops 30 feet again. It then bounces back 75 percent of 30 feet, or 22.5 feet, and repeats the process. Except for the initial 40 feet, all the measures double to account for going upward and then downward.

The question is: How far does the superball travel in 10 bounces, plus that first drop? To figure this out by using a geometric sequence (see the section "Summing geometrically"), you let the first term equal 30, the ratio equal 0.75, and the number of terms equal 10. You double that sum of the sequence to account for the distance the ball travels both up and down, and you add the first 40 feet to get the total distance traveled by the ball.

First, you find the value of the sum of the sequence:

$$S_{10} = \frac{30\left(1 - 0.75^{10}\right)}{1 - 0.75}$$
$$= \frac{30\left(1 - 0.0563135147\right)}{0.25}$$
$$= \frac{30\left(0.9436864853\right)}{0.25}$$
$$= 113.2423782$$

Now you double the sum of the sequence and add 40 feet:

$2(113.2423782) + 40 = 266.4847565$ feet

The ball travels over 266 feet in 10 bounces. Well done!

The formula I use in *Algebra For Dummies* that caused all the questions was Distance $= 40 + 240\left[1 - 0.75^n\right]$. I created this formula by doubling the terms in the formula for the sum of the sequence and adding 40:

$$\text{Distance} = 40 + 2\left[\frac{30\left(1 - 0.75^n\right)}{1 - 0.75}\right]$$
$$= 40 + \frac{60\left(1 - 0.75^n\right)}{0.25}$$
$$= 40 + \frac{60\left(1 - 0.75^n\right)}{0.25} \cdot \frac{4}{4}$$
$$= 40 + \frac{4\left(60\right)\left(1 - 0.75^n\right)}{4\left(0.25\right)}$$
$$= 40 + \frac{240\left(1 - 0.75^n\right)}{1}$$
$$= 40 + 240\left(1 - 0.75^n\right)$$

Whew! I'm glad I have that off my chest. Now when people contact me about the formula, I can refer them to this book for a complete explanation.

Highlighting Special Formulas

Algebra offers several special types of sequences and series that you may use frequently in higher mathematics, such as calculus, and in financial and physics applications. For these applications, you have formulas for the sums of the terms in the sequences. Adding consecutive integers is a task made easier because you have formulas at your disposal. Counting tiles to be used in a floor or mosaic, computing the total amount of money in an annuity, and other such applications also use sums of sequences of numbers.

Here are some of the special formulas you may see:

The sum of the first n positive integers:

$$1 + 2 + 3 + \ldots + n = \frac{n(n+1)}{2}$$

The sum of the first n squares of the positive integers:

$$1^2 + 2^2 + 3^2 + \ldots + n^2 = \frac{n(n+1)(2n+1)}{6}$$

The sum of the first n cubes of the positive integers:

$$1^3 + 2^3 + 3^3 + \ldots + n^3 = \frac{n^2(n+1)^2}{4}$$

The sum of the first n odd positive integers:

$$1 + 3 + 5 + 7 + \ldots + (2n-1) = n^2$$

For example, if you want the sum of the first 10 squares of positive integers, you use the formula to get the following:

$$1 + 4 + 9 + \ldots 100 = \frac{10(10+1)(20+1)}{6}$$
$$= \frac{10(11)(21)}{6} = 385$$

The n is the number of the term, not the term itself.

To use the formula for the sum of odd-numbered positive integers, you need to determine the number of the term — which odd number signifies the end of the sequence. For instance, if you want to add all the positive odd numbers from 1 to 49, you determine the 49th term. Using the equation $2n - 1$ for 49, you see that $n = 25$, because $2(25) - 1 = 50 - 1 = 49$. So, you use $n = 25$ in the formula:

$$1 + 3 + 5 + \ldots + 49 = 25^2 = 625$$

Chapter 17

Everything You Wanted to Know about Sets

Back in the 1970s, a book titled *Everything You Always Wanted to Know About Sex — But Were Afraid to Ask* (Bantam) hit the shelves of stores everywhere. The book caused quite a buzz. Nowadays, however, people don't seem to be afraid to ask about — or inform you about — anything. So, in the modern spirit, this chapter lets it all hang out. Here you find out all the deep, dirty secrets concerning sets. I cover the union and intersection of sets, complementary sets, counting on sets, and drawing some very revealing pictures. Can you handle it?

Revealing Set Notation

A *set* is a collection of items. The items may be people, pairs of shoes, or numbers, for example, but they usually have something in common — even if the only characteristic that ties them together is the fact that they appear in the same set. The items in a set are called the set's *elements*.

Mathematicians have developed some specific notation and rules to help you maneuver through the world of sets. The symbols and vocabulary aren't difficult to master; you just have to remember what they mean. Familiarizing yourself with set notation is like learning a new language.

Listing elements with a roster

The name of a set appears as a capital letter (A in the following example, for instance) to distinguish it from other sets. To organize the elements of a set, you put them in *roster notation,* which just means to list the elements. For example, if a set A contains the first five whole numbers, you write the set as follows:

A = {0, 1, 2, 3, 4}

You list the elements in the set inside the braces and separate them by commas. The order in which you list the elements doesn't matter. You could also say that A = {0, 2, 4, 1, 3}, for example.

Building sets from scratch

A simple way to describe a set (other than set notation; see the previous section) is to use *rule* or *set builder notation.* For example, you can write the set A = {0, 1, 2, 3, 4} as follows:

A = {$x \mid x \, \varepsilon W, x < 5$}

You read the set builder notation as "A is the set containing all x such that x is an element of W, the whole numbers, and x is less than 5." The vertical bar, |, separates the variable from its rule, and the epsilon, ε, means *is an element of.* This is all wonderful math shorthand.

You may be wondering why in the world anyone would want to use this long-winded notation when listing the elements is easy enough. You'd be right to wonder; it doesn't make sense in the previous example, but what if you want to talk about a set B that contains all the odd numbers between 0 and 100? Do you really want to list all the elements?

When dealing with sets that have huge numbers of elements, using set builder notation can save you time and busywork. And if the pattern of elements is obvious (always a tricky word in math), you can use an ellipsis. For the set containing all the odd numbers between 0 and 100, for example, you can write B = {1, 3, 5, 7, ..., 99}, or you can use the set builder notation: B = {$x \mid x$ is odd, and $0 < x < 100$}. Both methods are easier than listing all the elements in the set.

Going for all (universal set) or nothing (empty set)

Consider the sets F = {Iowa, Ohio, Utah} and I = {Idaho, Illinois, Indiana, Iowa}. Set F contains three elements, and set I contains four elements. Within these sets, you can branch out to a set that you call the *universal set* for F and I. You can also distinguish a set called the *empty set,* or *null set.* This all or nothing business lays the foundation for performing set operations (see the section "Operating on Sets" later in this chapter).

The following list presents the characteristics of these all or nothing sets:

- ✓ **Universal set:** A universal set for one or more sets contains all the possible elements in a particular category. The writer of the situation must decide how many elements need to be considered in a particular problem. But one characteristic is pretty standard: The universal set is denoted U.

    For example, you could say that the universal set for F = {Iowa, Ohio, Utah} and I = {Idaho, Illinois, Indiana, Iowa} is U = {states in the United States}. The universal set for F and I doesn't have to be a set containing all the states; it can just be all the states that start with a vowel.

- ✓ **Empty (or null) set:** The opposite of the universal set is the empty set (or null set). The empty set doesn't contain anything (no kidding!). The two types of notation used to indicate the empty set are ø and { }. The first notation resembles a zero with a slash through it, and the second is empty braces. You must use one notation or the other, not both at the same time, to indicate that you have the empty set.

    For example, if you want to list all the elements in set G, where G is the set that contains all the states starting with the letter Q, you write G = { }. You have the empty set because no states in the United States start with the letter Q.

Subbing in with subsets

The real world provides many special titles for the little guy. Apartments can have sublets, movies can have subtitles, and ships can be submarines, so of course sets can have subsets. A *subset* is a set completely contained within another set — no element in a subset is absent from the set it's a subset of. Whew! That's a mouthful, and my sentence ends with a preposition. For the sake of good English, I'll start calling the sets *subsets* and *supersets* — a *superset* representing what the subset is a subset of (there I go again!).

Indicating subsets with notation

The set B = {2, 4, 8, 16, 32} is a subset of C = {$x | x = 2^n$, $n \in Z$}. Set B is a subset of set C, because B is completely contained in C. Set C consists of all the numbers that are powers of 2, where the powers are all elements of the set of integers (Z). The notation for *subset of* is ⊂, and you write B ⊂ C to say that B is a subset of C.

The letter Z usually represents the integers, the positive and negative whole numbers, and zero.

Another way you can write the superset C is

$$C = \left\{ ..., \frac{1}{8}, \frac{1}{4}, \frac{1}{2}, 1, 2, 4, 8, 16, 32, 64, 128, ... \right\}$$

using ellipses to indicate that the set continues on forever.

When one set is a subset of another, and the two sets aren't equal (meaning they don't contain the same elements), the subset is called a *proper subset*, indicating that the subset has fewer elements than the superset. Technically, any set is its own subset, so you can say that a set is an *improper subset* of itself. You write that statement with the subset notation and a line under it to indicate "subset and, also, equal to." To say that set B is its own subset, you write B ⊆ B. This may seem like a silly thing to do, but, as with all mathematical rules, you have a good reason for doing so. One of the reasons has to do with the number of subsets of any given set.

Counting the number of subdivisions

Have a look at the following listings of some selected sets and all their subsets. Notice that I include the empty set in each list of subsets. I do so because the empty set fulfills the definition that no element of the subset is absent from the superset. Here's the first set:

The subsets of A = {3, 8} are {3}, {8}, ø, {3, 8}.

The set A has four subsets: two subsets with one element, one with no elements, and one with both elements from the original set.

Here's a pet-themed set:

The subsets of B = {dog, cat, mouse} are {dog}, {cat}, {mouse}, {dog, cat}, {dog, mouse}, {cat, mouse}, ø, {dog, cat, mouse}.

The set B has eight subsets: three subsets with one element, three with two elements, one with no elements, and one with all the elements from the original set.

Time to get a bit bigger:

The subsets of set C = {r, s, t, u} are {r}, {s}, {t}, {u}, {r, s}, {r, t}, {r, u}, {s, t}, {s, u}, {t, u}, {r, s, t}, {r, s, u}, {r, t, u}, {s, t, u}, ø, {r, s, t, u}.

The set C has 16 subsets: four subsets with one element, six with two elements, four with three elements, one with no elements, and one with all the elements.

Have you determined a pattern yet? See if the following information helps:

A set with two elements has four subsets.

A set with three elements has eight subsets.

A set with four elements has sixteen subsets.

The number of subsets produced by a set is equal to a power of 2.

If a set A has *n* elements, it has 2^n subsets.

You can apply this rule to the set Q = {1, 2, 3, 4, 5, 6}, for example. The set has six elements, so it has 2^6 = 64 subsets. Knowing the number of subsets a set has doesn't necessarily help you list them all, but it lets you know if you've missed any. (Check out the section "Mixing up sets with combinations" to find out how many subsets of each type [number of elements] a set has.)

Operating on Sets

Algebra provides three basic operations that you can perform on sets: union, intersection, and complement (negation). The operations union and intersection take two sets at a time to perform — much like addition and subtraction take two numbers. Finding the complement of a set (or negating it) is like finding the opposite of the set, so you perform it on only one set at a time.

An additional process, which isn't really an operation, is counting up the number of elements contained in a set. This process has its own special notation to tell you to do the counting, just like the union, intersection, and negation operations have their own special symbols.

Celebrating the union of two sets

Finding the union of two sets is like merging two companies. You put them together to make one big set (forming unions probably isn't as lucrative, however).

To find the union of sets A and B, denoted A ∪ B, you combine all the elements of both sets by writing them into one set. You don't duplicate any elements that the sets have in common. Here's an example:

> Set A = {10, 20, 30, 40, 50, 60}
>
> Set B = {15, 30, 45, 60}
>
> A ∪ B = {10, 15, 20, 30, 40, 45, 50, 60}

You can also say that each set is a subset of the union of the sets.

You can apply the union operation on more than two sets at a time (but you have to have at least two sets). For instance, if you have sets R = {rabbit, bunny, hare}, S = {bunny, egg, basket, spring}, and T = {summer, fall, winter, spring}, the union R ∪ S ∪ T = {rabbit, bunny, hare, egg, basket, spring, summer, fall, winter}. Again, you mention each element only once in the union of the sets.

A couple of special unions involve subsets and the empty set. What happens when you perform the union of two sets, and one set is the subset of the other? For instance, say you have G = {1, 2, 3, 4, 5, 6} and H = {2, 4, 6}. You can write the union as G ∪ H = G, because the union of the sets is just the set G. H is a subset of G — every element in H is contained in G. Also, because H is a subset of G, H must also be a subset of the union of the two sets, G ∪ H.

Because the empty set is a subset of every set, you can say G ∪ ø = G, H ∪ ø = H, T ∪ ø = T, and so on. Think of how adding zero to a number affects the number — it doesn't.

Looking both ways for set intersections

The intersection of two sets is like the intersection of two streets. If Main Street runs east and west and University Street runs north and south, they share the street where they intersect. The street department doesn't have to pave the street twice, because the two streets share that little part.

To find the intersection of sets A and B, denoted A ∩ B, you list all the elements that the two sets have in common. If the sets have nothing in common, their intersection is the empty set. Here's an example:

> Set A = {a, e, i, o, u}
>
> Set B = {v, o, w, e, l}
>
> A ∩ B = {e, o}

If set A is a subset of another set, the intersection of A and its superset is just the set A. For instance, if set C = {a, c, e, g, i, k, m, o, q, s, u, w, z} and A = {a, e, i, o, u}, A ∩ C = {a, e, i, o, u} = A.

The intersection of any set with the empty set is just the empty set.

Feeling complementary about sets

The complement of set A, written A', contains every element that doesn't appear in A. Every item that doesn't appear in a set can be plenty of things — unless you limit your search.

To determine the complement of a set, you need to know the universal set. If A = {p, q, r}, and the universal set, U, contains all the letters of the alphabet, A' = {a, b, c, d, e, f, g, h, i, j, k, l, m, n, o, s, t, u, v, w, x, y, z}. You still deal with a lot of elements, but you limit your search to letters of the alphabet.

The number of elements in a set plus the number of elements in its complement always equals the number of elements in the universal set. You write this rule as $n(A) + n(A') = n(U)$. This relationship is very useful when you have to deal with large amounts of elements and you want some easy ways to count them.

Counting the elements in sets

Situations often arise when you need to be able to tell how many elements a set contains. When sets appear in probability, logic, or other mathematical problems, you don't always care what the elements are — just how many of them sit in the sets. The notation indicating that you want to know the number of elements contained in a set is $n(A) = k$, meaning "The number of elements in set A is k." The number k will always be some whole number: 0, 1, 2, and so on.

The union and intersection of sets have an interesting relationship concerning the number of elements:

$$n(A \cup B) = n(A) + n(B) - n(A \cap B)$$

Consider the sets A = {2, 4, 6, 8, 10, 12, 14, 16, 18, 20} and B = {3, 6, 9, 12, 15, 18, 21}, for example. The union of A and B, A ∪ B, = {2, 3, 4, 6, 8, 9, 10, 12, 14, 15, 16, 18, 20, 21}, and the intersection, A ∩ B, = {6, 12, 18}. You can apply the previous formula to count the number of elements in each set and the results of the operations. Here are the different bits of input: $n(A) = 10$, $n(B) = 7$, $n(A \cup B) = 14$, and $n(A \cap B) = 3$.

Filling in the numbers, you find $14 = 10 + 7 - 3$. This is a true statement. You want to check your arithmetic to be sure that you've figured the problem correctly.

Drawing Venn You Feel Like It

Venn diagrams are pictures that show the relationships between two or more sets and the elements in those sets. The adage "a picture is worth a thousand words" is never truer than with these diagrams. Venn diagrams can help you sort out a situation and come to a conclusion. Many problems that you solve by using Venn diagrams come in paragraphs — plenty of words and numbers and confusing relationships. Labeling the circles in the diagrams and filling in numbers helps you determine how everything works together — and allows you to see if you've forgotten anything.

You usually draw Venn diagrams with intersecting circles. You label the circles with the names of the sets, and you encase the circles with the universal set (a rectangle around the circles). The elements shared by the sets are shared by the overlapping parts of the respective circles.

For instance, the Venn diagram in Figure 17-1 shows set A, which contains the letters of the alphabet that spell *encyclopedia,* and set B, which contains the letters in the alphabet that rhyme with "see." Both sets are encased in the universal set — all the letters in the alphabet.

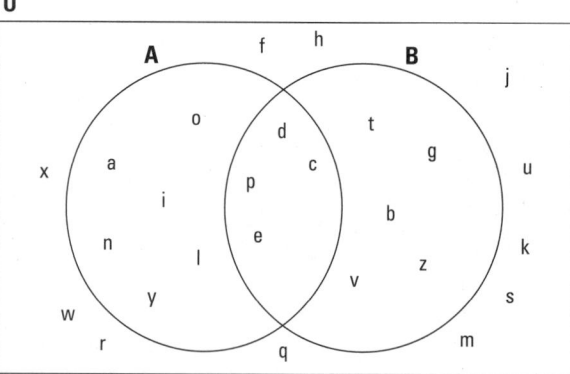

Figure 17-1:
A Venn diagram with two sets enclosed by the universal set.

Notice that the letters *c, d, e,* and *p* appear in both sets. The Venn diagram makes it easy to see where the elements are and what characteristics they have.

Applying the Venn diagram

The business of sorting letters based on their placement in a word or what they rhyme with, which you see in examples in the previous section, may not seem very fulfilling. The actual applications get a little more complicated, but the examples here show you the basics. Some actual uses of Venn diagrams appear in the world of advertising (charting types of advertising and the results), of politics (figuring out who has what opinions on issues and how to make use of their votes), of genetics and medicine (looking at characteristics and reactions based on symptoms and results), and so on. The following example shows, in a simplified version, how useful Venn diagrams can be.

A Chicago-area newspaper interviewed 40 people to determine if they were Chicago White Sox fans and/or if they cheered for the Chicago Bears. (For those of you who could care less about sports, the White Sox are a baseball team, and the Bears are a football team.) Of the people interviewed, 25 are White Sox fans, 9 like both the Sox and the Bears, and 7 don't care for either team. How many people are Bears fans?

As you can see, 25 + 9 + 7 = 41, which is more than the 40 people interviewed. The process has to have some overlap. You can sort out the overlap with a Venn diagram. You create one circle for White Sox fans, another for Bears fans, and a rectangle for all the people interviewed. Start by putting the seven fans who don't cheer for either team outside the circles (but inside the rectangle). You then put the nine who like both teams in the intersection of the two circles (see Figure 17-2).

Surveyed

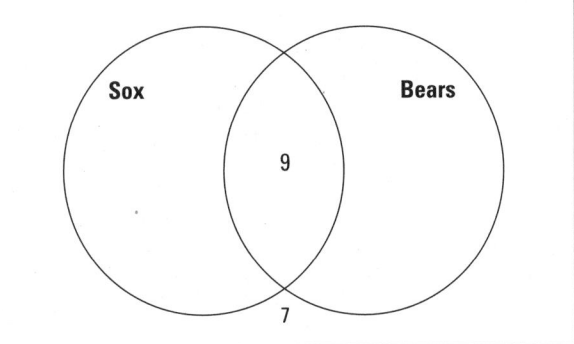

Figure 17-2:
Watch for the overlap created by combining two groups.

The number of White Sox fans is 25, so, if you put 16 in only the Sox's circle, the number in the entire circle (including the overlap) sums to 25. The only area missing is the circle where people are Bears fans but not Sox fans. So far, you have a total of 25 + 7 = 32 people. Therefore, you can say that 8 people root for the Bears but not the Sox. You put an 8 in that area, and you see that the number of Bears fans totals 17 people (see Figure 17-3).

Surveyed

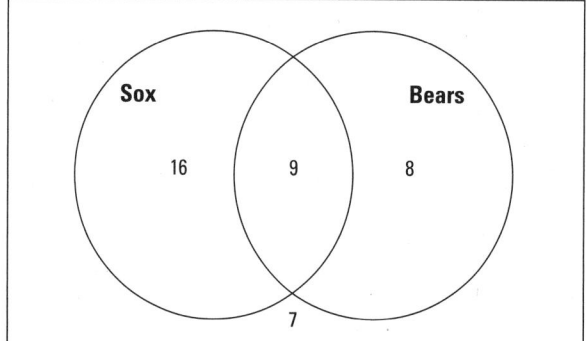

Figure 17-3:
You find
that 17
Chicagoans
root for
Da Bears.

Using Venn diagrams with set operations

You often have to use the set operations union, intersection, and complement (see the section "Operating on Sets") in different combinations when doing algebra problems. In these situations, Venn diagrams are useful for sorting some of the more complex statements.

For instance, is it true that $(C \cup D)' = C' \cap D'$ (translated as "the complement of the union of C and D equals the intersection of the complement of C and the complement of D")? You can sketch each of these situations to compare them by using Venn diagrams. You use shading to signify parts of a particular statement.

In Figure 17-4a, you see $C \cup D$ shaded in — every element that appears in both C and D. In Figure 17-4b, you see $(C \cup D)'$ shaded. The complement represents every element that is *not* in the specified set (but is in the universal set), so it's like the negative of a photograph — all the opposite areas are shaded.

Now you can deal with the other half of the equation. First, in Figure 17-5a, you see C' shaded in. The shaded area represents every element that isn't in C. In Figure 17-5b, you see both C' and D' shaded in, and their intersection (the elements they share) is much darker to show where they overlap. Does the overlap match the sketch for $(C \cup D)'$? Yes, Figures 17-4b and 17-5b are the same — the same areas are shaded in, so you've confirmed the equation.

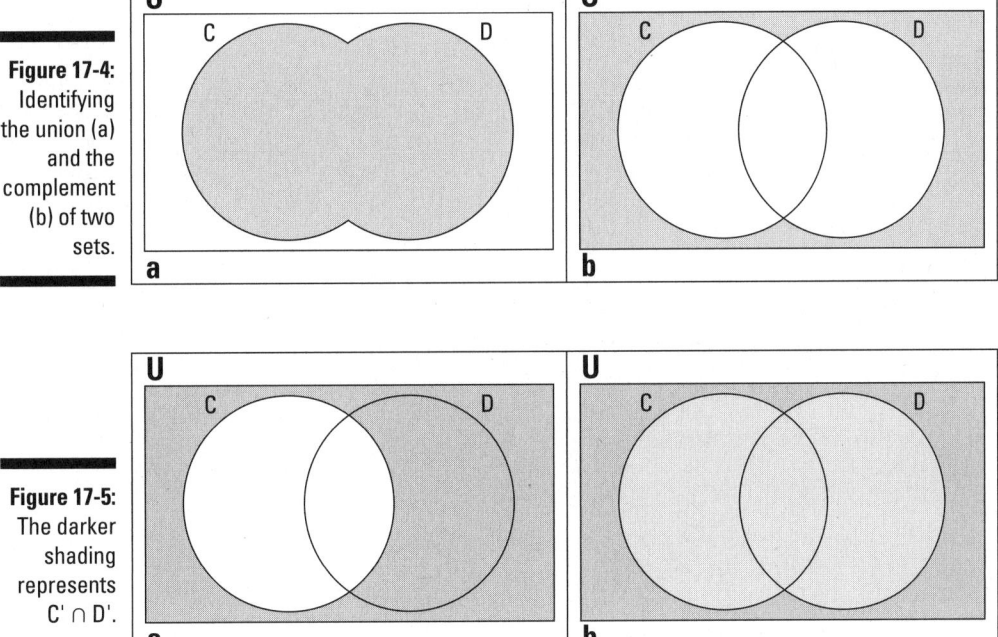

Figure 17-4: Identifying the union (a) and the complement (b) of two sets.

Figure 17-5: The darker shading represents C' ∩ D'.

The Venn diagrams show that two completely different statements can be equal. Sets and set operations are hard to work with in terms of algebra, which is why using a picture approach, like Venn diagrams, goes a long way toward showing or proving that an equation or other statement is true. Using Venn diagrams to pick apart compound statements gives an accurate and visual verification.

Adding a set to a Venn diagram

Showing the relationship between the elements of two sets with a Venn diagram is pretty straightforward. But, as with most mathematical processes, you can take the diagram one step further to illustrate the relationship between three sets. You can also handle four sets, but the diagram gets pretty hairy and isn't all that useful because of the way you have to draw the figure.

When two sets overlap, you divide the picture into four distinct areas: outside both circles, the overlap of the circles, and the two parts that have no overlap but appear in one circle or the other. When you overlap three circles in a Venn diagram, you create eight distinct areas. Figure 17-6 shows you how to complete the process.

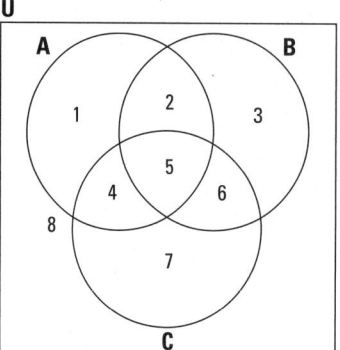

You can describe the eight different areas determined by the intersection of three circles as follows:

1. All the elements in A only

2. All the elements shared by A and B but not by C

3. All the elements in B only

4. All the elements shared by A and C but not by B

5. All the elements shared by A, B, and C

6. All the elements shared by B and C but not by A

7. All the elements in C only

8. All the elements not in A, B, or C

You may have to use a setup like this to sort out information and answer questions. For instance, say that a club with 25 members decides to order pizza for its next meeting; the secretary takes a poll to see what people like:

14 people like sausage.

10 people like pepperoni.

13 people like mushrooms.

5 people like both sausage and pepperoni.

10 people like both sausage and mushrooms.

7 people like both pepperoni and mushrooms.

4 people like all three toppings.

Here's the big question: How many people don't like any of the three toppings on their pizzas?

As you can see, the preferences sum to many more than 25, so you definitely have to account for some overlaps. You can answer the question by drawing three intersecting circles, labeling them Sausage, Pepperoni, and Mushrooms, and filling in numbers, starting with the last on the given list. Figure 17-7a shows the initial Venn diagram you can use.

Figure 17-7:
With a Venn diagram, you can tell how many people want a plain cheese pizza.

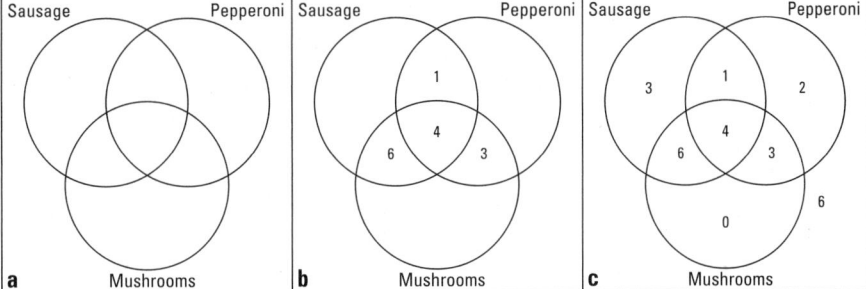

You put the 4, representing people who like all three toppings, in the middle section. Seven people like both pepperoni and mushrooms, but you already account for four of them in the middle, so put a 3 in the area for pepperoni and mushrooms, but not sausage. Ten people like both sausage and mushrooms, but you already put four of those people in the middle area, so put a 6 in the area for sausage and mushrooms, but not pepperoni. Five people like both sausage and pepperoni, but you already account for four; put a 1 in the sausage and pepperoni section, but not mushrooms. Figure 17-7b shows all these entries.

Now you can fill in the rest of the circles. Thirteen people like mushrooms only, and you already have 13 people in that circle, so put 0 for people who like mushrooms only. Ten people like pepperoni only; you already have 8 in that circle, so put a 2 for pepperoni only. Fourteen people like sausage only, so put a 3 in the section for sausage only to account for the difference. Look at all those filled-in numbers in Figure 17-7c!

You finish by adding all the numbers in Figure 17-7c. The numbers add to 19 people. The club has 25 members, so you conclude that 6 of them must not like sausage, pepperoni, or mushrooms. You can order a plain cheese pizza for these picky folk.

Focusing on Factorials

When you perform a *factorial* operation, you multiply the number you're operating on times every positive integer smaller than that number. The factorial operation is denoted by an exclamation mark, !. I cover the factorial operation in Chapter 16 and use it in some of the sequences in that chapter, but it doesn't truly come into its own until you use it with permutations, combinations, and probability problems (as you can see in the "How Do I Love Thee? Let Me Count Up the Ways" section later in this chapter).

One of the main reasons you use the factorial operation in conjunction with sets is to count the number of elements in the sets. When the numbers are nice, discrete, small values, you have no problem. But when the sets get very large — such as the number of handshakes that occur when everyone in a club of 40 people shakes hands with everyone else — you want a way to do a systematic count. Factorials are built into the formulas that allow you to do such counting.

Making factorial manageable

When you apply the factorial operation to a counting problem or algebra expansion of binomials, you run the risk of producing a very large number. Just look at how fast the factorial grows:

$$1! = 1$$
$$2! = 2 \cdot 1 = 2$$
$$3! = 3 \cdot 2 \cdot 1 = 6$$
$$4! = 4 \cdot 3 \cdot 2 \cdot 1 = 24$$
$$5! = 5 \cdot 4 \cdot 3 \cdot 2 \cdot 1 = 120$$
$$6! = 6 \cdot 5 \cdot 4 \cdot 3 \cdot 2 \cdot 1 = 720$$
$$7! = 7 \cdot 6 \cdot 5 \cdot 4 \cdot 3 \cdot 2 \cdot 1 = 5,040$$
$$8! = 8 \cdot 7 \cdot 6 \cdot 5 \cdot 4 \cdot 3 \cdot 2 \cdot 1 = 40,320$$
$$9! = 9 \cdot 8 \cdot 7 \cdot 6 \cdot 5 \cdot 4 \cdot 3 \cdot 2 \cdot 1 = 362,880$$
$$10! = 10 \cdot 9 \cdot 8 \cdot 7 \cdot 6 \cdot 5 \cdot 4 \cdot 3 \cdot 2 \cdot 1 = 3,628,800$$

Check out the last two numbers in the list. You see that 10! has the same digits as 9! — only with an extra zero. This observation illustrates one of the properties of the factorial operation. Another property is the definition of 0! (zero factorial).

The two properties (or rules) for the factorial operation are as follows:

- ✔ **$n! = n \cdot (n - 1)!$:** To find $n!$, you multiply the number n times the factorial that comes immediately before it.

- ✔ **$0! = 1$:** The value of zero factorial is one. You just have to trust this one.

If you know that $13! = 6,227,020,800$, and you want $14!$, you can try your calculator first. Most hand-held calculators go into scientific notation when you get to numbers this large. The calculator won't give you the exact value — it rounds off to eight decimal places. But, if you're industrious, you can find $14!$ with the first property from the previous list: $14 \cdot 13! = 14(6,227,020,800) = 87,178,291,200$ (multiplied by hand!).

Simplifying factorials

The process of simplifying factorials in fractions, multiplication problems, or formulas is simple enough, as long as you keep in mind how the factorial operation works. For instance, if you want to simplify the fraction $\frac{8!}{4!}$, you can't just divide the 8 by the 4 to get $2!$. The factorials just don't work that way. You'll find that reducing the fraction is much easier and more accurate than trying to figure out the values of the huge numbers created by the factorials and then working with them in a fraction.

You have a string of numbers multiplied together, so the product of all those numbers lends itself to factorization or reduction of a fraction. Look at the two factorials written out:

$$\frac{8!}{4!} = \frac{8 \cdot 7 \cdot 6 \cdot 5 \cdot 4 \cdot 3 \cdot 2 \cdot 1}{4 \cdot 3 \cdot 2 \cdot 1}$$

You have two options to simplify this factorial:

- ✔ You can cancel out all the factors of $4!$ and multiply what's left:

$$\frac{8!}{4!} = \frac{8 \cdot 7 \cdot 6 \cdot 5 \cdot \cancel{4} \cdot \cancel{3} \cdot \cancel{2} \cdot \cancel{1}}{\cancel{4} \cdot \cancel{3} \cdot \cancel{2} \cdot \cancel{1}} = \frac{8 \cdot 7 \cdot 6 \cdot 5}{1} = 1,680$$

- ✔ You can take advantage of the rule $n! = n(n - 1)!$ (see the previous section) when writing the larger factorial:

$$\frac{8!}{4!} = \frac{8 \cdot 7 \cdot 6 \cdot 5 \cdot 4!}{4!} = \frac{8 \cdot 7 \cdot 6 \cdot 5 \cdot \cancel{4!}}{\cancel{4!}} = \frac{8 \cdot 7 \cdot 6 \cdot 5}{1} = 1,680$$

You see how useful this process is when you have to answer a problem with factorials such as $\frac{40!}{37!3!}$. The value of $40!$ is huge, and so is the value of $37!$.

But the reduced form of the example fraction is quite nice. You can reduce the fraction by using the first property of factorials:

$$\frac{40!}{3!37!} = \frac{40 \cdot 39 \cdot 38 \cdot 37!}{3 \cdot 2 \cdot 1 \cdot 37!}$$

$$= \frac{40 \cdot 39 \cdot 38 \cdot \cancel{37!}}{3 \cdot 2 \cdot 1 \cdot \cancel{37!}}$$

$$= \frac{40 \cdot \cancel{39}^{13} \cdot \cancel{38}^{19}}{\cancel{3} \cdot \cancel{2} \cdot 1}$$

$$= 40 \cdot 13 \cdot 19 = 9{,}880$$

The values 39 and 38 in the numerator are each divisible by one of the factors in the denominator. You multiply what you have left for your answer.

How Do I Love Thee? Let Me Count Up the Ways

You may think that you know everything there is to know about counting. After all, you've been counting since your parents asked you, a 3-year-old, how many toes you have and how many cookies are on the plate. Counting toes and cookies is a breeze. Counting very, very large sets of elements is where the challenge comes in. Fortunately, algebra provides you with some techniques for counting large sets of elements more efficiently: the multiplication principle, permutations, and combinations (for more in-depth info on these topics, check out *Probability For Dummies* [Wiley], by Deborah Rumsey, PhD). You use each technique in a decidedly different situation, although deciding which technique to use is often the biggest challenge.

Applying the multiplication principle to sets

The *multiplication principle* is true to its name: It calls for you to multiply the number of elements in different sets to find the total number of ways that tasks or other things can be done. If you can do Task 1 in m_1 ways, Task 2 in m_2 ways, Task 3 in m_3 ways, and so on, you can perform all the tasks in a total of $m_1 \cdot m_2 \cdot m_3 \cdot \ldots$ ways.

For instance, if you have six shirts, four pairs of pants, eight pairs of socks, and two pairs of shoes, you have $6 \cdot 4 \cdot 8 \cdot 2 = 384$ different ways to get dressed. Of course, this doesn't take into account any color or style issues.

How many different license plates are available in a state with the following rules:

- ✔ All the plates have three letters followed by two numbers.

- ✔ The first letter can't be O.

- ✔ The first number can't be zero or one.

Think of this problem as a prime candidate for the multiplication principle. Here's the first rule for the license plate:

# of ways letter 1 · # of ways letter 2 · # of ways letter 3 ·
# of ways number 1 · # of ways number 2

You can't use O for the first letter, so that leaves 25 ways to choose the first letter. The second and third letters have no restrictions, so you can choose any one of the 26 letters for them. The first number can't be zero or one, so that leaves you eight choices. The second number has no restrictions, so you have all ten choices. You simply multiply these choices together to get your answer: $25 \cdot 26 \cdot 26 \cdot 8 \cdot 10 = 1,352,000$. You can assume that this isn't a very big state if it has only a million or so license-plate possibilities. And trust me, using the multiplication principle to find this answer is faster than going to the DMV for help!

Arranging permutations of sets

The *permutation* of some set of elements is the rearrangement of the order of those elements. For example, you can rearrange the letters in the word "act" to spell six different words (well, not actual *words;* they could possibly be acronyms for something): act, cat, atc, cta, tac, tca. Therefore, you say that for the word "act," you have six permutations of three elements.

Counting the permutations

Knowing how many permutations a set of elements has doesn't tell you what those permutations are, but you at least know how many permutations to look for. When you find the permutations for a set of elements, you select where you get the elements from and how many you need to rearrange. You assume that you don't replace any of the elements you select before you choose again.

You find the number of permutations, P, of n elements taken r at a time with the formula $_nP_r = \dfrac{n!}{(n-r)!}$ (see the section "Focusing on Factorials" earlier in this chapter for an explanation of !).

If you want to arrange four of your six vases on a shelf, for example, how many different arrangements (orders) are possible? If you label your vases A, B, C, D, E, and F, some of the arrangements are ABCD, ABCE, ABCF, BCFD, BFDC, and so on.

Using the formula, you let $n = 6$ and $r = 4$:

$$_6P_4 = \frac{6!}{(6-4)!} = \frac{6!}{2!} = \frac{6 \cdot 5 \cdot 4 \cdot 3 \cdot \cancel{2} \cdot \cancel{1}}{\cancel{2} \cdot \cancel{1}} = 6 \cdot 5 \cdot 4 \cdot 3 = 360$$

With that number, I hope you don't plan on taking a picture of each arrangement!

Here's an example that illustrates some broader observations of permutations. Say that the seven dwarfs want to line up for a family photo. How many different ways can the dwarfs line up for the picture? In algebraic terms, you want a permutation of seven things taken seven at a time. Using the previous formula, you find the following:

$$_7P_7 = \frac{7!}{(7-7)!} = \frac{7!}{0!} = \frac{7 \cdot 6 \cdot 5 \cdot 4 \cdot 3 \cdot 2 \cdot 1}{1} = 7 \cdot 6 \cdot 5 \cdot 4 \cdot 3 \cdot 2 \cdot 1 = 5,040$$

Now you can see why mathematicians had to declare that $0! = 1$ (which I explain in the "Making factorial manageable" section earlier in the chapter). In the formula, you end up with a 0 in the denominator, because the number of items chosen is the same as the number of items available. By saying that $0! = 1$, the denominator becomes a 1, and the answer becomes the value in the numerator of the fraction.

When using the formula for permutations of n elements taken r at a time,

✔ $_nP_n = \dfrac{n!}{(n-n)!} = \dfrac{n!}{0!} = n!$. When you use all the elements in the arrangements, you just need to find $n!$.

✔ $_nP_1 = \dfrac{n!}{(n-1)!} = \dfrac{n \cdot \cancel{(n-1)}\cancel{(n-2)}\cancel{(n-3)} \dots \cancel{3} \cdot \cancel{2} \cdot \cancel{1}}{\cancel{(n-1)}\cancel{(n-2)}\cancel{(n-3)} \dots \cancel{3} \cdot \cancel{2} \cdot \cancel{1}} = n$. When you take only one of the elements from all the choices, you have only n different arrangements.

✔ $_nP_0 = \dfrac{n!}{(n-0)!} = \dfrac{n!}{n!} = 1$. When you don't use any of the elements from the set, you have only one way to do that.

Some problems call for a mixture of permutations and the multiplication principle (see the previous section). Say, for example, that you want to make up a new password for your computer account. The first rule for the system is that the first two entries must be letters — with no repeats, and you can't use O or I. The next four entries must be digits — with no restrictions. You can then enter two more letters with no restrictions. Finally, the last three entries must be three digits with no repeats. How many different passwords are possible?

Break down the problem into its four different parts, setting up the multiplication principle: ×

   2 letters, no repeats, no O or I × 4 digits × 2 letters × 3 digits, no repeats

You can now set up the permutations:

   ✔ The arrangement of the first two letters is a permutation of 24 elements taken 2 at a time.

   ✔ The next four digits give you 10 choices each time — multiplied together.

   ✔ The following two letters give you 26 choices each time — multiplied together.

   ✔ The final three digits represent a permutation of 10 elements taken 3 at a time.

The computation goes as follows:

$$_{24}P_2 \times 10 \cdot 10 \cdot 10 \cdot 10 \times 26 \cdot 26 \times {}_{10}P_3$$
$$= \frac{24!}{22!} \times 10^4 \times 26^2 \times \frac{10!}{7!}$$
$$= 552 \times 10,000 \times 676 \times 720$$
$$= 2,686,694,400,000$$

You calculate over two-and-a-half trillion different passwords. Hopefully, the rules will keep the hackers out. Now you just have to remember your password!

Distinguishing one permutation from another

If you're lining up the books on your bookshelf based on color — for eye appeal — rather than subject or author, you aren't *distinguishing* between the different red books and the different blue books. And if you want to know the total number of permutations of the letters in the word "cheese," you can't *distinguish* between "cheese" and "cheese," where you switch the e's around. You can't tell the permutations apart, so you don't count rearrangements of the same letter as different permutations. The way you counter this effect is to use the formula for *distinguishable permutations*.

If a set of n objects has k_1 elements that are alike, k_2 elements that are alike, and so on, you find the number of distinguishable permutations of those n objects with the following formula:

$$\frac{n!}{k_1!k_2!k_3!\ldots}$$

Using this formula, you can determine the number of distinguishable permutations of the word "cheese" (see the section "Focusing on Factorials" for the division part of this formula):

$$\frac{6!}{3!} = \frac{6 \cdot 5 \cdot 4 \cdot \cancel{3!}}{3!} = 120$$

Cheese has six letters altogether, and the three e's are all the same.

You can also apply the formula to the books on the bookshelf. Say that you have ten blue books, five red books, six black books, one green book, and one gray book. You have over 8×10^{10} possible distinguishable arrangements of the books. You can determine this number by using the formula for distinguishable permutations:

$$\frac{23!}{10!5!6!} = 8.245512475 \times 10^{10}$$

Now say you want to put all the books of the same color together, and you don't care about the order of the books in each color grouping. How many different arrangements are possible? This problem is a permutation of the five colors (blue, red, black, green, and gray). A permutation of 5 elements taken 5 at a time is 5! = 120 different arrangements (see the section "Focusing on Factorials" for more on this calculation).

Being the finicky decorator that you are, you now decide that the arrangements within each color matter, and you want the books arranged in groups of the same color. How many different arrangements are possible? This problem involves the permutations of the different colors first and then permutations of the books in the blue, red, and black groups. Doing the math, you find the following:

$$\text{Colors} \times \text{Blues} \times \text{Reds} \times \text{Blacks}$$
$$= \left(_5P_5 \right) \times \left(_{10}P_{10} \right) \times \left(_5P_5 \right) \times \left(_6P_6 \right)$$
$$= 5! \times 10! \times 5! \times 6!$$
$$= 3.76233984 \times 10^{13}$$

You have quite a few ways to arrange the books on the bookshelf. Maybe alphabetical order makes more sense.

Proving the Four-Color Problem

One famous multiple-color problem was around a long time before a computer finally proved it to be true. The Four-Color Problem states that it never takes more than four different colors to color a map so that no two states or countries with a common border ever have the same color. Mathematicians struggled with the proof of this problem for decades. They couldn't find a counterexample, but they couldn't prove the problem, either. It took a modern-day computer to run through all the different possibilities and prove the problem true. The following figure shows a possible configuration of bordering states. It doesn't take more than four different colors to complete this map so that no two states with a common border share the same color. Can you do it?

Mixing up sets with combinations

Combination is a precise mathematical term that refers to how you can choose a certain number of elements from a set. With combinations, the order in which the elements appear doesn't matter. Combinations, for example, allow you to determine how many different ways you can choose three people to serve on a committee (the order in which you choose them doesn't matter — unless you want to make one of them the chairperson).

You find the number of possible ways to choose r elements from a set containing n elements, C, with the following formula (see the section "Focusing on Factorials" for an explanation of !):

$$_nC_r = \frac{n!}{(n-r)!\,r!}$$

The bulk of this formula should look familiar; it's the formula for the number of permutations of n elements taken r at a time (see the previous section) — with an extra factor in the denominator. The formula is actually the permutations divided by $r!$.

You can also indicate combinations with two other notations:

$C(n, r)$

$\binom{n}{r}$

The different notations all mean the same thing and have the same answer. For the most part, the notation you use is a matter of personal preference or what you can find on your calculator. You read all the notations as "n choose r."

For instance, say you have tickets to the theatre, and you want to choose three people to join you. How many different ways can you choose those three people if you have five close friends to choose from? Your friends are Violet, Wally, Xanthia, Yvonne, and Zeke. You can take

{Violet, Wally, Xanthia}

{Violet, Wally, Yvonne}

{Violet, Wally, Zeke}

{Wally, Xanthia, Yvonne}

{Wally, Xanthia, Zeke}

{Xanthia, Yvonne, Zeke}

You find six different subsets. Is that all? Have you forgotten any?

You can check your work with the formula for the number of combinations of five elements taken three at a time:

$$_5C_3 = \frac{5!}{(5-3)!3!} = \frac{5 \cdot \cancel{4}^{2} \cdot \cancel{3} \cdot \cancel{2} \cdot \cancel{1}}{\cancel{2} \cdot 1 \cdot \cancel{3} \cdot \cancel{2} \cdot \cancel{1}} = 10$$

Oops! You must have forgotten some combinations. (See the section "Drawing a tree diagram for a combination" to see what you missed.)

Branching Out with Tree Diagrams

Combinations and permutations (see the previous section) tell you how many subsets or groupings to expect within certain sets, but they don't shine a light on what elements are in those groupings or subsets. A nice, orderly way to write out all the combinations or permutations is to use a tree diagram. The name *tree diagram* comes from the fact that the drawing looks something like a family tree — branching out from the left to the right. The left-most entries are your "first choices," the next column of entries shows what you can choose next, and so on.

You can't use tree diagrams in all situations dealing with permutations and combinations — the diagrams get too large and spread out all over the place — but you can use tree diagrams in many situations involving counting items. Tree diagrams also appear in the worlds of statistics, genetics, and other studies.

Picturing a tree diagram for a permutation

When you choose two different letters from a set of four letters to form different "words," you can find the number of different words possible with the permutation $_4P_2 = \dfrac{4!}{(4-2)!} = \dfrac{4 \cdot 3 \cdot \cancel{2} \cdot \cancel{1}}{\cancel{2} \cdot \cancel{1}} = 12$ (see the section "Arranging permutations of sets"). You can form 12 different two-letter words. To make a list of the words — and not repeat yourself or leave any out — you should create a tree diagram.

Go out on a limb and follow these simple steps:

1. **Start out with the designated number of items and list them vertically — one under the other — leaving some space for your tree to grow to the right.**

2. **Connect the next set of entries to the first set by drawing short line segments — the branches.**

3. **Keep connecting more entries until you complete the task as many times as needed.**

For example, say you want to create permutations out of the word "seat." Figure 17-8 shows the tree diagram that organizes these permutations.

First Choice	Second Choice	Two-Letter Words
S	E	SE
	A	SA
	T	ST
E	S	ES
	A	EA
	T	ET
A	S	AS
	E	AE
	T	AT
T	S	TS
	E	TE
	A	TA

Figure 17-8: Creating two-letter words (permutations) from *seat*.

Drawing a tree diagram for a combination

In the section "Mixing up sets with combinations," you receive the job of choosing three friends from a set of five to accompany you to the theatre, and you find that you have ten different ways to choose them. The order in which you choose them doesn't matter; they get to go or they don't. The tree diagram in Figure 17-9 shows how you determine all the arrangements without repeating yourself.

You figure out the different arrangements by starting out with any three people. If you do your tree diagram correctly, it doesn't matter who you start with — you get the same number of groupings with the same combinations of people in the groupings. (I didn't include poor Yvonne and Zeke in the first column of entries because their names come last in the alphabet.)

Because the order doesn't matter, you first figure out all the arrangements with Violet and Wally, and then Violet and Xanthia, and then Violet and Yvonne. The number of branches decreases because you can't repeat groupings. You then move to Wally and Xanthia and then Wally and Yvonne. You finally get to Xanthia and stop there, because you've accounted for all the groupings.

Figure 17-9:
The branches of the tree diagram get smaller as you account for all the possibilities.

Do you see which sets I missed in the section "Mixing up sets with combinations"? You can easily miss some combinations if you don't have an orderly method for listing them. And you may be concerned that Violet has a much better chance for selection than her peers, but that isn't the case. If you look through all ten sets of three people, you see that each name appears six times.

Part V
The Part of Tens

The 5th Wave   By Rich Tennant

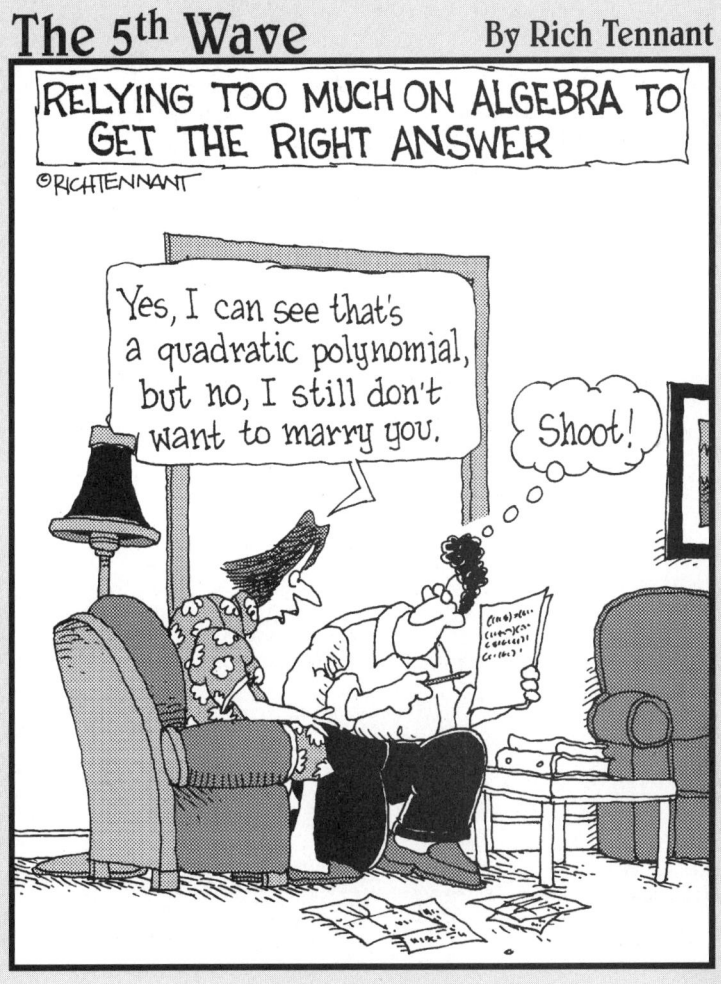

In this part . . .

In the age of technology and calculators, you may think you have numbers all figured out. Want to know pi? Just hit the correct key, right? Well, you may not know all there is to know about numbers just yet. In Part V, I give you some refreshing ways to look at numbers in a handy list format. I present some multiplication tricks you can do in your head, which is much faster than hauling out a calculator, replacing its batteries, entering the numbers, and getting an answer. I also provide a list of special numbers that gives you insight into the common characteristics of numbers, as well as the differences.

Chapter 18

Ten Multiplication Tricks

In This Chapter

▶ Saving time on tests and homework problems with fast multiplication

▶ Avoiding calculators and division (and sometimes multiplication itself)

You studied your multiplication tables way back in the third grade. You can quickly rattle off the product of 7 and 9 or 8 and 6 (can't you?). In this chapter, you discover some new patterns and tricks to help you with products and processes — tricks that save you time. You're also more apt to get the right answer when using these tricks (not that you ever have the wrong answer, of course). And think of how you can amaze your friends and colleagues by pulling seemingly tough answers out of thin air.

But before you get moving, I have a confession to make: I've snuck in an addition trick as well because I know that you'll love it.

Squaring Numbers that End in 5

You know that 5 squared is 25. But what about 15^2, 25^2, 35^2, and so on? Don't have your calculator on you? No worries. To square a number that ends in 5, follow these steps:

1. **Write down the last digits of the answer: 25.**

   The squared form of a number that ends in 5 always ends with 25.

2. **Take the digit or digits in front of the original 5 and multiply them by the next bigger number.**

3. **Put the product of Step 2 in front of the 25, and you have the square of the number you want.**

To square 35, for example, you write down the last two digits, 25. Now you multiply the 3 times the next bigger number, 4, to get 12. Put the 12 in front of the 25, and you have your answer: $35^2 = 1,225$. Squaring 65, you know that $6 \cdot 7 = 42$, so $65^2 = 4,225$.

This trick even works with three-digit numbers. You find the square of 105 by multiplying $10 \cdot 11$, giving you 110: $105^2 = 11,025$.

Finding the Next Perfect Square

In the previous section, you discover how to find the squares of numbers that end in 5. But what about all the other squares? A very nice property you can use deals with the next square in any list of the squares of whole numbers. The property states that you can get to the next square in the list by taking the square of the number you already have and adding the root plus the next number (the root of the square you want).

For example, if you know that $25^2 = 625$, and you want 26^2, you just add 625 (the square of 25) + 25 (the root of 625) + 26 (the next highest number), which equals $650 + 26 = 676$. To find the square of 81, you take $80^2 = 6,400$ and then add up the ingredients: $6,400 + 80 + 81 = 6,400 + 161 = 6,561$.

Recognizing the Pattern in Multiples of 9

Have you ever noticed something special about the first ten multiples of 9? Here's the list (notice that I've made the first multiple into a two-digit number): 09, 18, 27, 36, 45, 54, 63, 72, 81, 90. Each of the multiples has two digits that add up to 9. Also, the first digit of each multiple is one less than its multiplier. So, when multiplying 7 times 9, you can start with one digit smaller than 7, the number 6. To find the sum, you take $9 - 6$. The product of 7 and 9 is 63.

Casting Out 9s

A wonderful way to do a quick check of your addition or multiplication is to *cast out 9s.* The easiest way to explain this method is to demonstrate it (I start with the rogue addition trick I shoehorn into this chapter).

Assume that you have to add the following column of numbers, and you want to check your work without having to add again (many people don't catch their errors by adding again because they make the same error again — it gets embedded in the mind):

```
  1492
  1984
  2006
  1776
+ 1812
  9070
```

To check your work, you look at all the digits in the numbers you're adding. Start by crossing out (*casting out*) all the 9s. And don't stop there! You can cast out sets of numbers that add up to 9 — the 1 and 8 in the second value and the 1 and 8 in the fifth value. You can also cross out the 2 and 6 in the third value and the 1 in the fourth value:

```
  1492
  1984
  2006
  1776
+ 1812
  9070
```

Now cross out the 1, 4, and 4 in the first and second values. Cross out the 2 and 7 in the first and fourth values and the 7 and 2 in the fourth and fifth values:

```
  1492
  1984
  2006
  1776
+ 1812
  9070
```

What do you have left? Not much! Add any digits you haven't crossed out. You get 0 + 0 + 6 + 1 = 7. If the sum you find is bigger than 9, you add the digits of that sum (and then add *that* sum if it's bigger than 9).

Now look at the answer in the addition problem. Cross out the 9, and you have 7 left. That matches the 7 from the digits you find in your check. The sum checks out because they match.

This check isn't foolproof. If you make an error that differs by 9 from the correct answer, you won't catch your mistake. However, the ease and quickness of this method make up for that potential failure.

Casting Out 9s: The Multiplication Moves

The previous section shows you how to cast out 9s for addition; here's how you cast out 9s in a multiplication problem. You cross out 9s or sums of 9s in the first value. (If you can't cross out any, add up the digits. If the sum is bigger than 9, add those digits.) Do the same with the second number. Repeat the process again with the answer. Here's an example:

$$\begin{array}{r} 4812 \\ \times\,7535 \\ \hline 36{,}258{,}420 \end{array}$$

In the first value, you cast out the sum of 9 (the 8 + 1) and then add the two remaining digits (4 + 2 to get 6). In the second value, you add the digits for a sum of 20. You then add those digits (2 + 0) for a sum of 2. In the answer, you find two sets of digits that sum to 9 (3 + 6 and 5 + 4). After crossing them out, you add the remaining digits and then their sum (2 + 8 + 2 + 0 = 12; 1 + 2 = 3).

You finish by multiplying the sum of 6 from the first value by the 2 from the second value to get a product of 12. Add those two digits to get 3. That 3 matches the 3 in the digits from the answer. The answer checks out.

Multiplying by 11

Multiplying single digits by 11 is simple, in-your-head math. You just take the single digit, make two of them, and you're done. Multiplying a larger number times 11 is a bit trickier. However, you can make it just as easy by bookending the value with zeros and adding adjacent digits.

For instance, when multiplying 142,327 · 11, you put a zero in front of the number and behind the number, and double every digit of the original number giving you 01144223322770. Now you add each pair of adjacent digits:

0 + 1, 1 + 4, 4 + 2, 2 + 3, 3 + 2, 2 + 7, 7 + 0

The sums are 1, 5, 6, 5, 5, 9, and 7, so the product of 142,327 · 11 is 1,565,597.

In the previous product, you see no carry-over. If one or more of the sums you find is greater than nine, you carry the tens digit over to the sum to the left of the digit in question.

For instance, when multiplying 56,429 · 11, you add the following:

$$0 + 5, 5 + 6, 6 + 4, 4 + 2, 2 + 9, 9 + 0$$

The sums are 5, 11, 10, 6, 11, and 9. Starting at the right side of the answer, you see that the last digit is 9. Now follow these steps:

1. **Write down the 1 from the ones place of the 11 in front of the 9, and carry the other 1 over to the 6 to make it 7. Write down the 7 in front of the 1.**

   You now have 719.

2. **Write down the 0 in front of the 7, and carry the 1 over to the left, adding it to the 11.**

   Now you have 0719.

   Adding the 1 to the 11 gives you 12.

3. **Write down the 2 in front of the 0, and carry the 1 in the tens place over to the 5 to make it 6. Write the 6 in front of the 2.**

   You now have 620,719.

So, 56,429 · 11 = 620,719.

Multiplying by 5

To multiply any number by five in your head, you can just halve the number you want to multiply and add a zero to the end of the number.

To multiply 14 · 5, for example, you take half of 14, which is 7, and put a 0 after the 7 — 14 · 5 = 70.

But what if the number you want to multiply is odd and halving gives you a decimal? In this case, you don't add a zero to the end of the halved number; you just drop the decimal point.

For example, if you want to multiply 43 · 5, you take half of 43, which is 21.5. Dropping the decimal point, you get 215 — 43 · 5 = 215.

Finding Common Denominators

When adding or subtracting fractions, you need a common denominator. In the problem $\frac{3}{16} + \frac{5}{24}$, for example, the common denominator should be the least common multiple of 16 and 24 (the smallest number that they both divide into evenly).

A quick way to find the common denominator of two fractions is to take the larger of the two denominators and check its multiples to see if the other denominator divides them evenly. In this case, you start at $24 \cdot 2 = 48$. The number 16 divides 48 evenly ($48 \div 3$), so 48 is the common denominator.

When subtracting the fractions $\frac{13}{15} - \frac{3}{20}$, you want the least common multiple of 15 and 20. Using the larger denominator, you try $20 \cdot 2$, which is 40. But 15 doesn't divide 40 evenly. You try $20 \cdot 3$, and you get 60. This time you have a winner; 15 divides 60 evenly ($60 \div 4$).

Determining Divisors

When reducing fractions or factoring numbers out of terms in an expression, you want the largest number that divides two different numbers evenly. For instance, if you want to reduce the fraction $\frac{36}{48}$, you want to find the number that divides both the numerator and denominator evenly. You know that two divides both because they're even numbers. But you should aim to reduce only once, so you want the GCD: the greatest common divisor.

To find the GCD, follow these steps:

1. **Divide the smaller number into the larger number and check out the remainder.**

   For the previous example, $48 \div 36 = 1$ with R 12 (the remainder is 12).

2. **Divide the smaller number by the remainder.**

   You find that $36 \div 12 = 3$, with R 0. You find no remainder, so the divisor — the 12 — is the GCD. The reduced fraction is ¾.

Other problems may involve more steps. To reduce the fraction ²⁰⁄₂₈, you find the GCD of the numbers 20 and 28, for example, you find $28 \div 20 = 1$ with R 8. Now you divide the 20 by the 8: $20 \div 8 = 2$ with R 4. You divide the 8 by the remainder: $8 \div 4 = 2$ with R 0. When you get a remainder of zero, the last number you divided by is the GCD. The 4 divides 20 and 28 evenly: ⁵⁄₇.

Multiplying Two-Digit Numbers

To multiply two 2-digit numbers in your head, you can use the FOIL method (First, Outer, Inner, Last; see Chapter 1).

For instance, to multiply 23 · 12, do the Last portion first by multiplying the 3 times the 2. Put a 6 in the right-most answer position. Now you cross-multiply. The 2 in 23 times the 2 in 12 is 4. You add that value to the value for 3 times 1, and you get 7, the Outer and Inner portion. Put the 7 in front of the 6 for the answer. Now you multiply the 2 in 23 by the 1 in 12. Put the 2 in front of the 7 and 6 for a complete answer: 276.

If any of the products you find are greater than nine, you carry over the number in the tens place and add it to the next product or cross product.

Chapter 19

Ten Special Types of Numbers

In This Chapter

▶ Associating numbers with shapes

▶ Characterizing numbers as perfect, narcissistic, and so on

*M*athematicians classify numbers in many ways, much like psychologists (and gossipers) classify people: even or odd, positive or negative, rational or irrational, and so on. In this chapter, you discover even more ways to pigeonhole numbers by putting them into interesting groupings that make them special.

Triangular Numbers

Triangular numbers are numbers in the sequence 1, 3, 6, 10, 15, 21, 28, and so on (for more on sequences, see Chapter 16). You may have noticed that each term in the sequence is larger than the previous number by one more value than the previous difference between the terms. Say what? Okay, let me put that another way: To find each term, you add 2 to the previous term, and then 3, and then 4, and then 5, and so on.

The formula for finding the *nth* triangular number is $\frac{n^2 + n}{2}$. Here's how you use the formula to find the eighth triangular number (one more than the number 28 in the previous list): $\frac{8^2 + 8}{2} = \frac{64 + 8}{2} = \frac{72}{2} = 36$. The number 36 is 8 more than the number 28, which is 7 more than 21, and so on. You can also depict the triangular numbers by counting the connected dots in triangular arrays:

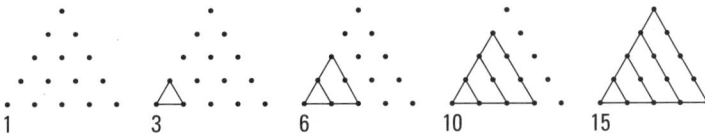

1   3   6   10   15

Square Numbers

The *squares* of numbers should be familiar to you: 1, 4, 9, 16, 25, 36, and so on. As the following figure shows, connected dots in arrays can illustrate squared numbers. Perfect squares show up in algebra and geometry all the time. You see the squares when you're working with quadratic equations (Chapter 3) and the conics (Chapter 11). In geometry, Pythagoras was dependent on squares to do his famous Pythagorean Theorem (the sum of the squares of the two shorter sides of a right triangle is equal to the square of the longest side).

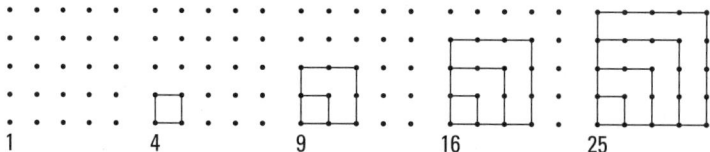

Hexagonal Numbers

Hexagonal numbers appear in the sequence 1, 7, 19, 37, 61, 91, and so on. You can illustrate these numbers by drawing hexagons, and then hexagons around the hexagons, and so on; you then count how many hexagons you have, as you see here:

Another way you can list hexagonal numbers, if you're not into drawing all the six-sided figures, is to use the formula for the *nth* hexagonal number: $n^3 - (n-1)^3$ or $3n^2 - 3n + 1$. You may have noticed from the pattern that you create the next hexagonal number in the sequence by adding six to the previous difference (7 to 19 is 12, and 19 to 37 is 18, for example). The numbers can get pretty large pretty quickly, though, so the formula helps. You can use this formula to find the tenth hexagonal number, for example: $10^3 - (10-1)^3 = 10^3 - 9^3 = 1,000 - 729 = 271$.

Perfect Numbers

A *perfect number* is a number in which the sum of its proper divisors (numbers smaller than the number in question that divide it evenly) is equal to the number in question. For instance, the number 6 is a perfect number, because 1 + 2 + 3 = 6. The next few perfect numbers are 28 (1 + 2 + 4 + 7 + 14), 496 (1 + 2 + 4 + 8 + 16 + 31 + 62 + 124 + 248), and 8,128. Notice how these proper divisors all have "pairs," except for the 1. The pairs multiply together to give you the number in question: 2 times 14, 4 times 7 in the case of the 28. Can you find all the proper divisors of 8,128? Have fun! There's no magic formula for finding these numbers. Just get out your calculator!

Amicable Numbers

Two numbers are *amicable* if each value is equal to the sum of the proper divisors of the other. The numbers 284 and 220 are amicable, for example, because the sum of the proper divisors of 284 is 220 (1 + 2 + 4 + 71 + 142), and the sum of the proper divisors of 220 is 284 (1 + 2 + 4 + 5 + 10 + 11 + 20 + 22 + 44 + 55 + 110). Other sets of amicable numbers include 1,184-1,210 and 17,296-18,416. Do you want to find more amicable numbers? Go for it, but don't expect a magic formula to help you.

Happy Numbers

A number is considered *happy* if the sum of the squares of its digits, or the sum of the squares of the sum of its squares, is equal to one.

For instance, the number 203 is happy because the sum of the squares of its digits is $2^2 + 0^2 + 3^2 = 4 + 9 = 13$; the sum of the squares of the digits in 13 is $1^2 + 3^2 = 1 + 9 = 10$; and the sum of the squares of the digits in 10 is $1^2 + 0^2 = 1$. Finally! You may be surprised to find that a bigger number such as 2,211 is happy in fewer steps! You take $2^2 + 2^2 + 1^2 + 1^2 = 4 + 4 + 1 + 1 = 10$, and $1^2 + 0^2 = 1$.

Abundant Numbers

An *abundant number* is a number with one rule: The sum of its divisors must be greater than twice the number. For instance, the number 12 is abundant because 1 + 2 + 3 + 4 + 6 + 12 = 28, and 28 is greater than twice 12. Some other abundant numbers include 18, 20, 24, 30, 36, and 100.

Deficient Numbers

A *deficient number* has almost the opposite rule of an abundant number (see the previous section). If the sum of the proper divisors of a number is less than the original number, the number is a deficient number. An example of a deficient number is 15, because the sum of its proper divisors is $1 + 3 + 5 = 9$, and 9 is less than 15. What about the number 32? Is it deficient? The sum of its proper divisors is $1 + 2 + 4 + 8 + 16 = 31$, so 32 is also deficient.

Narcissistic Numbers

A *narcissistic number* is a number that you can write by using operations involving all the digits in the number. For instance, the number 371 is narcissistic because if you take the digits — 3, 7 and 1 — raise each to the third power, and add the powers together, you get the original number: $3^3 + 7^3 + 1^3 = 27 + 343 + 1 = 371$. The digits of the number look in the mirror with operations, so to speak, and see only themselves. Another narcissistic number is 2,427, because you can raise the digits to successively larger powers, add up the powers, and get the original number: $2^1 + 4^2 + 2^3 + 7^4 = 2 + 16 + 8 + 2,401 = 2,427$. The possibilities for self-absorbed numbers are endless.

Prime Numbers

A *prime number* is divisible only by the number 1 and itself. The first 15 prime numbers are as follows: 2, 3, 5, 7, 11, 13, 17, 19, 23, 29, 31, 37, 41, 43, and 47. Even though mathematicians have been working on prime numbers for centuries, they haven't found a formula or pattern that finds or predicts all the prime numbers.

A famous *conjecture* (mathematicians haven't proven a conjecture, but they haven't disproved it either) involving prime numbers is Goldbach's Conjecture. Goldbach theorized that you can write every even number greater than two as the sum of two primes. For instance, $8 = 3 + 5$, $20 = 7 + 13$, and so on.

People armed with powerful computers and lots of computing time are always coming up with new prime numbers, but it doesn't happen too frequently. If you're interested in trying to make some money from finding a prime number, look up Mersenne Primes on the Internet. There's a big money bounty on finding the next Mersenne Prime (a prime that's one less than the number two raised to a prime power).

Index

• *G* •

• *H* •

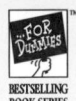

Algebra II Workbook
For Dummies®

Changing the Absolute Value Format

$|ax + b| = c \rightarrow ax + b = c$ or $ax + b = -c$

$|ax + b| > c \rightarrow ax + b > c$ or $ax + b < -c$

$|ax + b| < c \rightarrow -c < ax + b < c$

Permutations and Combinations

Permutations of r things taken from a group of n things:

$P(n, r) = \dfrac{n!}{(n-r)!}$

Combinations of r things taken from a group of n things:

$C(n, r) = \dbinom{n}{r} = \dfrac{n!}{r!(n-r)!}$

Rules for Exponents

$b^x \cdot b^y = b^{x+y}$ $\qquad \dfrac{b^x}{b^y} = b^{x-y}$

$(b^x)^y = b^{xy}$ $\qquad b^0 = 1$

$\dfrac{1}{b} = b^{-1}$ $\qquad \dfrac{1}{b^x} = b^{-x}$

$\left(\dfrac{a}{b}\right)^{-1} = \dfrac{b}{a}$ $\qquad$ If $b^x = b^y$, then $x = y$

Laws of Logarithms

$\log_b xy = \log_b x + \log_b y$

$\log_b \dfrac{x}{y} = \log_b x - \log_b y$

$\log_b x^n = n \log_b x$

$\log_b 1 = 0$

$\log_b b = 1$

$\log_b \dfrac{1}{x} = -\log_b x$

Conics: Standard Equations

Putting the equations of conic sections in standard form can tell you quite a bit about what the figures look like. Take a look at the formulas:

- **Parabola:** $a(x - h)^2 = y - k$ or $a(y - k)^2 = x - h$, where *(h, k)* is the vertex
- **Circle:** $(x - h)^2 + (y - k)^2 = r^2$, where *(h, k)* is the center and *r* is the radius
- **Ellipse:** $\dfrac{(x - h)^2}{a^2} + \dfrac{(y - k)^2}{b^2} = 1$, where *(h, k)* is the center
- **Hyperbola:** $\dfrac{(x - h)^2}{a^2} - \dfrac{(y - k)^2}{b^2} = 1$ or $\dfrac{(y - k)^2}{b^2} - \dfrac{(x - h)^2}{a^2} = 1$, where *(h, k)* is the center

The Quadratic Formula

If $ax^2 + bx + c = 0$, then $x = \dfrac{-b \pm \sqrt{b^2 - 4ac}}{2a}$

Proportions

If $\dfrac{a}{b} = \dfrac{c}{d}$, then $\dfrac{b}{a} = \dfrac{d}{c}$ and $\dfrac{a+b}{b} = \dfrac{c+d}{d}$ and $\dfrac{a-b}{b} = \dfrac{c-d}{d}$ and $\dfrac{a-b}{a+b} = \dfrac{c-d}{c+d}$

Rules for Radicals

$\sqrt{x} = x^{1/2}$ $\qquad \sqrt[b]{x^a} = x^{a/b}$

$\sqrt{xy} = \sqrt{x}\sqrt{y}$ $\qquad \sqrt{\dfrac{x}{y}} = \dfrac{\sqrt{x}}{\sqrt{y}}$

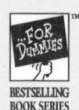

Algebra II Workbook For Dummies®

Cheat Sheet

Graphing Formulas

Some formulas pop up again and again when you're working with the x-y coordinate system. Consider the two points (x_1, y_1) and (x_2, y_2):

- **Slope of the line through the points:** $m = \dfrac{y_2 - y_1}{x_2 - x_1}$
- **Slope-intercept form of line:** $y = mx + b$
- **Point-slope form of line with slope m:** $y - y_1 = m(x - x_1)$
- **Distance formula:** $d = \sqrt{(x_2 - x_1)^2 + (y_2 - y_1)^2}$
- **Midpoint formula:** $M = \left(\dfrac{x_1 + x_2}{2}, \dfrac{y_1 + y_2}{2} \right)$

Number Systems

- **Counting numbers (natural numbers):** 1, 2, 3, 4, 5, . . .
- **Whole numbers:** 0, 1, 2, 3, 4, 5, . . .
- **Integers:** . . ., −4, −3, −2, −1, 0, 1, 2, 3, 4, . . .
- **Rational numbers:** $\dfrac{p}{q}$, where p and q are integers and $q \neq 0$
- **Irrational numbers:** Can't be written as $\dfrac{p}{q}$; decimal equivalents never repeat and never terminate
- **Complex numbers:** $a + bi$, where a and b are real numbers and $i = \sqrt{-1}$

Powers of $(a+b)$

$(a + b)^0 = 1$

$(a + b)^1 = a + b$

$(a + b)^2 = a^2 + 2ab + b^2$

$(a + b)^3 = a^3 + 3a^2 b + 3ab^2 + b^3$

$(a + b)^4 = a^4 + 4a^3 b + 6a^2 b^2 + 4ab^3 + b^4$

$(a + b)^5 = a^5 + 5a^4 b + 10a^3 b^2 + 10a^2 b^3 + 5ab^4 + b^5$

Cramer's Rule

In the linear system of equations

$$a_1 x + b_1 y = c_1$$
$$a_2 x + b_2 y = c_2$$

let

$$d = a_1 b_2 - b_1 a_2$$
$$n_x = c_1 b_2 - b_1 c_2$$
$$n_y = a_1 c_2 - c_1 a_2$$

Then $x = \dfrac{n_x}{d} = \dfrac{c_1 b_2 - b_1 c_2}{a_1 b_2 - b_1 a_2}$ and $y = \dfrac{n_y}{d} = \dfrac{a_1 c_2 - c_1 a_2}{a_1 b_2 - b_1 a_2}$

Binomial Theorem

$$(a + b)^n = a^n + na^{n-1} b^1 + \frac{n(n-1)}{2} a^{n-2} b^2 + \cdots + na^1 b^{n-1} + b^n$$

$$= \binom{n}{0} a^n + \binom{n}{1} a^{n-1} b^1 + \cdots + \binom{n}{k} a^{n-k} b^k + \cdots \binom{n}{n-1} a^1 b^{n-1} + \binom{n}{n} b^n$$

For Dummies: Bestselling Book Series for Beginners

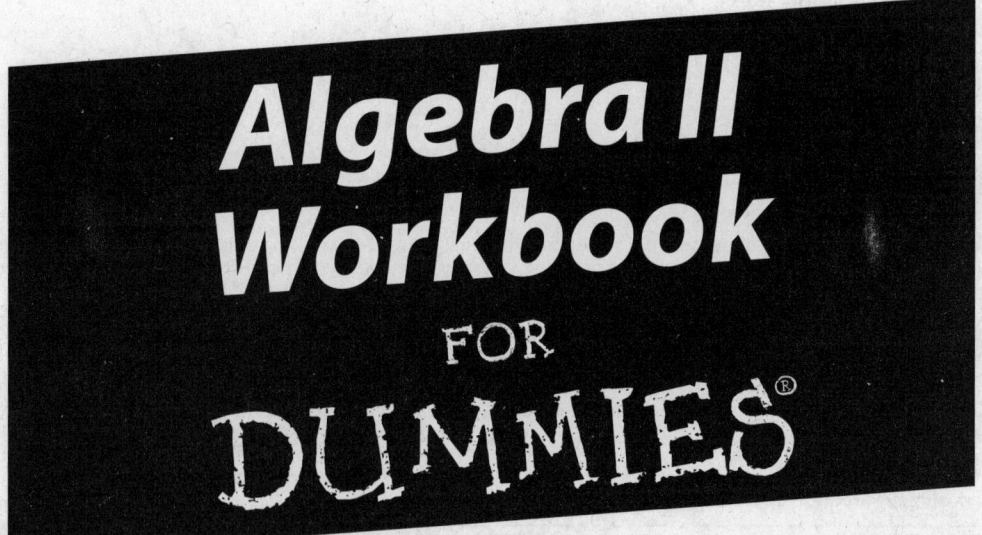

Algebra II Workbook

FOR DUMMIES®

by Mary Jane Sterling

BICENTENNIAL
1807
WILEY
2007
BICENTENNIAL

Wiley Publishing, Inc.

Algebra II Workbook For Dummies®

Published by
Wiley Publishing, Inc.
111 River St.
Hoboken, NJ 07030-5774
www.wiley.com

Copyright © 2007 by Wiley Publishing, Inc., Indianapolis, Indiana

Published by Wiley Publishing, Inc., Indianapolis, Indiana

Published simultaneously in Canada

No part of this publication may be reproduced, stored in a retrieval system, or transmitted in any form or by any means, electronic, mechanical, photocopying, recording, scanning, or otherwise, except as permitted under Sections 107 or 108 of the 1976 United States Copyright Act, without either the prior written permission of the Publisher, or authorization through payment of the appropriate per-copy fee to the Copyright Clearance Center, 222 Rosewood Drive, Danvers, MA 01923, 978-750-8400, fax 978-646-8600. Requests to the Publisher for permission should be addressed to the Legal Department, Wiley Publishing, Inc., 10475 Crosspoint Blvd., Indianapolis, IN 46256, 317-572-3447, fax 317-572-4355, or online at http://www.wiley.com/go/permissions.

Trademarks: Wiley, the Wiley Publishing logo, For Dummies, the Dummies Man logo, A Reference for the Rest of Us!, The Dummies Way, Dummies Daily, The Fun and Easy Way, Dummies.com and related trade dress are trademarks or registered trademarks of John Wiley & Sons, Inc. and/or its affiliates in the United States and other countries, and may not be used without written permission. All other trademarks are the property of their respective owners. Wiley Publishing, Inc., is not associated with any product or vendor mentioned in this book.

LIMIT OF LIABILITY/DISCLAIMER OF WARRANTY: THE PUBLISHER AND THE AUTHOR MAKE NO REPRESENTATIONS OR WARRANTIES WITH RESPECT TO THE ACCURACY OR COMPLETENESS OF THE CONTENTS OF THIS WORK AND SPECIFICALLY DISCLAIM ALL WARRANTIES, INCLUDING WITHOUT LIMITATION WARRANTIES OF FITNESS FOR A PARTICULAR PURPOSE. NO WARRANTY MAY BE CREATED OR EXTENDED BY SALES OR PROMOTIONAL MATERIALS. THE ADVICE AND STRATEGIES CONTAINED HEREIN MAY NOT BE SUITABLE FOR EVERY SITUATION. THIS WORK IS SOLD WITH THE UNDERSTANDING THAT THE PUBLISHER IS NOT ENGAGED IN RENDERING LEGAL, ACCOUNTING, OR OTHER PROFESSIONAL SERVICES. IF PROFESSIONAL ASSISTANCE IS REQUIRED, THE SERVICES OF A COMPETENT PROFESSIONAL PERSON SHOULD BE SOUGHT. NEITHER THE PUBLISHER NOR THE AUTHOR SHALL BE LIABLE FOR DAMAGES ARISING HEREFROM. THE FACT THAT AN ORGANIZATION OR WEBSITE IS REFERRED TO IN THIS WORK AS A CITATION AND/OR A POTENTIAL SOURCE OF FURTHER INFORMATION DOES NOT MEAN THAT THE AUTHOR OR THE PUBLISHER ENDORSES THE INFORMATION THE ORGANIZATION OR WEBSITE MAY PROVIDE OR RECOMMENDATIONS IT MAY MAKE. FURTHER, READERS SHOULD BE AWARE THAT INTERNET WEBSITES LISTED IN THIS WORK MAY HAVE CHANGED OR DISAPPEARED BETWEEN WHEN THIS WORK WAS WRITTEN AND WHEN IT IS READ.

For general information on our other products and services, please contact our Customer Care Department within the U.S. at 800-762-2974, outside the U.S. at 317-572-3993, or fax 317-572-4002.

For technical support, please visit www.wiley.com/techsupport.

Wiley also publishes its books in a variety of electronic formats. Some content that appears in print may not be available in electronic books.

Library of Congress Control Number: 2006936749

ISBN: 978-0-470-05238-9

Manufactured in the United States of America

10 9 8 7 6 5 4

1O/RS/RS/QW/IN

WILEY

About the Author

Mary Jane Sterling is also the author of *Algebra For Dummies, Trigonometry For Dummies, Algebra Workbook For Dummies, Trigonometry Workbook For Dummies, Algebra II For Dummies, Algebra I CliffsStudySolver,* and *Algebra II CliffsStudySolver.* She has been teaching at Bradley University in Peoria, Illinois, for over 25 years. Mary Jane enjoys working with her students both in the classroom and out, and she spends her free time on Kiwanis Family projects.

Dedication

The author dedicates this book to her children, Jon, Jim and Jane. Mary Jane and her husband Ted raised their children to be independent and adventuresome — and, boy, did it work. Jon is making his mark with his professional endeavors in California. Jim has seen the world — the good and the bad — with the Marine Corps. And Jane is poised to make her mark in her chosen profession in New York. At least their parents will have interesting places to visit when they go to see them — if the kids stay put long enough.

Author's Acknowledgments

Thanks to Tim Gallan and Danielle Voirol for seeing me through another project — and for your patience and professionalism. And thank you to the technical editor, David Herzog, who also worked on an earlier project and continues to keep me on the straight and narrow! Finally, thank you to Kathy Cox for lining up this project. Best wishes to you for a wonderful new career.

Publisher's Acknowledgments

We're proud of this book; please send us your comments through our Dummies online registration form located at www.dummies.com/register/.

Some of the people who helped bring this book to market include the following:

Acquisitions, Editorial, and Media Development

Senior Project Editor: Tim Gallan

Acquisitions Editors: Kathleen M. Cox, Lindsay Lefevere

Copy Editor: Danielle Voirol

Technical Editor: David Herzog

Editorial Manager: Christine Meloy Beck

Editorial Assistants: Erin Calligan, Joe Niesen, David Lutton

Cartoons: Rich Tennant (www.the5thwave.com)

Composition Services

Project Coordinator: Jennifer Theriot

Layout and Graphics: Denny Hager, Stephanie D. Jumper

Anniversary Logo Design: Richard Pacifico

Proofreaders: Melissa D. Buddendeck, John Greenough, Christine Pingleton, Dwight Ramsey

Indexer: Valerie Haynes Perry

Publishing and Editorial for Consumer Dummies

Diane Graves Steele, Vice President and Publisher, Consumer Dummies

Joyce Pepple, Acquisitions Director, Consumer Dummies

Kristin A. Cocks, Product Development Director, Consumer Dummies

Michael Spring, Vice President and Publisher, Travel

Kelly Regan, Editorial Director, Travel

Publishing for Technology Dummies

Andy Cummings, Vice President and Publisher, Dummies Technology/General User

Composition Services

Gerry Fahey, Vice President of Production Services

Debbie Stailey, Director of Composition Services

Contents at a Glance

Table of Contents

Introduction

Here you are, pencil in hand, ready to take on the challenges of doing Algebra II problems. How did you get here? Are you taking an Algebra II class and just not getting enough homework? Or have you found a few gaps in the instruction and you want to fill them in before you end up with a flood of questions? Maybe you've been away from algebra for a while and want a review. Or perhaps you need material for those graphing parties you started hosting after movie night got old. If you're looking for some good-natured, clear explanations on how to do some standard and challenging algebra problems, then you've come to the right place.

I hope you can find everything you need in this book to practice the concepts of Algebra II. You'll find some basic (to get you in the mood) and advanced algebra topics. But not all the basics are here — that's where Algebra I comes in. The topics that aren't here are referenced for your investigation or further study.

Calculus and other more-advanced math drive Algebra II. Algebra is the passport to studying calculus and trigonometry and number theory and geometry and all sorts of good mathematics and science. Algebra is basic, and the algebra here can help you grow in your skills and knowledge.

About This Book

You don't have to do the problems in this book in the order they're presented. You can go to the topics you want or need and refer back to earlier problems if necessary. You can jump back and forth and up and down, if so inclined (but please, not on the furniture). The organization allows you to move freely about and find what you need.

Use this book as a review or to supplement your study of Algebra II. Each section has a short explanation and an example or two — enough information to allow you to do the problems. If you want more background or historical information on a topic, you can refer to the companion book, *Algebra II For Dummies,* where I go more deeply into what's involved with each type of problem. (If you need more-basic information, you can try *Algebra For Dummies* and *Algebra Workbook For Dummies*). In this workbook, I get to the point quickly but with enough detail to see you through. The answers to the problems, at the end of each chapter, provide even more step-by-step instruction.

Conventions Used in This Book

When you're reading this book, you don't have to go very far to figure out what a mathematical word means. The specialized terms are *italicized* and defined right then and there (italics may also indicate emphasis or mark variables). I really like to use the correct mathematical terms in my explanations — the exact terms are quicker and aren't ambiguous. You know exactly what you're getting.

Also, if you need additional information to understand the material you're looking at or if further discussion may shed light on a subject, I include clear cross-references to tell you in what chapter or section you can find a particular topic.

Foolish Assumptions

Algebra II is basically a continuation of Algebra I, so I made some assumptions about you:

- ✔ In a previous study of algebra, you mastered the arithmetic of signed numbers — how to combine positive and negative numbers and come out with the correct sign. You also know the properties you need for working with algebraic expressions and the rules of order (order of operations).

- ✔ You know how to solve equations and do some of the easier graphs on the coordinate system. Even though I offer some light review of these equations and graphs in this book, I still assume that you have a general knowledge of the procedures.

- ✔ You can read the math in the examples without a lot of wordy explanation. I give step-by-step explanations in the examples and worked-out problems, but I tend to write the solutions to equations horizontally to save space. So you may see the solution of an equation as $4x = 12$; $x = 3$, with a semicolon separating lines of the equation. In a beginning algebra course, you usually write lines of the equations under one another, but I assume that you're just as comfortable with spreading out across the line.

- ✔ You have access to a simple graphing calculator. For the most part, you can do the arithmetic manipulations in your head, but some problems involve bigger-than-life numbers.

How This Book Is Organized

This book begins with some basic chapters involving equation-solving skills, and then it gets into the different kinds of functions. The next two parts consist of topics that are relatively loosely connected, but they really seem to belong together. The organization should help you locate what you want, whether you're jumping into the middle or going step by step from beginning to end.

The practice problems are packed full of challenges. I try to cover most, if not all, the eventualities that you'd face working with a particular topic. That's why you'll see so many chances to practice good mathematics within each of these problems. Read on to find out how the parts of this book break down.

Part 1: Homing In on the Basics

The first four chapters give you a crash course on performing the simple algebraic manipulations necessary to solve equations and inequalities. This part focuses on some basic factoring and equation-solving skills — skills that you use throughout the whole book and topics that are quick and easy to refer back to. You solve quadratic, rational, and radical equations. You also delve into the world of the absolute value. A quick look at graphing is here as well.

Part II: Functions

The five chapters in this part deal with the special properties that carry through all the different types of functions (linear, polynomial, exponential, and so on). I discuss the characteristics that make the functions alike, as well as those that make the functions different from one another.

A function's domain is linked to a rational function's asymptotes. A function's inverse is essential to exponential and logarithmic functions. The problems allow you to discover and confirm what you read. I also throw in a few graphing tips where appropriate.

Part III: Conics and Systems of Equations

This part focuses on conic sections, their graphs, and then systems of equations and inequalities. The conic sections are four special kinds of curves. They come with a language of their own, but they also intermingle with the other curves and functions to form intersections — common solutions. You can practice the methods for solving systems of equations (by graphing or performing algebraic manipulations) with the problems in this part. These topics can help open the door to calculus, engineering, physics, astronomy, and other types of science and mathematics besides algebra.

Part IV: Other Good Stuff: Lists, Arrays, and Imaginary Numbers

What makes Algebra II so darned interesting is that you have all the skills and practice necessary from your exposure to Algebra I that you can appreciate and be successful with these new topics. The complex numbers (which incorporate the square root of –1), matrices, sequences, and series, and sets are all present in this part.

Part V: The Part of Tens

The two chapters in the Part of Tens cover two completely different topics. The first chapter offers my choices for the ten most frequent graphs. Think of it as being like a David Letterman countdown. Check out the pictures and descriptions of the basic graphs, knowing you can modify those basic graphs as you like. The second list of ten discusses common sequences of numbers and their sums.

Icons Used in This Book

The icons in this book are great for calling attention to what you need to remember or what you need to be careful about when doing algebra. Don't let the icons throw you. Think of them as being faithful guides, helping you along the way.

You can read the word *rules* as a noun or a verb. Sometimes it's hard to differentiate. But usually, in this book, *rules* is a noun. This icon marks a formula or theorem or law from algebra that pertains to the subject at hand. The rule applies at that moment and at any moment in algebra.

You see this icon when I present an example problem whose solution I walk you through step by step. You get a problem and a detailed answer.

This word can have several meanings, too. Does it refer to the *tip* of an iceberg, the dollar bills you leave the server at the restaurant, or a hint at the racetrack? The last is closest to the correct interpretation, although I'm in no way promoting gambling. (Chapter 16 can show you how to calculate the dismal odds of winning the lottery.) When you see this icon, pay attention to what I offer to help you through the topic. This info can save you time and frustration.

This icon marks a rule or process that's one for the scrapbook. You want to remember this information even after you close the book. These concepts can come in handy when you need to take a trip down memory lane (say, when you're taking a test or working at your night job as an aerospace engineer).

The Tip and Remember icons serve as light taps on the shoulder. This icon is more in-your-face. Pay close attention when you see this icon. You can get in big trouble if you don't heed what it has to offer.

Where to Go from Here

Even though this workbook is designed so you can jump right into the middle or go to the end or start anywhere you want, why would you want to miss out on all the fun and satisfaction of working through the problems as you get them?

Okay, so you may not do math problems just for your amusement. In that case, you want to find a section you're interested in or want practice on, try a couple problems, and check how you do. See whether you can work the example problems first and then go on to the practice problems provided. Time to get started!

Part I
Homing In on the Basics

The 5th Wave   By Rich Tennant

"David's using algebra to calculate the tip.
Barbara— would you mind being a fractional
exponent?"

In this part . . .

Reviewing the basics in algebra is like preparing the crust on a pizza. You need the flour and water and yeast and oil — basic ingredients to hold the whole thing together. Some people do a lot of preparation, kneading their dough quite a bit. Others like their basics thin — just enough to do the job. This part lets you pick and choose from information on factoring, graphing, and solving equations. Concentrate heavily on the basics in this part or make your study crispy and light.

Chapter 1
Going Beyond Beginning Algebra

In This Chapter

▶ Applying order of operations and algebraic properties

▶ Using FOIL and other products

▶ Solving linear and absolute value equations

▶ Dealing with inequalities

The nice thing about the rules in algebra is that they apply no matter what level of mathematics or what area of math you're studying. Everyone follows the same rules, so you find a nice consistency and orderliness. In this chapter, I discuss and use the basic rules to prepare you for what comes next in Algebra II.

Good Citizenship: Following the Order of Operations and Other Properties

The order of operations in mathematics deals with what comes first (much like the chicken and the egg). When faced with multiple operations, this *order* tells you the proper course of action.

The *order of operations* states that you use the following sequence of events:

1. **Raise to powers or find roots.**

2. **Multiply or divide.**

3. **Add or subtract.**

Special groupings can override the normal order of operations. For instance, a^{b+c} asks you to add $b + c$ before raising a to a power. If groupings are a part of the expression, first perform whatever's in the grouping symbol. The most common grouping symbols are parentheses, ( ); brackets, [ ]; braces, { }; fraction bars, —; absolute value bars, | |; and radical signs, $\sqrt{\ }$.

If you find more than one operation from the same level, move from left to right performing those operations.

The commutative, associative, and distributive properties allow you to rewrite expressions and not change their value. So what do these properties say? Great question! And here's the answer:

✔ **Commutative property of addition and multiplication:** $a + b = b + a$, and $a \cdot b = b \cdot a$; the order doesn't matter.

Tip: Rewrite subtraction problems as addition problems so you can use the commutative (and associative) property. In other words, think of $a - b$ as $a + -b$.

> ✔ **Associative property of addition and multiplication:** $a + (b + c) = (a + b) + c$, and $a(b \cdot c) = (a \cdot b)c$; the order is the same, but the grouping changes.
>
> ✔ **Distributive property of multiplication over addition (or subtraction):** $a(b + c) = a \cdot b + a \cdot c$, and $a(b - c) = a \cdot b - a \cdot c$.

The *multiplication property of zero* states that if the product of $a \cdot b = 0$, then either a or b (or both) must be equal to 0.

0.   Use the order of operations and other properties to simplify the expression $\dfrac{12\left(\left(\frac{5}{6} + \frac{3}{4}\right) - \frac{1}{6}\right)}{5\sqrt{2(3)^2 + 7}}$

A. $\frac{17}{25}$. The big fraction bar is a grouping symbol, so you deal with the numerator and denominator separately. Use the commutative and associative properties to rearrange the fractions in the numerator; square the 3 under the radical in the denominator. Next, in the numerator, combine the fractions that have a common denominator; below the fraction bar, multiply the two numbers under the radical. Reduce the first fraction in the numerator; add the numbers under the radical. Distribute the 12 over the two fractions; take the square root in the denominator. Simplify the numerator and denominator.

Here's what the process looks like: $\dfrac{12\left(\left(\frac{5}{6} - \frac{1}{6}\right) + \frac{3}{4}\right)}{5\sqrt{2(9) + 7}} = \dfrac{12\left(\left(\frac{4}{6}\right) + \frac{3}{4}\right)}{5\sqrt{18 + 7}} = \dfrac{12\left(\frac{2}{3} + \frac{3}{4}\right)}{5\sqrt{25}} = \dfrac{8 + 9}{5(5)} = \dfrac{17}{25}$

---

1.   Simplify $-3\left[4 - 2\left(3^2\right)\right] =$

Solve It

2.   Simplify $4^3(3^2 + 11)(7^4 - 7^4)(10^{10}) =$

Solve It

3.   Simplify
$153 + 187 - 153 + 270 - 471 - 270 + 471 =$

Solve It

4.   Simplify $\left(6^2 + 4\right)\sqrt{5^2 - 3^2}$

Solve It

5. Simplify $\dfrac{\sqrt[3]{4\,(3^2+7)}}{\sqrt{121}-\sqrt{49}}$

Solve It

6. Simplify $7+3\left|2\,(8)-5^2\right|$

(For info on absolute value, see the upcoming "Dealing with Linear Absolute Value Equations" section.)

Solve It

Specializing in Products and FOIL

Multiplying algebraic expressions together can be dandy and nice or downright gruesome. Taking advantage of patterns and processes makes the multiplication quicker, easier, and more accurate.

When multiplying two binomials together, you have to multiply the two terms in the first binomial times the two terms in the second binomial — you're actually *distributing* the first terms over the second. The FOIL acronym describes a way of multiplying those terms in an organized fashion, saving space and time. *FOIL* refers to multiplying the two *First* terms together, then the two *Outer* terms, then the two *Inner* terms, and finally the two *Last* terms. The *Outer* and *Inner* terms usually combine. Then you add all the products together. So if you have *(ax + b)(cx + d)*, you can do the multiplication of the terms, or FOIL, like so:

Terms	Product
First	*ax(cx)*
Outer	*ax(d)*
Inner	*b(cx)*
Last	*b(d)*
	$= acx^2 + adx + bcx + bd$
	$= acx^2 + (ad + bc)x + bd$

The following examples show some multiplication patterns to use when multiplying *binomials* (expressions with two terms).

Q. Find the square of the binomial: $(2x + 3)^2$

A. $4x^2 + 12x + 9$. When squaring a binomial, you square both terms and put twice the product of the two original terms between the squares: $(a + b)^2 = a^2 + 2ab + b^2$. So $(2x + 3)^2 = (2x)^2 + 2(2x)(3) + 3^2 = 4x^2 + 12x + 9$.

Q. Multiply the two binomials together using FOIL: $(3x - 5)(4x + 7)$

A. $12x^2 + x - 35$. Find the products: *First*, $3x(4x) = 12x^2$, plus *Outer*, $3x(7) = 21x$, plus *Inner*, $(-5)(4x) = -20x$, plus *Last*, $(-5)(7) = -35$. So the product of $(3x - 5)(4x + 7)$ is

$$(3x)(4x) + (3x)(7) + (-5)(4x) + (-5)(7)$$

$$= 12x^2 + 21x - 20x - 35$$

$$= 12x^2 + x - 35$$

EXAMPLE

Q. Find the product of the binomial and the trinomial: $(2x + 7)(3x^2 + x - 5)$

A. $6x^3 + 23x^2 - 3x - 35$. Distribute the $2x$ over the terms in the trinomial, and then distribute the 7 over the same terms. Combine like terms to simplify. The product of $(2x + 7)(3x^2 + x - 5)$ is

$$= 2x(3x^2 + x - 5) + 7(3x^2 + x - 5)$$
$$= 6x^3 + 2x^2 - 10x + 21x^2 + 7x - 35$$
$$= 6x^3 + 23x^2 - 3x - 35$$

---

7. Square the binomial: $(4x - 5)^2$

Solve It

8. Multiply $(5y - 6)(5y + 6)$

Solve It

9. Multiply $(8z - 3)(2z + 5)$

Solve It

10. Multiply $(2x - 5)(4x^2 + 10x + 25)$

Solve It

Variables on the Side: Solving Linear Equations

A linear equation has the general format $ax + b = c$, where x is the variable and a, b, and c are constants. When you solve a linear equation, you're looking for the value that x takes on to make the linear equation a true statement. The general game plan for solving linear equations is to isolate the term with the variable on one side of the equation and then multiply or divide to find the solution.

Q. Solve for x in the equation
$\frac{3(x+7)-5}{4} = 2x+9$

A. $x = -4$. First multiply each side by 4 to get rid of the fraction. Then distribute the 3 over the terms in the parentheses. Combine the like terms on the left. Next, you want all variable terms on one side of the equation, so subtract $8x$ and 16 from each side. Finally, divide each side by -5.

$$\frac{\cancel{4}}{1} \cdot \frac{3(x+7)-5}{\cancel{4}} = 4(2x+9)$$
$$3(x+7)-5 = 8x+36$$
$$3x+21-5 = 8x+36$$
$$3x+16 = 8x+36$$
$$-5x = 20$$
$$x = -4$$

---

11. Solve for x: $4x-5 = 3(3x+10)$

Solve It

12. Solve for x: $5[9-x(x+8)] = 5x(1-x)$

Solve It

13. Solve for x: $\frac{x}{4} + \frac{x-6}{5} = \frac{x+2}{3}$

Solve It

14. Solve for x: $(x-1) + 2(x-2) + 3(x-3) + 4(x-4) + 5(x-5) = 0$

Solve It

Dealing with Linear Absolute Value Equations

The absolute value of a number is the number's distance from 0. The formal definition of *absolute value* is $|a| = \begin{cases} a, a \geq 0 \\ -a, a < 0 \end{cases}$. In other words, the absolute value of a number is exactly that number unless the number is negative; when the number is negative, its absolute value is the opposite, or a positive. The absolute value of a number, then, is the number's value without a sign; it's never negative.

When solving linear absolute value equations, you have two possibilities: one that the quantity *inside* the absolute value bars is positive, and the other that it's negative. Because you have to consider both situations, you usually get two different answers, one from each scenario. The two answers come from setting the quantity first equal to a positive value and then equal to a negative value.

Before setting the quantity equal to the positive and negative values, first isolate the absolute value term on one side of the equation by adding or subtracting the other terms (if you have any) from each side of the equation.

If you find more than one absolute value expression in your problem, you have to get *down and dirty* — consider all the possibilities. A value inside absolute value bars can be either positive or negative, so you look at all the different combinations: Both values within the bars are positive, or the first is positive and the second is negative, or the first is negative and the second positive, or both are negative. Whew!

Q. Solve for x in $|3x + 7| = 11$

A. $x = \frac{4}{3}$, **–6.** First rewrite the absolute value equation as two separate linear equations. In the first equation, assume that the $3x + 7$ is positive and set it equal to 11. In the second equation, also equal to 11, assume that the $(3x + 7)$ is negative. For that one, *negate* (multiply by –1) the whole binomial, and then solve the equation.

$$3x + 7 = 11; \ 3x = 4; \ x = \frac{4}{3}.$$

$$-(3x + 7) = 11; \ -3x - 7 = 11; \ -3x = 18; \ x = -6$$

---

15. Solve for x in $|2x + 1| = 83$

Solve It

16. Solve for x in $5|4 - 3x| + 6 = 11$

Solve It

---

17. Solve for x in $|x + 3| + |2x - 1| = 6$

Solve It

18. Solve for x in $3|4x - 5| + 10 = 7$

Solve It

Greater Math Skills: Equalizing Linear Inequalities

A linear inequality resembles a linear equation — except for the relationship. The basic form for a linear inequality is $ax + b > c$. The > can be replaced by ≥ or < or ≤, but the methods used to solve the inequality stay the same. When the extra bar appears under the inequality symbol, it means *or equal to,* so you read ≤ as "is less than or equal to."

The main time to watch out when solving inequalities is when you multiply or divide each side of the inequality by a negative number. When you do that — and yes, you're allowed — you have to reverse the sense or the relationship. The inequality > becomes <, and vice versa.

When solving *absolute value inequalities* (see the preceding section for more on absolute values), you first drop the absolute value bars. You then apply one of two separate rules for absolute value inequalities, depending on which way the inequality symbol faces:

- ✔ Solve $|ax + b| > c$ using the two inequalities $ax + b > c$ and $ax + b < -c$.
- ✔ Solve $|ax + b| < c$ using the single compound inequality $-c < ax + b < c$.

Two ways of writing your answers are inequality notation and interval notation:

- ✔ **Inequality notation:** This notation is just what it says it is: If your answer is all x's greater than or equal to 3, you write $x \geq 3$. To say that the answer is all x values between –5 and +5, including the –5 but not the positive 5, you write $-5 \leq x < 5$.

- ✔ **Interval notation:** Some mathematicians prefer interval notation because it's so short and sweet. You simply list the starting and stopping points of the numbers you want to use. When you see this notation, you just have to recognize that you're discussing intervals of numbers (and not, for instance, the coordinates of a point). The rule is that you use a bracket, [or], when you want to include the number and use a parenthesis, (or), when you don't want to include the number. You always use a parenthesis with ∞ or –∞. For example, to write $x \geq 3$ in interval notation, you use $[3, \infty)$. Writing $-5 \leq x < 5$ in interval notation, you have $[-5, 5)$.

Q. Solve the inequality $8x - 15 \geq 10x + 7$. Write the answer in both inequality and interval notation.

A. $x \leq -11$. First subtract $10x$ from each side; then add 15 to each side. This step moves the variable terms to the left and the constants to the right: $-2x \geq 22$. Now divide each side by –2. Because you're dividing by a negative number, you need to reverse the inequality sign: $x \leq -11$. That's the answer in inequality notation. The solution is that x can be any number either equal to or smaller than –11. In interval notation, you write this $(-\infty, -11]$.

Q. Solve the inequality $|5 - 6x| < 7$. Write the answer in both inequality and interval notation.

A. $\frac{1}{3} < x < 2$ **and** $\left(-\frac{1}{3}, 2\right)$. Rewrite the absolute value inequality as the inequality $-7 < 5 - 6x < 7$.

Subtract 5 from each of the three sections, or *intervals*, of the inequality to put the variable term by itself: $-12 < -6x < 2$. Now divide each section by -6, reversing the inequality symbols. Then, after you've gone to all the trouble of reversing the inequalities, rewrite the statement again

with the smaller number on the left to correspond to numbers on the number line. This step requires reversing the inequalities again. Here are the details:

$$-12 < -6x < 2$$
$$2 > x > -\frac{1}{3}$$
$$-\frac{1}{3} < x < 2$$

In interval notation, this answer is $\left(-\frac{1}{3}, 2\right)$.

This answer looks very much like the coordinates of a point. In instances like this, be very clear about what you're trying to convey with the interval notation.

---

19. Solve the inequality $5x + 7 \le 22$

Solve It

20. Solve the inequality $8(y - 4) \ge 5(3y - 12)$

Solve It

---

21. Solve the inequality $|3x + 7| > 4$

Solve It

22. Solve the inequality $4|5 - 2x| + 3 < 7$

Solve It

Answers to Problems on Going Beyond Beginning Algebra

This section provides the answers (in bold) to the practice problems in this chapter.

1 Simplify $-3[4 - 2(3^2)]$. The answer is **42.**

$$-3[4 - 2(9)] = -3[4 - 18] = -3[-14] = 42$$

2 Simplify $4^3(3^2 + 11)(7^4 - 7^4)(10^{10})$. The answer is **0.**

The third factor is 0. This makes the whole product equal to 0. Remember, the *multiplication property of zero* says that if any factor in a product is equal to 0, then the entire product is equal to 0.

3 Simplify $153 + 187 - 153 + 270 - 471 - 270 + 471$. The answer is **187.**

Use the associative and commutative properties to write the numbers and their opposites together:

$$187 + 153 - 153 + 270 - 270 - 471 + 471 = 187 + 0 + 0 + 0 = 187$$

4 Simplify $(6^2 + 4)\sqrt{5^2 - 3^2}$. The answer is **160.**

$$(36 + 4)\sqrt{25 - 9} = (40)\sqrt{16} = (40)(4) = 160$$

5 Simplify $\dfrac{\sqrt[3]{4(3^2 + 7)}}{\sqrt{121} - \sqrt{49}}$. The answer is **1.**

$$\frac{\sqrt[3]{4(9 + 7)}}{\sqrt{121} - \sqrt{49}} = \frac{\sqrt[3]{4(16)}}{11 - 7} = \frac{\sqrt[3]{64}}{4} = \frac{4}{4} = 1$$

6 Simplify $7 + 3|2(8) - 5^2|$. The answer is **34.**

$$7 + 3|16 - 25| = 7 + 3|-9| = 7 + 3(9) = 7 + 27 = 34$$

7 Square the binomial: $(4x - 5)^2$. The answer is $\mathbf{16x^2 - 40x + 25.}$

When squaring a binomial, the two terms are each squared, and the term between them is twice the product of the original terms.

A common error in squaring binomials is to forget the middle term and just use the squares of the two terms in the binomial. If you tend to forget the middle term, you can get the same answer through FOIL — $(4x - 5)(4x - 5) = 16x^2 - 20x - 20x + 25 = 16x^2 - 40x + 25$.

8 Multiply $(5y - 6)(5y + 6)$. The answer is $\mathbf{25y^2 - 36.}$

The product of two binomials that contain the sum and difference of the same two terms results in a binomial that's the difference between the squares of the terms.

9 Multiply $(8z - 3)(2z + 5)$. The answer is $\mathbf{16z^2 + 34z - 15.}$

Using FOIL, the first term in the answer is the product of the first two terms: $(8z)(2z)$. The middle term is the sum of the products of the *Outer* and *Inner* terms: $(8z)(5)$ and $(-3)(2z)$. The final term is the product of the two last terms: $(-3)(5)$.

10 Multiply $(2x - 5)(4x^2 + 10x + 25)$. The answer is $\mathbf{8x^3 - 125.}$

Distribute the first term in the binomial $(2x)$ over the terms in the trinomial, and then distribute the second term in the binomial (-5) over the terms in the trinomial. After that, combine like terms:

$$= 2x(4x^2 + 10x + 25) - 5(4x^2 + 10x + 25)$$
$$= 8x^3 + 20x^2 + 50x - 20x^2 - 50x - 125$$
$$= 8x^3 + 20x^2 - 20x^2 + 50x - 50x - 125$$
$$= 8x^3 - 125$$

11 Solve for x: $4x - 5 = 3(3x + 10)$. The answer is $\boldsymbol{x = -7}$.

Distribute the terms on the right to get $4x - 5 = 9x + 30$. Subtract $9x$ and add 5 to each side, which gives you $-5x = 35$. Then divide each side by -5: $x = -7$.

12 Solve for x: $5[9 - x(x + 8)] = 5x(1 - x)$. The answer is $\boldsymbol{x = 1}$.

First distribute the x in the bracket on the left and the $5x$ on the right. Then you can distribute the 5 outside the bracket on the left to see what the individual terms are. The squared terms disappear when you add $5x^2$ to each side of the equation. Solve for x:

$$5[9 - x^2 - 8x] = 5x - 5x^2$$
$$45 - 5x^2 - 40x = 5x - 5x^2$$
$$45 - 40x = 5x$$
$$45 = 45x$$
$$1 = x$$

13 Solve for x: $\dfrac{x}{4} + \dfrac{x-6}{5} = \dfrac{x+2}{3}$. The answer is $\boldsymbol{x = 16}$.

First multiply each fraction by 60, the least common multiple. Then distribute and simplify:

$$\overset{15}{\cancel{60}}\left(\frac{x}{\cancel{4}_1}\right) + \overset{12}{\cancel{60}}\left(\frac{x-6}{\cancel{5}_1}\right) = \overset{20}{\cancel{60}}\left(\frac{x+2}{\cancel{3}_1}\right)$$
$$15x + 12(x - 6) = 20(x + 2)$$
$$15x + 12x - 72 = 20x + 40$$
$$7x = 112$$
$$x = 16$$

14 Solve for x: $(x - 1) + 2(x - 2) + 3(x - 3) + 4(x - 4) + 5(x - 5) = 0$. The answer is $\boldsymbol{x = \dfrac{11}{3}}$.

Distribute over the terms. Combine like terms and solve for x.

$$x - 1 + 2x - 4 + 3x - 9 + 4x - 16 + 5x - 25 = 0$$
$$15x - 55 = 0$$
$$15x = 55$$
$$x = \frac{11}{3}$$

15 Solve for x in $|2x + 1| = 83$. The answer is $\boldsymbol{x = 41, -42}$.

First let the value in the absolute value be positive, and solve $2x + 1 = 83$. You get $2x = 82$, or $x = 41$. Next, let the value in the absolute value be negative, and solve $-(2x + 1) = 83$. (If you multiply each side by -1, you don't have to distribute the negative sign over two terms. Instead, you have $2x + 1 = -83$.) Solving this, you get $2x = -84$, or $x = -42$.

16 Solve for x in $5|4 - 3x| + 6 = 11$. The answer is $\boldsymbol{x = 1, \dfrac{5}{3}}$.

First subtract 6 from each side, and then divide by 5. (You can't apply the rule for changing an absolute value equation into linear equations unless the absolute value is isolated on one side of the equation.) Now you have $|4 - 3x| = 1$. Letting the expression inside the absolute value be positive, you have $4 - 3x = 1$, which means $-3x = -3$ and $x = 1$. Now let the expression inside the absolute value be negative, which means that you negate the whole thing: $-(4 - 3x) = 1$ gives you $4 - 3x = -1$, which means $-3x = -5$; $x = \frac{5}{3}$.

17 Solve for x in $|x + 3| + |2x - 1| = 6$. The answer is $\boldsymbol{x = \frac{4}{3}, -2.}$

You need four different equations to solve this problem. Consider that both absolute values may be positive; then that the first is positive and the second, negative; then that the first is negative and the second, positive; and lastly that both are negative.

Unfortunately, not every equation gives you an answer that really works. Perhaps no value of x can make the first absolute value negative and the second positive. An *extraneous* solution to an equation or inequality is a false or incorrect solution. It occurs when you change the original format of the equation to a form that's more easily solved. The extraneous solution may be a solution to the new form, but it doesn't work in the original. After you change the format, simply solve each of the equations that's produced and then check the answer in the original equation.

Here are the situations, one at a time:

- **Positive/Positive:** $(x + 3) + (2x - 1) = 6$; $x + 3 + 2x - 1 = 6$; $3x + 2 = 6$; $3x = 4$; $x = \frac{4}{3}$. When you put this answer back into the original equation, it works — you get a true statement.

- **Positive/Negative:** $(x + 3) + -(2x - 1) = 6$; $x + 3 - 2x + 1 = 6$; $-x + 4 = 6$; $-x = 2$; $x = -2$. Putting this value back into the original equation, you get a true statement, so this answer is valid, too.

- **Negative/Positive:** $-(x + 3) + (2x - 1) = 6$; $-x - 3 + 2x - 1 = 6$; $x - 4 = 6$; $x = 10$. When you put this back into the original equation, you get $13 + 19 = 6$. That isn't true, so this solution is extraneous. It doesn't work.

- **Negative/Negative:** $-(x + 3) + -(2x - 1) = 6$; $-x - 3 - 2x + 1 = 6$; $-3x - 2 = 6$; $-3x = 8$; $x = -\frac{8}{3}$.

  You get another extraneous solution. When you put this value for x back into the original equation, you get

  $$\left| -\frac{8}{3} + 3 \right| + \left| 2\left(-\frac{8}{3}\right) - 1 \right| = 6$$
  $$\left| \frac{1}{3} \right| + \left| -\frac{19}{3} \right| = 6$$
  $$\frac{20}{3} = 6$$

  Not so! This solution doesn't work.

18 Solve for x in $3|4x - 5| + 10 = 7$. **No solution.**

This absolute value equation has no solution, because it asks you to find some number whose absolute value is negative. When you subtract 10 from each side and divide each side by 3, you get $|4x - 5| = -1$. That's impossible. The absolute value of any number is either positive or 0; it's never negative.

19 Solve the inequality $5x + 7 \le 22$. The answer is $\boldsymbol{x \le 3.}$

First subtract 7 from each side to get $5x \le 15$. Then divide each side by 5. In interval notation, write the solution as $(-\infty, 3]$.

20 Solve the inequality $8(y - 4) \ge 5(3y - 12)$. The answer is $\boldsymbol{y \le 4.}$

Distribute the 8 and 5 over their respective binomials to get $8y - 32 \ge 15y - 60$. Subtract $15y$ from each side and add 32 to each side: $-7y \ge -28$. When you divide each side by -7, you reverse the inequality to get $y \le 4$. Write this answer as $(-\infty, 4]$ in interval notation.

21 Solve the inequality $|3x + 7| > 4$. The answer is $x > -1$ or $x < -\dfrac{11}{3}$.

Rewrite the absolute value inequality as two separate linear inequalities: $3x + 7 > 4$ and $3x + 7 < -4$. Solving the first, you get $3x > -3$, or $x > -1$. Solving the second, you get $3x < -11$, or $x < -\dfrac{11}{3}$. In interval notation, you write this solution as $(-1, \infty)$ or $\left(-\infty, -\dfrac{11}{3}\right)$.

22 Solve the inequality $4|5 - 2x| + 3 < 7$. The answer is $2 < x < 3$.

First subtract 3 from each side and then divide each side by 4 to get $|5 - 2x| < 1$. Then rewrite the absolute value inequality as $-1 < 5 - 2x < 1$. Subtract 5 from each interval: $-6 < -2x < -4$. Dividing each interval by -2 in the next step means reversing the inequality signs: $3 > x > 2$. The numbers should be written from smallest to largest, so switch the numbers and their inequality signs to get $2 < x < 3$. In interval notation, you write this answer as $(2, 3)$.

Chapter 2

Handling Quadratic (and Quadratic-Like) Equations and Inequalities

. . . .

In This Chapter

▶ Finding solutions with radicals

▶ Solving quadratic equations with factoring and the quadratic formula

▶ Completing the square

▶ Changing equations with higher powers to quadratic form

▶ Dealing with quadratic inequalities using number lines

. . . .

$\mathcal{Q}$uadratic equations and inequalities include variables that have powers, or exponents, of 2. The power 2 opens up the possibilities for more solutions than do linear equations (whose variables have powers of 1 — see Chapter 1). For instance, the linear equation $2x + 3 = 5$ has one solution, $x = 1$, but the quadratic equation $2x^2 + 3 = 5$ has two solutions, $x = 1$ and $x = -1$. You can solve quadratic equations through factoring, employing the quadratic formula, completing the square, or using the nifty square root rule when possible. Quadratic inequalities, on the other hand, are best solved by looking at intervals on a number line.

Some of the equations in this chapter go beyond the second degree (power or exponent on the variable) — they start out with a degree of 3 or 4 or more, but a little tweaking brings you back to the basic quadratic equation and its many possibilities. (For information on higher-powered equations that you can't change to quadratic form, see Chapter 7, on polynomials.)

Finding Reasonable Solutions with Radicals

When a quadratic equation consists of just a squared term and a constant, you use the *square root rule* to quickly solve the equation:

Solve the equation $ax^2 = k$ by dividing each side of the equation by a and then taking the square root of each side: $ax^2 = k$; $x^2 = \frac{k}{a}$; $x = \pm\sqrt{\frac{k}{a}}$. Note that you end up with two roots — one positive and one negative.

Q. Solve $9x^2 + 11 = 27$ using the square root rule.

A. $x = \pm\dfrac{4}{3}$. Subtract 11 from each side, and you get $9x^2 = 16$. Then divide each side of the equation by 9, take the square root of each side, and simplify: $x^2 = \dfrac{16}{9}$; $x = \pm\sqrt{\dfrac{16}{9}} = \pm\dfrac{\sqrt{16}}{\sqrt{9}} = \pm\dfrac{4}{3}$. The equation gives you one positive and one negative solution.

Q. Solve $3x^2 = 36$ using the square root rule.

A. $x = \pm 2\sqrt{3}$. Divide each side by 3, and then take the square root of each side: $x^2 = 12$; $x = \pm\sqrt{12}$. You can simplify the radical because of the following property:
$$\sqrt{a \cdot b} = \sqrt{a} \cdot \sqrt{b}$$
Thus, the equation becomes
$$x = \pm\sqrt{4 \cdot 3} = \pm\sqrt{4}\sqrt{3} = \pm 2\sqrt{3}.$$

1. Solve for x: $4x^2 = 25$

Solve It

2. Solve for x: $3x^2 + 8 = 56$

Solve It

3. Solve for x: $x^2 = 7$

Solve It

4. Solve for x: $5x^2 = 200$

Solve It

UnFOILed Again! Successfully Factoring for Solutions

The quickest, easiest way to solve a quadratic equation is to factor it and set the individual factors equal to 0. Of course, the expression in the equation has to be factorable. If it isn't, you can rely on that old standby, the quadratic formula (see "Resorting to the Quadratic Formula," later in this chapter). In either case, make sure your quadratic equation is in the correct form before you begin.

You can factor a quadratic equation in the form $ax^2 + bx + c = 0$ if you can find two binomials $(dx + e)(fx + g)$ whose product is $ax^2 + bx + c$. Factoring is like working out a puzzle to figure out what the coefficients and constants are.

If $ax^2 + bx + c = (dx + e)(fx + g)$, then $d \cdot f = a$ and $e \cdot g = c$; the sum of $d \cdot g$ and $e \cdot f$ is b. This process is essentially undoing FOIL. (*FOIL* is the acronym for remembering how to multiply the terms in two binomials together: *F*irst, *O*uter, *I*nner, *L*ast — see Chapter 1 for details.)

TIP

Knowing whether to use + or − signs in the binomials can really ease the factoring process. Table 2-1 shows you how the order of the + and − signs in the quadratic equation can give you clues about the signs that show up in the factors.

Table 2-1	Signs to Use in the Binomials When Factoring	
Quadratic Equation	*Signs in the Binomials*	*Example*
$ax^2 + bx + c$	(+)(+)	$x^2 + 5x + 6 = (x + 2)(x + 3)$
$ax^2 - bx + c$	(−)(−)	$x^2 - 5x + 6 = (x - 2)(x - 3)$
$ax^2 + bx - c$	(+)(−) or (−)(+)	$x^2 + x - 6 = (x + 3)(x - 2)$
$ax^2 - bx - c$	(+)(−) or (−)(+)	$x^2 - x - 6 = (x + 2)(x - 3)$

EXAMPLE

Q. Solve by factoring: $8x^2 - 2x - 3 = 0$

A. $x = \frac{3}{4}$ or $x = -\frac{1}{2}$. Factor the left side using unFOIL. The two coefficients of x in your final, factored equation need to have a product of 8, so first determine the factors of 8; you'll use $8 \cdot 1$ or $4 \cdot 2$. The product of the two constants in the binomials has to be 3, so you'll need $3 \cdot 1$. Through trial and error, you find that the 8 and 1 just don't work with this problem. Arranging the factors as $(4x \ \ 3)(2x \ \ 1)$, you see that the product of the two *O*uter terms is $4x$ and that the product of the two *I*nner terms is $6x$. The difference between $4x$ and $6x$ is $2x$, which is the middle term. Placing + and − in the correct positions, you have $8x^2 - 2x - 3 = (4x - 3)(2x + 1) = 0$. The multiplication property of zero (see Chapter 1) tells you that at least one of the binomials equals 0; $4x - 3 = 0$ gives you $x = \frac{3}{4}$, and $2x + 1 = 0$ gives you $x = -\frac{1}{2}$.

Q. Solve by factoring: $12x^2 + 44x + 24 = 0$

A. $x = -\frac{2}{3}$ or $x = -3$. First factor a 4 out of each term to get $4(3x^2 + 11x + 6) = 0$. Now factor the trinomial to get $4(3x + 2)(x + 3) = 0$. Setting each binomial equal to 0, you get $x = -\frac{2}{3}$ and $x = -3$. Don't bother setting the 4 equal to 0, because $4 = 0$ has no solution.

5. Solve by factoring: $x^2 - 15x + 56 = 0$

Solve It

6. Solve by factoring: $6x^2 + 19x + 15 = 0$

Solve It

7. Solve by factoring: $x^2 + x - 30 = 0$

Solve It

8. Solve by factoring: $5x^2 + 9x - 2 = 0$

Solve It

9. Solve by factoring: $6x^2 - 45x - 24 = 0$

Solve It

10. Solve by factoring: $4x^2 + 28x + 49 = 0$

Solve It

11. Solve by factoring: $a^2x^2 - 2abx + b^2 = 0$

Solve It

12. Solve by factoring: $25x^2 - 36 = 0$

Solve It

Your Bag of Tricks: Factoring Multiple Ways

Factoring expressions in equations is a quick, efficient way of solving the equation, because you can obtain solutions by setting the individual factors equal to 0. However, using unFOIL isn't your only option for breaking down an equation. Here's how to tackle quadratic and higher-degree equations that take more than one method of factoring:

✔ **Binomials:** When you have two terms in a quadratic equation, look for

- A variable you can factor out, such as x^2 in $x^4 - x^2 = 0$
- The difference of squares — if $x^2 - a^2 = 0$, then $(x + a)(x - a) = 0$

✔ **Trinomials:** When you find three terms in the quadratic equation, look for

- A common factor
- A way to write the expression as the product of two binomials (unFOIL)

Q. Solve the equation using two or more factoring methods: $2x^4 + 5x^3 - 3x^2 = 0$

A. $x = 0$, $x = \frac{1}{2}$, **or** $x = -3$. First factor x^2 out of each term to get $x^2(2x^2 + 5x - 3) = 0$. Then factor the trinomial in the parentheses to get $x^2(2x - 1)(x + 3) = 0$. Setting $x^2 = 0$, the solution is $x = 0$. When $2x - 1 = 0$, $x = \frac{1}{2}$. And when $x + 3 = 0$, $x = -3$.

Q. Solve the equation using two or more factoring methods: $5x^4 - 405 = 0$

A. $x = 3$ **or** $x = -3$. The two terms are divisible by 5, so you get $5(x^4 - 81) = 0$. The binomial in the parentheses is the difference between two squares, which factors into $5(x^2 - 9)(x^2 + 9) = 0$. The first binomial factors, but the second doesn't, so you have $5(x - 3)(x + 3)(x^2 + 9) = 0$. Setting $x - 3 = 0$, you get $x = 3$; and when $x + 3 = 0$, $x = -3$; $x^2 + 9$ and 5 can't equal 0, so you have only two real answers.

13. Factor and solve for x: $5x^2 - 20 = 0$

Solve It

14. Factor and solve for x: $x^5 - x^3 = 0$

Solve It

15. Factor and solve for x: $2x^4 + 5x^3 - 3x^2 = 0$

Solve It

16. Factor and solve for x: $400x^2 - 6{,}100x + 22{,}500 = 0$

Solve It

Keeping Your Act Together: Factoring by Grouping

Grouping is a type of factoring in which you separate terms into collections that have the same factor. When grouping works, collecting terms and factoring puts common factors among the different collections or groups.

Q. Solve the equation by grouping the terms to factor: $x^3 + 3x^2 - 16x - 48 = 0$.

A. $x = -3$, **4, or** -4. The first two terms have a common factor of x^2, and the second two terms have a common factor of -16. Factoring, you have $x^2(x + 3) - 16(x + 3) = 0$.

The two new terms now have $(x + 3)$ as the common factor. Factor out that grouping to get $(x + 3)(x^2 - 16) = 0$. The second binomial is a quadratic that's the difference of two squares. Factoring that second binomial, $(x + 3)(x - 4)(x + 4) = 0$. Setting the factors equal to 0, you get $x = -3$, $x = 4$, and $x = -4$.

17. Use factoring by grouping to solve: $x^2 - 3x - 6x + 18 = 0$

Solve It

18. Use factoring by grouping to solve: $8x^3 + 12x^2 - 2x - 3 = 0$

Solve It

19. Use factoring by grouping to solve:
$x^3 - 4x^2 - 25x + 100 = 0$

Solve It

20. Use factoring by grouping to solve:
$x^3(x-2)^2 + 7x^2(x-2)^2 - x(x-2)^2 - 7(x-2)^2 = 0$

Solve It

Resorting to the Quadratic Formula

The quadratic formula gives you the values of the variable x in the quadratic equation $ax^2 + bx + c = 0$. When using the formula, be sure to write the quadratic equation in exactly the correct format so that you correctly substitute the numbers for a, b, and c. People use the quadratic formula on quadratic equations that don't factor, but you can also use it on quadratic equations that do factor. Factoring is often quicker and more efficient, but sometimes the factorization just doesn't come to you. Having this fallback is nice.

 ALGEBRA RULES

The solutions of the quadratic equation $ax^2 + bx + c = 0$ are $x = \dfrac{-b \pm \sqrt{b^2 - 4ac}}{2a}$.

When the quadratic formula yields a negative number under the radical, there's no real solution — no real number that x can be. Chapter 13 covers this type of situation. For now, your answer is simply *no solution*.

 EXAMPLE

Q. Use the quadratic formula to solve the equation $4x^2 + 7x - 5 = 0$

A. $x = \dfrac{-7 + \sqrt{129}}{8} \approx \textbf{0.545}$ **or** $x = \dfrac{-7 - \sqrt{129}}{8} \approx$ **−2.295.** The values of a, b, and c in the formula are 4, 7, and −5, respectively. Substituting into the formula,

$$x = \frac{-7 \pm \sqrt{7^2 - 4(4)(-5)}}{2(4)}$$

$$= \frac{-7 \pm \sqrt{49 + 80}}{8} = \frac{-7 \pm \sqrt{129}}{8}$$

So the values of x are $x = \dfrac{-7 + \sqrt{129}}{8} \approx$ 0.545 and $x = \dfrac{-7 - \sqrt{129}}{8} \approx -2.295$.

Q. Use the quadratic formula to solve the equation $x^2 - 8x - 10 = 0$

A. $x = \mathbf{4 \pm \sqrt{26}}$. The values of a, b, and c are 1, −8, and −10, respectively. Here's what you get when you substitute into the formula:

$$x = \frac{-(-8) \pm \sqrt{(-8)^2 - 4(1)(-10)}}{2(1)}$$

$$= \frac{8 \pm \sqrt{64 + 40}}{2} = \frac{8 \pm \sqrt{104}}{2} = \frac{8 \pm 2\sqrt{26}}{2}$$

$$= \frac{\overset{4}{8} \pm \cancel{2}\sqrt{26}}{\cancel{2}} = 4 \pm \sqrt{26}$$

The decimal equivalents for the solutions are $x \approx 9.099$ and $x \approx -1.099$.

21. Use the quadratic formula to solve
$x^2 - 6x - 4 = 0$

Solve It

22. Use the quadratic formula to solve
$2x^2 + 3x - 14 = 0$

Solve It

23. Use the quadratic formula to solve
$3x^2 - 8x + 2 = 0$

Solve It

24. Use the quadratic formula to solve
$x^2 + 4x - 14 = 0$.

Solve It

Solving Quadratics by Completing the Square

The two most popular methods for solving a quadratic equation are factoring and using the quadratic formula. The last method you'd ever choose to use is completing the square. Why, you ask, do you have to know this method, then? That's a good question. And rather than say that it's good for you, I'll tell it like it is: The methods you use to solve quadratic equations by completing the square are the same methods you use to rewrite equations of conics in their standard form (which I cover in Chapter 10). Doing these rewrites here, in which you get the answer to a problem, is good practice. And convincing you that the method works for conics is easier if you've seen it before. Think of this technique as a small step toward a brighter future.

Here's how to solve the quadratic equation $ax^2 + bx + c = 0$ by completing the square:

1. **Divide every term in the equation by a, the coefficient of the x^2 term.**

2. **Move the constant term to the opposite side of the equation by adding or subtracting.**

3. **Find half the value of the coefficient on the x term; square the result of the halving; then add that amount to each side of the equation.**

4. **Factor the perfect square trinomial that you've created.**

   The factoring always has a binomial whose first term is an x and whose second term is half the coefficient of the x term on the left.

5. **Find the square root of each side of the equation.**

6. **Solve for the x by adding or subtracting a constant from each side.**

Q. Solve the quadratic equation $3x^2 + 4x - 2 = 0$ by completing the square.

A. $x = \dfrac{\left(-2 \pm \sqrt{10}\right)}{3}$. Using the steps outlined, first divide each term by 3:

$$\frac{3x^2}{3} + \frac{4x}{3} - \frac{2}{3} = 0$$

$$x^2 + \frac{4}{3}x - \frac{2}{3} = 0$$

Now add $\frac{2}{3}$ to each side to get $x^2 + \frac{4}{3}x = \frac{2}{3}$.

Half of the coefficient on the x term, $\frac{4}{3}$, is $\frac{2}{3}$. The square of that is $\frac{4}{9}$. Add that square to both sides of the equation: $x^2 + \frac{4}{3}x + \frac{4}{9} = \frac{2}{3} + \frac{4}{9} = \frac{10}{9}$.

Next, factor the trinomial on the left; you get the square of a binomial. Then take the square root of each side:

$$x^2 + \frac{4}{3}x + \frac{4}{9} = \frac{10}{9}$$

$$\left(x + \frac{2}{3}\right)^2 = \frac{10}{9}$$

$$\sqrt{\left(x + \frac{2}{3}\right)^2} = \pm\sqrt{\frac{10}{9}}$$

$$x + \frac{2}{3} = \pm\frac{\sqrt{10}}{3}$$

Now subtract $\frac{2}{3}$ from each side to solve for

x: $x = -\frac{2}{3} \pm \frac{\sqrt{10}}{3} = \frac{-2 \pm \sqrt{10}}{3}$.

---

25. Solve by completing the square:
$x^2 - 8x - 9 = 0$

Solve It

26. Solve by completing the square:
$2x^2 + 7x - 4 = 0$

Solve It

---

27. Solve by completing the square:
$x^2 + 10x - 3 = 0$

Solve It

28. Solve by completing the square:
$ax^2 + bx + c = 0$

Solve It

Working with Quadratic-Like Equations

When you solve equations with degrees higher than 2, you don't have the huge advantage of knowing a formula that always gives you the answer whether the terms factor or not. Because of this, you try to use every convenience to your fullest. One of these "conveniences" is recognizing a higher degree equation that resembles a quadratic equation.

A *quadratic-like* equation has three terms, two with variables in them and one constant term. Also, the power on one of the variable terms is twice that of the other. The general format is: $ax^{2n} + bx^n + c = 0$. When you can, factor these equations as the product of two binomials using the same unFOIL format as with the quadratics (see the earlier section "UnFOILed Again! Successfully Factoring for Solutions").

 EXAMPLE

Q. Solve the equation $8x^6 + 7x^3 - 1 = 0$

A. $x = \frac{1}{2}$ or $x = -1$. Consider the quadratic equation that sort of looks like this equation: $8y^2 + 7y - 1 = 0$. This quadratic equation factors into $(8y - 1)(y + 1) = 0$. Use the same factorization pattern, replacing each y with x^3. You get $(8x^3 - 1)(x^3 + 1) = 0$. Using the multiplication property of zero, you determine that $8x^3 - 1 = 0$; $x = \sqrt[3]{\frac{1}{8}} = \frac{1}{2}$.

When you set the other factor equal to 0, you get $x^3 + 1 = 0$; $x = -1$.

You can do the same thing more formally with a process called *substitution*. You actually replace all the higher-degree terms with the lower and solve the new problem. The only trick is remembering to substitute back in when you're done.

Q. Solve the equation $x^4 + 4x^2 - 14 = 0$.

A. $x = \pm\sqrt{-2 + 3\sqrt{2}}$. First substitute y for x^2 and y^2 for x^4, giving you $y^2 + 4y - 14 = 0$. This quadratic equation requires the quadratic formula:

$$y = \frac{-4 \pm \sqrt{4^2 - 4(1)(-14)}}{2(1)}$$

$$= \frac{-4 \pm \sqrt{16 + 56}}{2} = \frac{-4 \pm \sqrt{72}}{2}$$

$$= \frac{-4 \pm \sqrt{36}\sqrt{2}}{2} = \frac{-\overset{2}{\cancel{4}} \pm \overset{3}{\cancel{6}}\sqrt{2}}{\cancel{2}}$$

$$= -2 \pm 3\sqrt{2}$$

You get two solutions, but wait! The solutions tell you what y is equal to, and $y = x^2$. Replacing the y with x^2 and solving for x using the square root property, you get

$$x^2 = -2 \pm 3\sqrt{2}$$
$$x = \pm\sqrt{-2 \pm 3\sqrt{2}}$$

A negative under a radical isn't a real number, so you can use only the solutions $x = \pm\sqrt{-2 + 3\sqrt{2}}$.

29. Solve for x: $x^4 - 17x^2 + 16 = 0$

Solve It

30. Solve for x: $x^8 - 15x^4 - 16 = 0$

Solve It

31. Solve for x: $x^6 - 7x^3 - 8 = 0$

Solve It

32. Solve for x: $x^8 - 626x^4 + 625 = 0$

Solve It

33. Solve for x: $3x^4 - 22x^2 + 7 = 0$

Solve It

34. Solve for x: $x^{10} + x^5 - 30 = 0$

Solve It

Checking Out Quadratic Inequalities

You can solve a quadratic inequality such as $x^2 - 2x - 8 < 0$ most efficiently with a number line. (This method is also ideal for inequalities that involve higher-degree polynomials, and a number line also works when you have inequalities with fractions that have a polynomial in the numerator and denominator.)

First write all the terms on one side of the inequality symbol and factor the expression. Then determine the *zeros* — the numbers that make the expression equal to 0. Put these zeros on a number line, from smallest to largest. The zeros determine the different intervals that divide where the expression is positive from where it's negative. You determine the sign in each interval by testing a number that falls between the zeros: Simply replace the x's in the factored expression with the test number. Then you can answer the question about whether the numbers in that interval are greater than or less than 0 (positive or negative).

Q. Solve the inequality $x^2 + 7x > 30$

A. **$x < -10$ or $x > 3$.** First subtract 30 from each side of the inequality: $x^2 + 7x - 30 > 0$. Factor the trinomial into $(x + 10)(x - 3) > 0$. The zeros of the factored form are $x = -10$ and $x = 3$. Put those zeros on a number line. Then determine whether the factored product is positive or negative in the three different intervals on either side of the zeros. Test a number below -10, a number between -10 and 3, and a number above 3.

In Figure 2-1, you see that both factors are negative whenever x is a number smaller than -10. Multiply the two negative factors together, and that makes the product positive. All the numbers less than -10 will result in a positive product — which satisfies the inequality, because you want the result to be greater than 0. Back to the figure, the product of the two factors is negative between -10 and 3, so none of these numbers work. The numbers greater than 3 all work in the inequality, because both factors are positive. The solution for the inequality is $x < -10$ or $x > 3$. Writing this solution in interval notation, you get $(-\infty, -10) \cup (3, \infty)$. (See the Chapter 1 for info on interval notation.)

Figure 2-1:
Signs indicate whether the factor is positive or negative.

$(x + 10)$ negative	$(x + 10)$ positive	$(x + 10)$ positive
$(x - 3)$ negative	$(x - 3)$ negative	$(x - 3)$ positive

-10 3

35. Solve the inequality: $x^2 + x - 42 \geq 0$

Solve It

36. Solve the inequality: $x^3 - 9x < 0$

Solve It

37. Solve the inequality: $x^2 + 4x + 7 \leq 0$

Solve It

38. Solve the inequality: $\dfrac{x^2 - 4x - 5}{x + 2} \geq 0$

Solve It

Answers to Problems on Quadratic (and Quadratic-Like) Equations and Inequalities

The following are the answers to the practice problems presented earlier in this chapter.

1 Solve for x: $4x^2 = 25$. The answer is $x = \pm\dfrac{5}{2}$.

Divide each side by 4. Then take the square root of each side. Both 25 and 4 are perfect squares, so the answer is a fraction — you don't have to find a decimal approximation, because the answer is a nice, rational number.

2 Solve for x: $3x^2 + 8 = 56$. The answer is $x = \pm 4$.

Subtract 8 from each side; then divide each side by 3. When you take the square root of each side, you get both +4 and –4.

3 Solve for x: $x^2 = 7$. The answer is $x = \pm\sqrt{7}$.

The number 7 isn't a perfect square, so leave the answer in *exact form* as plus or minus the square root of 7.

4 Solve for x: $5x^2 = 200$. The answer is $x = \pm 2\sqrt{10}$.

First divide each side by 5 to get $x^2 = 40$. When you take the square root of each side, you have $\pm\sqrt{40}$, which you can simplify: $\pm\sqrt{40} = \pm\sqrt{4}\sqrt{10} = \pm 2\sqrt{10}$.

 ALGEBRA RULES $\dfrac{1}{+\dfrac{1}{2}}$

Use this rule to simplify radicals in which one of the factors of the number under the radical is a perfect square: $\sqrt{a \cdot b} = \sqrt{a}\sqrt{b}$.

5 Solve by factoring: $x^2 - 15x + 56 = 0$. The answer is $x = 7, 8$.

The trinomial factors into $(x - 7)(x - 8) = 0$. Setting the factors equal to 0, you get $x = 7$ or 8.

6 Solve by factoring: $6x^2 + 19x + 15 = 0$. The answer is $x = -\dfrac{5}{3}, -\dfrac{3}{2}$.

The trinomial factors into $(3x + 5)(2x + 3) = 0$. Set the factors equal to 0. Subtract and divide in each case to get the two solutions.

7 Solve by factoring: $x^2 + x - 30 = 0$. The answer is $x = -6, 5$.

The trinomial factors into $(x + 6)(x - 5) = 0$. Set the individual factors equal to 0 and solve for x in each case.

8 Solve by factoring: $5x^2 + 9x - 2 = 0$. The answer is $x = \dfrac{1}{5}, -2$.

The trinomial factors into $(5x - 1)(x + 2) = 0$. Set the factors equal to 0 and solve for x in each case.

9 Solve by factoring: $6x^2 - 45x - 24 = 0$. The answer is $x = -\dfrac{1}{2}, 8$.

The terms share a common factor of 3; factor it out first to get $3(2x^2 - 15x - 8) = 0$. Continue factoring to get $3(2x + 1)(x - 8) = 0$. Set each factor equal to 0 to get the answers. When you set the first factor equal to 0, you get $3 = 0$. This statement is never true, so you don't get a solution from that factor — just the other two.

10 Solve by factoring: $4x^2 + 28x + 49 = 0$. The answer is $x = -\frac{7}{2}$.

This equation has a perfect square trinomial on the left. Factored, you get $(2x + 7)^2 = 0$. Setting that factor equal to 0, you get the sole answer. This answer is called a *double root,* because the answer technically appears twice in the factorization: $(2x + 7)(2x + 7)$.

11 Solve by factoring: $a^2x^2 - 2abx + b^2 = 0$. The answer is $x = \frac{b}{a}$.

The trinomial factors as the square of the binomial *(ax − b)*. The trinomial is a perfect square trinomial. Set it equal to 0 to solve for the value of x you need.

12 Solve by factoring: $25x^2 - 36 = 0$. The answer is $x = \frac{6}{5}, -\frac{6}{5}$.

This equation includes the difference between two perfect squares, which factors into the difference and sum of the same two values. The binomial factors into $(5x - 6)(5x + 6) = 0$. Setting the two factors equal to 0, you get the two answers (two opposites).

13 Factor and solve for x: $5x^2 - 20 = 0$. The answer is $x = 2, -2$.

First factor 5 out of each term to get $5(x^2 - 4) = 0$. The binomial in the parentheses factors into the difference and sum of the same two numbers, $5(x - 2)(x + 2) = 0$. Setting the factors equal to 0, you get the two solutions. Of course, setting $5 = 0$ doesn't yield a solution to the equation.

14 Factor and solve for x: $x^5 - x^3 = 0$. The answer is $x = 0, 1, -1$.

First factor out the x^3 to get $x^3(x^2 - 1) = 0$. The binomial does factor, so you have $x^3(x - 1)(x + 1) = 0$. Setting the three factors equal to 0, you get the three solutions. Actually, the x^3 is a *triple root,* giving you the answer 0 three times.

People commonly miss the solution obtained by setting x or a power of x equal to 0 after factoring it out of the terms in an equation. Be careful to set $x^n = 0$ to get the answer that $x = 0$.

15 Factor and solve for x: $2x^4 + 5x^3 - 3x^2 = 0$. The answer is $x = 0, \frac{1}{2}, -3$.

First factor out x^2 from each term to get $x^2(2x^2 + 5x - 3) = 0$. The quadratic in the parentheses factors, giving you $x^2(2x - 1)(x + 3) = 0$. Set the three factors equal to 0 to get the three solutions. Note that you have a double root at $x = 0$ because of the exponent on the x.

16 Factor and solve for x: $400x^2 - 6{,}100x + 22{,}500 = 0$. The answer is $x = \frac{25}{4}, 9$.

The numbers here seem enormous, but they're all divisible by 100. Factor that out first to get $100(4x^2 - 61x + 225) = 0$. The numbers used in the quadratic in the parentheses are also large, but think about the possible factors of each. The number 4 can be $4 \cdot 1$ or $2 \cdot 2$. The number 225 is 15^2, but it's also $225 \cdot 1$, $75 \cdot 3$, $45 \cdot 5$, and $25 \cdot 9$. It's this last pair, coupled with $4 \cdot 1$, that works to give a sum of 61 for the middle term. The quadratic factors to give you $100(4x - 25)(x - 9) = 0$. Set the two binomials equal to 0 and solve.

17 Use factoring by grouping to solve: $x^2 - 3x - 6x + 18 = 0$. The answer is $x = 3, 6$.

This equation is actually quadratic, so you can solve it by combining the two middle terms and then using unFOIL. To factor this expression by grouping, though, factor the x out of the first two terms and a −6 out of the second two terms to get $x(x - 3) - 6(x - 3) = 0$. The grouping and factoring results in two terms — each with the common factor of $(x - 3)$. Now factor that binomial out of the two terms and you get $(x - 3)(x - 6) = 0$. Setting the factors equal to 0 gives you the two solutions.

18 Use factoring by grouping to solve: $8x^3 + 12x^2 - 2x - 3 = 0$. The answer is $x = -\frac{3}{2}, \frac{1}{2}, -\frac{1}{2}$.

The first two terms have a common factor of $4x^2$, and the second two terms have a common factor of -1 (you have to divide by -1 so that the binomials match). After doing the two factorizations, you end up with $4x^2(2x + 3) - 1(2x + 3) = 0$. Now, factoring out $(2x + 3)$, you write the equation as $(2x + 3)(4x^2 - 1) = 0$. The second binomial factors into the difference and sum of the same two numbers, $(2x + 3)(2x - 1)(2x + 1) = 0$. Set each factor equal to 0 to find the solutions.

19 Use factoring by grouping to solve: $x^3 - 4x^2 - 25x + 100 = 0$. The answer is $x = 4, 5, -5$.

Factor x^2 out of the first two terms and -25 out of the second two terms to get $x^2(x - 4) - 25(x - 4) = 0$. The common factor $(x - 4)$ comes out of the two terms to give you $(x - 4)(x^2 - 25) = 0$. The second binomial factors into the difference and sum of the same two numbers (or if you prefer, when you set the two binomials equal to 0, you can rewrite $x^2 - 25 = 0$ as $x^2 = 25$ and use the square root rule). Set each binomial equal to 0 and solve.

20 Use factoring by grouping to solve: $x^3(x - 2)^2 + 7x^2(x - 2)^2 - x(x - 2)^2 - 7(x - 2)^2 = 0$. The answer is $x = 2, -7, 1, -1$.

Notice that each of the four terms has a factor of $(x - 2)^2$. After factoring that out, you can factor by grouping by taking x^2 out of the first two terms in the brackets and -1 out of the second two terms: $(x - 2)^2[x^3 + 7x^2 - x - 7] = (x - 2)^2[x^2(x + 7) - 1(x + 7)] = 0$. Factor out the $(x + 7)$ within the brackets to get $(x - 2)^2[(x + 7)(x^2 - 1)] = (x - 2)^2(x + 7)(x - 1)(x + 1) = 0$. Set the individual factors equal to 0 to find the solutions. The solution $x = 2$ is a double root because of the power on its binomial.

21 Use the quadratic formula to solve $x^2 - 6x - 4 = 0$. The answer is $x = 3 \pm \sqrt{13}$.

The values of a, b, and c are 1, -6, and -4, respectively. Substituting these values into the quadratic formula, you get

$$x = \frac{-(-6) \pm \sqrt{(-6)^2 - 4(1)(-4)}}{2(1)}$$
$$= \frac{6 \pm \sqrt{36 + 16}}{2} = \frac{6 \pm \sqrt{52}}{2}$$

You can simplify the radical, because 52 is the product of a perfect square (4) and another number (13). After simplifying the radical, reduce the fraction:

$$x = \frac{6 \pm \sqrt{4}\sqrt{13}}{2} = \frac{6 \pm 2\sqrt{13}}{2}$$
$$= \frac{\overset{3}{\cancel{6}} \pm \cancel{2}\sqrt{13}}{\cancel{2}} = 3 \pm \sqrt{13}$$

22 Use the quadratic formula to solve $2x^2 + 3x - 14 = 0$. The answer is $x = 2, -\frac{7}{2}$.

The values of a, b, and c are 2, 3, and -14, respectively. Substituting these values into the quadratic formula, you get

$$x = \frac{-3 \pm \sqrt{3^2 - 4(2)(-14)}}{2(2)}$$
$$= \frac{-3 \pm \sqrt{9 + 112}}{4} = \frac{-3 \pm \sqrt{121}}{4}$$
$$= \frac{-3 \pm 11}{4}$$

Note that the answer doesn't contain a radical. The final answers are both rational numbers; you get them by first using the $+$ and then the $-$ in the fraction. Having no radical in the answer means that you could have factored the quadratic to find the solution.

23 Use the quadratic formula to solve $3x^2 - 8x + 2 = 0$. The answer is $x = \dfrac{4 \pm \sqrt{10}}{3}$.

The values of a, b, and c are 3, –8, and 2, respectively. Substituting these values into the quadratic formula, you get

$$x = \frac{-(-8) \pm \sqrt{(-8)^2 - 4(3)(2)}}{2(3)}$$

$$= \frac{8 \pm \sqrt{64 - 24}}{6} = \frac{8 \pm \sqrt{40}}{6}$$

Simplify the radical by rewriting it as the product of the root of a perfect square (4) times the root of the other factor (10). Then reduce the fraction:

$$x = \frac{8 \pm \sqrt{4}\sqrt{10}}{6} = \frac{8 \pm 2\sqrt{10}}{6}$$

$$= \frac{\overset{4}{\cancel{8}} \pm \overset{}{\cancel{2}}\sqrt{10}}{\underset{3}{\cancel{6}}} = \frac{4 \pm \sqrt{10}}{3}$$

24 Use the quadratic formula to solve $x^2 + 4x - 14 = 0$. The answer is $x = x = -2 \pm 3\sqrt{2}$.

Use 1, 4, and –14 for a, b, and c, respectively:

$$x = \frac{-4 \pm \sqrt{4^2 - 4(1)(-14)}}{2(1)}$$

$$= \frac{-4 \pm \sqrt{16 + 56}}{2} = \frac{-4 \pm \sqrt{72}}{2}$$

$$= \frac{-4 \pm \sqrt{36}\sqrt{2}}{2} = \frac{\overset{2}{\cancel{-4}} \pm \overset{3}{\cancel{6}}\sqrt{2}}{\cancel{2}}$$

$$= -2 \pm 3\sqrt{2}$$

25 Solve by completing the square: $x^2 - 8x - 9 = 0$. The answer is $x = 9, -1$.

The coefficient on the x^2 term is 1, so you can skip the first step. To solve the equation by completing the square, you first add 9 to each side and then add 16 to each side (16 is –4 squared — half of the –8 coefficient squared). Write the terms on the left as the square of a binomial, and then take the square root of each side:

$$x^2 - 8x = 9$$
$$x^2 - 8x + 16 = 9 + 16$$
$$(x - 4)^2 = 25$$
$$\sqrt{(x - 4)^2} = \pm\sqrt{25}$$
$$x - 4 = \pm 5$$

Now solve for x by adding 4 to each side: $x = 4 \pm s\ 5$. Obtain the two solutions by applying the + and the –: $x = 4 + 5 = 9$ and $x = 4 - 5 = -1$.

26 Solve by completing the square: $2x^2 + 7x - 4 = 0$. The answer is $x = \dfrac{1}{2}, -4$.

You first divide each term by 2 and then add 2 to each side to put the constant on the right side of the equation. Take half of the coefficient on the x term and square it (half of $\frac{7}{2}$ is $\frac{1}{2} \cdot \frac{7}{2} = \frac{7}{4}$).

Then add that square to each side. Factor the terms on the left and simplify on the right, and then take the square root of each side:

$$\frac{2}{2}x^2 + \frac{7}{2}x - \frac{4}{2} = 0$$

$$x^2 + \frac{7}{2}x - 2 = 0$$

$$x^2 + \frac{7}{2}x = 2$$

$$x^2 + \frac{7}{2}x + \left(\frac{7}{4}\right)^2 = 2 + \left(\frac{7}{4}\right)^2$$

$$\left(x + \frac{7}{4}\right)^2 = \frac{32 + 49}{16} = \frac{81}{16}$$

$$\sqrt{\left(x + \frac{7}{4}\right)^2} = \pm\sqrt{\frac{81}{16}}$$

$$x + \frac{7}{4} = \pm\frac{9}{4}$$

Now subtract $\frac{7}{4}$ from each side and apply the + and – to get the answers:

$$x = -\frac{7}{4} \pm \frac{9}{4}$$

$$x = -\frac{7}{4} + \frac{9}{4} = \frac{2}{4} = \frac{1}{2}$$

or

$$x = -\frac{7}{4} - \frac{9}{4} = -\frac{16}{4} = -4$$

27 Solve by completing the square: $x^2 + 10x - 3 = 0$. The answer is $x = -5 \pm 2\sqrt{7}$.

Add 3 to each side of the equation. Then take half of 10 and square it — and add the 25 to each side. Factor on the left, and then take the square root of each side:

$$x^2 + 10x = 3$$

$$x^2 + 10x + 25 = 3 + 25$$

$$(x + 5)^2 = 28$$

$$\sqrt{(x + 5)^2} = \pm\sqrt{28}$$

$$x + 5 = \pm\sqrt{4}\sqrt{7} = \pm 2\sqrt{7}$$

Now solve for x by subtracting 5 from each side. The answer is $x = -5 + 2\sqrt{7}$.

28 Solve by completing the square: $ax^2 + bx + c = 0$. The answer is $x = \dfrac{-b \pm \sqrt{b^2 - 4ac}}{2a}$.

Look familiar? Well, it should. It's the quadratic formula. You can solve for the quadratic formula by taking the standard quadratic equation and solving for x by completing the square. You first divide each term by a. Subtract $\frac{c}{a}$ from each side. Then take half of the coefficient of the x term, $\frac{1}{2} \cdot \frac{b}{a} = \frac{b}{2a}$, square it, and add that squared value to both sides of the equation. Factor on the left and simplify on the right. Then take the square root of each side. Subtract $\frac{b}{2a}$ from each side and simplify:

$$ax^2 + bx + c = 0$$

$$\frac{a}{a}x^2 + \frac{b}{a}x + \frac{c}{a} = 0$$

$$x^2 + \frac{b}{a}x = -\frac{c}{a}$$

$$x^2 + \frac{b}{a}x + \left(\frac{b}{2a}\right)^2 = \left(\frac{b}{2a}\right)^2 - \frac{c}{a}$$

$$x^2 + \frac{b}{a}x + \frac{b^2}{4a^2} = \frac{b^2}{4a^2} - \frac{4ac}{4a^2}$$

$$\left(x + \frac{b}{2a}\right)^2 = \frac{b^2 - 4ac}{4a^2}$$

$$\sqrt{\left(x + \frac{b}{2a}\right)^2} = \pm\sqrt{\frac{b^2 - 4ac}{4a^2}}$$

$$x + \frac{b}{2a} = \pm\frac{\sqrt{b^2 - 4ac}}{\sqrt{4a^2}}$$

$$= -\frac{b}{a} \pm \frac{\sqrt{b^2 - 4ac}}{2a}$$

$$= \frac{-b \pm \sqrt{b^2 - 4ac}}{2a}$$

29 Solve for x: $x^4 - 17x^2 + 16 = 0$. The answer is $x = 4, -4, 1, -1$.

Factor into the product of two quadratic binomials, $(x^2 - 16)(x^2 - 1) = 0$. Now, each binomial factors into the difference and sum of the same numbers, $(x - 4)(x + 4)(x - 1)(x + 1) = 0$. Set each factor equal to 0 for the solutions.

30 Solve for x: $x^8 - 15x^4 - 16 = 0$. The answer is $x = 2, -2$.

The left side factors into $(x^4 - 16)(x^4 + 1) = 0$. The first binomial factors, giving you $(x^2 - 4)(x^2 + 4)(x^4 + 1) = 0$. Only the first factor yields a real solution. When $x^2 - 4 = 0$, $x = 2$ or -2.

31 Solve for x: $x^6 - 7x^3 - 8 = 0$. The answer is $x = 2, -1$.

The trinomial factors into $(x^3 - 8)(x^3 + 1) = 0$. Setting the two binomials equal to 0, you get the two solutions after taking the cube root of each side.

32 Solve for x: $x^8 - 626x^4 + 625 = 0$. The answer is $x = 5, -5, 1, -1$.

First factor the trinomial into $(x^4 - 625)(x^4 - 1) = 0$. Each binomial also factors: $(x^2 - 25)(x^2 + 25)(x^2 - 1)(x^2 + 1) = 0$. Now factor the two binomials that are differences of perfect squares to get $(x - 5)(x + 5)(x^2 + 25)(x - 1)(x + 1)(x^2 + 1) = 0$. Four of the factors produce solutions when you set them equal to 0. The other factors yield answers that are complex numbers. (See Chapter 13 for the scoop on complex numbers.)

33 Solve for x: $3x^4 - 22x^2 + 7 = 0$. The answer is $x = \pm\frac{\sqrt{3}}{3}, \pm\sqrt{7}$.

First factor the trinomial into $(3x^2 - 1)(x^2 - 7) = 0$. The binomials don't factor, but you can find the solutions by setting the biomials equal to 0 and using the square root rule:

$$3x^2 - 1 = 0, \; 3x^2 = 1, \; x^2 = \frac{1}{3}, \; x = \pm\sqrt{\frac{1}{3}} = \pm\frac{\sqrt{3}}{3}$$

$$x^2 - 7 = 0, \; x^2 = 7, \; x = \pm\sqrt{7}$$

34 Solve for x: $x^{10} + x^5 - 30 = 0$. The answer is $x = -\sqrt[5]{6}, \sqrt[5]{5}$.

Factor the left side into $(x^5 + 6)(x^5 - 5) = 0$. Setting the two factors equal to 0 and taking the fifth root of each side, you get the two solutions.

35 Solve the inequality: $x^2 + x - 42 \geq 0$. The answer is $x \leq -7$ **or** $x \geq 6$.

First, factor the quadratic into $(x + 7)(x - 6)$. When you set the factors equal to 0, you get the two zeros for the number line. Place the -7 and 6 on the number line and test numbers on either side and between the zeros to determine whether the values in the interval are positive or negative.

The product of the two factors is positive when x is less than -7 or greater than 6. Include the two zeros in your answer. In interval notation, write the answer as $(-\infty, -7] \cup [6, \infty)$.

36 Solve the inequality: $x^3 - 9x < 0$. The answer is $x < -3$ **or** $0 < x < 3$.

Factoring the expression on the left, you get $x(x - 3)(x + 3)$. Setting those factors equal to 0, the three zeros for the number line are 0, 3, and -3. None of these zeros appear in the answer, because you want values of x that make the expression negative. Place the zeros on the number line in order from the smallest to the largest. Then test the factors in the different intervals determined by the zeros.

The product is negative when an odd number of the factors is negative. The product of the three factors is negative when x is smaller than -3 and when x is between 0 and $+3$. In interval notation, write the answer as $(-\infty, -3) \cup (0, 3)$.

37 Solve the inequality: $x^2 + 4x + 7 \leq 0$. **No solution.**

The quadratic doesn't factor, and when you use the quadratic formula, you get a negative under the radical. This tells you that there's no real solution. So back to the drawing board. Just try some x's in the expression on the left to see whether you can make it negative. When you let $x = -1$, you get $1 - 4 + 7 = 4$. That doesn't work. When you let $x = -10$, you get $100 - 40 + 7 = 67$. Might as well stop. No solutions make the quadratic 0, so the expression never changes signs. Looks like you always get a positive number, no matter what you put in for x, so this inequality has no solution.

38 Solve the inequality: $\dfrac{x^2 - 4x - 5}{x + 2} \geq 0$. The answer is $-2 < x \leq -1$ **or** $x \geq 5$.

Factoring the numerator of the fraction, you get $(x - 5)(x + 1)$. The two zeros from the numerator are 5 and -1, and the zero from the denominator is -2. Put those zeros on the number line and test to see what the signs of the factors are in the intervals.

$(x - 5)$ negative	$(x - 5)$ negative	$(x - 5)$ negative	$(x - 5)$ positive
$(x + 1)$ negative	$(x + 1)$ negative	$(x + 1)$ positive	$(x + 1)$ positive
$(x + 2)$ negative	$(x + 2)$ positive	$(x + 2)$ positive	$(x + 2)$ positive

-2	-1	5	

The result is positive when an even number of factors is negative. The fraction is positive between –2 and –1 and then, again, when x is greater than 5. You can include the zeros at 5 and –1, because they make the numerator equal to 0. That's allowed. You can't include the –2, because that makes the denominator equal to 0, and numbers divided by 0 are undefined. Written in interval notation, the solution is $(-2, -1] \cup [5, \infty)$.

Chapter 3

Rooting Out the Rational, the Radical, and the Negative

*R*ational numbers are those you can write as a fraction with an integer in both the numerator and denominator. Rational numbers have the added attraction of having decimal equivalents that either terminate or repeat in a regular pattern. (This is what we call *behaving rationally*.)

Then you come to those pesky radicals. A *radical* indicates a root. The square root, cube root, fourth root and so on are numbers whose repeated products give you the number under the radical sign. Numbers that are stuck under a radical sign — and have no exact value — don't behave very well. The decimal equivalents of numbers under a radical never end and never have a pattern or repeat. Consequently, numbers in this category are called *irrational numbers*.

You can use a negative exponent to show that you've moved a factor from the denominator of a fraction to the numerator. For instance, if you don't like the fraction ½, you can move the 2 in the denominator up to the numerator by writing the ½ as 2^{-1}. See! The fraction notation is gone! Writing and dealing with numbers and variables with negative exponents is often easier than working with fractions. The numbers with negative exponents combine more easily.

In this chapter, you see rational numbers, irrational numbers, and negative exponents. I put these exciting types of numbers into equations and then ask you to solve for any real solutions to the equations. You often need to use some special handling to solve these problems. Rest assured that the sections in this chapter can acquaint you with the rules involved.

Doing Away with Denominators with an LCD

What in the world is an LCD? It almost sounds like something out of the '60s. (Oh, no, that's LSD. Sorry!) An *LCD* is a least common denominator. When you have fractions and need to add or subtract them, you need a common denominator for the fractions. And of course, the most preferable is the *least* of these — you don't want the numbers to get too big.

A good process to use when solving equations that contain fractions is to change all the fractions so that they have the same denominator. Ideally, the common denominator is the LCD. So what do you do after you go to all the trouble of changing all the fractions so they have the same denominator? You get rid of the denominators in the fractions by multiplying both sides of the equation by that LCD. The end result is a linear or quadratic or some other equation that you can solve more easily. The only catch is that you may introduce an extraneous solution.

An *extraneous solution* is an answer that doesn't work in the original equation. Extraneous solutions arise when you change the original equation to another, more-convenient form in order to solve the equation. Check for extraneous solutions by substituting any solutions you find back into the original equation.

Q. Solve for x in $\dfrac{3x}{x+2} - \dfrac{3}{x-1} = 1$ by finding an LCD and simplifying the equation.

A. $3x^2 - 3x - 3x - 6 = x^2 + x - 2$

$2x^2 - 7x - 4 = 0$

You first find the LCD of the fractions, which is $(x+2)(x-1)$. Then multiply all the terms in the equation by this LCD:

$$\frac{3x}{x+2} \cdot \frac{(x+2)(x-1)}{1} - \frac{3}{x-1} \cdot \frac{(x+2)(x-1)}{1} = \frac{1}{1} \cdot \frac{(x+2)(x-1)}{1}$$

$$\frac{3x}{x+2} \cdot \frac{(x+2)(x-1)}{1} - \frac{3}{x-1} \cdot \frac{(x+2)(x-1)}{1} = \frac{1}{1} \cdot \frac{(x+2)(x-1)}{1}$$

$$3x(x-1) - 3(x+2) = (x+2)(x-1)$$

Multiply and distribute. Then set the equation equal to 0:

$3x^2 - 3x - 3x - 6 = x^2 + x - 2$

$2x^2 - 7x - 4 = 0$

This quadratic equation factors into $(2x+1)(x-4) = 0$. Setting the two factors equal to 0, you get that $x = -\frac{1}{2}, 4$. You need to check both answers. Neither creates a situation in which you have a 0 in the denominator, and both do work in the original equation. So both answers work — neither solution is extraneous.

---

1. Solve for x: $\dfrac{x}{2} - \dfrac{3x}{4} = 2$

Solve It

2. Solve for x: $\dfrac{4}{x+1} + \dfrac{6}{x-1} = 4$

Solve It

3. Solve for x: $\dfrac{5}{x} - 6 = -\dfrac{3}{x+2}$

Solve It

4. Solve for x: $\dfrac{2}{x-2} + \dfrac{8}{x+2} = \dfrac{8}{x^2-4}$

Solve It

Simplifying and Solving Proportions

A proportion is an equation involving two ratios (fractions) set equal to each other. The equation $\dfrac{14}{21} = \dfrac{24}{36}$ is a proportion. Both fractions in that proportion reduce to $\dfrac{2}{3}$, so it's fairly easy to see how this statement is true.

 Proportions have some interesting, helpful, and easy-to-use properties:

In the proportion $\dfrac{a}{b} = \dfrac{c}{d}$

- ✔ The cross-products are equal: $a \cdot d = b \cdot c$.
- ✔ The reciprocals are equal (you can flip the fractions): $\dfrac{b}{a} = \dfrac{d}{c}$.
- ✔ You can reduce the fractions vertically or horizontally: You can divide out factors that are common to both numerators or both denominators or the left fraction or the right fraction. (You can't, however, divide out a factor from the numerator of one fraction and the denominator of the other.)

The properties of proportions come in useful when solving equations involving fractions. When you can, change an equation with fractions in it to a proportion for ease in solving.

Q. Solve the equation using the properties of proportions: $\dfrac{5}{x+3} + \dfrac{x}{10} = \dfrac{3x+6}{10}$

A. $x = -8$ or $x = 2$. First subtract $\dfrac{x}{10}$ from each side of the equation to create the proportion:

$$\dfrac{5}{x+3} + \dfrac{x}{10} - \dfrac{x}{10} = \dfrac{3x+6}{10} - \dfrac{x}{10}$$

$$\dfrac{5}{x+3} = \dfrac{2x+6}{10}$$

Then reduce the right-hand fraction by factoring out a 2:

$$\dfrac{5}{x+3} = \dfrac{2(x+3)}{10}$$

$$\dfrac{5}{x+3} = \dfrac{\cancel{2}(x+3)}{\cancel{10}^{5}}$$

$$\dfrac{5}{x+3} = \dfrac{x+3}{5}$$

Now cross-multiply and simplify the resulting quadratic equation. Factor the quadratic and solve for the values of x:

$$5 \cdot 5 = (x+3)(x+3)$$
$$25 = x^2 + 6x + 9$$
$$0 = x^2 + 6x - 16$$
$$0 = (x+8)(x-2)$$
$$x = -8 \ \text{or} \ x = 2$$

To check for extraneous solutions, you need to substitute the two different values in for x to see whether they work in the original equation. When $x = -8$, you get

$$\frac{5}{-8+3} + \frac{-8}{10} = \frac{3(-8)+6}{10}$$
$$\frac{5}{-5} + \frac{-4}{5} = \frac{-24+6}{10}$$
$$\frac{-5}{5} - \frac{4}{5} = \frac{-18}{10}$$
$$\frac{-9}{5} = \frac{-9}{5}$$

This solution works. Now, substituting in $x = 2$,

$$\frac{5}{2+3} + \frac{2}{10} = \frac{3(2)+6}{10}$$
$$\frac{5}{5} + \frac{1}{5} = \frac{6+6}{10}$$
$$\frac{6}{5} = \frac{12}{10}$$
$$\frac{6}{5} = \frac{6}{5}$$

Both solutions work.

---

5. Solve for x: $\dfrac{x+1}{9} = \dfrac{x+6}{24}$

Solve It

6. Solve for x: $\dfrac{3x-2}{20} + \dfrac{7}{x+1} = \dfrac{12}{x+1}$

Solve It

---

7. Solve for x: $\dfrac{144x}{x+71} = \dfrac{64(x+2)}{12(x+7)}$

Solve It

8. Solve for x: $\dfrac{x+6}{5-2x} = \dfrac{x+2}{x-9}$

Solve It

Wrangling with Radicals

Radical equations contain square roots or cube roots or some other roots of algebraic expressions. The most efficient way of solving these equations is to get rid of the radical signs. Squaring both sides once or twice gets rid the square root radicals, but doing so may introduce an extraneous root. You need to check your solutions carefully.

Q. Solve for x: $\sqrt{5x-1} + 2 = x - 1$

A. **$x = 10$.** First subtract 2 from each side to get the radical by itself on one side of the equation. Then square both sides of the equation:

$$\sqrt{5x-1} = x - 3$$
$$\left(\sqrt{5x-1}\right)^2 = (x-3)^2$$
$$5x - 1 = x^2 - 6x + 9$$

Now set the quadratic equation equal to 0 and factor. Set the two factors equal to 0 to solve for x:

$$0 = x^2 - 11x + 10$$
$$= (x - 10)(x - 1)$$
$$x = 10 \text{ or } x = 1$$

To check the solutions in the *original* equation, first let $x = 10$, and you see that 10 is a solution:

$$\sqrt{5(10) - 1} + 2 = 10 - 1$$
$$\sqrt{50 - 1} + 2 = 9$$
$$\sqrt{49} + 2 = 9$$
$$7 + 2 = 9$$

Trying the $x = 1$, you get

$$\sqrt{5(1) - 1} + 2 = 1 - 1$$
$$\sqrt{5 - 1} + 2 = 0$$
$$\sqrt{4} + 2 = 0$$
$$2 + 2 \neq 0$$

Thus, the solution $x = 1$ is extraneous; it's a solution of the quadratic equation, but it doesn't work in the original equation.

Q. Solve for x: $\sqrt{x+5} + \sqrt{2x+3} = 9$

A. **$x = 11$.** This problem requires you to square the sides of the equation twice. You first move one of the radicals to the right and then square both sides. The result still has a radical in it, so you isolate that radical term on one side and square both sides again. Solve the resulting equation for x:

$$\sqrt{x+5} = 9 - \sqrt{2x+3}$$
$$\left(\sqrt{x+5}\right)^2 = \left(9 - \sqrt{2x+3}\right)^2$$
$$x + 5 = 81 - 18\sqrt{2x+3} + 2x +$$
$$x + 5 = 84 + 2x - 18\sqrt{2x+3}$$
$$-x - 79 = -18\sqrt{2x+3}$$
$$(-x - 79)^2 = \left(-18\sqrt{2x+3}\right)^2$$
$$x^2 + 158x + 6241 = 324(2x+3)$$
$$x^2 + 158x + 6241 = 648x + 972$$
$$x^2 - 490x + 5269 = 0$$
$$(x - 11)(x - 479) = 0$$
$$x = 11 \text{ or } x = 479$$

Substituting in $x = 11$, you find that 11 is a solution:

$$\sqrt{11 + 5} + \sqrt{2(11) + 3} = 9$$
$$\sqrt{16} + \sqrt{25} = 9$$
$$4 + 5 = 9$$

However, $x = 479$ is extraneous:

$$\sqrt{479 + 5} + \sqrt{2(479) + 3} = 9$$
$$\sqrt{484} + \sqrt{961} = 9E$$
$$22 + 31 \neq 9$$

9. Solve for x: $\sqrt{2x+5} - 2 = 1$

Solve It

10. Solve for x: $\sqrt{3x+7} + 1 = x$

Solve It

11. Solve for x: $\sqrt{2x+7} - 2x = 1$

Solve It

12. Solve for x: $\sqrt{2x-3} - \sqrt{x-2} = 1$

Solve It

Changing Negative Attitudes toward Negative Exponents

Negative exponents are really quite useful. They allow you to combine factors of numbers and variables with simple addition, subtraction, and multiplication. For instance, if you want to multiply $\frac{1}{x^3} \cdot \frac{3}{x^5} \cdot x^9$, just change the two denominators to numbers with negative exponents $(x^{-3} \cdot 3x^{-5} \cdot x^9)$ and multiply the factors with the same base x by *adding the exponents* $(3x^{-3-5+9} = 3x^1)$.

Here's the rule for changing fractions to numbers with negative exponents (and back).

$$\frac{1}{x} = x^{-1} \text{ and } \frac{1}{x^n} = x^{-n}$$

When negative exponents occur in equations, you can factor out the terms containing those negative exponents or treat those equations like other equations with a similar format. You usually have to get rid of the negative exponents (convert the terms with negative exponents to fractions) at the end of the problem to solve for the value of x.

Q. Solve for x: $x^{-1} + 3x^{-2} = 0$

A. $x = -3$. First factor out x^{-2} from each term, giving you $x^{-2}(x + 3) = 0$. Now set each factor equal to 0. The factor x^{-2} will never equal 0, because fractions (like $\frac{1}{x^2}$) can't equal 0 unless the numerator is 0. The solution for $x + 3 = 0$ is $x = -3$.

Q. Solve for x: $3x^{-2} + 2x^{-1} = 1$

A. $x = 3$ **or** $x = -1$. Rewrite the equation by setting it equal to 0: $3x^{-2} + 2x^{-1} - 1 = 0$. This gives you a *quadratic-like* equation (see Chapter 2 for more on this type of equation). The left side factors into the product of two binomials, $(3x^{-1} - 1)(x^{-1} + 1) = 0$. Setting the first factor equal to 0, you get

$$3x^{-1} - 1 = 0$$
$$\frac{3}{x} = 1$$
$$x = 3$$

The second factor yields you

$$x^{-1} + 1 = 0$$
$$\frac{1}{x} = -1$$
$$x = -1$$

These two solutions work in the original equation.

---

13. Solve for x: $x^{-2} = \dfrac{1}{9}$

Solve It

14. Solve for x: $4x^{-2} - x^{-1} = 0$

Solve It

---

15. Solve for x: $x^{-2} - 13x^{-1} + 36 = 0$

Solve It

16. Solve for x: $8x^{-2} = 27x$

Solve It

Divided Powers: Solving Equations with Fractional Exponents

You use fractional exponents to replace radicals and combinations of radicals and powers.

Here's how fractional exponents work:

$$\sqrt{a} = a^{1/2}, \text{ and more generally, } \sqrt[m]{a^n} = a^{n/m}$$

When solving equations with fractional exponents, use much the same technique as with negative exponents: Look for common factors, and look for quadratic-like or other equation patterns (see Chapter 2 for more info on equation patterns).

Q. Solve for x: $x^{5/3} + 2x^{4/3} - 3x = 0$

A. $x = 0$ or $x = -27$ or $x = 1$. Each term has a variable x in it, and the lowest power is x^1. Factor that out of each term, and then factor the trinomial using the same pattern as when factoring $y^2 + 2y - 3 = 0$:

$$x^{5/3} + 2x^{4/3} - 3x = 0$$
$$x\left(x^{2/3} + 2x^{1/3} - 3\right) = 0$$
$$x\left(x^{1/3} + 3\right)\left(x^{1/3} - 1\right) = 0$$

Set each of the three factors equal to 0, and solve the three equations for x. When $x = 0$, you have a solution right there. When $x^{1/3} + 3 = 0$, subtract 3 from each side and then raise each side of the equation to the third power to get

$$x^{1/3} = -3$$
$$\left(x^{1/3}\right)^3 = (-3)^3$$
$$x = -27$$

Use the same technique with $x^{1/3} = 1$ to get that $x = 1$.

---

17. Solve for x: $x^{2/3} = 4$

Solve It

18. Solve for x: $x^{2/3} - 13x^{1/3} + 36 = 0$

Solve It

---

19. Solve for x: $x + x^{1/2} = 6$

Solve It

20. Solve for x: $x^{1/2} = 4x^{1/6}$

Solve It

Answers to Problems on Rooting Out the Rational, the Radical, and the Negative

The following are the answers to the practice problems presented in this chapter.

1 Solve for x: $\frac{x}{2} - \frac{3x}{4} = 2$. The answer is $x = -8$.

Clear the equation of fractions by multiplying each term by the LCD (least common denominator), which is 4: $\frac{x}{2} \cdot \frac{4^2}{1} - \frac{3x}{4} \cdot \frac{4}{1} = 2 \cdot 4$. The equation simplifies to $2x - 3x = 8$. Combine the terms on the left, and the equation becomes $-x = 8$, or $x = -8$.

2 Solve for x: $\frac{4}{x+1} + \frac{6}{x-1} = 4$. The answer is $x = -\frac{1}{2}, 3$.

Multiply each fraction by the LCD, $(x+1)(x-1)$, to clear the equation of fractions. Then distribute the remaining values in each term:

$$\frac{4}{x+1} \cdot \frac{(x+1)(x-1)}{1} + \frac{6}{x-1} \cdot \frac{(x+1)(x-1)}{1} = 4 \cdot (x+1)(x-1)$$

$$4(x-1) + 6(x+1) = 4(x^2-1)$$

$$4x - 4 + 6x + 6 = 4x^2 - 4$$

Rewrite the equation as a quadratic set equal to 0, and factor for the solutions.

$$10x + 2 = 4x^2 - 4$$
$$0 = 4x^2 - 10x - 6$$
$$0 = 2(2x^2 - 5x - 3)$$
$$0 = 2(2x+1)(x-3)$$
$$x = -\frac{1}{2} \text{ or } x = 3$$

Both answers work — neither is extraneous (neither causes the denominator to be 0).

3 Solve for x: $\frac{5}{x} - 6 = -\frac{3}{x+2}$. The answer is $x = -\frac{5}{3}, 1$.

Multiply each fraction by the LCD, $x(x+2)$. Then distribute the remaining values in each term:

$$\frac{5}{x} \cdot \frac{x(x+2)}{1} - 6 \cdot x(x+2) = -\frac{3}{x+2} \cdot \frac{x(x+2)}{1}$$

$$5(x+2) - 6x(x+2) = -3x$$

$$5x + 10 - 6x^2 - 12x = -3x$$

Rewrite the equation as a quadratic set equal to 0, and factor for the solutions. Neither answer is extraneous (neither causes the denominator to be 0):

$$10 - 6x^2 - 7x = -3x$$
$$0 = 6x^2 + 4x - 10$$
$$0 = 2(3x^2 + 2x - 5)$$
$$0 = 2(3x+5)(x-1)$$
$$x = -\frac{5}{3} \text{ or } x = 1$$

4 Solve for x: $\dfrac{2}{x-2} + \dfrac{8}{x+2} = \dfrac{8}{x^2-4}$. **No solution.**

Multiply each fraction by the LCD, $(x-2)(x+2) = x^2-4$. Then distribute the remaining values in each term. What's left is a linear equation that you can solve for x:

$$\frac{2}{x-2} \cdot \frac{(x-2)(x+2)}{1} + \frac{8}{x+2} \cdot \frac{(x-2)(x+2)}{1} = \frac{8}{x^2-4} \cdot \frac{x^2-4}{1}$$

$$2(x+2) + 8(x-2) = 8$$
$$2x + 4 + 8x - 16 = 8$$
$$10x - 12 = 8$$
$$10x = 20$$
$$x = 2$$

The solution $x = 2$ looks like a perfectly respectable answer. Unfortunately, it isn't. If you replace all the x's in the original equation with 2, you get 0 in the denominators of the first and last terms. The value 2 is a solution of the linear equation that you created to solve the problem, but it's extraneous. It doesn't solve the original equation, and nothing else works there, either.

5 Solve for x: $\dfrac{x+1}{9} = \dfrac{x+6}{24}$. The answer is $x = 2$.

You can first simplify the proportion by reducing horizontally across the bottom; both denominators are divisible by 3. Then cross-multiply, distribute the factors, and solve the resulting linear equation:

$$\frac{x+1}{9_3} = \frac{x+6}{24_8}$$
$$8(x+1) = 3(x+6)$$
$$8x + 8 = 3x + 18 \quad E$$
$$5x = 10$$
$$x = 2$$

6 Solve for x: $\dfrac{3x-2}{20} + \dfrac{7}{x+1} = \dfrac{12}{x+1}$. The answer is $x = \dfrac{17}{3}, -6$.

Combine the two terms with the same denominator by subtracting $\dfrac{7}{x+1}$ from each side. Then cross-multiply and rewrite the resulting equation as a quadratic set equal to 0. Factor the quadratic and solve for x:

$$\frac{3x-2}{20} = \frac{12}{x+1} - \frac{7}{x+1}$$
$$\frac{3x-2}{20} = \frac{5}{x+1}$$
$$(3x-2)(x+1) = 100$$
$$3x^2 + x - 2 = 100$$
$$3x^2 + x - 102 = 0$$
$$(3x-17)(x+6) = 0$$
$$x = \frac{17}{3} \text{ or } x = -6$$

7 Solve for x: $\dfrac{144x}{x+71} = \dfrac{64(x+2)}{12(x+7)}$. The answer is $x = -\dfrac{17}{13}, 1$.

You can reduce the fraction on the right by dividing the numerator and denominator by 4. Then you can reduce horizontally, across the top, by dividing both numerators by 16. Cross-multiply what's left and simplify to form a quadratic equation you can set equal to 0 and factor for the solution:

$$\frac{144x}{x+71} = \frac{\overset{16}{\cancel{64}}(x+2)}{\underset{3}{\cancel{12}}(x+7)}$$

$$\frac{\overset{9}{\cancel{144}}x}{x+71} = \frac{\overset{1}{\cancel{16}}(x+2)}{3(x+7)}$$

$$9x(3)(x+7) = (x+71)(x+2)$$

$$27x^2 + 189x = x^2 + 73x + 142$$

$$26x^2 + 116x - 142 = 0$$

$$2(13x^2 + 58x - 71) = 0$$

$$2(13x + 71)(x - 1) = 0$$

$$x = -\frac{71}{13} \text{ or } x = 1$$

8 Solve for x: $\frac{x+6}{5-2x} = \frac{x+2}{x-9}$. The answer is $x = \frac{16}{3}, -4$.

Cross-multiply and then simplify the terms by distributing. Rewrite the equation as a quadratic equation set equal to 0. Factor and find for the solutions:

$$\frac{x+6}{5-2x} = \frac{x+2}{x-9}$$

$$(x+6)(x-9) = (x+2)(5-2x)$$

$$x^2 - 3x - 54 = -2x^2 + x + 10$$

$$3x^2 - 4x - 64 = 0$$

$$(3x - 16)(x + 4) = 0$$

$$x = \frac{16}{3} \text{ or } x = -4$$

9 Solve for x: $\sqrt{2x+5} - 2 = 1$. The answer is $x = 2$.

Add 2 to each side of the equation, and then square both sides, which gives you $2x + 5 = 9$. Subtract 5 from each side to get $2x = 4$. Dividing by 2, the answer is $x = 2$. Substituting this number into the original equation, you end up with $3 - 2 = 1$, so the answer checks.

10 Solve for x: $\sqrt{3x+7} + 1 = x$. The answer is $x = 6$.

Subtract 1 from each side, and then square both sides of the equation to get $3x + 7 = x^2 - 2x + 1$. Rewrite this as a quadratic equation set equal to 0: $x^2 - 5x - 6 = 0$. This equation factors into $(x-6)(x+1) = 0$. When you try $x = 6$ in the original equation, you get $5 + 1 = 6$. When you try $x = -1$ in the original equation, you get $2 + 1 = -1$, which isn't true; $x = -1$ is an extraneous solution.

11 Solve for x: $\sqrt{2x+7} - 2x = 1$. The answer is $x = 1$.

Add $2x$ to each side of the equation, and then square both sides to get $2x + 7 = 4x^2 + 4x + 1$. Set this quadratic equation equal to 0 and factor: $0 = 4x^2 + 2x - 6 = 2(2x^2 + x - 3) = 2(2x + 3)(x - 1)$. The first solution, $-\frac{3}{2}$, doesn't work, because it makes the original equation read $2 + 3 = 1$. The second solution, $x = 1$, does work; you get $3 - 2 = 1$.

12 Solve for x: $\sqrt{2x-3} - \sqrt{x-2} = 1$. The answer is $x = 6, 2$.

You have to square both sides of this equation twice to get rid of the radicals. First, add $\sqrt{x-2}$ to each side, and then square both sides. Simplify the terms, and then isolate the remaining radical expression by putting the rest of the terms on the other side. Square both sides, and then solve the resulting quadratic equation:

$$\sqrt{2x-3} = 1 + \sqrt{x-2}$$
$$\left(\sqrt{2x-3}\right)^2 = \left(1 + \sqrt{x-2}\right)^2$$
$$2x - 3 = 1 + 2\sqrt{x-2} + x - 2$$
$$2x - 3 = -1 + x + 2\sqrt{x-2}$$
$$x - 2 = 2\sqrt{x-2}$$
$$(x-2)^2 = \left(2\sqrt{x-2}\right)^2$$
$$x^2 - 4x + 4 = 4(x-2)$$
$$x^2 - 4x + 4 = 4x - 8$$
$$x^2 - 8x + 12 = 0$$
$$(x-6)(x-2) = 0$$
$$x = 6 \text{ or } x = 2$$

Both the 6 and the 2 work in the original equation.

13 Solve for x: $x^{-2} = \frac{1}{9}$. The answer is $x = 3, -3$.

First rewrite the equation without the negative exponent: $\frac{1}{x^2} = \frac{1}{9}$. You have a proportion, so you can "flip" the fractions and set the reciprocals equal to one another. (If you need a refresher on proportion properties, refer to the section called "Simplifying and Solving Proportions," earlier in this chapter.) The new equation reads: $x^2 = 9$. Use the square root property (take the square root of both sides), and you get $x = \pm 3$.

14 Solve for x: $4x^{-2} - x^{-1} = 0$. The answer is $x = 4$.

Factor x^{-2} out of each term to get $x^{-2}(4-x) = 0$. Setting x^{-2} equal to 0 doesn't yield you an answer, because the equation $\frac{1}{x^2} = 0$ has no solution. The only way a fraction can equal 0 is if the numerator equals 0, and with a 1 in the numerator of this fraction, that can never happen. The other factor gives you the solution $x = 4$.

15 Solve for x: $x^{-2} - 13x^{-1} + 36 = 0$. The answer is $x = \frac{1}{9}, \frac{1}{4}$.

This trinomial can be considered to be *quadratic-like* (see Chapter 2), because the power on the first term is twice that of the second and the last term is a constant. Factoring the trinomial into the product of two binomials, you get $(x^{-1} - 9)(x^{-1} - 4) = 0$. Setting these factors equal to 0, use the equations $\frac{1}{x} - 9 = 0$ or $\frac{1}{x} - 4 = 0$ to solve for the solutions:

$$\frac{1}{x} - 9 = 0 \text{ or } \frac{1}{x} - 4 = 0$$
$$\frac{1}{x} = 9 \quad \text{ or } \quad \frac{1}{x} = 4$$
$$\frac{1}{x} = \frac{9}{1} \quad \text{ or } \quad \frac{1}{x} = \frac{4}{1}$$
$$\frac{x}{1} = \frac{1}{9} \quad \text{ or } \quad \frac{x}{1} = \frac{1}{4}$$

16 Solve for x: $8x^{-2} = 27x$. The answer is $x = \frac{2}{3}$.

Subtract $27x$ from each side and then factor x^{-2} out of each term to get $x^{-2}(8 - 27x^3) = 0$. The first factor, x^{-2}, doesn't yield a solution, but setting the second factor equal to 0, you get

$$8 = 27x^3$$

$$\frac{8}{27} = x^3$$

$$\sqrt[3]{\frac{8}{27}} = x$$

$$\frac{2}{3} = x$$

When you have only two terms in an equation with negative exponents, consider rewriting the equation as a proportion. You can do problems 14 and 16 this way. For problem 14, you write the equation as

$$4x^{-2} = x^{-1}$$

$$\frac{4}{x^2} = \frac{1}{x}$$

and then cross-multiply. The downside to this method is that you're apt to introduce extraneous solutions, so you have to be especially wary. Likewise, problem 16 can appear as $\frac{8}{x^2} = \frac{27x}{1}$. Simply cross-multiply and solve for x.

17 Solve for x: $x^{2/3} = 4$. The answer is **$x = 8, -8$.**

First cube each side of the equation (raise each side to the third power): $\left(x^{2/3}\right)^3 = 4^3$. That gives you the equation $x^2 = 64$. Now use the square root property, and you end up with $x = 8$ or $x = -8$.

18 Solve for x: $x^{2/3} - 13x^{1/3} + 36 = 0$. The answer is **$x = 64, 729$.**

This trinomial factors using the same factorization pattern as in $y^2 - 13y + 36 = (y - 4)(y - 9) = 0$. The factorization with the fractional exponents is $\left(x^{1/3} - 4\right)\left(x^{1/3} - 9\right) = 0$. Setting each factor equal to 0 and solving by cubing both sides of the equations, you get $x^{1/3} - 4 = 0$, $x^{1/3} = 4$, $\left(x^{1/3}\right)^3 = 4^3$, $x = 64$ and $x^{1/3} - 9 = 0$, $x^{1/3} = 9$, $\left(x^{1/3}\right)^3 = 9^3$, $x = 729$.

19 Solve for x: $x + x^{1/2} = 6$. The answer is **$x = 4$.**

Set the equation equal to 0, giving you the trinomial $x + x^{1/2} - 6 = 0$. Next, factor the trinomial into $(x^{1/2} - 2)(x^{1/2} + 3) = 0$. Setting the first factor equal to 0 and solving for x, you get $x^{1/2} - 2 = 0$, $x^{1/2} = 2$, $\left(x^{1/2}\right)^2 = (2)^2$, $x = 4$. But when you try to solve $x^{1/2} + 3 = 0$, you get $x^{1/2} = -3$, which is an impossible statement; the only answer is 4.

20 Solve for x: $x^{1/2} = 4x^{1/6}$. The answer is **$x = 0, 64$.**

First set the equation equal to 0, and then factor out the common factor of $x^{1/6}$. You get $x^{1/6} - 4x^{1/6} = 0$, $x^{1/6}\left(x^{1/3} - 4\right) = 0$. When you set $x^{1/6} = 0$, you get $x = 0$. After setting $x^{1/3} - 4 = 0$, move the 4 to the right and cube each side to get $x = 64$.

Chapter 4

Graphing for the Good Life

A graph is a mathematical picture. In statistics or other topics that deal with numbers of things, you see bar graphs, scatter plots, line graphs, and so on. In algebra, you use graphs to represent all the sets of numbers that work in a particular equation. A graph is very useful — it gives you all sorts of information right up front. For instance, it can tell you how high or how low the points go. It can show you, quickly, if the graph represents points that spread all over the place or points that are limited in their scope.

In this chapter, you become familiar with the coordinate graphing system and its many facets. Your goal is to quickly graph equations, so in this chapter, you see how to use the properties of the different types of equations in order to make the process easier and faster.

I also provide a bit of information on graphing with some technological help. The graphing calculator information near the end of the chapter is purposely generic so you can see the technological graphing possibilities and then go to your particular calculator manual for all the gory details.

Coordinating Axes, Coordinates of Points, and Quadrants

The *Cartesian coordinate system* is the standard graphing arrangement set up for mathematical points and equations. In this system, the *x*- and *y-axes* (singular: axis) are perpendicular lines that divide a plane into four sections called *quadrants*. The *origin* is the point at the center of the plane where the axes intersect. *Coordinates* are ordered pairs of numbers, such as (–3, 4) or (7, 0); any ordered pair *(x, y)* tells you how many units from the origin a particular point is in the *x* direction (left or right) and the *y* direction (up or down). Thus, a point on one of the axes always has a 0 for one of the coordinates, and the origin has a 0 for both coordinates. Figure 4-1 shows you the coordinate system's axes, origin, quadrants, and a few select points.

The quadrants are labeled in a counterclockwise direction, starting with Quadrant I in the upper right-hand corner. Given which quadrant a point is in, you can tell whether the point's *x*- and *y*-values are positive or negative. A negative coordinate indicates that you move the point left or down from the origin; positive coordinates go to the right or upward. Therefore, all the points in a particular quadrant have the same signs on their *x*- and *y*-coordinates. For instance, all points in Quadrant IV have a positive *x*-coordinate and a negative *y*-coordinate.

Figure 4-1:
The point
(–3, 4) is 3
units left
and 4 units
up from the
origin.

Q. Which quadrant do the points (–3, 4), (–1, 6) and (–5, 5) lie in?

A. **Quadrant II.** The *x*-coordinates are all negative, so that means that the points are in either Quadrant II or Quadrant III. The *y*-coordinates are positive, which is true of points in Quadrant I or Quadrant II. The only quadrant satisfying both of these conditions is Quadrant II.

Use Figure 4-2 to answer questions 1–4.

Figure 4-2:
Points
in their
quadrants.

1. Name the points in Figure 4-2 that lie in Quadrant II or on the *x*-axis

Solve It

2. Name the points in Figure 4-2 that lie in Quadrant IV or on the *y*-axis

Solve It

3. Identify the quadrant that contains all points with a negative *x*-coordinate and a negative *y*-coordinate

Solve It

4. If you draw a segment connecting points (–4, 4) and (2, –2), what point lies on that line segment that has *x*- and *y*-coordinates with the same value (the numbers are the same)?

Solve It

Crossing the Line: Using Intercepts and Symmetry to Graph

The *intercepts* of a graph are where the graph of an equation crosses an axis. The x-intercepts all have the general form $(h, 0)$ — the x-coordinate is some number, and the y-coordinate is 0. The y-intercepts all have the general form $(0, k)$, where the x-coordinate is 0 and the y-coordinate is some number. Replacing one of the variables in an equation with a 0 makes the equation simpler, so finding intercepts by putting in 0s also simplifies the graphing.

To find the x-intercepts of the graph of an equation, let $y = 0$ and solve for x. (If you have the y isolated on one side of the equation, you solve for x-intercepts the same way you'd solve any equation set equal to 0 — the x-intercepts are *solutions* or *roots* of the equation.) To find the y-intercepts, let $x = 0$ and solve for y.

A graph has *symmetry* if half of it looks like a mirror reflection or 180-degree rotation of the other half. Instead of plotting points to check whether a graph is symmetrical, you can try the following algebraic tests.

Here are the types of symmetry and how to look for them:

✔ A graph is *symmetric with respect to the y-axis* if the graph is a mirror image on either side of the vertical (y) axis. If you replace all the x's in the equation with $-x$, the equation remains unchanged.

✔ A graph is *symmetric with respect to the x-axis* if it's a mirror image on either side of the horizontal (x) axis. If you replace all the y's in the equation with $-y$, the equation remains unchanged.

✔ A graph is *symmetric about the origin* if a 180-degree turn around the origin makes the parts of the graph coincide. If you replace all the x's with $-x$ and the y's with $-y$, the equation remains unchanged.

Q. Use intercepts and symmetry to help graph the equation $x^2 + (y - 3)^2 = 25$

A. Letting $y = 0$, you get $x^2 + 9 = 25$, so $x^2 = 16$ and $x = \pm 4$. The two x-intercepts are $(4, 0)$ and $(-4, 0)$. When you let $x = 0$, you get $(y - 3)^2 = 25$. Taking the square root of each side, $y - 3 = \pm 5$, so $y = 3 + 5 = 8$ or $y = 3 - 5 = -2$. The two y-intercepts are $(0, 8)$ and $(0, -2)$. When you replace x with $-x$ in the equation, the original equation doesn't change, so the graph is symmetric with respect to the y-axis. The graph in Figure 4-3a shows the four intercepts graphed on the coordinate axes. You may recognize this equation as the standard form for a circle (see Chapter 10). Even if you don't immediately recognize that, you can just plug and plot a few more points and use the fact that the graph is symmetric about the y-axis to fill in the rest of the graph. Figure 4-3b shows the whole graph drawn in.

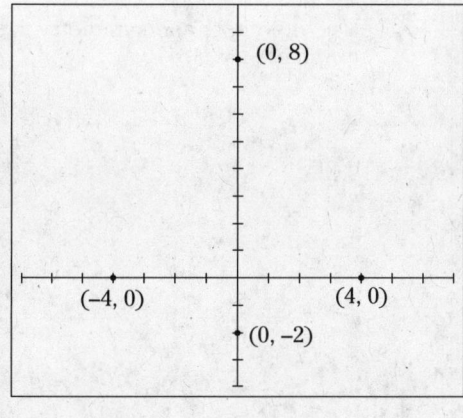

Figure 4-3:
The four intercepts drawn in and the rest of the graph, showing that the equation is a circle.

a

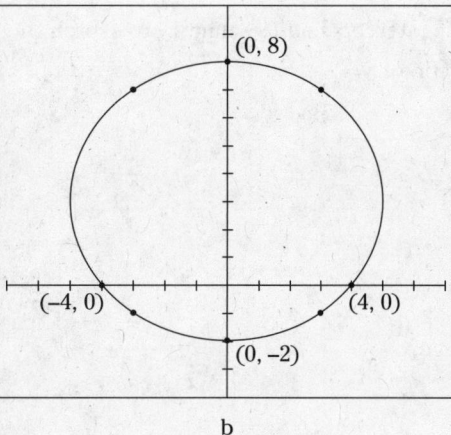

b

5. Use intercepts and symmetry to sketch the graph of $y = x^2 - 9$

Solve It

6. Use intercepts and symmetry to sketch the graph of $y = 3 - |x|$

Solve It

7. Use intercepts and symmetry to sketch the graph of $y = \dfrac{1}{x^2}$

Solve It

8. Use intercepts and symmetry to sketch the graph of $y = \sqrt{16 - x^2}$

Solve It

Graphing Lines Using Slope-Intercept and Standard Forms

REMEMBER

Just two points determine a line; only one distinct line can go through those two points. People usually indicate a line with one of two common forms of the equation: *slope-intercept* or *standard*.

- **Slope-intercept form:** $y = mx + b$. The m is a constant that equals the slope of the line, and the b is the y-coordinate of the y-intercept of the line.

- **Standard form:** $Ax + By = C$. When using this form, the slope is equal to $-\dfrac{A}{B}$, the x-intercept is $\left(\dfrac{C}{A}, 0\right)$, and the y-intercept is $\left(0, \dfrac{C}{B}\right)$.

EXAMPLE

Q. Use the equation of the line $4x - 3y = 12$ to find the slope and two intercepts of the line. Then change this equation to slope-intercept form and verify the slope and y-intercept. Sketch a graph of this line.

A. The slope is $-\dfrac{A}{B} = -\dfrac{4}{-3} = \dfrac{4}{3}$, the x-intercept is $\left(\dfrac{C}{A}, 0\right) = \left(\dfrac{12}{4}, 0\right) = (3, 0)$, and the y-intercept is

$\left(0, \dfrac{C}{B}\right) = \left(0, \dfrac{12}{-3}\right) = (0, -4)$. You get these answers from the fractions I give you in the second bullet point. To write this standard-form equation in slope-intercept form, subtract $4x$ from each side $(-3y = -4x + 12)$ and then divide each term by -3 to get

$$\frac{-3y}{-3} = \frac{-4x}{-3} + \frac{12}{-3}$$

$$y = \frac{4}{3}x - 4$$

You see that the slope and y-intercept match what you found earlier. To sketch the line, you can use either the two intercepts or the y-intercept and the slope. To use the y-intercept and slope, start at the y-intercept and look at the denominator of the fraction representing the slope (if the slope is an integer, then the denominator is 1). The denominator tells you how many units to move to the right from the y-intercept. Then look at the numerator of the slope. Count up as many units as are in the numerator if the slope is positive; count down if it's negative. Check out Figure 4-4 to see how to graph this line using both methods.

Figure 4-4:
Using the intercepts and slope-intercept to graph a line.

9. Sketch the graph of $y = \frac{2}{3}x - 4$ using the slope and y-intercept

Solve It

10. Sketch the graph of $y = -5x + 1$ using the slope and y-intercept

Solve It

11. Sketch the graph of $2x + 3y = 6$ using the two intercepts; determine the slope of the line

Solve It

12. Sketch the graph of $9x - 2y = 18$ using the two intercepts; determine the slope of the line

Solve It

Graphing Basic Polynomial Curves

A *polynomial* is a smooth curve that has x-values ranging from negative infinity to positive infinity. The equation representing the curve determines the y-values for points on the curve. There's just one y-value for every x-value, because polynomials are types of functions.

A *function* is a special type of relationship or equation in which you find no more than one y-value for every x-value.

The degree (exponent) of a polynomial tells you the maximum number of x-intercepts the polynomial can have and the maximum number of *turning points* (where the graph changes direction from going upward to downward or vice versa). If the degree of the polynomial is n, then you can find a maximum of n x-intercepts and a maximum of $n - 1$ turning points. You use the intercepts and symmetry and a few extra plug-and-plot points to sketch in the graph of the polynomial.

A polynomial may have fewer intercepts or turning points but never more than the maximum (makes sense, huh?). A polynomial has fewer x-intercepts if the equation has a double or triple root — the same solution appearing more than once. You also find fewer x-intercepts when roots are imaginary. Refer to Chapter 13 for information on imaginary and complex numbers. You can find out more about graphing and working with polynomials in Chapter 7.

Q. Sketch the graph of the polynomial
$y = x^3 - 9x$

A. Let $y = 0$ to solve for the x-intercepts:
$0 = x^3 - 9x = x(x^2 - 9) = x(x - 3)(x + 3)$. The
x-intercepts are $(0, 0)$, $(3, 0)$, and $(-3, 0)$.
When you let $x = 0$ to find the y-intercept,
you get $(0, 0)$, an x-intercept. This repeated
intercept always happens when one of the
intercepts is the origin. This graph is symmetric with respect to the origin, because
if you replace each x with $-x$ and each y

with $-y$, you get $(-y) = (-x)^3 - 9(-x)$. This
equation simplifies to $-y = -x^3 + 9x$. Multiplying each side of the equation by -1, you
get the original equation — there was no
change. (See the previous section, "Crossing
the Line: Using Intercepts and Symmetry to
Graph," for more info on symmetry tests.)
Figure 4-5 shows the intercepts and a few
other points plotted. These points should
be enough, along with the symmetry, to
sketch in the whole curve. You can see that
this graph contains two turning points.

Figure 4-5:
Connect the
dots to
sketch the
graph of
the curve.

---

13. Use intercepts, turning points, symmetry,
and a few select points to sketch the graph
of $y = x^3 - 4x^2$

Solve It

14. Use intercepts, turning points, symmetry,
and a few select points to sketch the graph
of $y = x^4 - x^2$

Solve It

15. Use intercepts, turning points, symmetry, and a few select points to sketch the graph of $y = x(x + 3)(x - 4)(x - 6)$

Solve It

16. Use intercepts, turning points, symmetry, and a few select points to sketch the graph of $y = (x^2 - 36)(x^2 - 25)$

Solve It

Grappling with Radical and Absolute Value Equations

Radical equations and absolute value equations have distinctive-looking curves. Radical equations are usually restricted in their domains (possible x-values — see Chapter 5), because negatives under even-powered radicals represent numbers that don't exist. Consequently, you usually see nice, smoothly rising (or falling) curves that start or end abruptly at one point in their domain. Absolute value equations have a distinctive V-shape to them, facing either upward or downward.

Intercepts and symmetry help with the sketches of these curves, too, but even more helpful are the "endpoint" in the case of the radical equation and the tip of the V in the case of an absolute value equation.

Q. Sketch the graph of $y = \sqrt{x^2 - 9}$.

A. The domain of this function doesn't include any numbers between –3 and 3. Any x between those two numbers gives you a negative value under the radical. The two x-intercepts are at (–3, 0) and (3, 0).

The graph has no y-intercept. The two intercepts are sort of "endpoints" of the two parts of the curve. This curve is symmetric with respect to the y-axis. Figure 4-6 shows you the intercepts, a few other selected points, and the curves that are to the left and right of the intercepts.

Figure 4-6:
The graph of
the radical
equation is
symmetric.

Q. Sketch the graph of $y = 3|x + 4| - 2$

A. The distinctive V-shape in this graph opens upward because the number multiplied by the absolute value operation is positive. The 3 multiplier also acts like slope, making the curve relatively steep. The lowest y-value this curve can have is –2; that's because the term involving the absolute value can never be less than 0, and subtracting 2 lowers the y-value (0) by 2 units. The V-shape therefore has its lowest point at (–4, –2). The –4 comes from the value that makes the absolute value term equal to 0.

The graph has a y-intercept of (0, 10). You get this value by letting $x = 0$ and solving for y. The graph also has two x-intercepts. You find them by letting $y = 0$ and then solving the absolute value equation $3|x + 4| = 2$.

This equation has two solutions, $x = -3\frac{1}{3}$, $x = -4\frac{2}{3}$. (Refer to Chapter 1 if you need info on solving absolute value equations.) Figure 4-7 shows the graph of this absolute value equation.

Figure 4-7:
The V is for
victory in
graphing
absolute
values!

17. Sketch the graph of $y = \sqrt{4-x}$

Solve It

18. Sketch the graph of $y = \sqrt[3]{3x-1}$

Solve It

19. Sketch the graph of $y = |2x+3| - 4$

Solve It

20. Sketch the graph of $y = 2|x-3| + 1$

Solve It

Enter the Machines: Using a Graphing Calculator

Graphing calculators are wonderful instruments — and not just for graphing equations. Most graphing calculators allow you to perform intricate calculations, analyze data statistically, work with matrices, and so on. The main drawback with graphing calculators, though, is that they give you what you ask for and not necessarily what you mean or want. It doesn't hurt to be too liberal with parentheses to be sure you've entered something into the calculator correctly (calculators strictly follow the order of operations — see Chapter 1).

 You need a fairly good idea of what the graph should look like so you can adjust the viewing window to get a good picture of what's going on. If you expect to get a U-shaped curve and the graph you get is a line, then you know that either the viewing window is wrong or you put in the equation incorrectly. It's helpful to know the general shape, where the intercepts are, and how high and low the graph goes before you even start.

Q. Use a graphing calculator to graph
$y = x(x - 13)^2(x + 11)$

A. If you type this function into the graphing menu and use a "standard graph" command (it graphs from –10 to +10 both left and right and up and down) you don't see much of anything. Of course, you know that something should be there, because you know about intercepts and turning points. Looking at the factored equation, you see that the x-intercepts are at 0, 13, and –11. This graph is a fourth degree polynomial, so you could have as many as three turning points (refer to "Graphing Basic Polynomial Curves" for details). So for starters, the standard graphing window that goes left

only to –10 isn't wide enough. Start out by changing the x-coordinates of your viewing window so that they go down to –13 (two less than the lowest x-intercept) and up to 15 (two larger than the highest x-intercept).

Then use the function on your calculator that automatically fits the curve to the width to figure out how high and low the y-values should be. You get a graph that looks very much like the one in Figure 4-8. The curve touches the x-axis at x = 13, because the power on that factor is an even number. This even power causes a *touch and go,* a point at which the graph touches the axis but doesn't cross it (you have a double root). Odd powers on the factors allow the curve to cross sides of the x-axis.

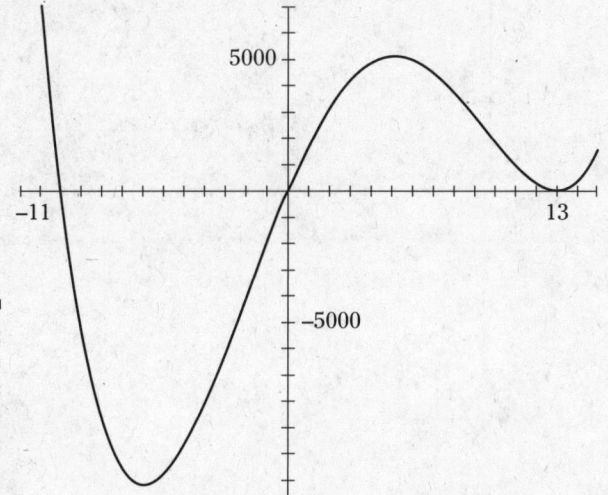

Figure 4-8: The curve crosses at (–11, 0) and touches at 13.

Q. Use a graphing calculator to graph $y = \dfrac{1}{x+2}$

A. This perfectly swell rational function should graph very nicely on your calculator. The catch in graphing, though, is that you need to enter the equation correctly. If you enter "$y = 1 \div x + 2$" into your calculator, you get the wrong graph — the calculator follows the order of operations and divides before adding. Both the x and the 2 need to be in the denominator of the fraction, so put parentheses around them and enter "$y = 1 \div (x + 2)$."

The second issue you need to be aware of is that calculators tend to connect ends of the curves — even when they shouldn't be connected. The correct graph doesn't have connections between where x is smaller than –2 and where x is larger than –2 because at $x = -2$, the denominator is 0. (For more on what's happening in those gaps or holes in the graph, refer to Chapter 8.) You can turn your calculator to dot mode to prevent these stray connections — or just be aware that they shouldn't be there and ignore them. Figure 4-9 shows the graph of the equation. The window shows the graph from –5 to 5; that's wide enough.

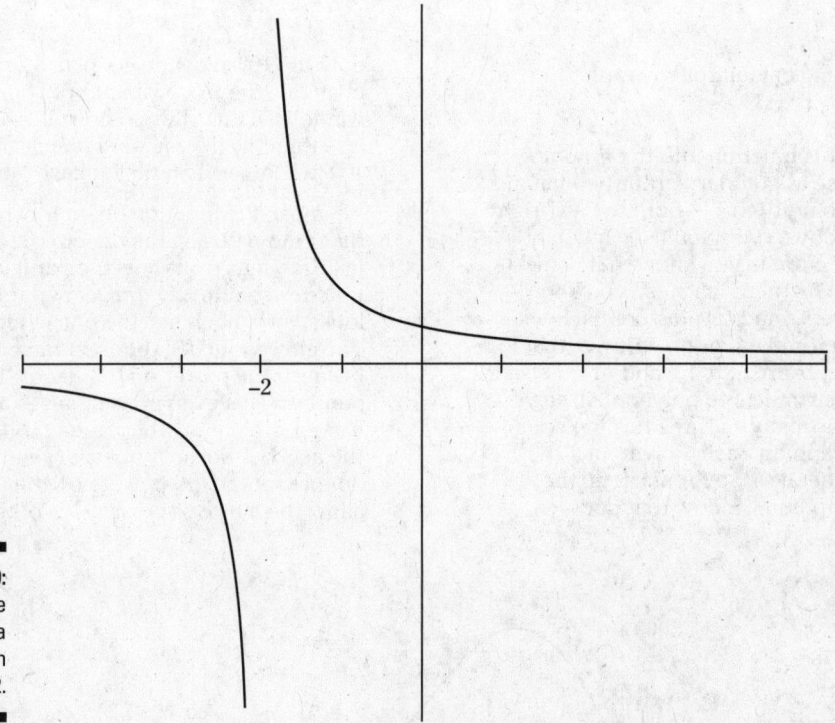

Figure 4-9:
The curve takes a break when $x = -2$.

21. Sketch the graph of $y = 3x + 12$

Solve It

22. Sketch the graph of $y = \sqrt{x^2 - 9}$

Solve It

23. Sketch the graph of $y = x^2(x^2 - 100)(x - 12)$

Solve It

24. Sketch the graph of $y = 2^{2x + 1}$

Solve It

Answers to Problems on Graphing for the Good Life

The following are the answers to the practice problems presented in this chapter.

1 Name the points in Figure 4-2 that lie in Quadrant II or on the *x*-axis. The answer is **(–3, 4), (–7, 3), (–2, 1), (–6, 0), (5, 0).**

The points that lie in Quadrant II are in the upper left-hand corner of the coordinate plane. The *x*-coordinate of these points is always negative, and the *y*-coordinate is positive. Points that lie on the *x*-axis always have a *y*-coordinate of 0.

2 Name the points in Figure 4-2 that lie in Quadrant IV or on the *y*-axis. The answer is **(1, –2), (5, –3), (0, 4), (0, –3).**

The points that lie in Quadrant IV are in the lower right-hand corner of the coordinate plane. The *y*-coordinate of these points is always negative, and the *x*-coordinate is positive. Points that lie on the *y*-axis always have an *x*-coordinate of 0.

3 Identify the quadrant that contains all points with a negative *x*-coordinate and a negative *y*-coordinate. The answer is **Quadrant III.**

Both the *x*- and *y*-coordinates are negative for points in the third quadrant.

4 If you draw a segment connecting points (–4, 4) and (2, –2), what point lies on that line segment that has *x*- and *y*-coordinates with the same value (the numbers are the same)? The answer is **(0, 0).**

The segment connecting the two points contains points in which the coordinates have opposite values, such as –4 and 4, 3 and –3, –1 and 1, and so on. The only point on the segment that wouldn't have two coordinates with opposite signs is the origin — (0, 0).

5 Use intercepts and symmetry to sketch the graph of $y = x^2 - 9$.

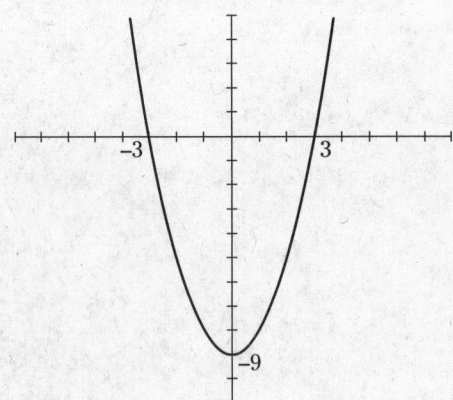

The intercepts are (0, –9), (3, 0), and (–3, 0). The curve is symmetric with respect to the *y*-axis.

6 Use intercepts and symmetry to sketch the graph of $y = 3 - |x|$.

The intercepts are $(0, 3)$, $(3, 0)$, and $(-3, 0)$. The V-shaped absolute value curve is symmetric with respect to the y-axis.

7 Use intercepts and symmetry to sketch the graph of $y = \dfrac{1}{x^2}$.

There are no intercepts. The curve is symmetric with respect to the y-axis. The y-values are always positive, and no curve is possible when $x = 0$.

8 Use intercepts and symmetry to sketch the graph of $y = \sqrt{16 - x^2}$.

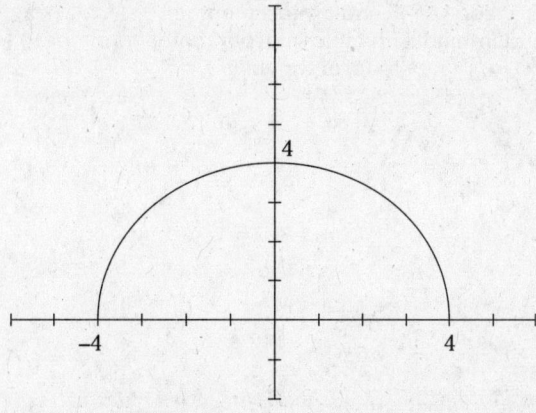

The intercepts are (0, 4), (4, 0), and (–4, 0). The curve is symmetric with respect to the y-axis. The value of x can't be greater than 4 or less than –4, so the curve exists only between and including those two numbers.

9 Sketch the graph of $y = \frac{2}{3}x - 4$ using the slope and y-intercept.

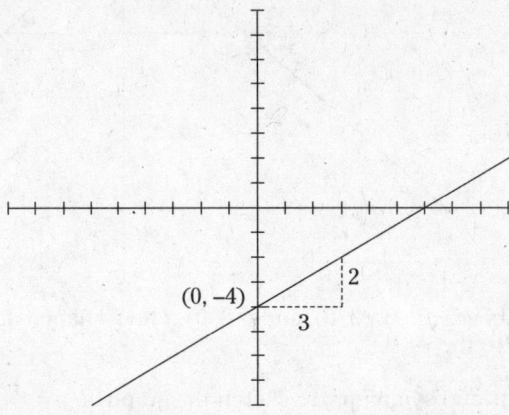

The line has a slope of $\frac{2}{3}$ and a y-intercept at (0, –4). Mark a point for the y-intercept on your graph. Then count 3 units to the right and 2 units up from that y-intercept to locate another point on the line. Connect the two points to form the line.

10 Sketch the graph of $y = -5x + 1$ using the slope and y-intercept.

The line has a slope of –5 and a y-intercept at (0, 1). Mark a point for the y-intercept. Then count 1 unit to the right and 5 units down from that y-intercept to locate another point on the line. Connect the two points to form the line.

11 Sketch the graph of $2x + 3y = 6$ using the two intercepts; determine the slope of the line.

The slope is $-\dfrac{2}{3}$**.** The intercepts are $(0, 2)$ and $(3, 0)$. Draw a line through those two intercepts.

12 Sketch the graph of $9x - 2y = 18$ using the two intercepts; determine the slope of the line.

The slope is $\dfrac{9}{2}$**.** The intercepts are $(0, -9)$ and $(2, 0)$. Draw a line through the two intercepts.

13 Use intercepts, turning points, symmetry, and a few select points to sketch the graph of $y = x^3 - 4x^2$.

The intercepts are $(0, 0)$ and $(4, 0)$, and the graph has no symmetry. The curve just touches at $x = 0$, because the power on the x-factor is an even number. The graph has two turning points.

14   Use intercepts, turning points, symmetry, and a few select points to sketch the graph of $y = x^4 - x^2$.

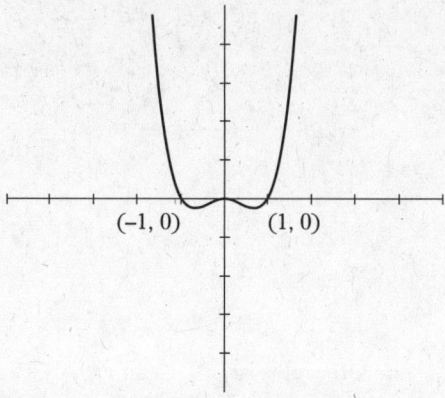

The intercepts are at $(0, 0)$, $(1, 0)$, and $(-1, 0)$. The curve is symmetric with respect to the y-axis. You have three turning points.

15   Use intercepts, turning points, symmetry, and a few select points to sketch the graph of $y = x(x + 3)(x - 4)(x - 6)$.

The intercepts are $(0, 0)$, $(-3, 0)$, $(4, 0)$, and $(6, 0)$. There's no symmetry. The graph has three turning points.

16   Use intercepts, turning points, symmetry, and a few select points to sketch the graph of $y = (x^2 - 36)(x^2 - 25)$.

The intercepts are (6, 0), (–6, 0), (5, 0), and (–5, 0). The curve is symmetric with respect to the y-axis. You can find three turning points.

17 Sketch the graph of $y = \sqrt{4-x}$.

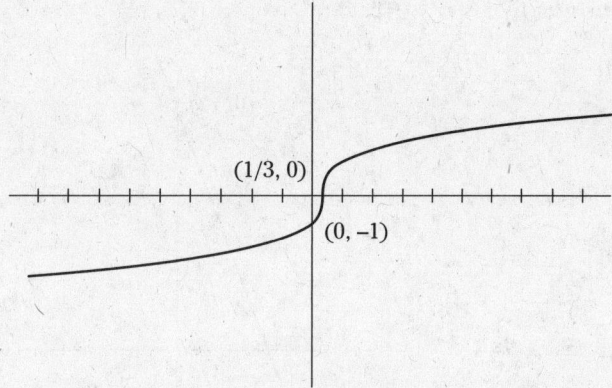

The intercepts are at (0, 2) and (4, 0). There are no points for when $x > 4$.

18 Sketch the graph of $y = \sqrt[3]{3x-1}$.

The intercepts are at (0, –1) and $\left(\frac{1}{3}, 0\right)$. There's no symmetry.

19 Sketch the graph of $y = |2x + 3| - 4$.

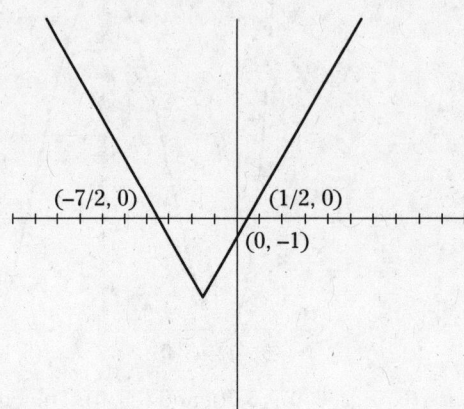

The intercepts are at $(0, -1)$, $\left(\frac{1}{2}, 0\right)$, and $\left(-\frac{7}{2}, 0\right)$.

20 Sketch the graph of $y = 2|x - 3| + 1$.

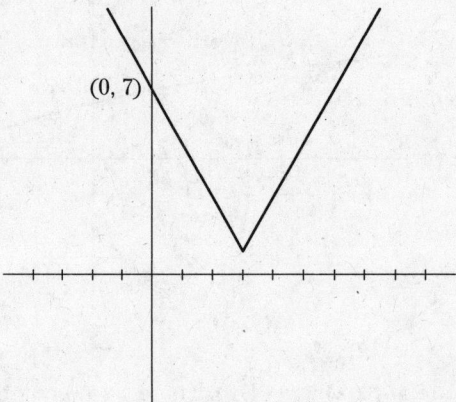

The only intercept is at $(0, 7)$.

21 Sketch the graph of $y = 3x + 12$.

The intercepts are at (0, 12) and (–4, 0). Set the window in the graphing screen to include a *y*-value that goes as high as the *y*-intercept, which is at 12.

22 Sketch the graph of $y = \sqrt{x^2 - 9}$.

(–3, 0)   (3, 0)

The intercepts are at (–3, 0) and (3, 0). You have no *y*-intercept. The curve is symmetric with respect to the *y*-axis. Be sure to enter the equation into the calculator very carefully, including both the x^2 term and –9 under the radical. You can usually do so by putting parentheses around the two terms. Instead of using a radical, you can raise the two terms in parentheses to the 0.5 power.

23 Sketch the graph of $y = x^2(x^2 - 100)(x - 12)$.

50,000

(–10, 0)   (10, 0)   (12, 0)

The intercepts are at (0, 0), (10, 0), (–10, 0), and (12, 0). Be sure to widen the graphing view to include all of the *x*-intercepts. Then use the feature that automatically fits the window to graph the curve. You can see that the *y*-values go to a high of 50,000 and a low of –5,000.

24 Sketch the graph of $y = 2^{2x+1}$.

The only intercept is at (0, 2). Be very careful when entering this equation into the calculator. You have to write the exponent, $2x + 1$, entirely in parentheses so the calculator can read it correctly.

Part II
Functions

The 5th Wave   By Rich Tennant

In this part . . .

Functions are carefully defined relationships in mathematics — just like you carefully describe your pizza order. Linear functions are like cheese pizzas — no extra lumps or bumps. Polynomial functions can twist and turn — just like some of the braided crusts on pizza. You have some functions with a multitude of characteristics, much like a supreme. With functions, their name or description tells you what the characteristics are and what you can do with them — see my drift? Great! Grab a bite to eat if you need to and read on.

Chapter 5

Formulating Functions

● ●

In This Chapter

▶ Determining how much functions are worth

▶ Checking the limits on input and output: Domain and range

▶ Defining functions by their characteristics

▶ Working with functions with multiple parts

▶ Performing operations on functions

▶ Looking at and creating inverse functions

● ●

A function is a very special type of relationship in mathematics. *Functions* are those relationships between input values and output values that guarantee that you'll never get more than one output for any input. For example, the equation $y = 8x + 2x^2 - 3^x$ is a function, because when you plug in any value for x, such as 2, you get only one value for y; in this case, $y = 15$. This characteristic of having just one answer when you put a number into an equation is essential when you're dealing with functions that model the cost of production or the number of amoebas in a culture. You don't want to be saying, "Well, there are either 16 or 16,000,000 amoebas in that dish." You'd prefer just one answer when you use the formula.

This chapter deals with how you can categorize a function and determine its characteristics. A function can be *even* or *odd* or even *piecewise*. Each label has its importance, and each has its place when studying and using functions.

Evaluating Functions

You read an equation like $y = 3x^3 - 4x^2 + 5x - 11$ with words such as "y is a function of x and is equal to three times x cubed minus four times. . . ." You get the point. From the equation, you determine that the y output is some function of the x input. You can plot the x's and y's in a graph. Another way of writing this function is $f(x) = 3x^3 - 4x^2 + 5x - 11$. This format is called *function notation*. The name of the function is f.

Using function notation when you need to list two or more functions is very handy, because you can name the functions f of x, g of x, h of x, and so on. Doing so helps to distinguish one function from another when you're working a bunch of them and they all do different things. When you use function notation, such as $f(x) = 3x^3 - 4x^2 + 5x - 11$, you *evaluate* the function for some x-value when you replace each x with the number in the parentheses and solve the equation. For example, using the rule $f(x)$ and evaluating for $x = 2$, you write that $f(2) = 3(2)^3 - 4(2)^2 + 5(2) - 11 = 3(8) - 4(4) + 10 - 11 = 24 - 16 - 1 = 7$.

Q. Find $f(2)$, $g(3)$, $h(-4)$ and $k(0)$ if $f(x) = 3x^3 - 4x^2 + 5x - 11$, $g(x) = |2x - 5| + 8$, $h(x) = \sqrt[3]{x + 31}$, and $k(x) = \dfrac{x^2 + 2}{x - 1}$

A. Replace the x in each particular function with the number that's in the parentheses. Solve the expressions.

✔ $f(2) = 3(2)^3 - 4(2)^2 + 5(2) - 11 = 3(8) - 4(4) + 10 - 11 = 24 - 16 + 10 - 11 = 7$

✔ $g(3) = |2(3) - 5| + 8 = |6 - 5| + 8 = |1| + 8 = 1 + 8 = 9$

✔ $h(-4) = \sqrt[3]{(-4) + 31} = \sqrt[3]{27} = 3$

✔ $k(0) = \dfrac{(0)^2 + 2}{(0) - 1} = \dfrac{2}{-1} = -2$

---

1. If $f(x) = 3x^2 - 4x + 2$, find $f(-3)$

Solve It

2. If $g(x) = 5|2x - 3| - 1$, find $g(-3)$

Solve It

3. If $h(x) = \dfrac{12}{x^2 - 3}$, find $h(-3)$

Solve It

4. If $k(x) = \sqrt{25 - x^2}$, find $k(-3)$

Solve It

Determining the Domain and Range of a Function

The domain of a function is where it lives. No, that's not strictly true, but it does give you an idea of what the concept of domain involves. A function's *domain* consists of all of the numbers that you can use as input values for a particular function. You always start out hoping or assuming that any real number is in the domain and can be used in the function's rule, but you have to discard numbers that cause inconsistencies or unreal numbers to appear when you perform the function's operations. If a function's rule involves a fraction, then you can't use any number in the function that would make the denominator of the fraction equal to 0. If a function's rule involves an even-powered radical, then you can't let the expression under the radical be a negative.

The range of a function is what you get after evaluating input values. The *range* consists of all the real numbers that are results of performing the function operations on all the domain or input values.

Two ways of writing about domain and range are inequality notation and interval notation (see Chapter 1 for details on these formats). You may also write the numbers that aren't part of the answer set, like $x \neq -3, 3$.

Q. Determine the domain and range of the function $f(x) = \dfrac{24}{x^2 - 1}$

A. **Domain: $x \neq -1, 1$; range: $f(x) \neq 0$.** The domain of $f(x)$ consists of all the real numbers except for $x = 1$ and $x = -1$. Both 1 and -1 make the denominator of the fraction equal to 0 — a situation that results in values that don't exist. You can't divide by 0. The way to describe this domain in interval notation is $(-\infty, -1) \cup (-1, 1) \cup (1, \infty)$ — the union symbol, $\cup$, means *or*. Interval notation is a bit cumbersome here, so I usually prefer just saying what's *not* allowed: $x \neq -1, 1$.

The range of $f(x)$ is all real numbers except for 0. When you input numbers into the rule, such as $f(2) = 8$, $f(-3) = 3$, $f(\frac{1}{2}) = -32$, and so on, you see, after a while, that you get big numbers and small numbers — both positive and negative — and the only

number you can never get is 0. The numerator of a fraction has to equal 0 for the value of the fraction to be 0, so this function's rule will never result in 0. Saying that the range of $f(x)$ is all real numbers except 0, in interval notation, you write $(-\infty, 0) \cup (0, \infty)$. Otherwise, you can simply write that $f(x) \neq 0$.

Q. Determine the domain and range of the function $g(x) = \sqrt{6 - x}$

A. **Domain: $x \leq 6$; range: $g(x) \geq 0$.** The domain of $g(x)$ is all real numbers smaller than or equal to 6, and the range of $g(x)$ is all positive numbers and 0 — domain: $(-\infty, 6]$; range: $[0, \infty)$. You can't have a negative number under a square root radical, so the input values can't be larger than 6. And when you take the square root of positive numbers or 0, you always get results that are either positive or 0.

---

5. Determine the domain and range of the function $f(x) = x^2 - 5$

Solve It

6. Determine the domain and range of the function $g(x) = |x + 7|$

Solve It

---

7. Determine the domain and range of the function $h(x) = \sqrt{x - 8}$

Solve It

8. Determine the domain and range of the function $k(x) = \dfrac{1}{x - 3}$

Solve It

Recognizing Even, Odd, and One-to-One Functions

The words *even, odd,* and *one-to-one* are ways of describing functions just like *perching, wading,* and *gull-like* are ways of describing birds. A perching bird still has many characteristics in common with other types of birds, but it has some unique characteristics that set it apart and make it special in a particular way.

Anyway, back to functions:

- An *even* function has the same output regardless of whether you put in x or $-x$. Inputting a number and its opposite yields the same output value: $f(-x) = f(x)$. The graphs of even functions are symmetric with respect to the y-axis (see Chapter 4 for more on symmetry).

- An *odd* function does the opposite: The result of inputting x is the opposite of when you input $-x$ into the function rule: $f(-x) = -f(x)$. The graphs of odd functions are symmetric with respect to the origin (like a 180-degree turn).

- With a *one-to-one* function, not only do you get one output for every input value, but each output value also occurs exactly once. There's just one input value for every output value. If $f(x_1) = f(x_2)$, then $x_1 = x_2$.

Q. Describe the following functions as *even, odd,* or *neither;* then determine whether they're *one-to-one*

  a. $f(x) = \dfrac{4}{x^2 + 1}$

  b. $g(x) = x + 2$

  c. $h(x) = 4x - x^3$

A. Here are the answers:

  a. Even, not one-to-one. Function f is an even function. For instance, you see that $f(1) = 2$ and $f(-1) = 2$. You get the same result. Substituting $-x$ into the function equation, you get $f(x) = \dfrac{4}{(-x)^2 + 1} = \dfrac{4}{x^2 + 1}$. The function value is the same as for $f(x)$. The function f is not one-to-one, though, for the same reason that made it even. The output value 2 has two different input values. One-to-one functions have just one input value for every output value.

  b. Neither, one-to-one. Function g is neither even nor odd. If you replace the x with $-x$, you get $g(-x) = (-x) + 2 = -x + 2$. This isn't the same as the original $g(x)$, and it isn't the opposite, either. To be the opposite, both terms would have to be negative. Function g is one-to-one, though. Every input value has just one output, and every output has just one input. You can confirm this conclusion by looking at the graph of g. It's a line that slants upward and doesn't double back on itself.

  c. Odd, not one-to-one. Function h is an odd function. Replacing each x with $-x$, you get $h(-x) = 4(-x) - (-x)^3 = -4x + x^3 = -(4x - x^3)$. Function h isn't one-to-one, though. Both $h(2)$ and $h(-2)$ give you the same output — they're both equal to 0.

9. Determine whether the function is *even, odd,* or *neither;* then determine whether it's one-to-one: $f(x) = x^2 - 5$

Solve It

10. Determine whether the function is *even, odd,* or *neither;* then determine whether it's one-to-one: $g(x) = |x|$

Solve It

11. Determine whether the function is *even, odd,* or *neither;* then determine whether it's one-to-one: $h(x) = x^3 + 1$

Solve It

12. Determine whether the function is *even, odd,* or *neither;* then determine whether it's one-to-one: $k(x) = \sqrt[5]{x}$

Solve It

They Come in Pieces: Working through Piecewise Functions

Piecewise functions are functions that may remind you of the people in front of you in the line at the bakery: They just can't seem to make up their minds. *Piecewise* functions have two or more definitions, depending on what part of their domain (input) you're talking about.

Piecewise functions have a huge place in business, where people charge different prices depending on how many items you're buying or where companies charge for utility usage based on the quantity that month. Even the income tax rate is piecewise, with the percentage rate determined by your income.

Q. The cost for one-quart geraniums is based on the number of plants you buy. The piecewise function

$$C(g) = \begin{cases} 3.00g & \text{if } 0 \le g < 20 \\ 2.50g + 10.00 & \text{if } 20 \le g < 40 \\ 2.00g + 30 & \text{if } 40 \le g \le 400 \end{cases}$$

gives you the breakdown of prices. How much do 10 plants cost, and how much is each one? How much for 25 plants, and what's the individual price?

A. **Ten cost $30.00, or $3.00 each; 25 cost $63.50, or $2.54 each.** The domain of the function starts at 0 and goes on up to 400 (that's all the plants they have). If you buy 10 plants, they'll cost you $3.00 each, because $C(10) = 3.00(10) = \$30.00$. If you buy 25 plants, they'll cost you $2.54 each, because $C(25) = 2.50(25) + 10.00 = \$63.50$, and $63.50 \div 25 = 2.54$. The rule you use to determine the total cost depends on the input value. You use the rule that corresponds to where the input value is in the definition of the function.

Q. Given the piecewise function

$$f(x) = \begin{cases} x^2 + 4 & \text{if } x \le -1 \\ 2x - 3 & \text{if } x > -1 \end{cases}$$

find $f(-2)$ and $f(0)$

A. **$f(-2) = 8$; $f(0) = -3$.** When x is less than or equal to -1, you use the top rule, so $f(-2) = (-2)^2 + 4 = 4 + 4 = 8$. The number 0 is greater than -1, so $f(0) = 2(0) - 3 = 0 - 3 = -3$. You use the bottom rule.

13. Given the piecewise function

$$f(x) = \begin{cases} 3x & \text{if } x < 0 \\ 4x & \text{if } x \ge 0 \end{cases}$$

find $f(-1)$

Solve It

14. Given the piecewise function

$$g(x) = \begin{cases} 2x - 1 & \text{if } x < -3 \\ x^2 - 2 & \text{if } x \ge -3 \end{cases}$$

find $g(0)$

Solve It

15. Given the piecewise function

$$f(x) = \begin{cases} x + 1 & \text{if } x < -2 \\ 4 & \text{if } -2 \le x < 3 \\ x + 5 & \text{if } x \ge 3 \end{cases}$$

find $f(0)$ and $f(-2)$

Solve It

16. Given the piecewise function

$$k(x) = \begin{cases} x - 1 & \text{if } x < a \\ 2x + 3 & \text{if } x \ge a \end{cases}$$

find a value for a so that $k(a)$ gives you the same answer using both rules

Solve It

Composing Functions and Simplifying the Difference Quotient

You can add, subtract, multiply, and divide two or more functions by just taking their algebraic rules and performing those operations. For instance, if you want to add functions f and g, where $f(x) = 3x^2 - 9x + 7$ and $g(x) = -x^2 - 6x - 12$, then you just combine the like terms: $f + g = 3x^2 - 9x + 7 + (-x^2 - 6x - 12) = 3x^2 - x^2 - 9x - 6x + 7 - 12 = 2x^2 - 15x - 5$. Not much more to it than that.

Functions do have an operation, though, that's special to them — plain old numbers don't use it. The operation I'm referring to is composition. You *compose* functions by making one function the input of another. The operation symbol for composition is ∘, and you write it between the two functions you're composing, such as $f \circ g$.

The composition of functions f and g is defined as $f \circ g = f(g)$. This function is not commutative, so in general, $f \circ g \neq g \circ f$.

You can use the composition of functions when working with difference quotients. The difference quotient is the main component of the definition of the derivative in calculus. Most algebra books include exercises on simplifying the difference quotient to get you to practice a problem that you'll use when you get to calculus.

The *difference quotient* for the function $f(x)$ is $\dfrac{f(x+h)-f(x)}{h}$.

Q. Find both $f \circ g$ and $g \circ f$ if the function $f(x) = 5x^2 - 2x + 1$ and $g(x) = 9 - 4x$

A. $f \circ g = 406 - 352x + 80x^2$; $g \circ f = 5 + 8x - 20x^2$. First, $f \circ g = f(g) = f(9-4x)$. Replacing each x in the definition of function f with the rule for g, you get $f(9-4x) = 5(9-4x)^2 - 2(9-4x) + 1 = 5(81 - 72x + 16x^2) - 2(9-4x) + 1 = 405 - 360x + 80x^2 - 18 + 8x + 1 = 406 - 352x + 80x^2$. Now, reversing the roles, $g \circ f = g(5x^2 - 2x + 1) = 9 - 4(5x^2 - 2x + 1) = 9 - 20x^2 + 8x - 4 = 5 + 8x - 20x^2$. As you can see, by changing the order, you get different answers.

Q. Find the difference quotient for the function $f(x) = x^3 + 2x^2 - 3$

A. $3x^2 + 3xh + h^2 + 4x + 2h$. You substitute $f(x+h)$ in for the first term in the numerator — you get the terms of $f(x+h)$ by replacing every x in the function's rule with an $x+h$. This step resembles composition if you think that $g(x) = x+h$ and that you're composing $f(g)$. Then substitute the original function $f(x)$ in for the second term. Be sure to put the values defining $f(x)$ in parentheses so that the negative sign gets distributed over each term:

$$\frac{f(x+h)-f(x)}{h} = \frac{(x+h)^3 + 2(x+h)^2 - 3 - (x^3 + 2x^2 - 3)}{h}$$

$$= \frac{x^3 + 3x^2h + 3xh^2 + h^3 + 2(x^2 + 2xh + h^2) - 3 - x^3 - 2x^2 + 3}{h}$$

$$= \frac{x^3 + 3x^2h + 3xh^2 + h^3 + 2x^2 + 4xh + 2h^2 - 3 - x^3 - 2x^2 + 3}{h}$$

After cubing and squaring the binomial and distributing over the parentheses, you find pairs of terms and their opposites in the numerator. You can eliminate them, and then you can factor out an h and reduce the fraction:

$$\frac{\cancel{x^3} + 3x^2h + 3xh^2 + h^3 + \cancel{2x^2} + 4xh + 2h^2 \cancel{-3} \cancel{-x^3} \cancel{-2x^2} \cancel{+3}}{h}$$

$$= \frac{3x^2h + 3xh^2 + h^3 + 4xh + 2h^2}{h} = \frac{\cancel{h}(3x^2 + 3xh + h^2 + 4x + 2h)}{\cancel{h}}$$

$$= 3x^2 + 3xh + h^2 + 4x + 2h$$

This answer probably doesn't seem like much, but it's one step short of finding the derivative. An important part of the result is that you get rid of the denominator with the h in it.

17. Find both $f \circ g$ and $g \circ f$ if $f(x) = 2x^2 + 3x - 4$ and $g(x) = x - 2$

Solve It

18. Find both $f \circ g$ and $g \circ f$ if $f(x) = 2x + 7$ and $g(x) = 5 - 3x$

Solve It

19. Find both $f \circ g$ and $g \circ f$ if $f(x) = x^2 - 9$ and $g(x) = x^3 + 3$

Solve It

20. Find both $f \circ g$ and $g \circ f$ if $f(x) = f(x) = 2x - 8$ and $g(x) = \frac{x + 8}{2}$

Solve It

21. Evaluate the difference quotient for $f(x) = 3x^2 + 2x - 5$

Solve It

22. Evaluate the difference quotient for $f(x) = \frac{3x + 1}{x - 2}$

Solve It

Solving for Inverse Functions

An *inverse operation* does the opposite of what the original operation does. For instance, the inverse of adding 6 to a number is subtracting 6. The inverse of dividing by 4 is multiplying by 4. An *inverse function* undoes or performs the opposite of what the original function did.

Consider the function $f(x) = \frac{3x + 2}{x - 4}$. Letting $x = 2$, you get $f(2) = -4$: $f(2) = \frac{3(2) + 2}{(2) - 4} = \frac{6 + 2}{-2} = \frac{8}{-2} = -4$. The inverse function for f, designated by f^{-1}, is $f^{-1}(x) = \frac{4x + 2}{x - 3}$. Look what happens when I put -4 in for x in the inverse: $f^{-1}(-4) = \frac{4(-4) + 2}{(-4) - 3} = \frac{-16 + 2}{-7} = \frac{-14}{-7} = 2$. The result is 2 the number that I initially put in function f. The inverse gets you back to the original input.

You may wonder where I found the inverse for the function *f*. I used the following process. But first, be aware that only one-to-one functions have inverses. When you put a result into an inverse function, it's important that you get just one answer. And that's the case only with one-to-one functions. Go to "Recognizing Even, Odd and One-to-One Functions," earlier in this chapter, if you need more information.

To solve for an inverse function:

1. **Change *f* (*x*) to *y* for convenience.**
2. **Switch all the *x*'s in the function equation to *y*'s and the *y* to an *x*.**
3. **Solve for *y* in this new equation.**
4. **Replace the *y* with *f*⁻¹(*x*).**

Q. Find the inverse for $f(x) = \dfrac{3x+2}{x-4}$

A. $f^{-1}(x) = \dfrac{4x+2}{x-3}$. Using the four steps, first replace the *f* (*x*) with *y* to get $y = \dfrac{3x+2}{x-4}$.

Next, switch all the *x*'s and *y* to get $x = \dfrac{3y+2}{y-4}$. To solve for *y*, first multiply each side of the equation by the denominator, *y* – 4. Then distribute the *x* over the binomial. Next, add 4*x* to each side and subtract 3*y* from each side to get all the terms with *y* in them on the left side of the equation:

$$x(y-4) = 3y+2$$
$$xy - 4x = 3y + 2$$
$$xy - 3y = 4x + 2$$

Now factor out the *y* on the left and divide each side of the equation by *x* – 3. Change the *y* to $f^{-1}(x)$:

$$y(x-3) = 4x+2$$
$$y = \frac{4x+2}{x-3}$$
$$f^{-1} = \frac{4x+2}{x-3}$$

(Yes, this is the inverse from the example in the main text. Thought you should see where it came from!)

Note: The composition of functions is usually *not* commutative, but when you compose a function and its inverse, the result of both compositions is *x*. So with this special combination, you do have commutativity.

88 **Part II: Functions**

23. Find the inverse function for $f(x) = 3x - 1$

Solve It

24. Find the inverse function for $f(x) = x^3 + 3$

Solve It

25. Find the inverse function for $f(x) = \frac{x-1}{2x+1}$

Solve It

26. Find the inverse function for $f(x) = \frac{2x-2}{3x-2}$

Solve It

Answers to Problems on Formulating Functions

The following are the answers to the practice problems presented earlier in this chapter.

1 If $f(x) = 3x^2 - 4x + 2$, find $f(-3)$. The answer is **$f(-3) = 41$.**

Replace all the x's in the function equation with -3 and simplify: $f(-3) = 3(-3)^2 - 4(-3) + 2 = 3(9) + 12 + 2 = 41$.

2 If $g(x) = 5|2x - 3| - 1$, find $g(-3)$. The answer is **$g(-3) = 44$.**

Replace the x inside the absolute value operation with the -3. Simplify what's in the absolute value operation, and then find the absolute value before multiplying by 5 and subtracting 1: $g(-3) = 5|2(-3) - 3| - 1 = 5|-9| - 1 = 5(9) - 1 = 45 - 1 = 44$.

3 If $h(x) = \dfrac{12}{x^2 - 3}$, find $h(-3)$. The answer is **$h(-3) = 2$.**

Replace the x in the denominator with -3. After squaring the -3 and subtracting 3, divide the numerator by the result: $\dfrac{12}{(-3)^2 - 3} = \dfrac{12}{9 - 3} = \dfrac{12}{6}$.

4 If $k(x) = \sqrt{25 - x^2}$, find $k(-3)$. The answer is **$k(-3) = 4$.**

Subtracting $(-3)^2 = 9$ from 25 under the radical, find the square root of 16, which is 4.

5 Determine the domain and range of the function $f(x) = x^2 - 5$. The answer is **domain: all real numbers, or $(-\infty, \infty)$; range: $y = f(x) \geq -5$, or $[-5, \infty)$.**

Substitute in any real number for x, and the result is always a real number, so the domain is all reals. The lowest number that's an output value is -5. That's because when you square x, it becomes positive. The smallest x^2 can be is 0, and when you subtract 5, you get -5. All the other values in the range are larger than that, and they go up without bound.

6 Determine the domain and range of the function $g(x) = |x + 7|$. The answer is **domain: all real numbers, or $(-\infty, \infty)$; range: $y = g(x) \geq 0$, or $[0, \infty)$.**

You can substitute in any real number for x, and the result will be a real number, so the domain is all reals. The lowest number that's an output value is 0, because the absolute value of any number is positive — except for 0, which is as low as the range can go.

7 Determine the domain and range of the function $h(x) = \sqrt{x - 8}$. The answer is **domain: $x \geq 8$, or $[8, \infty)$; range: $y = h(x) \geq 0$, or $[0, \infty)$.**

You can't have a negative under the square root radical, so the smallest number that you can substitute in for x is 8. The results of taking a square root are always positive or at least 0.

8 Determine the domain and range of the function $k(x) = \dfrac{1}{x - 3}$. The answer is **domain: $x \neq 3$, or $(-\infty, 3) \cup (3, \infty)$; range: $y = k(x) \neq 0$, or $(-\infty, 0) \cup (0, \infty)$.**

You can't have a 0 in the denominator, so x can't be 3. The function output is any real number except 0, because a fraction can equal 0 only if the numerator is equal to 0. The numerator is always the number 1.

9 Determine whether the function is *even, odd,* or *neither;* then determine whether it's *one-to-one:* $f(x) = x^2 - 5$. The function is **even; not one-to-one.**

The function is an even function, because when you replace the x's with $-x$'s, you get the same function definition: $f(-x) = (-x)^2 - 5 = x^2 - 5 = f(x)$. The function isn't one-to-one, because you find more than one x-value for all but one y-value. For instance, if $f(x) = 11$, x can be either 4 or -4.

10 Determine whether the function is *even, odd,* or *neither;* then determine whether it's *one-to-one:* $g(x) = |x|$. The function is **even; not one-to-one.**

The function is an even function, because when you replace the x's with $-x$'s, you get the same function definition: $g(-x) = |-x| = |x| = g(x)$. The function isn't one-to-one because it has more than one x-value for all but one y-value. For instance, if $g(x) = 3$, x can be either 3 or -3.

11 Determine whether the function is *even, odd,* or *neither;* then determine whether it's *one-to-one:* $h(x) = x^3 + 1$. The function is **neither even nor odd; one-to-one.**

The function isn't even or odd, because when you replace the x's with $-x$'s, you don't get the same function definition and you don't get the opposite: $h(-x) = (-x)^3 + 1 = -x^3 + 1 \neq x^3 + 1$. Also, $-x^3 + 1 \neq -(x^3 + 1)$. The function is one-to-one because every y-value has only one x-value and every x-value has only one y-value. The cube on the x doesn't change the sign of the result.

12 Determine whether the function is *even, odd,* or *neither;* then determine whether it's *one-to-one:* $k(x) = \sqrt[5]{x}$. The function is **odd; one-to-one.**

The function is an odd function, because when you replace the x's with $-x$'s, you get the opposite of the function definition: $k(-x) = \sqrt[5]{-x} = -\sqrt[5]{x}$. It's also one-to-one, because it's a fifth root, and when you take the fifth root of a number, its sign doesn't change.

13 Given the piecewise function
$$f(x) = \begin{cases} 3x & \text{if } x < 0 \\ 4x & \text{if } x \geq 0 \end{cases}$$
find $f(-1)$. The answer is $\boldsymbol{f(-1) = -3.}$

Use the top rule, because the input value, -1, is less than 0.

14 Given the piecewise function
$$g(x) = \begin{cases} 2x - 1 & \text{if } x < -3 \\ x^2 - 2 & \text{if } x \geq -3 \end{cases}$$
find $g(0)$. The answer is $\boldsymbol{g(0) = -2.}$

Use the bottom rule, because the input value, 0, is greater than -3.

15 Given the piecewise function
$$f(x) = \begin{cases} x + 1 & \text{if } x < -2 \\ 4 & \text{if } -2 \leq x < 3 \\ x + 5 & \text{if } x \geq 3 \end{cases}$$
find $f(0)$ and $f(-2)$. The answer is $\boldsymbol{f(0) = 4; f(-2) = 4.}$

Use the middle rule for both problems. The input value 0 is between -2 and 3. And the rule here just says that the answer is 4 — you have no x to substitute the 0 into. For the second part of the problem, you want to input -2. This value uses the second rule as well, because the -2 is included in the interval with the less-than-or-equal-to sign.

16 Given the piecewise function
$$k(x) = \begin{cases} x - 1 & \text{if } x < a \\ 2x + 3 & \text{if } x \geq a \end{cases}$$
find a value for a so that $k(a)$ gives you the same answer using both rules. The answer is $\boldsymbol{a = -4.}$

If you want to have the second rule pick up where the first one left off, you want the y or function values to be the same. Replace the x's in both rules with a's and set the rules equal to one another. You get $a - 1 = 2a + 3$. Solving for a, you get $a = -4$. So the function rule can now read

$$k(x) = \begin{cases} x - 1 & \text{if } x < -4 \\ 2x + 3 & \text{if } x \geq -4 \end{cases}$$

the function value is the same, $k(-4) = -5$, in the bottom rule (where you'd evaluate the -4) as it would be if you were to inadvertently substitute into the top rule.

17 Find both $f \circ g$ and $g \circ f$ if $f(x) = 2x^2 + 3x - 4$ and $g(x) = x - 2$. The answer is $f \circ g = 2x^2 - 5x - 2$; $g \circ f = 2x^2 + 3x - 6$.

First, to compute the composition $f \circ g$, you follow these steps:

$$\begin{aligned} f \circ g &= 2(x - 2)^2 + 3(x - 2) - 4 \\ &= 2(x^2 - 4x + 4) + 3(x - 2) - 4 \\ &= 2x^2 - 8x + 8 + 3x - 6 - 4 \\ &= 2x^2 + 5x - 2 \end{aligned}$$

Next, the composition $g \circ f$ is

$$\begin{aligned} g \circ f &= (2x^2 + 3x - 4) - 2 \\ &= 2x^2 + 3x - 6 \end{aligned}$$

18 Find both $f \circ g$ and $g \circ f$ if $f(x) = 2x + 7$ and $g(x) = 5 - 3x$. The answer is $f \circ g = 17 - 6x$; $g \circ f = -16 - 6x$.

First compute the composition $f \circ g$

$$\begin{aligned} f \circ g &= 2(5 - 3x) + 7 \\ &= 10 - 6x + 7 \\ &= 17 - 6x \end{aligned}$$

Then, the composition $g \circ f$ is

$$\begin{aligned} g \circ f &= 5 - 3(2x + 7) \\ &= 5 - 6x - 21 \\ &= -16 - 6x \end{aligned}$$

19 Find both $f \circ g$ and $g \circ f$ if $f(x) = x^2 - 9$ and $g(x) = x^3 + 3$. The answer is $f \circ g = x^6 + 3x^3$; $g \circ f = x^6 - 27x^4 + 243x^2 - 726$.

The composition $f \circ g$ is

$$\begin{aligned} f \circ g &= (x^3 + 3)^2 - 9 \\ &= x^6 + 3x^3 + 9 - 9 \\ &= x^6 + 3x^3 \end{aligned}$$

Here are the steps to find $g \circ f$:

$$\begin{aligned} g \circ f &= (x^2 - 9)^3 + 3 \\ &= x^6 - 27x^4 + 243x^2 - 729 + 3 \\ &= x^6 - 27x^4 + 243x^2 - 726 \end{aligned}$$

20 Find both $f \circ g$ and $g \circ f$ if $f(x) = f(x) = 2x - 8$ and $g(x) = \dfrac{x + 8}{2}$. The answer is $f \circ g = x$; $g \circ f = x$.

First compute $f \circ g$: $f \circ g = 2\left(\dfrac{x + 8}{2}\right) - 8 = x + 8 - 8 = x$. The composition $g \circ f$ is $g \circ f = \dfrac{(2x - 8) + 8}{2} = \dfrac{2x}{2} = x$. In this case, the composition is commutative; the answers are the same.

You get this result when the two functions are inverses of one another.

21 Evaluate the difference quotient for $f(x) = 3x^2 + 2x - 5$. The answer is **$6x + 3h + 2$**.

Compute the difference quotient:

$$\frac{f(x+h) - f(x)}{h} = \frac{3(x+h)^2 + 2(x+h) - 5 - (3x^2 - 2x - 5)}{h}$$

$$= \frac{3(x^2 + 2xh + h^2) + 2(x+h) - 5 - (3x^2 + 2x - 5)}{h}$$

$$= \frac{3x^2 + 6xh + 3h^2 + 2x + 2h - 5 - 3x^2 - 2x + 5}{h}$$

$$= \frac{3x^2 + 6xh + 3h^2 + 2x + 2h - 5 - 3x^2 - 2x + 5}{h}$$

$$= \frac{6xh + 3h^2 + 2h}{h} = \frac{h(6x + 3h + 2)}{h} = 6x + 3h + 2$$

22 Evaluate the difference quotient for $f(x) = \dfrac{3x+1}{x-2}$. The answer is $\dfrac{-7}{(x+h-2)(x-2)}$.

Compute the difference quotient, $\dfrac{f(x+h) - f(x)}{h} = \dfrac{\dfrac{3(x+h)+1}{(x+h)-2} - \dfrac{3x+1}{x-2}}{h}$. You have a complex fraction — a fraction within a fraction. The best approach here is to find a common denominator for the two fractions in the numerator, write them each with that common denominator, and combine them; then multiply the numerator times the reciprocal of the denominator:

$$\frac{\dfrac{3(x+h)+1}{(x+h)-2} \cdot \dfrac{x-2}{x-2} - \dfrac{3x+1}{x-2} \cdot \dfrac{(x+h)-2}{(x+h)-2}}{h}$$

$$= \frac{\dfrac{(3x+3h+1)(x-2)}{(x+h-2)(x-2)} - \dfrac{(3x+1)(x+h-2)}{(x-2)(x+h-2)}}{h}$$

$$= \frac{\dfrac{3x^2 - 6x + 3xh - 6h + x - 2}{(x+h-2)(x-2)} - \dfrac{3x^2 + 3xh - 6x + x + h - 2}{(x-2)(x+h-2)}}{h}$$

$$= \frac{\dfrac{3x^2 - 6x + 3xh - 6h + x - 2 - 3x^2 - 3xh + 6x - x - h + 2}{(x+h-2)(x-2)}}{h}$$

$$= \frac{-7h}{(x+h-2)(x-2)} \cdot \frac{1}{h} = \frac{-7}{(x+h-2)(x-2)}$$

23 Find the inverse function for $f(x) = 3x - 1$. The answer is $f^{-1}(x) = \dfrac{x+1}{3}$.

Use the steps to solve for the inverse function:

$$y = 3x - 1$$
$$x = 3y - 1$$
$$x + 1 = 3y$$
$$\frac{x+1}{3} = y$$

24 Find the inverse function for $f(x) = x^3 + 3$. The answer is $f^{-1}(x) = \sqrt[3]{x-3}$.

Use the steps to solve for the inverse function:

$$y = x^3 + 3$$
$$x = y^3 + 3$$
$$x - 3 = y^3$$
$$\sqrt[3]{x-3} = \sqrt[3]{y^3}$$
$$\sqrt[3]{x-3} = y$$

25 Find the inverse function for $f(x) = \frac{x-1}{2x+1}$. The answer is $f^{-1}(x) = \frac{x+1}{1-2x}$.

Using the steps to solve for the inverse function:

$$y = \frac{x-1}{2x+1}$$
$$x = \frac{y-1}{2y+1}$$
$$x(2y+1) = y-1$$
$$2xy + x = y - 1$$
$$y(2x-1) = -x-1$$
$$y = \frac{-x-1}{2x-1} = \frac{x+1}{1-2x}$$

In the last step, I multiplied both numerator and denominator by –1 to reverse all the signs.

26 Find the inverse function for $f(x) = \frac{2x-2}{3x-2}$. The answer is $f^{-1}(x) = \frac{2x-2}{3x-2}$.

Notice that this function is its own inverse! Sorta neat, don't you think? You get this result when the graph of a function is symmetric with respect to the line $y = x$. To find this answer, use the steps to solve for the inverse function:

$$y = \frac{2x-2}{3x-2}$$
$$x = \frac{2y-2}{3y-2}$$
$$x(3y-2) = 2y-2$$
$$3xy - 2x = 2y - 2$$
$$3xy - 2y = 2x - 2$$
$$y(3x-2) = 2x-2$$
$$y = \frac{2x-2}{3x-2}$$

Chapter 6

Specializing in Quadratic Functions

Quadratic functions are relatively recognizable, easily manipulated, and surprisingly common in the realm of the functions that model what goes on in the real world — perfect pawns for all your brilliant plots (in algebra)!

The standard form for the quadratic function is $f(x) = ax^2 + bx + c$ where a, b, and c are real numbers that are constants, $a \neq 0$, and x is the variable.

From this simple format, you can tell a lot about the nature of the function. Solving for intercepts or other function values is always possible because of the factoring properties of the quadratics and the quadratic formula (see Chapter 2 for more info on solving quadratic equations). Saying anything bad about these functions is just really hard — so I won't!

In this chapter, I show you how to find the x- and y-intercepts of a quadratic function. Every quadratic function has a y-intercept, but not all of them have x-intercepts — here, you can find out how to recognize which case you have. You can also see what wonderful practical models quadratic functions can be in the real world. Finally, I tell you how to graph quadratic functions to get those smooth, U-shaped curves.

Finding Intercepts and the Vertex of a Parabola

The *intercepts* of a function are where the curve or graph of the function crosses an axis. The x-intercepts of a function always have a 0 for the y-coordinate. They look like (–3, 0), (4, 0), and so on. A function has only one y-intercept (because of the definition of a function), and the y-intercept always has a 0 for its x-coordinate.

To solve for the x-intercepts of a quadratic function, you set $f(x)$ equal to 0 (this is the same as setting y equal to 0) and then solve for x. To solve for the y-intercept of a quadratic function, you set x equal to 0 and solve for $f(x)$.

The graph of a quadratic function is a parabola, a U-shaped curve; the parabola's *vertex* is a point that's at the highest or lowest position of the graph of the function. The vertex is a parabola's only turning point. Here are a couple things to note:

✔ When the constant a is positive, the parabola opens upward and the vertex is in the lowest position (it's the *minimum*).

✔ When a is negative, the parabola opens downward, and the vertex is in the highest position (the *maximum*).

You can find the x-coordinate of the vertex with $x = -\dfrac{b}{2a}$. Solve for the x-coordinate, and then substitute that number into the equation of the function to solve for the y-coordinate of the vertex.

Q. Find the intercepts and vertex of the function $f(x) = 4x^2 + 8x - 21$

A. **y-intercept: (0, –21); x-intercepts: $\left(-\dfrac{7}{2}, 0\right)$, $\left(\dfrac{3}{2}, 0\right)$; vertex: (–1, –25).** To find the y-intercept, let all the x's equal 0 in the equation; you get $y = 4(0)^2 + 8(0) - 21 = -21$, so the y-intercept is (0, –21). For the x-intercepts, let $f(x) = 0$ and solve $0 = 4x^2 + 8x - 21$. Factoring, $0 = (2x + 7)(2x - 3)$,

which has the solutions $x = -\dfrac{7}{2}$, $x = \dfrac{3}{2}$. The x-intercepts are $\left(-\dfrac{7}{2}, 0\right)$, $\left(\dfrac{3}{2}, 0\right)$. You can find the x-coordinate of the the vertex using the 4 and 8 in the equation, where $x = \dfrac{-8}{2(4)} = \dfrac{-8}{8} = -1$. Substitute the –1 in for the x's to solve for the y-coordinate of the vertex: $y = 4(-1)^2 + 8(-1) - 21 = 4 - 8 - 21 = -25$. The coordinates of the vertex are (–1, –25).

1. Find the intercepts and vertex of the function $f(x) = 2x^2 - 9x - 18$

Solve It

2. Find the intercepts and vertex of the function $f(x) = 25 - 4x^2$

Solve It

3. Find the intercepts and vertex of the function $f(x) = x^2 + 10x + 7$

Solve It

4. Find the intercepts and vertex of the function $f(x) = -x^2 - 8x - 16$

Solve It

Applying Quadratics to Real-Life Situations

Quadratic equations lend themselves to modeling situations that happen in real life, such as the rise and fall of profits from selling goods, the decrease and increase in the amount of time it takes to run a mile based on your age, and so on. The wonderful part of having something that can be modeled by a quadratic is that you can easily solve the equation and predict the patterns in the function values.

The vertex and x-intercepts are especially useful. These intercepts tell you where numbers change from positive to negative or negative to positive, so you know, for instance, where the ground is located in a physics problem or when you'd start making a profit or losing money in a business venture. The vertex tells you where you can find the absolute maximum or minimum cost, profit, speed, height, time, or whatever you're modeling.

Q. In 1972, you could buy a Mercury Comet for about $3,200 dollars. Cars can depreciate in value pretty quickly, but a 1972 Comet in pristine condition may be worth a lot of money to a collector today. Let the value of one of these Comets be modeled by the quadratic function $v(t) = 18.75t^2 - 450t + 3,200$, where t is the number of years since 1972. When is the value of the function equal to 0 (what is an x-intercept), what was the car's lowest value, and what was its value in 2000?

A. **The car's value never dropped to 0, the lowest value was $500, and the car was worth $5,300 in the year 2000.** In this model, the y-intercept represents the initial value. When $t = 0$, the function is $v(0) = 3,200$, which corresponds to the purchase price. Find the x-intercepts by solving

$18.75t^2 - 450t + 3,200 = 0$. Using the quadratic formula (you could try factoring, but it's a bit of a challenge and, as it turns out, the equation doesn't factor), you get $-37,500$ under the radical in the formula. You can't get a real-number solution, so the graph has no x-intercept. The value of the Comet doesn't ever get down to 0. Find the lowest value by determining the vertex. Using the formula, $t = \dfrac{-(-450)}{2(18.75)} = \dfrac{450}{37.50} = 12$. This coordinate tells you that 12 years from the beginning (1984 — add 12 to 1972) the value of the Comet is at its lowest. Replace the t's in the formula with 12, and you get $v(12) = 18.75(12)^2 - 450(12) + 3,200 = 500$. The Comet was worth $500 in 1984. To find the value of the car in 2000, you let $t = 28$, because the year 2000 is 28 years after 1972. The value of the car in 2000 is $v(28) = 18.75(28)^2 - 450(28) + 3,200 = \$5,300$.

5. The height of a ball t seconds after it's thrown into the air from the top of a building can be modeled by $h(t) = -16t^2 + 48t + 64$, where $h(t)$ is height in feet. How high is the building, how high does the ball rise before starting to drop downward, and after how many seconds does the ball hit the ground?

Solve It

6. The profit function telling Georgio how much money he will net for producing and selling x speciality umbrellas is given by $P(x) = -0.00405x^2 + 8.15x - 100$. What is Georgio's loss if he doesn't sell any of the umbrellas he produces, how many umbrellas does he have to sell to break even, and how many does he have to sell to earn the greatest possible profit?

Solve It

7. Chip ran through a maze in less than a minute the first time he tried. His times got better for a while with each new try, but then his times got worse (he took longer) due to fatigue. The amount of time Chip took to run through the maze on the *a*th try can be modeled by $T(a) = 0.5a^2 - 9a + 48.5$. How long did Chip take to run the maze the first time, and what was his best time?

Solve It

8. A highway underpass is parabolic in shape. If the curve of the underpass can be modeled by $h(x) = 50 - 0.02x^2$, where x and $h(x)$ are in feet, then how high is the highest point of the underpass, and how wide is it?

Solve It

Graphing Parabolas

The graph of a quadratic function is a smooth, U-shaped curve that opens either upward or downward, depending on the sign of the coefficient of the x^2 term. The vertex and intercepts offer the quickest, easiest points to help with the graph of the parabola. You can resort to solving for other points if the graph has no *x*-intercepts or if you need additional information to determine more about the shape.

Another aid to use when graphing parabolas is the *axis of symmetry;* a parabola is symmetric about a vertical line that runs through the vertex. Points on either side of the axis of symmetry that have the same *y*-value are equal distances from the axis. The equation of the *axis of symmetry* is $x = h$, where *(h, k)* is the vertex of the parabola.

Q. Sketch the graph of the parabola $f(x) = -x^2 + 6x + 40$, labeling any intercepts and the vertex and showing the axis of symmetry

A. As you can see in Figure 6-1, the *y*-intercept is (0, 40); you can find it by letting all the *x*'s equal 0 and simplifying. Find the *x*-intercepts by setting $-x^2 + 6x + 40$ equal to 0 and factoring: $0 = -(x^2 - 6x - 40) = -(x + 4)(x - 10)$; $x = -4$ and 10, so the intercepts are at (−4, 0) and (10, 0). The vertex is at (3, 49): You find the *x*-value using the formula (in the section "Finding Intercepts and the Vertex of a Parabola") and then replace the *x*'s with 3s and simplify for the *y*-coordinate.

Figure 6-1:
The
parabola
opens
downward
because the
x^2 term is
negative.

9. Sketch the graph of the parabola $f(x) = 4x^2$, labeling any intercepts and the vertex and showing the axis of symmetry

Solve It

10. Sketch the graph of the parabola $f(x) = -\frac{1}{3}x^2 + 3$, labeling any intercepts and the vertex and showing the axis of symmetry

Solve It

11. Sketch the graph of the parabola $f(x) = 3x^2 - 6x - 9$, labeling any intercepts and the vertex and showing the axis of symmetry

Solve It

12. Sketch the graph of the parabola $f(x) = -2x^2 + 10x - 8$, labeling any intercepts and the vertex and showing the axis of symmetry

Solve It

13. Sketch the graph of the parabola $f(x) = \frac{1}{2}x^2 + 3x + 1$, labeling any intercepts and the vertex and showing the axis of symmetry

Solve It

14. Sketch the graph of the parabola $f(x) = -x^2 + 6x$, labeling any intercepts and the vertex and showing the axis of symmetry

Solve It

Answers to Problems on Quadratic Functions

The following are the answers to the practice problems presented earlier in this chapter.

1 Find the intercepts and vertex of the function $f(x) = 2x^2 - 9x - 18$. The answer is **y-intercept: $(0, -18)$; x-intercepts: $(6, 0)$ and $\left(-\frac{3}{2}, 0\right)$; vertex: $\left(\frac{9}{4}, -28\frac{1}{8}\right)$.**

You find the y-intercept by letting all the x's equal 0 and simplifying. Find the x-intercepts by setting $2x^2 - 9x - 18$ equal to 0, factoring that equation into $(2x + 3)(x - 6) = 0$, and solving for x. The x-value of the vertex is $x = \frac{-(-9)}{2(2)} = \frac{9}{4}$. Substituting the x-value in for all the x's, you get

$$f\left(\frac{9}{4}\right) = 2\left(\frac{9}{4}\right)^2 - 9\left(\frac{9}{4}\right) - 18 = 2\left(\frac{81}{16}\right) - \frac{81}{4} - 18$$

$$= \frac{162}{16} - \frac{324}{16} - \frac{288}{16} = \frac{-450}{16} = -\frac{225}{8} = -28\frac{1}{8}$$

2 Find the intercepts and vertex of the function $f(x) = 25 - 4x^2$. The answer is **y-intercept: $(0, 25)$; x-intercepts: $\left(\frac{5}{2}, 0\right)$, $\left(-\frac{5}{2}, 0\right)$; vertex: $(0, 25)$.**

You find the y-intercept by letting all the x's equal 0 and simplifying. Find x-intercepts by setting $25 - 4x^2$ equal to 0; the left side of the equation factors into $(5 - 2x)(5 + 2x) = 0$. You can then solve for x. The x-value of the vertex is $x = \frac{-0}{2(-4)} = 0$. You have no x term in the original formula (only x^2), so the coefficient b is equal to 0. Because the x-value of the coordinate for the vertex is 0, you get the y-intercept as the vertex.

3 Find the intercepts and vertex of the function $f(x) = x^2 + 10x + 7$. The answer is **y-intercept: $(0, 7)$; x-intercepts: $\left(-5 + 3\sqrt{2}, 0\right)$, $\left(-5 - 3\sqrt{2}, 0\right)$; vertex: $(-5, -18)$.**

You find the y-intercept by letting all the x's equal 0 and simplifying. Find the x-intercepts by setting $x^2 + 10x + 7$ equal to 0 and solving with the quadratic formula (the trinomial doesn't factor):

$$x = \frac{-10 \pm \sqrt{10^2 - 4(1)(7)}}{2(1)} = \frac{-10 \pm \sqrt{100 - 28}}{2}$$

$$= \frac{-10 \pm \sqrt{72}}{2} = \frac{-10 \pm 6\sqrt{2}}{2} = -5 \pm 3\sqrt{2}$$

These x-intercepts end up being at about $(-0.757, 0)$ and $(-9.243, 0)$. The x-value of the vertex is $x = \frac{-(10)}{2(1)} = \frac{-10}{2} = -5$. Substituting the x-value in for all the x's, you get $f(-5) = (-5)^2 + 10(-5) + 7 = 25 - 50 + 7 = -18$. Thus, the vertex is at $(-5, -18)$.

4 Find the intercepts and vertex of the function $f(x) = -x^2 - 8x - 16$. The answer is **y-intercept: $(0, -16)$; x-intercept: $(-4, 0)$; vertex: $(-4, 0)$.**

You find the y-intercept by letting all the x's equal 0 and simplifying. Find the x-intercepts by setting $-x^2 - 8x - 16$ equal to 0, factoring that into $-1(x + 4)^2 = 0$, and solving for x. You find only one x-intercept; the graph of the parabola doesn't cross the axis at that intercept — it just touches (is tangent to) the axis. The x-value of the vertex is $x = \frac{-(-8)}{2(-1)} = \frac{8}{-2} = -4$. Substituting the x-value in for all the x's, you get $f(-4) = -(-4)^2 - 8(-4) - 16 = -16 + 32 - 16 = 0$. The vertex is at the x-intercept.

5 The height of a ball t seconds after it's thrown into the air from the top of a building can be modeled by $h(t) = -16t^2 + 48t + 64$, where $h(t)$ is height in feet. How high is the building, how high does the ball rise before starting to drop downward, and after how many seconds does the ball hit the ground? **The building is 64 feet tall, the ball peaks at 136 feet, and it takes 4 seconds to hit the ground.**

The ball is thrown from the top of the building, so you want the height of the ball when $t = 0$. This number is the initial t value (the y-intercept). When $t = 0$, $h = 64$, so the building is 64 feet high. The ball is at its highest at the vertex of the parabola. Solving for the t value, you get that the vertex occurs where $t = 1.5$ seconds. Substituting $t = 1.5$ into the formula, you get that $h = 136$ feet. The ball hits the ground when $h = 0$. Solving $-16t^2 + 48t + 64 = 0$, you factor to get $-16(t - 4)(t + 1) = 0$. The solution $t = 4$ tells you when the ball hits the ground. The $t = -1$ represents going backward in time, or in this case, where the ball would have started if it had been launched from the ground — not the top of a building.

6   The profit function telling Georgio how much money he will net for producing and selling x speciality umbrellas is given by $P(x) = -0.00405x^2 + 8.15x - 100$. What is Georgio's loss if he doesn't sell any of the umbrellas he produces, how many umbrellas does he have to sell to break even, and how many does he have to sell to earn the greatest possible profit? **Georgio loses $100 (earns –$100) if he sells 0, needs to sell 13 to break even, and can maximize profits if he sells 1,006 umbrellas.**

If Georgio sells no umbrellas, then $x = 0$, and he makes a negative profit (loss) of $100. The break-even point comes when the profit changes from negative to positive, at an x-intercept. Using the quadratic formula, you get that $x = 2,000$ or about 12.35. The first (smaller) x-intercept is where the function changes from negative to positive. The second is where the profit becomes a loss again (too many umbrellas, too much overtime?). So 13 umbrellas would yield a positive profit — he'd break even. The maximum profit occurs at the vertex. Using the formula for the x-value of the vertex, you get that $x \approx 1,006.17$. Substituting 1,006 into the formula and then 1,007 into the formula, you see that Georgio gets slightly more profit with 1,006 umbrellas. He'd make about $4,000.

7   Chip ran through a maze in less than a minute the first time he tried. His times got better for a while with each new try, but then his times got worse (he took longer) due to fatigue. The amount of time Chip took to run through the maze on the ath try can be modeled by $T(a) = 0.5a^2 - 9a + 48.5$. How long did Chip take to run the maze the first time, and what was his best time? **Chip took 40 seconds the first time; his best time was 8 seconds.**

Because the variable a represents the number of the attempt, find $T(1)$ for the time of the first attempt. $T(1) = 40$ seconds. The best (minimum) time is at the vertex. Solving for the a value (which is the number of the attempt), $a = \dfrac{-(-9)}{2(0.5)} = 9$. He had the best time on the ninth attempt, and $T(9) = 8$.

8   A highway underpass is parabolic in shape. If the curve of the underpass can be modeled by $h(x) = 50 - 0.02x^2$, where x and $h(x)$ are in feet, then how high is the highest point of the underpass, and how wide is it? **The underpass is 50 feet high and 100 feet wide.**

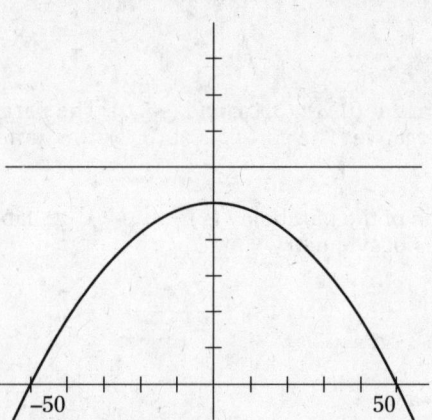

The highest point occurs at the vertex: $x = \dfrac{-0}{2(-0.02)} = 0$. The x-coordinate of the vertex is 0, so the vertex is also the y-intercept, at $(0, 50)$. The two x-intercepts represent the endpoints of the width of the overpass. Setting $50 - 0.02x^2$ equal to 0, you solve for x and get $x = 50$ and $x = -50$. These two points are 100 units apart — the width of the underpass.

9 Sketch the graph of the parabola $f(x) = 4x^2$, labeling any intercepts and the vertex and showing the axis of symmetry.

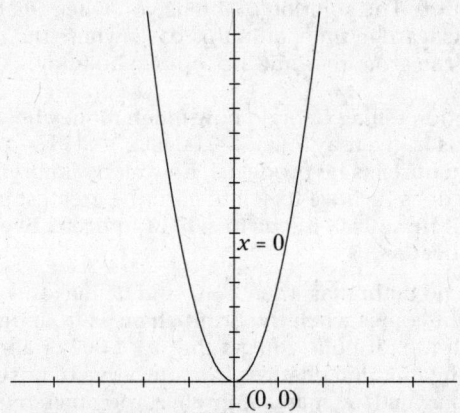

The only intercept is at $(0, 0)$. The parabola opens upward, because 4 is positive. The vertex is at $(0, 0)$, and the equation of the axis of symmetry is $x = 0$.

10 Sketch the graph of the parabola $f(x) = f(x) = -\dfrac{1}{3}x^2 + 3$, labeling any intercepts and the vertex and showing the axis of symmetry.

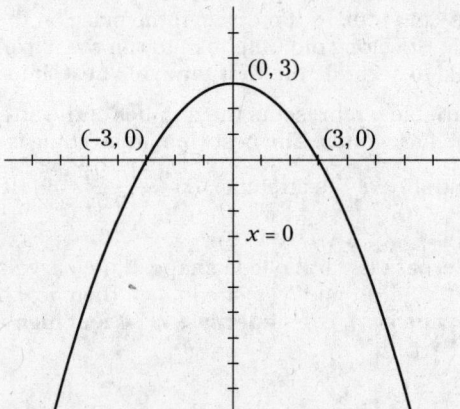

The intercepts are at $(0, 3)$, $(3, 0)$, and $(-3, 0)$. The parabola opens downward, because the coefficient of x^2 is negative. The vertex is at $(0, 3)$, the y-intercept, and the equation of the axis of symmetry is $x = 0$.

11 Sketch the graph of the parabola $f(x) = 3x^2 - 6x - 9$, labeling any intercepts and the vertex and showing the axis of symmetry.

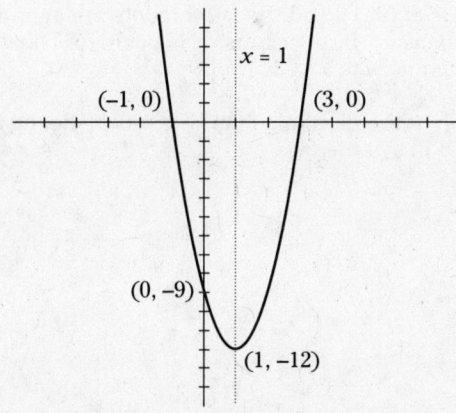

The intercepts are at $(0, -9)$, $(3, 0)$, and $(-1, 0)$. The parabola opens upward, because 3 is positive. The vertex is at $(1, -12)$, and the equation of the axis of symmetry is $x = 1$.

12 Sketch the graph of the parabola $f(x) = -2x^2 + 10x - 8$, labeling any intercepts and the vertex and showing the axis of symmetry.

The intercepts are at $(0, -8)$, $(4, 0)$, and $(1, 0)$. The parabola opens downward because -2 is negative. The vertex is at $(2.5, 12)$, and the equation of the axis of symmetry is $x = 2.5$.

13 Sketch the graph of the parabola $f(x) = f(x) = \frac{1}{2}x^2 + 3x + 1$, labeling any intercepts and the vertex and showing the axis of symmetry.

The y-intercept is at (0, 1), and the x-intercepts are at approximately (–0.354, 0) and (–5.646, 0). The parabola opens upward, because ½ is positive. The vertex is at (–3, 3.5), and the equation of the axis of symmetry is $x = -3$.

14 Sketch the graph of the parabola $f(x) = -x^2 + 6x$, labeling any intercepts and the vertex and showing the axis of symmetry.

The intercepts are at (0, 0) and (6, 0). The parabola opens downward because of the –1 coefficient on the x^2 term. The vertex is at (3, 9), and the equation of the axis of symmetry is $x = 3$.

Chapter 7

Plugging In Polynomials

- -

In This Chapter

▶ Locating the intercepts in polynomials

▶ Figuring out where a function changes signs

▶ Assembling all the information you need to graph a polynomial curve

▶ Looking for roots with clues from Descartes — and others

▶ Doing division synthetically and taking other shortcuts

▶ Building up polynomials from their roots

- -

A polynomial function is a well-mannered function with nice, predictable habits. The graph of a polynomial function is always a smooth curve with no abrupt changes in direction or shape. The equation of a polynomial function has only whole-number exponents on the variables.

A *polynomial function* has the form $f(x) = a_n x^n + a_{n-1} x^{n-1} + a_{n-2n} x^{n-2} + \ldots a_1 x^1 + a_0 x^0$, where n is a whole number and any a_{n-i} is a real number. The x^0 is actually equal to 1, so that term is a constant; I include the final x and superscript here to complete the pattern. A polynomial function has at most n x-intercepts and at most $n - 1$ turning points (where the curve of the graph changes direction).

In this chapter, you graph polynomial functions after finding their intercepts and determining where the function is positive or negative. More-advanced tools used to find the x-intercepts (roots) of the polynomial equation include the Rational Root Theorem and Descartes's Rule of Sign. Finally, the Remainder Theorem, along with synthetic division, lets you quickly evaluate functions, and the Factor Theorem tells you how to reconstruct polynomials if you know only their roots.

Finding Basic Polynomial Intercepts

A polynomial function has exactly one y-intercept (where it crosses the y-axis). You find the y-intercept by replacing all the x's in the function formula with 0 and solving.

Polynomial functions may or may not have x-intercepts. If a polynomial function has an odd degree (the highest power is an odd number), then the polynomial has to have at least one x-intercept; the curve has to cross the x-axis. A polynomial function with an even degree may or may not cross the x-axis. But in any case, a polynomial won't have more x-intercepts than its degree. You can bank on that.

To solve for the x-intercepts, set the function equation equal to 0 and find out which values of x are solutions. Those answers are the x-values of the intercepts. The biggest challenge often comes when trying to find that solution of the function equation. The problems in this first section are easily solvable. The next section and "Possible Roots and Where to Find Them: The Rational Root Theorem and Descartes's Rule" show you how to handle equations that aren't as cooperative with their solutions.

Q. Find the intercepts of the polynomial
$y = x^3 - 16x$

A. **y-intercept: (0, 0); x-intercepts: (0, 0), (4, 0), and (–4, 0).** Note the repeat in the answers. The intercept (0, 0) always appears in both intercept categories (or in neither one). To find the y-intercept, replace all the x's in the function equation with 0 and solve: $y = 0^3 - 16(0) = 0$. To find the x-intercepts, set the function equation equal to 0 and solve: $0 = x^3 - 16x = x(x^2 - 16) = x(x - 4)(x + 4)$. The solutions are $x = 0$, 4, and –4.

Q. Find the intercepts of the polynomial
$y = x^5 + 3x^4 - 5x^3 - 15x^2$

A. **y-intercept: (0, 0); x-intercepts: (0, 0), (–3, 0), $\left(\sqrt{5}, 0\right)$, $\left(-\sqrt{5}, 0\right)$.** You could find as many as five x-intercepts for this function, but as it turns out, you have only four. The rule gives you the maximum number of x-intercepts possible, not always exactly how many. Replacing all the x's with 0's and simplifying tells you that the y-intercept is 0. To solve for the x-intercepts, you have to factor the function rule. First pull out the x^2 factor, and then factor the four terms in the parentheses by grouping (Chapter 2 explains this factoring technique): $x^5 + 3x^4 - 5x^3 - 15x^2 = x^2(x^3 + 3x^2 - 5x - 15) = x^2[x^2(x + 3) - 5(x + 3)] = x^2[(x + 3)(x^2 - 5)]$. The exponent 2 on the x term indicates a double root at $x = 0$ (I explain how a double root affects the graph in "Graphing Polynomials," later in this chapter). The factor $x^2 - 5$ yields two roots — one positive and one negative — because you have to take a square root to solve for x.

1. Find the intercepts of the polynomial
$y = (x + 4)(x + 2)(x^2 - 1)$

Solve It

2. Find the intercepts of the polynomial
$y = x(x - 3)^2(x - 6)(x^2 + 4)$

Solve It

3. Find the intercepts of the polynomial
$y = -2x^3 - 6x^2 + 8x$

Solve It

4. Find the intercepts of the polynomial
$y = x^4 - 26x^2 + 25$

Solve It

Digging Up More-Difficult Polynomial Roots with Factoring

When solving for roots (x-intercepts of a polynomial), you usually need to factor the function rule and set it equal to 0. The factorization can be simple and obvious or complicated and obscure. You always hope for the simple and obvious, move to challenging and doable, and resort to the "big guns" when the factors are more obscure. You can find the "big guns" later in this chapter in "Possible Roots and Where to Find Them: The Rational Root Theorem and Descartes's Rule."

Before you go to those lengths, though, you need to exhaust other methods. The other methods of factoring include

- Dividing by a greatest common factor (GCF)
- Factoring a perfect square binomial
- Factoring by grouping
- Factoring quadratic-like trinomials

Flip to Chapter 2 if you need a review of these methods.

Q. Find the roots (solutions) of the polynomial $x^7 - 82x^5 + 81x^3 = 0$

A. $x = 0, 0, 0, 9, -9, 1, -1$. This equation technically has seven solutions, but the 0 is a multiple root, so you end up with only five different numbers. To find these solutions, you first factor x^3 out of each term to get $x^3(x^2 - 81)(x^2 - 1) = 0$. The two binomials are both the difference of perfect squares, so you can factor them into the difference and sum of the roots of the terms. You get $x^3(x - 9)(x + 9)(x - 1)(x + 1) = 0$. Setting each of the factors equal to 0, you find the roots.

Q. Find the roots (solutions) of the polynomial $x^3 - 16x^2 + 100x - 1,600 = 0$

A. $x = 16$. The polynomial doesn't have a common factor in the four terms, but you can group the terms for the pairs of common factors. You get $x^2(x - 16) + 100(x - 16) = 0$, which factors into $(x - 16)(x^2 + 100) = 0$. The second binomial is the sum of squares, which doesn't factor. Setting these two factors equal to 0, you get $x = 16$ from the first factor, but the second factor doesn't produce any real answers. Even though you started with a third-degree polynomial, which can yield up to three solutions, this polynomial has only one real root. The solution is just $x = 16$.

5. Find the roots (solutions) of the polynomial $3x^4 - 12x^3 - 27x^2 + 108x = 0$

Solve It

6. Find the roots (solutions) of the polynomial $x^5 - 16x^3 + x^2 - 16 = 0$

Solve It

7. Find the roots (solutions) of the polynomial $x^6 + 9x^3 + 8 = 0$

Solve It

8. Find the roots (solutions) of the polynomial $36x^5 - 13x^3 + x = 0$

Solve It

Determining Where a Function Is Positive or Negative

A polynomial function is positive when the function value (or *y*-value) is a positive number. Graphically, you can see where a polynomial function is positive by where the curve is above the *x*-axis. When you don't have a graph to look at, one way to determine where a polynomial function is positive is to just plug numbers in for *x* and see what you get. Not only is plugging in random numbers time consuming, but it also can lead to inaccuracies.

The best method for determining on what intervals (between what *x*-values) a polynomial is positive or negative is to find the *zeros* or *roots* (*x*-intercepts; values for which the polynomial is 0), and then plug in a number between each pair of intercepts; this test number lets you check the positive or negative nature of the function there. Make sure you also test numbers in the intervals from $-\infty$ to the lowest *x*-intercept and from the highest *x*-intercept to ∞.

To write the positive and negative natures of a function in "mathspeak," you can use inequalities or interval notation. Remember that at *x*-intercepts, the function equals 0. Because 0 is neither positive nor negative, you can't use $\leq$, $\geq$, or brackets when you write your answers; answers simply include < and > or parentheses. (For details on interval notation, see Chapter 1.)

Q. Determine where the polynomial $y = x(x-3)(x-7)(x+2)^2(x+6)$ is positive and where it's negative

A. **Positive: $(-\infty, -6) \cup (0, 3) \cup (7, \infty)$; negative: $(-6, -2) \cup (-2, 0) \cup (3, 7)$.** You first determine the x-intercepts (zeros or roots) by setting the function equal to 0. The polynomial $x(x-3)(x-7)(x+2)^2(x+6) = 0$ when $x = 0$, $x = 3$, $x = 7$, $x = -2$, and $x = -6$. Draw a number line with the x-values representing the zeros shown on the line — in their correct order from smallest to largest. Figure 7-1 shows the bare number line on the top and the "worked through" number line underneath that first one.

You test each interval sectioned off by the zeros on the number line by choosing some number that lies in each interval and plugging that number in for each x in the function rule. You determine whether each individual factor is positive or negative, record that fact on the number line, and then see whether the product of all those factors comes out to be positive or negative. For instance, if you look at the interval between the -2 and 0 on the number line, the most obvious choice for x is -1. (Any number works — pick a nice one when you can.) Replacing all the x's in the function

rule with -1, you see that the first factor, x, is negative, the factor $(x-3)$ is negative, the $(x-7)$ is negative, the $(x+2)^2$ is positive, and the $(x+6)$ is positive. You write $(-)(-)(-)(+)(+)$ in the interval and determine that the product is negative (if you have an odd number of negative signs, the product is positive; an even number, and the product is positive). So the function is negative for all the numbers between -2 and 0.

Reading the signs (sort of like astrology?) on the number line, you determine that the value of the function equation is positive when the x's are less than -6, between 0 and 3, or greater than 7. Between -6 and -2, between -2 and 0, or between 3 and 7, the function is negative. Even though you don't have a sign change at -2, you can't include -2 in the numbers that give a negative function value. At -2, the function is 0. Here's the answer in inequality notation: the function is positive for $x < -6$ or $0 < x < 3$ or $x > 7$, and the function is negative for $-6 < x < -2$ or $-2 < x < 0$ or $3 < x < 7$. In interval notation, you write that the function is positive for $(-\infty, -6) \cup (0, 3) \cup (7, \infty)$, and it's negative for $(-6, -2) \cup (-2, 0) \cup (3, 7)$. The $\cup$ symbol means *union* and is standard set notation. It replaces the word *or*.

Figure 7-1: A number line shows the positive and negative intervals.

9. Determine where the polynomial $y = 6x^2 - 6x$ is positive and where it's negative

Solve It

10. Determine where the polynomial $y = -2x^2 - 7x + 15$ is positive and where it's negative

Solve It

11. Determine where the polynomial $y = (x + 3)(x - 2)(x - 5)^2$ is positive and where it's negative

Solve It

12. Determine where the polynomial $y = x^2(x + 3)^2(x - 4)^4$ is positive and where it's negative

Solve It

Graphing Polynomials

The graph of a polynomial is a smooth, winding curve that starts somewhere to the left, moves across the graph paper, crossing or touching or avoiding the x-axis, and then disappears to the right. You can determine the points that make up the graph by inserting x-values into the function rule, but graphing a polynomial that way can be a long, tedious chore. A more efficient method involves finding the intercepts, plotting a few additional points for clarity, and boldly sketching in the rest.

Q. Sketch the graph of $y = (x^2 - 1)(x + 3)^2$

A. First, determine the intercepts and place them on the graph. When $x = 0$, $y = -9$; that's the y-intercept. When $y = 0$, $x = 1, -1$, or -3; those are the three x-intercepts. The intercept at $(-3, 0)$ is a double root. The curve only touches the axis at that point — it doesn't cross the x-axis. Next, use a few test points between the intercepts to figure out where the function is positive or negative. The function is negative between $x = -1$ and $x = 1$. Other than that (and at the x-intercepts), the function is positive — above the x-axis. Figure 7-2 shows you the completed graph. The graph shows the intercepts and a few arbitrary points.

Q. Sketch the graph of $y = -x^3 - x^2 + 25x + 25$

A. The y-intercept is at $(0, 25)$. You find that answer by letting all the x's be 0. To solve for the x-intercepts, set the function rule equal to 0 and factor by grouping: $0 = -x^2(x + 1) + 25(x + 1) = (x + 1)(-x^2 + 25) = (x + 1)(5 - x)(5 + x)$. The x-intercepts are $(-1, 0)$, $(5, 0)$, and $(-5, 0)$. The function is negative on the intervals $(-5, -1)$ and $(5, \infty)$ and positive on the intervals $(-\infty, -5)$ and $(-1, 5)$. Figure 7-3 shows the graph of the function with the intercepts and a few select points for clarity.

Figure 7-2:
The curve
has three
turning
points.

Figure 7-3:
The graph
goes to neg-
ative infinity
as *x* gets
really large.

13. Sketch the graph of $y = (x + 4)(x + 2)(x - 3)$

Solve It

14. Sketch the graph of $y = x(x - 3)^2(x - 6)$

Solve It

15. Sketch the graph of $y = -2x^3 - 6x^2 + 8x$

Solve It

16. Sketch the graph of $y = x^4 - 26x^2 + 25$

Solve It

Possible Roots and Where to Find Them: The Rational Root Theorem and Descartes's Rule

You may come across problems that you simply can't figure out how to factor. However, that doesn't mean you're stuck graphing the function on a calculator, hoping you can zoom in enough to guess where the roots are. You have a couple more theorems at your disposal, and together they can give you a good idea of what the exact solutions may be.

So what's the Rational Root Theorem and what does it have to with Descartes's Rule? Check it out:

✔ **Rational Root Theorem:** If the polynomial $f(x) = a_n x^n + a_{n-1} x^{n-1} + a_{n-2} x^{n-2} + \ldots + a_1 x^1 + a_0 x^0$ has any rational roots, then they all meet the requirement that they can be written as a fraction equal to $\dfrac{\text{factor of } a_0}{\text{factor of } a_n}$.

✔ **Descartes's Rule of Sign:** By looking at which terms in the polynomial are positive and which ones are negative and counting how many times the signs change, you can figure out how many positive or negative roots you're likely to find:

- **Positive roots:** The polynomial $f(x) = a_n x^n + a_{n-1} x^{n-1} + a_{n-2} x^{n-2} + \ldots + a_1 x^1 + a_0 x^0$ has at most n real roots. Count the number of times the sign changes in f, and call it p. The value of p is the maximum number of *positive* real roots of f. If the number of positive roots is not p, then it is $p-2$, $p-4$, $p-6$, and so on.

- **Negative roots:** In the polynomial $f(x) = a_n x^n + a_{n-1} x^{n-1} + a_{n-2} x^{n-2} + \ldots + a_1 x^1 + a_0 x^0$ find $f(-x)$. Then count the number of times the sign changes in $f(-x)$ and call it q. The value of q is the maximum number of *negative* roots of f. If the number of negative roots is not q, then it is $q-2$, $q-4$, $q-6$, and so on.

If the polynomial doesn't have a constant term, you first have to factor out the greatest power of x possible before applying the Rational Root Theorem or Descartes's Rule of Sign.

Q. Determine all the possibilities for rational roots of the polynomial $4x^4 - 25x^2 + 36 = 0$. Then determine how many of the real roots of the polynomial may be positive and how many may be negative. Factor the polynomial to confirm your results.

A. **Possible real roots:** ± 1, ± 2, ± 3, ± 4, ± 6, ± 9, ± 12, ± 18, ± 36, $\pm\frac{1}{2}$, $\pm\frac{3}{2}$, $\pm\frac{9}{2}$, $\pm\frac{1}{4}$, $\pm\frac{3}{4}$, $\pm\frac{9}{4}$; **number of positive roots: two or zero; number of negative roots: two or zero; actual solution:** $x = \frac{3}{2}$, $-\frac{3}{2}$,

2, -2. Any rational roots of this polynomial equation have to be numbers that are the results of dividing factors of the constant term, 36, by factors of the lead coefficient, 4. First list the factors of the constant term. Use both + and − for each factor. The list reads: ± 1, ± 2, ± 3, ± 4, ± 6, ± 9, ± 12, ± 18, ± 36. These numbers are all rational, and they're all possible solutions of the equation. Now divide each of the factors of 36 by the factors of 4. The factors of 4 are ± 1, ± 2, ± 4. Dividing the factors of 36 by ± 1 won't change any of them, so don't bother. Dividing the factors by ± 2 and ± 4 gives you $\frac{\pm 1}{\pm 2}$, $\frac{\pm 2}{\pm 2}$, $\frac{\pm 3}{\pm 2}$, $\frac{\pm 4}{\pm 2}$, $\frac{\pm 6}{\pm 2}$, $\frac{\pm 9}{\pm 2}$, $\frac{\pm 12}{\pm 2}$, $\frac{\pm 18}{\pm 2}$, $\frac{\pm 36}{\pm 2}$ and $\frac{\pm 1}{\pm 4}$, $\frac{\pm 2}{\pm 4}$, $\frac{\pm 3}{\pm 4}$, $\frac{\pm 4}{\pm 4}$, $\frac{\pm 6}{\pm 4}$, $\frac{\pm 9}{\pm 4}$, $\frac{\pm 12}{\pm 4}$, $\frac{\pm 18}{\pm 4}$, $\frac{\pm 36}{\pm 4}$. Reducing these fractions and eliminating the duplicates, the remaining fractions are $\pm\frac{1}{2}$, $\pm\frac{3}{2}$, $\pm\frac{9}{2}$, $\pm\frac{1}{4}$, $\pm\frac{3}{4}$, $\pm\frac{9}{4}$. The list of integers plus these fractions are all the possibilities for rational roots. No other rational numbers can work.

Now for the positive and negative natures of the real roots: First look at $f(x) = 4x^4 - 25x^2 + 36$ and count the number of sign changes from beginning to end. The terms change from positive (+4) to negative (−25) back to positive (+36). You find two sign changes, so the equation has either two or zero positive real roots. Notice that the two exponents are both even (and technically, the 36 is multiplied by x raised to the 0 power — which is also an even exponent). Now look at $f(-x) = 4(-x)^4 - 25(-x)^2 + 36 = 4x^4 - 25x^2 + 36$. The signs didn't change; you still have two sign changes, so there are either two or zero negative real roots. Now, to check the answer, the factorization is $4x^4 - 25x^2 + 36 = (4x^2 - 9)(x^2 - 4) = (2x - 3)(2x + 3)(x - 2)(x + 2) = 0$. The solutions are $\frac{3}{2}$, $-\frac{3}{2}$, 2, −2, so you find exactly two positive and two negative real roots.

---

17. Determine all the possibilities for rational roots of the polynomial $9x^4 - 37x^2 + 4 = 0$. Then determine how many of the real roots of the polynomial may be positive and how many may be negative. Factor the polynomial to confirm your results.

Solve It

18. Determine all the possibilities for rational roots of the polynomial $2x^4 - 3x^3 - 18x^2 + 27x = 0$. Then determine how many of the real roots of the polynomial may be positive and how many may be negative. Factor the polynomial to confirm your results.

Solve It

19. Determine all the possibilities for rational roots of the polynomial $x^4 - 4x^3 + 6x^2 - 4x + 1 = 0$. Then determine how many of the real roots of the polynomial may be positive and how many may be negative. Factor the polynomial to confirm your results.

Solve It

20. Determine all the possibilities for rational roots of the polynomial $x^5 + 8x^4 + 7x^3 - 25x^3 - 200x^2 - 175x = 0$. Then determine how many of the real roots of the polynomial may be positive and how many may be negative. Factor the polynomial to confirm your results.

Solve It

21. Determine all the possibilities for rational roots of the polynomial $3x^3 + 4x^2 + 30x + 40 = 0$. Then determine how many of the real roots of the polynomial may be positive and how many may be negative. Factor the polynomial to confirm your results.

Solve It

22. Determine all the possibilities for rational roots of the polynomial $x^6 - 16x^3 + 64 = 0$. Then determine how many of the real roots of the polynomial may be positive and how many may be negative. Factor the polynomial to confirm your results.

Solve It

Getting Real Results with Synthetic Division and the Remainder Theorem

Synthetic division is a method for dividing polynomials by binomials. Because this type of division occurs frequently in algebra and other mathematics courses, the process is especially useful. Synthetic division, along with the Remainder Theorem, can help you quickly evaluate a polynomial function at any value.

Basically, synthetic division lets you cut out all the "middlemen." You deal with the coefficients of the terms in the polynomial and the constant in the binomial and ignore (well, ignore just momentarily) the variables while doing the process. The easiest way to explain synthetic division is just to show you an example.

The Remainder Theorem says that when the polynomial $f(x) = a_n x^n + a_{n-1}x^{n-1} + a_{n-2}x^{n-2} + \ldots + a_1 x^1 + a_0$ is divided by the binomial $x - c$, the remainder of the division is equal to $f(c)$.

For instance, when you divide $f(x) = x^3 + 2x^2 - 5x + 4$ by the binomial $x - 1$, the remainder is 2, and $f(1) = 2$. You can find $f(1)$ by using 1 as your divisor in synthetic division.

Q. Do the division problem $(4x^5 + 3x^4 - 6x^2 + 5x - 2) \div (x + 3)$ using synthetic division.

A. $4x^4 - 9x^3 + 27x^2 + 266 - \dfrac{800}{x + 3}$. Here's how synthetic division works:

1. **Set up the problem.**

   Put the polynomial in decreasing order of the exponents, inserting a 0 for any missing term. Then write the coefficients in a horizontal row and put a division symbol, a little half-box, to the left of that row. The number you put in front of the division symbol (the divisor) is the *opposite* of the number in the binomial you're dividing by. So if you divide $4x^5 + 3x^4 - 6x^2 + 5x - 2$ by the binomial $x + 3$, you use –3 in synthetic division. Here's the initial setup of the division problem:

   $$\underline{-3\,|}\ 4\ \ 3\ \ 0\ \ -6\ \ 5\ \ -2$$

   Notice that you write a 0 where the x^3 term would be. The horizontal line leaves enough room beneath the numbers to write products.

2. **Drop the lead coefficient (the 4) below the horizontal line.**

3. **Multiply the number below the horizontal line by the divisor; write this product beneath the next coefficient, and add the coefficient and product together; put your answer below the horizontal line, and repeat the process.**

   Multiply the 4 by –3 and put that product, –12, beneath the next coefficient, the 3. Add the 3 and –12, and put that result, –9, below the horizontal line. Multiply that number by –3, put the product under the next coefficient, the 0, and add the 0 and 27. Keep repeating the multiply-add process all the way down the line until you're done:

   $$
   \begin{array}{r|rrrrrr}
   -3 & 4 & 3 & 0 & -6 & 5 & -2 \\
      &   & -12 & 27 & -81 & 261 & -798 \\
   \hline
      & 4 & -9 & 27 & -87 & 266 & -800
   \end{array}
   $$

4. **Put the variables back in as you write the answer.**

The last entry on the bottom is –800. This number is the remainder of the division, and the other numbers are coefficients of the quotient (answer). To write out the answer, start with a power that's one lower than the original polynomial's power, and use all the coefficients, dropping the power on the variable by 1 with each successive coefficient. The last number is the remainder, so write that over the divisor. In this case, write the answer of the division problem as

$$4x^4 - 9x^3 + 27x^2 + 266 - \frac{800}{x + 3}.$$

Q. Using the Remainder Theorem, find $f(-9)$ if $f(x) = x^5 + 10x^4 + 7x^3 - 20x^2 - 18x + 4$

A. $f(-9) = 4$. If you substitute the –9 in for the x's and simplify, you get 4 — if you're lucky. The numbers get really large and unwieldy. Using the Remainder Theorem and synthetic division, you have a much easier task: Use –9 as your divisor when you do synthetic division; your answer is the remainder, the last entry on the bottom:

$$
\begin{array}{r|rrrrrr}
-9 & 1 & 10 & 7 & -20 & -18 & 4 \\
   &   & -9 & -9 & 18 & 18 & 0 \\
\hline
   & 1 & 1 & -2 & -2 & 0 & 4
\end{array}
$$

---

23. Use synthetic division to divide $(9x^4 - 37x^2 + 4) \div (x - 2)$

Solve It

24. Use synthetic division to divide $(2x^4 + 3x^3 + x^2 - 3x + 5) \div (x + 1)$

Solve It

---

25. Use synthetic division to divide $(3x^3 + 4x^2 - 2x + 10) \div (x + 2)$

Solve It

26. Use synthetic division to divide $(x^4 - 12x^3 + 54x^2 - 108x + 81) \div (x - 3)$

Solve It

27. Find $f(2)$ if $f(x) = x^5 - 3x^4 + 4x^3 - 6x^2 + 5x - 1$

Solve It

28. Find $f(-2)$ if $f(x) = 6x^{10} - 16x^6 + 150x^5 - 80x^2$

Solve It

Connecting the Factor Theorem with a Polynomial's Roots

Factors of polynomials and roots of polynomial equations are closely linked. Whenever a variable's involved, factors produce roots, and roots identify factors.

The Factor Theorem is special because it works forward and backward. Technically, the theorem is an *if and only if* type of statement. *If* inputting some number c makes the function equal to 0, *then* it's true that $x - c$ is a factor of the function. And *if* you know that $x - c$ is a factor of some function, *then* inputting c must make the function equal 0.

The Factor Theorem says that if $x = \frac{b}{a}$ is a root of the polynomial $f(x)$, then the binomial $(ax - b)$ is a factor of the polynomial function $f(x)$. Also, if a polynomial factors into $f(x) = k(x - a)(x - b) \cdots$ where k is some constant factor, then a is a root, b is a root, and so on.

For instance, the polynomial $f(x) = 2x^3 - x^2 - 18x + 9$ has the roots $\frac{1}{2}, 3, -3$, and it factors into $f(x) = (2x - 1)(x - 3)(x + 3)$.

Q. If the roots of $f(x)$ are 2, –4, 6, –8, –8, and $\frac{2}{3}$, write a function rule for $f(x)$

A. Using the Factor Theorem, $f(x) = k(x - 2)(x + 4)(x - 6)(x + 8)^2(3x - 2)$. You're probably wondering what that k is for. With the information given, you can write only the factors that involve variables. Each factor, when set equal to 0, gives you one of the roots (solutions) of the polynomial equation. But the original function could contain some constant multiplier that doesn't show up in the roots. The value of the constant really doesn't really matter unless you're graphing the function. To figure out what k is, you need more information. For instance, if you know that $f(0) = 6{,}144$, you can determine that $k = 1$. For now, just put the k in as a place holder and as a reminder that something may be missing from the big picture.

29. If the roots of $f(x)$ are 1, 2, –2, 3, 3, 3, 4, and 4, then write a function rule for $f(x)$

Solve It

30. If the roots of $f(x)$ are 0, 0, 0, 1, 1, 1, 1, –2, –3, and –4, then write a function rule for $f(x)$

Solve It

31. If the roots of $f(x)$ are –1, –1, 1, $\frac{1}{2}$, $\frac{3}{4}$, $-\frac{7}{8}$, and $-\frac{7}{8}$, then write a function rule for $f(x)$

Solve It

Answers to Problems on Plugging In Polynomials

The following are the answers to the practice problems presented earlier in this chapter.

1 Find the intercepts of the polynomial $y = (x + 4)(x + 2)(x^2 - 1)$. The answer is **y-intercept: (0, –8); x-intercepts: (–4, 0), (–2, 0), (1, 0), (–1, 0).**

Replace all the x's with 0 to get the y-intercept, (0, –8). You can obtain the x-intercepts by letting $y = 0$ and setting each factor equal to 0 to get the solutions.

2 Find the intercepts of the polynomial $y = x(x - 3)^2(x - 6)(x^2 + 4)$. The answer is **y-intercept: (0, 0); x-intercepts: (0, 0), (3, 0), (6, 0).**

Replace all the x's with 0 to get the y-intercept, (0, 0). This answer is also, technically, an x-intercept. Obtain the x-intercepts by letting $y = 0$ and setting each factor equal to 0. The factor $(x^2 + 4)$ never equals 0, so you don't get an x-intercept from that factor.

3 Find the intercepts of the polynomial $y = -2x^3 - 6x^2 + 8x$. The answer is **y-intercept: (0, 0); x-intercepts: (0, 0), (–4, 0), (1, 0).**

First, factor the polynomial to get $y = -2x(x + 4)(x - 1)$. Replacing the x's with 0 gives you the y-intercept (0, 0), which is also an x-intercept. Setting the binomials equal to 0 gives you the other two x-intercepts.

4 Find the intercepts of the polynomial $y = x^4 - 26x^2 + 25$. The answer is **y-intercept: (0, 25); x-intercepts: (5, 0), (–5, 0), (1, 0), (–1, 0).**

Factoring the polynomial, you get $y = (x^2 - 25)(x^2 - 1) = (x - 5)(x + 5)(x - 1)(x + 1)$. Replacing all the x's with 0 gives you the y-intercept (0, 25). Setting each binomial factor equal to 0 gives you all the x-intercepts.

5 Find the roots (solutions) of the polynomial $3x^4 - 12x^3 - 27x^2 + 108x = 0$. The answer is **x = 0, 3, –3, 4.**

First factor $3x$ out of each term to get $3x(x^3 - 4x^2 - 9x + 36) = 0$. You can then factor the terms in the parentheses by grouping: $3x[x^2(x - 4) - 9(x - 4)] = 3x[(x - 4)(x^2 - 9)] = 3x[(x - 4)(x - 3)(x + 3)] = 0$. Set each factor equal to 0 to solve for the roots.

6 Find the roots (solutions) of the polynomial $x^5 - 16x^3 + x^2 - 16 = 0$. The answer is **x = 4, –4, –1.**

The polynomial factors by grouping: $x^3(x^2 - 16) + 1(x^2 - 16) = (x^2 - 16)(x^3 + 1) = (x - 4)(x + 4)(x + 1)(x^2 - x + 1) = 0$. The first three factors give you the real roots. The last factor is a quadratic that has no real solution when you set it equal to 0.

Note: You don't really have to factor $x^3 + 1$ in problem 6 to find the root. If you just set $x^3 + 1$ equal to 0, you get $x^3 = -1$, and taking the cube root of both sides gives you the solution –1. The factored form simply shows you how this problem could have had five roots, but not all are real numbers in this case.

7 Find the roots (solutions) of the polynomial $x^6 + 9x^3 + 8 = 0$. The answer is **x = –2, –1.**

The polynomial is quadratic-like (see Chapter 2 if you need more information on factoring quadratic-like trinomials). It factors into $(x^3 + 8)(x^3 + 1) = 0$. Setting each factor equal to 0, you get the two roots.

8 Find the roots (solutions) of the polynomial $36x^5 - 13x^3 + x = 0$. The answer is $x = 0, \frac{1}{3}, -\frac{1}{3}, \frac{1}{2}, -\frac{1}{2}$.

First factor x out of each term to get $x(36x^4 - 13x^2 + 1) = 0$. The trinomial factors, giving you $x(9x^2 - 1)(4x^2 - 1) = 0$. Each binomial is the difference of squares (see Chapter 2), so all the binomials factor. For the final factorization, you end up with $x(3x - 1)(3x + 1)(2x - 1)(2x + 1) = 0$. Setting each factor equal to 0 gives you the five different solutions.

9 Determine where the polynomial $y = 6x^2 - 6x$ is positive and where it's negative. The answer is **positive: (–∞, 0) ∪ (1, ∞); negative: (0, 1).**

First factor the polynomial to get $6x(x - 1)$. The zeros are at 0 and 1. Set up a number line with the 0 and 1, placed in that order. Check the signs of the factors in the three different intervals: less than 0, between 0 and 1, and then greater than 1. The products of the factors are negative for all values of x between 0 and 1 and positive everywhere else.

10 Determine where the polynomial $y = -2x^2 - 7x + 15$ is positive and where it's negative. The answer is **positive:** $\left(-5, \frac{3}{2}\right)$**; negative:** $\left(-\infty, -5\right) \cup \left(\frac{3}{2}, \infty\right)$**.**

Factor to get $y = -(2x - 3)(x + 5)$. Create a number line with the roots in their correct order. Checking the signs in the three intervals, you find that the function is positive only between the two zeros and negative elsewhere.

11 Determine where the polynomial $y = (x + 3)(x - 2)(x - 5)^2$ is positive and where it's negative. The answer is **positive: (–∞,–3) ∪ (2, 5) ∪ (5, ∞); negative: (–3, 2).**

Place the numbers representing the roots or zeros on the number line; then check out the signs of the factors and products of the factors to determine the sign of the product in each interval.

12 Determine where the polynomial $y = x^2(x + 3)^2(x - 4)^4$ is positive and where it's negative. The answer is **positive: (–∞, –3) ∪ (–3, 0) ∪ (0, 4) ∪ (4, ∞); negative: never.**

It's probably easiest just to say that this polynomial is positive *except* at the three zeros. The even powers make the binomials positive, except when they're 0.

13 Sketch the graph of $y = (x + 4)(x + 2)(x - 3)$.

The graph continues on to get higher and higher as x gets larger and larger.

14 Sketch the graph of $y = x(x - 3)^2(x - 6)$.

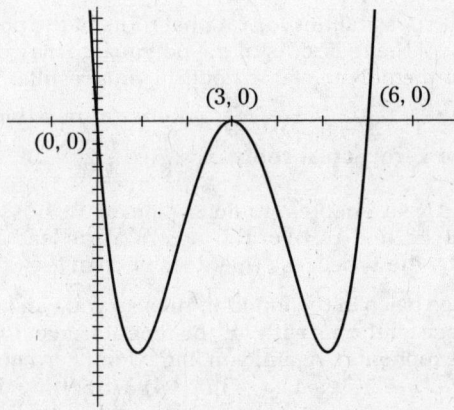

A double root is at $x = 3$, which is why the curve doesn't cross the axis at that point.

15 Sketch the graph of $y = -2x^3 - 6x^2 + 8x$.

The y-values approach negative infinity as the x-values get larger and larger.

16 Sketch the graph of $y = x^4 - 26x^2 + 25$.

The curve is symmetric with respect to the y-axis because it's an even function. (Chapter 5 can tell you more about even and odd functions.)

17 Determine all the possibilities for rational roots of the polynomial $9x^4 - 37x^2 + 4 = 0$. Then determine how many of the real roots of the polynomial may be positive and how many may be negative. Factor the polynomial to confirm your results. The answer is **possible rational roots:** $\pm 1, \pm 2, \pm 4, \pm\frac{1}{3}, \pm\frac{2}{3}, \pm\frac{4}{3}, \pm\frac{1}{9}, \pm\frac{2}{9}, \pm\frac{4}{9}$; **number of possible real roots — positive: two or zero, negative: two or zero; actual roots:** $x = \frac{1}{3}, -\frac{1}{3}, 2, -2$.

Use the Rational Root Theorem to determine all the possible rational roots. Put the factors of the constant, $\pm 1, \pm 2$, and ± 4, over the factors of the lead coefficent, 1, 3, and 9. You don't really have to divide by the 1, because that doesn't change the value.

The signs change twice in the function rule, so you can have two or zero positive real roots. When you replace all the x's with $-x$, the function rule doesn't change at all (because of the even-powered exponents). Again, you find two or zero negative real roots. Factoring the polynomial, you get $(3x - 1)(3x + 1)(x - 4)(x + 4) = 0$. Setting the factors equal to 0, you get the roots $x = \frac{1}{3}, -\frac{1}{3}, 2, -2$. This function has exactly two positive and two negative real roots. Notice that all four roots are in the list created using the Rational Root Theorem.

18 Determine all the possibilities for rational roots of the polynomial $2x^4 - 3x^3 - 18x^2 + 27x = 0$. Then determine how many of the real roots of the polynomial may be positive and how many may be negative. Factor the polynomial to confirm your results. The answer is **possible rational roots:** $\pm 1, \pm 3, \pm 9, \pm 27, \pm\frac{1}{2}, \pm\frac{3}{2}, \pm\frac{9}{2}, \pm\frac{27}{2}$; **number of possible real roots — positive: two or zero, negative: one; actual roots:** $x = 0, \frac{3}{2}, 3, -3$.

First factor out the x from each term to get $x(2x^3 - 3x^2 - 18x + 27) = 0$. Then use the Rational Root Theorem on the terms in the parentheses to determine all the possible rational roots. Put the factors of the constant, $\pm 1, \pm 3, \pm 9$, and ± 27, over the factors of the lead coefficient, 1 and 2.

You have are two sign changes on the terms within the parentheses. Now look at the polynomial inside the parentheses, and change each x to a $-x$. You get $2(-x)^3 - 3(-x)^2 - 18(-x) + 27 = -2x^3 - 3x^2 + 18x + 27$. You find only one sign change. Factoring the polynomial inside the parentheses by grouping, $x[x^2(2x - 3) - 9(2x - 3)] = x[(2x - 3)(x^2 - 9)] = x[(2x - 3)(x - 3)(x + 3)] = 0$. The roots of the entire polynomial are $0, \frac{3}{2}, 3, -3$. Two of the roots are positive; one is negative.

The number 0 is also a root, but 0 has no sign — it's neither positive nor negative, so it doesn't show up when you count possible positive and negative roots.

19 Determine all the possibilities for rational roots of the polynomial $x^4 - 4x^3 + 6x^2 - 4x + 1 = 0$. Then determine how many of the real roots of the polynomial may be positive and how many may be negative. Factor the polynomial to confirm your results. The answer is **possible rational roots:** ± 1; **number of possible real roots — positive: four or two or zero, negative: zero; actual roots:** $x = 1, 1, 1, 1$ (a quadruple root).

Using the Rational Root Theorem, you divide the factors of the constant, 1, by the factors of the lead coefficient, also a 1. That step gives you only two different possibilities for rational roots: 1 and -1.

The signs change four times in the original polynomial. Replacing each x with $-x$, you get $x^4 + 4x^3 + 6x^2 + 4x + 1 = 0$. The signs never change. The polynomial is the fourth power of the binomial $(x - 1)$, so it factors into $(x - 1)^4 = 0$, and the roots are 1, 1, 1, and 1. There are four positive roots (all the same number, of course).

20 Determine all the possibilities for rational roots of the polynomial $x^5 + 8x^4 + 7x^3 - 25x^3 - 200x^2 - 175x = 0$. Then determine how many of the real roots of the polynomial may be positive and how many may be negative. Factor the polynomial to confirm your results. The answer is **possible rational roots:** $\pm 1, \pm 5, \pm 7, \pm 25, \pm 35, \pm 175$; **number of possible real roots — positive: one, negative: three or one; actual roots:** $x = -7, -1, 0, 0, 0, 5, -5$.

After combining like terms and factoring out x, you get $x(x^4 + 8x^3 - 18x^2 - 200x - 175) = 0$. Apply the Rational Root Theorem to the terms in the parentheses. The lead coefficient is 1, so the only rational possibilities come from the factors of the constant, $\pm 1, \pm 5, \pm 7, \pm 25, \pm 35$, and ± 175.

The polynomial in the parentheses has just one sign change. When you replace each x in the parentheses with $-x$, you get $x^4 - 8x^3 - 18x^2 + 200x - 175$, and you find three sign changes. To factor this polynomial, go back to the original format where a "convenient" arrangement allows you to factor x^3 out of each of the first three terms and $-25x$ out of the last three terms. You get $x^3(x^2 + 8x + 7) - 25x(x^2 + 8x + 7) = (x^2 + 8x + 7)(x^3 - 25x) = (x + 7)(x + 1)x(x^2 - 25) = (x + 7)(x + 1)$ $x(x - 5)(x + 5)$. Setting this factorization equal to 0, you get roots of $-7, -1, 0, 0, 0, 5, -5$. Three roots are negative, one is positive, three are 0.

21 Determine all the possibilities for rational roots of the polynomial $3x^3 + 4x^2 + 30x + 40 = 0$. Then determine how many of the real roots of the polynomial may be positive and how many may be negative. Factor the polynomial to confirm your results. The answer is **possible rational roots: $\pm 1, \pm 2, \pm 4, \pm 5, \pm 8, \pm 10, \pm 20, \pm 40, \pm\frac{1}{3}, \pm\frac{2}{3}, \pm\frac{4}{3}, \pm\frac{5}{3}, \pm\frac{8}{3}, \pm\frac{10}{3}, \pm\frac{20}{3}, \pm\frac{40}{3}$; number of possible**

real roots: positive: zero, negative: three or one; actual root: $x = -\frac{4}{3}$.

Find the possible rational roots using the Rational Root Theorem. Divide the factors of the constant, $\pm 1, \pm 2, \pm 4, \pm 5, \pm 8, \pm 10, \pm 20$, and ± 40 by the factors of the lead coefficient, 1 and 3.

The original polynomial has no sign changes, but replacing each x with $-x$, you get $-3x^3 + 4x^2 - 30x + 40 = 0$. The sign changes three times, so you can have three or one negative roots. Checking the roots, you factor by grouping to get $x^2(3x + 4) + 10(3x + 4) = (3x + 4)(x^2 + 10) = 0$. The only real root is $x = -\frac{4}{3}$. The second factor has no solution when you set it equal to 0.

22 Determine all the possibilities for rational roots of the polynomial $x^6 - 16x^3 + 64 = 0$. Then determine how many of the real roots of the polynomial may be positive and how many may be negative. Factor the polynomial to confirm your results. The answer is **possible rational roots: $\pm 1, \pm 2, \pm 4, \pm 8, \pm 16, \pm 32, \pm 64$; number of possible real roots — positive: two or zero, negative: zero; actual roots: $x = 2, 2$.**

Using the Rational Root Theorem, you divide the factors of the constant, $\pm 1, \pm 2, \pm 4, \pm 8, \pm 16, \pm 32$, and ± 64, by the factors of the lead coefficient. The lead coefficient is 1, so dividing doesn't yield you any more possibilities for rational roots.

The sign changes twice in the polynomial, so you can have up to two positive roots. Replacing each x with $-x$, you get $x^6 + 16x^3 + 64$. You have no sign change here, so the roots can't be negative. The polynomial factors into $(x^3 - 8)^2 = 0$. The only solutions are 2 and 2 (two of the same 2s, too).

23 Use synthetic division to divide $(9x^4 - 37x^2 + 4) \div (x - 2)$. The answer is $9x^3 + 18x^2 - x - 2$.

The factor $x - 2$ divides evenly into the polynomial. There's no remainder.

$$
\begin{array}{r|rrrrr}
2 & 9 & 0 & -37 & 0 & 4 \\
  &   & 18 & 36 & -2 & -4 \\
\hline
  & 9 & 1 & -1 & -2 & 0
\end{array}
$$

24 Use synthetic division to divide $(2x^4 + 3x^3 + x^2 - 3x + 5) \div (x + 1)$. The answer is $2x^3 + x^2 - 3x + \dfrac{8}{x + 1}$.

Be sure to change the dividing number to -1. The remainder is 8, so you write it over the divisor.

$$
\begin{array}{r|rrrrr}
-1 & 2 & 3 & 1 & -3 & 5 \\
   &   & -2 & -1 & 0 & 3 \\
\hline
   & 2 & 1 & 0 & -3 & 8
\end{array}
$$

25 Use synthetic division to divide $(3x^3 + 4x^2 - 2x + 10) \div (x + 2)$. The answer is $3x^2 - 2x + 2 + \dfrac{6}{x + 2}$. The remainder is 6. Write it over the divisor, $x + 2$.

$$
\begin{array}{r|rrrr}
-2 & 3 & 4 & -2 & 10 \\
   &   & -6 & 4 & -4 \\
\hline
   & 3 & -2 & 2 & 6
\end{array}
$$

26 Use synthetic division to divide $(x^4 - 12x^3 + 54x^2 - 108x + 81) \div (x - 3)$. The answer is $x^3 - 9x^2 + 27x - 27$.

The binomial divides evenly into the polynomial, so you get no remainder in this division.

$$
\begin{array}{r|rrrrr}
3 & 1 & -12 & 54 & -108 & 81 \\
  &   & 3 & -27 & 81 & -81 \\
\hline
  & 1 & -9 & 27 & -27 & 0
\end{array}
$$

27 Find $f(2)$ if $f(x) = x^5 - 3x^4 + 4x^3 - 6x^2 + 5x - 1$. The answer is $f(2) = 1$.

Use the Remainder Theorem and synthetic division to find the value of the function. The divisor in the synthetic division is 2; your answer is the remainder, the last element on the bottom:

$$
\begin{array}{r|rrrrrr}
2 & 1 & -3 & 4 & -6 & 5 & -1 \\
  &   & 2 & -2 & 4 & -4 & 2 \\
\hline
  & 1 & -1 & 2 & -2 & 1 & 1
\end{array}
$$

28 Find $f(-2)$ if $f(x) = 6x^{10} - 16x^6 + 150x^5 - 80x^2$. The answer is $f(-2) = 0$.

Use the Remainder Theorem and synthetic division. Don't forget to put 0's in for the missing terms in the polynomial — all the way down to the constant term.

$$
\begin{array}{r|rrrrrrrrrrr}
-2 & 6 & 0 & 0 & 0 & -16 & 150 & 0 & 0 & -80 & 0 & 0 \\
   &   & -12 & 24 & -48 & 96 & -160 & 20 & -40 & 80 & 0 & 0 \\
\hline
   & 6 & -12 & 24 & -48 & 80 & -10 & 20 & -40 & 0 & 0 & 0
\end{array}
$$

29 If the roots of $f(x)$ are 1, 2, –2, 3, 3, 3, 4, and 4, then write a function rule for $f(x)$. The answer is $f(x) = k(x - 1)(x - 2)(x + 2)(x - 3)^3(x - 4)^2$.

The root 3 appears three times, and the root 4 appears twice. Write the corresponding factors with exponents. The k represents some constant multiplier — actually, any constant multiplier except 0. The roots will be the same no matter what constant you multiply the polynomial by.

30 If the roots of $f(x)$ are 0, 0, 0, 1, 1, 1, 1, –2, –3, and –4, then write a function rule for $f(x)$. The answer is $f(x) = kx^3(x - 1)^4(x + 2)(x + 3)(x + 4)$.

The root 0 appears three times, and the root 1 appears four times. Use exponents to reflect these multiple roots.

31 If the roots of $f(x)$ are $-1, -1, 1, \dfrac{1}{2}, \dfrac{3}{4}, -\dfrac{7}{8}$, and $-\dfrac{7}{8}$, then write a function rule for $f(x)$. The answer is $f(x) = k(x + 1)^2(2x - 1)(4x - 3)(8x + 7)^2$.

The root –1 appears twice. Use the exponent to reflect this. The fractional roots have factors that use the denominator of the fraction as a coefficient of x.

Another way to look at this problem is to simplify the equation related to the fractional root. For instance, you can simplify the equation for the root $x = \dfrac{3}{4}$ by multiplying each side of the equation by 4 to get $4x = 3$. Next, subtract 3 from each side to get $4x - 3 = 0$. The binomial on the left is the factor of the polynomial.

Chapter 8

Acting Rationally with Functions

- -

In This Chapter

▶ Showing domain and range with asymptotes of rational functions

▶ Checking out removable discontinuities

▶ Going to the limit with limits

▶ Graphing rational functions

- -

A *rational number* is a number that has a fractional basis — you can write it as a fraction with an integer in both the numerator and the denominator. When the denominator of a rational number is a 1, then the number is an integer, too. For example, $\frac{4}{3}, \frac{10}{2}$ or $-\frac{6}{1}$ are rational numbers. The 10 divided by 2 is equal to the integer 5, and the negative 6 divided by 1 is an integer. All integers are rational numbers, but not all rational numbers are integers.

You can construct a *rational function* by putting a polynomial in both the numerator and the denominator. For example, $\frac{x^2 - 3}{x^3 + 2x^2 - 5x - 1}$ and $\frac{4}{x^2 - 3x + 1}$ are rational functions. As you see, a rational function may alternately have a constant in the numerator and a polynomial in the denominator — mathematicians consider the constant in the numerator to be a polynomial with just the constant term in it.

In this chapter, you use the properties of rational functions that arise because of their fractional nature. These functions have issues of domain and range, and the graphs can get quite interesting with asymptotes (lines the functions follow but generally aren't allowed to cross) here, there, and everywhere.

Determining Domain and Intercepts of Rational Functions

The *domain* of a function consists of all the real numbers that you can use as input values. You're rather selective about domain values when working with rational functions because you can't let the denominator equal 0. The number 0 comes in very handy in many cases, but it can cause problems in rational functions.

To find the intercepts of rational functions, you use the same processes as with any other function. Replace all the x's with 0 to solve for the y-intercept; set the function rule equal to 0 and solve for x for the x-intercepts.

Q. Determine the domain and intercepts of the function $y = \dfrac{x^2 - 5x - 6}{x^3 - 4x}$

A. **Domain: all real numbers except for 0, 2, and –2 (in symbols, $x \neq 0, 2, -2$); x-intercepts: (6, 0) and (–1, 0); y-intercept: none.** To determine all these answers, first factor the numerator and denominator of the function rule. You get $y = \dfrac{(x - 6)(x + 1)}{x(x - 2)(x + 2)}$. To find the domain, set the denominator equal to 0. Yes, that's generally a no-no, but this technique is a good way to solve for what you *don't* want. The denominator is 0 when x is 0, 2, or –2, so you have to eliminate those numbers from the domain. Another way to write the domain — in a more positive way than to say what it *can't* be — is to use interval notation (see Chapter 1 for details). The domain is $(-\infty, -2) \cup (-2, 0) \cup (0, 2) \cup (2, \infty)$. You can now probably see why the "everything except" wording is so popular.

To solve for the y-intercept, you usually replace all the x's with 0. But you aren't allowed to do that in this function, because 0 isn't in the domain. That's why this function has no y-intercept. To solve for the x-intercepts, set the function rule equal to 0: $0 = \dfrac{(x - 6)(x + 1)}{x(x - 2)(x + 2)}$. You can multiply each side of the equation by the denominator, and you have the equation $(x - 6)(x + 1) = 0$, whose solutions are $x = 6$ and –1. Those solutions are the x-values of the two x-intercepts.

1. Determine the domain and intercepts of the function $y = \dfrac{x^2 - 16}{x^2 - 3x + 2}$

Solve It

2. Determine the domain and intercepts of the function $y = \dfrac{2x^2 - 2}{(x + 1)^2}$

Solve It

3. Determine the domain and intercepts of the function $y = \dfrac{x(x - 3)(x + 4)(x + 6)}{(x^2 - 25)(x^2 - 1)}$

Solve It

4. Determine the domain and intercepts of the function $y = \dfrac{5x^3}{x^4 + 1}$

Solve It

Introducing Vertical and Horizontal Asymptotes

An asymptote is something that isn't physically there — sort of like your conscience or your make-believe playmate. This line helps determine actions (much like a conscience) or the shapes of things, but it isn't really a part of the graph (or the reality of the moment). An *asymptote* is usually a ghostlike line that helps you graph a rational function. As the graph moves along toward an asymptote, it gets closer and closer to the asymptote but never touches it. Thus, the asymptote helps determine where the graph of the function can go or not go. It gives a general shape or direction for what the graph does way out there where you can't see it.

Vertical asymptotes mark places where the function has no domain. They're vertical lines drawn in lightly or with dashes to show they're not part of the graph. You solve for the equations of the vertical asymptotes by setting the denominator of the rational function equal to 0.

Note: If the numerator and denominator have a common factor, you may have a hole in the graph rather than an asymptote. For more information, see the upcoming "Removing Discontinuities" section.

Horizontal asymptotes indicate what happens to the curve as the *x*-values get very large or very small. To find the equation of a horizontal asymptote, use the following rule:

- ✔ If the highest variable power in the function equation occurs in the numerator only, the function has no horizontal asymptote (though it may have an oblique asymptote — see the next section).

- ✔ If the highest variable power in the function equation occurs in the denominator only, the equation of the horizontal asymptote is $y = 0$ (the *x*-axis).

- ✔ If both the numerator and denominator have terms with the highest variable power in them, take the coefficients of those terms with the highest power and make a fraction of them. The equation of the horizontal asymptote is $y = $ *the fraction you formed*.

Q. What are the asymptotes of the function $y = \dfrac{3x^2 + 9x - 12}{x^2 - 25}$?

A. **Vertical asymptotes:** $x = 5$ **and** $x = -5$; **horizontal asymptote:** $y = 3$. To solve for these equations, first factor the denominator and set it equal to 0: $x^2 - 25 = (x - 5)(x + 5) = 0$. The two solutions are $x = 5$ and $x = -5$. These numbers are the two values that you can't include in the domain, and the equations are the vertical asymptotes. To determine the horizontal asymptote, look at the original equation. The highest variable power is 2. You find squared terms in both the numerator and the denominator, so make a fraction of the coefficients of the x^2 terms: $\dfrac{\text{coefficient of } 3x^2}{\text{coefficient of } x^2} = \dfrac{3}{1}$. Put a *y* in front of the reduced equation, and you get $y = 3$.

Q. What are the asymptotes of the function $y = \dfrac{x^2 - 1}{x^4 - 81}$?

A. **Vertical asymptotes:** $x = 3$ **and** $x = -3$; **horizontal asymptote:** $y = 0$. Factor the denominator and set it equal to 0: $x^4 - 81 = (x^2 - 9)(x^2 + 9) = (x - 3)(x + 3)(x^2 + 9) = 0$. The two solutions are $x = 3$ and $x = -3$. The last factor never equals 0. You can't include the numbers 3 and –3 in the domain, so the solutions are the equations of the vertical asymptotes. To determine the horizontal asymptote, look at the original equation. The highest power is 2 in the numerator and 4 in the denominator. The highest variable power in the function rule is 4, and it occurs only in the denominator, so the equation of the horizontal asymptote is $y = 0$.

5. Determine the equations of the vertical and horizontal asymptotes of the function

$$y = \frac{x^2 - 16}{x^2 - 3x + 2}$$

Solve It

6. Determine the equations of the vertical and horizontal asymptotes of the function

$$y = \frac{9x}{x^2 - 9}$$

Solve It

7. Determine the equations of the vertical and horizontal asymptotes of the function

$$y = \frac{2x^2 - x - 3}{3x^2 - 7x - 20}$$

Solve It

8. Determine the equations of the vertical and horizontal asymptotes of the function

$$y = \frac{x^2 - 3x - 10}{x^2 - 5x - 6}$$

Solve It

Getting a New Slant with Oblique Asymptotes

An *oblique* or slant asymptote acts much like its cousins, the vertical and horizontal asymptotes. In other words, it helps you determine the ultimate direction or shape of the graph of a rational function. An oblique asymptote sometimes occurs when you have no horizontal asymptote. This kind of asymptote takes special circumstances, but the equations of these asymptotes are relatively easy to find when they do occur.

The rule for oblique asymptotes is that if the highest variable power in a rational function occurs in the numerator — and if that power is *exactly* one more than the power in the denominator — then the function has an oblique asymptote. You can find the equation of the oblique asymptote by dividing the numerator of the function rule by the denominator and using the first two terms in the quotient as the equation of the line.

Q. Find the equation of the oblique asymptote in the function $y = \frac{x^3 - 3x^2 + 3x + 1}{x^2 - 5x + 6}$

A. **$y = x + 2$.** To find this equation, you have to divide the denominator of the function rule into the numerator. This step requires *long division*. You can't use synthetic division (which I cover in Chapter 7) because the divisor isn't a binomial in the form $x - a$. Here's what the long division looks like:

$$
\begin{array}{r}
x + 2 \\
x^2 - 5x + 6 \overline{)\, x^3 - 3x^2 + 3x - 1} \\
\underline{x^3 - 5x^2 + 6x} \\
+ 2x^2 - 3x - 1 \\
\underline{2x^2 - 10x + 12} \\
7x - 13
\end{array}
$$

Ignore the remainder, and just use the first two terms in the quotient.

9. Find the equation of the oblique asymptote in the function $y = \dfrac{x^2 - 5x + 4}{x - 3}$

Solve It

10. Find the equation of the oblique asymptote in the function $y = \dfrac{x^2}{x - 1}$

Solve It

11. Find the equation of the oblique asymptote in the function $y = \dfrac{x^3 - 3x^2 - 4x + 1}{x^2 - 2x + 1}$

Solve It

12. Find the equation of the oblique asymptote in the function $y = \dfrac{-3x^4 + 4x^3 + 6x^2 + 4x + 2}{x^3 + 3x^2 + 3x + 1}$

Solve It

Removing Discontinuities

A function is *discontinuous* at a point or value of x where the function doesn't have a domain value. Discontinuities can come in the form of huge intervals of numbers, and they can also occur at just one particular number while everything on either side of that number works. Refer to the earlier section "Determining Domain and Intercepts" to see how you can have a discontinuity at any x-value that makes the denominator of a rational function equal to 0.

A *removable discontinuity,* which looks like a hole in the graph, is a bit of a misnomer. You can rewrite a function so that the function appears to no longer have a discontinuity at a particular point — in other words, you can simplify the function rule so you couldn't possibly get a 0 in the denominator. But the new form is just for appearances. Some functions have one or more removable discontinuities, but the discontinuity is still present, even after you've factored it out. You can't forget or ignore it. If you've factored out a discontinuity, you have to limit the domain of the new function rule ($x \neq ?$). The changed form in the function is just for convenience in evaluating and graphing.

If $y = \dfrac{f(x)}{g(x)}$ is equal to $\dfrac{0}{0}$ for some value of $x = c$, and if $(x - c)$ is a factor of both $f(x)$ and $g(x)$, then c is a *removable discontinuity* of the rational function.

Q. Determine where the function $y = \dfrac{x^2 + 4x + 4}{x^2 - 4}$ has a removable discontinuity. Remove the discontinuity and rewrite the function rule.

A. **Removable discontinuity: $x = -2$; revised function rule:** $y = \dfrac{x+2}{x-2}$, $x \neq -2$. If you let $x = -2$, you get $y = \dfrac{(-2)^2 + 4(-2) + 4}{(-2)^2 - 4} =$

$\dfrac{4 - 8 + 4}{4 - 4} = \dfrac{0}{0}$. Factor the numerator and denominator to get $y = \dfrac{(x+2)(x+2)}{(x-2)(x+2)}$. The factor $(x + 2)$ occurs in both, so you find a removable discontinuity at $x = -2$. Divide the numerator and denominator by the common factor to get $y = \dfrac{(x+2)\cancel{(x+2)}}{(x-2)\cancel{(x+2)}} = \dfrac{x+2}{x-2}$. You can write the original function as $y = \dfrac{x+2}{x-2}$, $x \neq -2$.

---

13. Determine where the function $y = \dfrac{x^2 - 7x - 18}{x^2 - 17x + 72}$ has a removable discontinuity. Remove the discontinuity and rewrite the function rule.

Solve It

14. Determine where the function $y = \dfrac{x^3 - 3x^2 + 2x}{x^3 - 4x}$ has a removable discontinuity. Remove the discontinuity and rewrite the function rule.

Solve It

---

15. Determine where the function $y = \dfrac{x^3 + 3x^2 - 4x - 12}{x^3 + 3x^2 - x - 3}$ has a removable discontinuity. Remove the discontinuity and rewrite the function rule.

Solve It

16. Determine where the function $y = \dfrac{x^3 - 3x^2 - 16x - 12}{x^3 - 4x^2 - 11x - 6}$ has a removable discontinuity. Remove the discontinuity and rewrite the function rule.

Solve It

Going the Limit: Limits at a Number and Infinity

Finding limits, which is especially important in calculus, is all about testing boundaries. The *limit* of a function is a y-value that the function approaches from both sides of a particular x-value. The x-value may or may not be in the domain of the function, but you can have values in the domain on either side of it (really close to it).

To find the limit of $f(x)$ at the x-value a, replace all the x's in the function rule with a and simplify. If you get a fraction with 0's in the numerator and denominator, remove the discontinuity, if possible (see the preceding section called "Removing Discontinuities") and evaluate $f(a)$ again. The notation for finding a limit at a is $\lim_{x \to a} f(x) = L$, where L is the y-value or limit.

To find the limit at infinity, replace all the x's with ∞ and simplify. If you get a fraction with ∞ in both numerator and denominator, then divide each term in the fraction by the highest variable power and evaluate again.

The key to solving the limits of infinity is usually to recall that the limit of a constant divided by infinity is 0.

Q. Find the limit of $f(x) = \dfrac{x^2 - 5x + 4}{2x^2 - 11x + 12}$ at $x = 4$

A. $\dfrac{3}{5}$. Replacing each x with 4 results in a fraction with 0's in both the numerator and denominator: $\lim_{x \to 4} \dfrac{x^2 - 5x + 4}{2x^2 - 11x + 12} =$.
$\dfrac{4^2 - 5(4) + 4}{2(4)^2 - 11(4) + 12} = \dfrac{16 - 20 + 4}{32 - 44 + 12} = \dfrac{0}{0}$.
factor and reduce to get $\lim_{x \to 4} \dfrac{x^2 - 5x + 4}{2x^2 - 11x + 12} =$
$\lim_{x \to 4} \dfrac{(x-4)(x-1)}{(2x-3)(x-4)} = \lim_{x \to 4} \dfrac{x-1}{2x-3} = \dfrac{4-1}{8-3} =$.
Even though the function is undefined at $x = 4$, it has a limit of $\dfrac{3}{5}$, which means that the function values get really close to that limit or y-value when you use x's slightly smaller and slightly larger than 4.

Q. Find the limit of $f(x) = \dfrac{3x}{x - 4}$ at $x = 4$

A. **No limit.** The function has no limit at 4:
$\lim_{x \to 4} \dfrac{3x}{x - 4} = \dfrac{3(4)}{4 - 4} = \dfrac{12}{0}$. You can't reduce the function rule. You can get a 0 in the denominator but not in the numerator.

Q. Find the limit of $f(x) = \dfrac{6x^2 - 2x + 3}{2x^2 + 5x - 2}$ as $x \to \infty$ by dividing each term by x^2

A. **3.** When you plug in ∞ for x, you get
$\lim_{x \to \infty} \dfrac{6x^2 - 2x + 3}{2x^2 + 5x - 2} = \dfrac{6(\infty)^2 - 2\infty + 3}{2(\infty)^2 + 5\infty - 2} = \dfrac{\infty}{\infty}$. A fraction with infinity over infinity doesn't tell you a thing — just that something big is dividing something else big. You need to be able to compare these huge numbers, so divide each term in the function by

x^2 and evaluate again: $\lim_{x \to \infty} \dfrac{\dfrac{6x^2}{x^2} - \dfrac{2x}{x^2} + \dfrac{3}{x^2}}{\dfrac{2x^2}{x^2} + \dfrac{5x}{x^2} - \dfrac{2}{x^2}} =$

$\lim_{x \to \infty} \dfrac{6 - \dfrac{2}{x} + \dfrac{3}{x^2}}{2 + \dfrac{5}{x} - \dfrac{2}{x^2}} = \dfrac{6 - \dfrac{2}{\infty} + \dfrac{3}{\infty^2}}{2 + \dfrac{5}{\infty} - \dfrac{2}{\infty^2}} = \dfrac{6 - 0 + 0}{2 + 0 - 0} =$

$\dfrac{6 - 0 + 0}{2 + 0 - 0} = \dfrac{6}{2} = 3$. Whenever you divide a constant by infinity, that fraction equals 0; your limit for this function is 3.

Q. Find the limit of $f(x) = \dfrac{4x^2 - 11x + 2}{2x + 47}$ as $x \to \infty$

A. **No limit.** This function has no limit as x approaches infinity:

$\lim_{x \to \infty} \dfrac{4x^2 - 11x + 2}{2x + 47} = \lim_{x \to \infty} \dfrac{\dfrac{4x^2}{x^2} - \dfrac{11x}{x^2} + \dfrac{2}{x^2}}{\dfrac{2x}{x^2} + \dfrac{47}{x^2}}$

$= \lim_{x \to \infty} \dfrac{4 - \dfrac{11}{x} + \dfrac{2}{x^2}}{\dfrac{2}{x} + \dfrac{47}{x^2}} = \dfrac{4 - \dfrac{11}{\infty} + \dfrac{2}{\infty^2}}{\dfrac{2}{\infty} + \dfrac{47}{\infty^2}} = \dfrac{4 - 0 + 0}{0 + 0} =$

You can't divide by 0.

17. Find the limit of $f(x) = \dfrac{x^2+5}{x^2-2}$ at $x = 3$

Solve It

18. Find the limit of $f(x) = \dfrac{3x^2-5x-2}{x^2-3x+2}$ at $x = 2$

Solve It

19. Find the limit of $f(x) = \dfrac{15x^2+6x+9}{3x^2+7x+11}$ as $x \to \infty$

Solve It

20. Find the limit of $f(x) = \dfrac{4x^2+3x-1}{2x^3+9x+11}$ as $x \to \infty$

Solve It

21. Find the limit of $f(x) = \dfrac{x^2-1}{x^2-2x+1}$ at $x = 1$

Solve It

22. Find the limit of $f(x) = \dfrac{2-8x^2-5x^3}{4x^2+3x+1}$ as $x \to \infty$

Solve It

Graphing Rational Functions

A rational function is a polynomial divided by another polynomial. The rational function rule makes the separate parts (the polynomials) of the graph pretty much smooth curves. What distinguishes the graphs of rational functions from those of polynomial functions is that rational functions have breaks or gaps or discontinuities shown with vertical asymptotes, as well as horizontal and oblique asymptotes that show other patterns of behavior.

To graph a rational function, follow these steps:

1. **Determine all intercepts.**

2. **Determine the domain and, at the same time, the equations of the vertical asymptotes.**

3. **Determine the equations of any horizontal or oblique asymptotes.**

4. **Sketch in all the intercepts and asymptotes.**

5. **Solve for any other additional points you need to finish drawing the the graph.**

Q. Sketch the graph of the function $y=\dfrac{x^2-3x-10}{x^2-9}$, indicating all intercepts and asymptotes

A. **The y-intercept is $\left(0, \dfrac{10}{9}\right)$, and the x-intercepts are (5, 0) and (–2, 0).** The domain consists of all real numbers except for $x = 3$ and $x = -3$, which are the equations of the vertical asymptotes. The equation of the horizontal asymptote is $y = 1$. Figure 8-1a shows the intercepts and asymptotes sketched in. As you can see, these points alone aren't too much help with the graph. But here are some observations you can make: The graph of the function to the left of $x = -3$ has to be above the horizontal asymptote, because in that interval, no x-intercepts allow the curve to cross the axis. The graph to the right of $x = 3$ has to be below the horizontal asymptote, crossing the x-axis at the intercept. And the curve between the vertical asymptotes probably rises from left to right, crossing the horizontal asymptote but hugging the two vertical asymptotes. Substituting some other values of x into the function rule, additional points include (–9, 1.361), (–5, 1.875), (–2.8, –5.38), (1, 1.5), (2, 2.4), (4, –0.857), and (7, 0.45), which you can use to complete the sketch of the function, shown in Figure 8-1b.

Q. Sketch the graph of the function $y=\dfrac{x^2-3x-4}{x+2}$, indicating all intercepts and asymptotes

A. **The y-intercept is (0, –2), and the x-intercepts are (4, 0) and (–1, 0).** You have a vertical asymptote of $x = -2$ and an oblique asymptote of $y = x - 5$. Figure 8-2a shows those points and asymptotes sketched in. From these points, you get a sense of what's happening in the upper right area, but you need more help with the rest of the graph. You can choose additional points to help with the rest of the sketch, like (–8, –14), (–4, –12), and (–3, –14), shown in Figure 8-2b.

Figure 8-1:
Start with
some
points and
asymptotes
and fill in
the details.

Figure 8-2:
The vertical
and slant
asymptotes
give shape
to the curve.

23. Sketch the graph of the function $y = \frac{x+3}{x-1}$, indicating all intercepts and asymptotes

Solve It

24. Sketch the graph of the function $y = \frac{x}{x^2 - 5x - 6}$, indicating all intercepts and asymptotes

Solve It

25. Sketch the graph of the function $y = \frac{x^2 - 16}{x^2 + 2x - 15}$, indicating all intercepts and asymptotes

Solve It

26. Sketch the graph of the function $y = \frac{x^2 - 5x + 4}{x - 3}$, indicating all intercepts and asymptotes

Solve It

27. Sketch the graph of the function $y = \dfrac{2x^2 - 6x - 20}{x^3 + x^2 - 12x}$, indicating all intercepts and asymptotes

Solve It

28. Sketch the graph of the function $y = \dfrac{x^2 - 7x - 18}{x^2 - 17x + 72}$, indicating all intercepts and asymptotes

Solve It

Answers to Problems on Rational Functions

The following are the answers to the practice problems presented earlier in this chapter.

1 Determine the domain and intercepts of the function $y = \dfrac{x^2 - 16}{x^2 - 3x + 2}$. The answer is **domain: $(-\infty, 1) \cup (1, 2) \cup (2, \infty)$; intercepts: (0, –8), (4, 0), (–4, 0).**

The denominator factors into $(x - 1)(x - 2)$. Therefore, the domain consists of all the real numbers except for 1 and 2. Those values make the denominator equal 0. You can find the y-intercept by letting all the x's in the function rule equal 0 and simplifying the fraction. Determine the x-intercepts by setting the function equation equal to 0, multiplying each side of the equation by the denominator, and then solving the quadratic equation that results.

2 Determine the domain and intercepts of the function $y = \dfrac{2x^2 - 2}{(x + 1)^2}$. The answer is **domain: $(-\infty, -1) \cup (-1, \infty)$; intercepts: (0, –2), (1, 0).**

The domain contains all real numbers except for the number –1, because –1 makes the denominator equal to 0. Calculate the y-intercept by letting all the x's in the function rule equal 0 and solving for y. You can find the x-intercept *after* factoring and reducing the function rule. The factor $x + 1$ divides out, leaving $2(x - 1)$ in the numerator.

3 Determine the domain and intercepts of the function $y = \dfrac{x(x - 3)(x + 4)(x + 6)}{(x^2 - 25)(x^2 - 1)}$. The answer is **domain: $(-\infty, -5) \cup (-5, -1) \cup (-1, 1) \cup (1, 5) \cup (5, \infty)$; intercepts: (0, 0), (3, 0), (–4, 0), (–6, 0).**

The domain consists of all real numbers except for –5, –1, 1, and 5. The y-intercept is also an x-intercept at (0, 0). Find the other x-intercepts by setting the function equation equal to 0 and solving for x.

4 Determine the domain and intercepts of the function $y = \dfrac{5x^3}{x^4 + 1}$. The answer is **domain: all real numbers; intercept: (0, 0).**

The domain contains all real numbers. No x-value can make the denominator equal to 0, so you can use all x's. When $x = 0$, $y = 0$. The origin is the only intercept.

5 Determine the equations of the vertical and horizontal asymptotes of the function $y = \dfrac{x^2 - 16}{x^2 - 3x + 2}$. The answer is **vertical: $x = 1$, $x = 2$; horizontal: $y = 1$.**

Determine the vertical asymptotes by setting the denominator equal to 0 and solving for x. Find the horizontal asymptote by making a fraction of the coefficients of the x^2 terms and setting the result equal to y.

6 Determine the equations of the vertical and horizontal asymptotes of the function $y = \dfrac{9x}{x^2 - 9}$. The answer is **vertical: $x = 3$, $x = -3$; horizontal: $y = 0$.**

Find the vertical asymptotes by setting the denominator equal to 0 and solving for x. The horizontal asymptote is $y = 0$, because the highest power of the variable is in the denominator of the fraction.

7 Determine the equations of the vertical and horizontal asymptotes of the function $y = \dfrac{2x^2 - x - 3}{3x^2 - 7x - 20}$. The answer is **vertical: $x = 4$, $x = -\dfrac{5}{3}$; horizontal: $y = \dfrac{2}{3}$.**

You can get the vertical asymptotes by setting the denominator equal to 0 and solving for x: $0 = 3x^2 - 7x - 20 = (3x + 5)(x - 4)$; $x = -\dfrac{5}{3}$, 4. For the horizontal asymptote, make a fraction of the coefficients of the x^2 terms and set the result equal to y.

8 Determine the equations of the vertical and horizontal asymptotes of the function $y = \dfrac{x^2 - 3x - 10}{x^2 - 5x - 6}$. The answer is **vertical: $x = 6$, $x = -1$; horizontal: $y = 1$.**

Find the vertical asymptotes by setting the denominator equal to 0 and solving for x. Determine the horizontal asymptote by making a fraction of the coefficients of the x^2 terms and setting the result equal to y.

9 Find the equation of the oblique asymptote in the function $y = \dfrac{x^2 - 5x + 4}{x - 3}$. The answer is $y = x - 2$.

Use synthetic division or long division to divide the denominator into the numerator:

```
3| 1  -5   4
        3  -6
   ─────────────
   1  -2  -2
```

The first two terms in the quotient are the slope and y-intercept of the oblique asymptote's equation.

10 Find the equation of the oblique asymptote in the function $y = \dfrac{x^2}{x - 1}$. The answer is $y = x + 1$.

Use synthetic division or long division to divide the denominator into the numerator:

```
1| 1  0  0
       1  1
   ──────────
   1  1  1
```

The first two terms in the quotient are the slope and y-intercept of the oblique asymptote's equation.

11 Find the equation of the oblique asymptote in the function $y = \dfrac{x^3 - 3x^2 - 4x + 1}{x^2 - 2x + 1}$. The answer is $y = x - 1$.

Use long division to divide the denominator into the numerator:

$$
\begin{array}{r}
x - 1 \\
x^2 - 2x + 1 \overline{)\, x^3 - 3x^2 - 4x + 1} \\
\underline{x^3 - 2x^2 + x} \\
-x^2 - 5x + 1 \\
\underline{-x^2 + 2x - 1} \\
-7x + 2
\end{array}
$$

The first two terms in the quotient are the slope and y-intercept of the oblique asymptote's equation.

12 Find the equation of the oblique asymptote in the function $y = \dfrac{-3x^4 + 4x^3 + 6x^2 + 4x + 2}{x^3 + 3x^2 + 3x + 1}$. The answer is $y = -3x + 13$.

Use long division to divide the denominator into the numerator:

$$
\begin{array}{r}
-3x + 13 \\
x^3 + 3x^2 + 3x + 1 \overline{)\, -3x^4 + 4x^3 + 6x^2 + 4x + 2} \\
\underline{-3x^4 - 9x^3 - 9x^2 - 3x} \\
13x^3 + 15x^2 + 7x + 2 \\
\underline{13x^3 + 39x^2 + 39x + 13} \\
-24x^2 - 32x - 11
\end{array}
$$

The first two terms in the quotient are the slope and y-intercept of the oblique asymptote's equation.

13 Determine where the function $y = \dfrac{x^2 - 7x - 18}{x^2 - 17x + 72}$ has a removable discontinuity. Remove the discontinuity and rewrite the function rule. The answer is **removable discontinuity: $x = 9$; revised function rule: $y = \dfrac{x + 2}{x - 8}$, $x \neq 9$.**

The function rule factors, and you have a common factor of $x - 9$ in the numerator and denominator. Divide through by $x - 9$: $y = \dfrac{x^2 - 7x - 18}{x^2 - 17x + 72} = \dfrac{(x - 9)(x + 2)}{(x - 8)(x - 9)}$.

14 Determine where the function $y = \dfrac{x^3 - 3x^2 + 2x}{x^3 - 4x}$ has a removable discontinuity. Remove the discontinuity and rewrite the function rule. The answer is **removable discontinuities: $x = 0, 2$; revised function rule: $y = \dfrac{x - 1}{x + 2}$, $x \neq 0, 2$.**

The function rule factors, and you find two common factors: x and $x - 2$. Divide through by both common factors: $y = \dfrac{x^3 - 3x^2 + 2x}{x^3 - 4x} = \dfrac{x(x - 1)(x - 2)}{x(x - 2)(x + 2)}$.

15 Determine where the function $y = \dfrac{x^3 + 3x^2 - 4x - 12}{x^3 + 3x^2 - x - 3}$ has a removable discontinuity. Remove the discontinuity and rewrite the function rule. The answer is **removable discontinuity: $x = -3$; revised function rule: $y = \dfrac{x^2 - 4}{x^2 - 1}$, $x \neq -3$.**

The function rule factors, and you have a common factor of $x + 3$ in the numerator and denominator. Divide through by $x + 3$: $y = \dfrac{x^3 + 3x^2 - 4x - 12}{x^3 + 3x^2 - x - 3} = \dfrac{x^2(x + 3) - 4(x + 3)}{x^2(x + 3) - 1(x + 3)} = \dfrac{(x + 3)(x^2 - 4)}{(x + 3)(x^2 - 1)} = \dfrac{x^2 - 4}{x^2 - 1}$.

16 Determine where the function $y = \dfrac{x^3 - 3x^2 - 16x - 12}{x^3 - 4x^2 - 11x - 6}$ has a removable discontinuity. Remove the discontinuity and rewrite the function rule. The answer is **removable discontinuity: $x = -1, 6$; revised function rule: $y = \dfrac{x + 2}{x + 1}$, $x \neq -1, 6$.**

The function rule factors, and the numerator and denominator have two common factors: $x + 1$ and $x - 6$. Divide through by both common factors: $y = \dfrac{x^3 - 3x^2 - 16x - 12}{x^3 - 4x^2 - 11x - 6} = \dfrac{(x + 1)(x + 2)(x - 6)}{(x + 1)(x + 1)(x - 6)} = \dfrac{x + 2}{x + 1}$.

17 Find the limit of $f(x) = \dfrac{x^2 + 5}{x^2 - 2}$ at $x = 3$. The answer is **2.**

Replace each x with 3 to get $\lim\limits_{x \to 3} \dfrac{x^2 + 5}{x^2 - 2} = \dfrac{9 + 5}{9 - 2} = \dfrac{14}{7} = 2$.

18 Find the limit of $f(x) = \dfrac{3x^2 - 5x - 2}{x^2 - 3x + 2}$ at $x = 2$. The answer is **7.**

Replacing each x with 2 gives you a 0 in both numerator and denominator. Factor the numerator and denominator, divide out the common factor, and substitute the 2 into the new form:

$$\lim\limits_{x \to 2} \dfrac{3x^2 - 5x - 2}{x^2 - 3x + 2} = \lim\limits_{x \to 2} \dfrac{(3x + 1)(x - 2)}{(x - 2)(x - 1)} = \lim\limits_{x \to 2} \dfrac{3x + 1}{x - 1}$$

$$= \dfrac{6 + 1}{2 - 1} = \dfrac{7}{1} = 7$$

19 Find the limit of $f(x) = \dfrac{15x^2 + 6x + 9}{3x^2 + 7x + 11}$ as $x \to \infty$. The answer is **5**.

Divide each term by x^2 and then replace each x with ∞:

$$\lim_{x \to \infty} \frac{15x^2 + 6x + 9}{3x^2 + 7x + 11} = \lim_{x \to \infty} \frac{\dfrac{15x^2}{x^2} + \dfrac{6x}{x^2} + \dfrac{9}{x^2}}{\dfrac{3x^2}{x^2} + \dfrac{7x}{x^2} + \dfrac{11}{x^2}} = \lim_{x \to \infty} \frac{15 + \dfrac{6}{x} + \dfrac{9}{x^2}}{3 + \dfrac{7}{x} + \dfrac{11}{x^2}}$$

$$= \frac{15 + \dfrac{6}{\infty} + \dfrac{9}{\infty^2}}{3 + \dfrac{7}{\infty} + \dfrac{11}{\infty^2}} = \frac{15 + 0 + 0}{3 + 0 + 0} = \frac{15}{3} = 5$$

20 Find the limit of $f(x) = \dfrac{4x^2 + 3x - 1}{2x^3 + 9x + 11}$ as $x \to \infty$. The answer is **0**.

Divide each term by x^3, and then replace each x with ∞:

$$\lim_{x \to \infty} \frac{4x^2 + 3x - 1}{2x^3 + 9x + 11} = \lim_{x \to \infty} \frac{\dfrac{4x^2}{x^3} + \dfrac{3x}{x^3} - \dfrac{1}{x^3}}{\dfrac{2x^3}{x^3} + \dfrac{9x}{x^3} + \dfrac{11}{x^3}} = \lim_{x \to \infty} \frac{\dfrac{4}{x} + \dfrac{3}{x^2} - \dfrac{1}{x^3}}{2 + \dfrac{9}{x^2} + \dfrac{11}{x^3}}$$

$$= \frac{\dfrac{4}{\infty} + \dfrac{3}{\infty^2} - \dfrac{1}{\infty^3}}{2 + \dfrac{9}{\infty^2} + \dfrac{11}{\infty^3}} = \frac{0 + 0 - 0}{2 + 0 + 0} = \frac{0}{2} = 0$$

21 Find the limit of $f(x) = \dfrac{x^2 - 1}{x^2 - 2x + 1}$ at $x = 1$. **No limit.**

Even after factoring the numerator and denominator and reducing the fraction, you still get a 0 in the denominator when replacing the x's with 1. This function has no limit at $x = 1$:

$$\lim_{x \to 1} \frac{x^2 - 1}{x^2 - 2x + 1} = \lim_{x \to 1} \frac{(x - 1)(x + 1)}{(x - 1)(x - 1)} = \lim_{x \to 1} \frac{x + 1}{x - 1} = \frac{2}{0}$$

22 Find the limit of $f(x) = \dfrac{2 - 8x^2 - 5x^3}{4x^2 + 3x + 1}$ as $x \to \infty$. **No limit.**

After dividing each term by x^3 and replacing each x with ∞, you end up with a fraction that has 0 in the denominator. This function has no limit as x approaches infinity:

$$\lim_{x \to \infty} \frac{2 - 8x^2 - 5x^3}{4x^2 + 3x + 1} = \lim_{x \to \infty} \frac{\dfrac{2}{x^3} - \dfrac{8x^2}{x^3} - \dfrac{5x^3}{x^3}}{\dfrac{4x^2}{x^3} + \dfrac{3x}{x^3} + \dfrac{1}{x^3}} = \lim_{x \to \infty} \frac{\dfrac{2}{x^3} - \dfrac{8}{x} - 5}{\dfrac{4}{x} + \dfrac{3}{x^2} + \dfrac{1}{x^3}}$$

$$= \frac{\dfrac{2}{\infty^3} - \dfrac{8}{\infty} - 5}{\dfrac{4}{\infty} + \dfrac{3}{\infty^2} + \dfrac{1}{\infty^3}} = \frac{0 - 0 - 5}{0 + 0 - 0} = \frac{-5}{0}$$

23 Sketch the graph of the function $y = \dfrac{x + 3}{x - 1}$, indicating all intercepts and asymptotes.

The intercepts are $(0, -3)$ and $(-3, 0)$. The vertical asymptote is $x = 1$, and the horizontal asymptote is $y = 1$.

24 Sketch the graph of the function $y = \dfrac{x}{x^2 - 5x - 6}$, indicating all intercepts and asymptotes.

The only intercept is $(0, 0)$. The vertical asymptotes are $x = 6$ and $x = -1$, and the horizontal asymptote is $y = 0$.

25 Sketch the graph of the function $y = \dfrac{x^2 - 16}{x^2 + 2x - 15}$, indicating all intercepts and asymptotes.

The intercepts are $\left(0, \frac{16}{15}\right)$, $(4, 0)$, and $(-4, 0)$. The vertical asymptotes are $x = -5$ and $x = 3$, and the horizontal asymptote is $y = 1$.

26 Sketch the graph of the function $y = \dfrac{x^2 - 5x + 4}{x - 3}$, indicating all intercepts and asymptotes.

The intercepts are $\left(0, -\frac{4}{3}\right)$, $(4, 0)$, and $(1, 0)$. The vertical asymptote is $x = 3$, and the oblique asymptote is $y = x - 2$.

27 Sketch the graph of the function $y = \dfrac{2x^2 - 6x - 20}{x^3 + x^2 - 12x}$, indicating all intercepts and asymptotes.

The intercepts are $(5, 0)$ and $(-2, 0)$. The vertical asymptotes are $x = 0$, $x = 3$, and $x = -4$, and the horizontal asymptote is $y = 2$. Notice that the curve crosses the horizontal asymptote at about $x = 1.5$ and then again at $x = 2$.

Horizontal asymptotes tell you what happens as you approach positive or negative infinity — the curve will flatten out above or below the asymptote. But remember, curves *never* cross vertical asymptotes.

28   Sketch the graph of the function $y = \dfrac{x^2 - 7x - 18}{x^2 - 17x + 72}$, indicating all intercepts and asymptotes.

The intercepts are $\left(0, -\dfrac{1}{4}\right)$ and (–2, 0). The vertical asymptote is $x = 8$, and the horizontal asymptote is $y = 1$. This function has a removable discontinuity at $x = 9$, which you graph by placing a hollow point (empty circle) at (9, 11).

Chapter 9

Exposing Exponential and Logarithmic Functions

*E*xponential and logarithmic functions are related to one another, but they're quite different from the algebraic functions such as polynomials, radicals, and rationals. Algebraic functions all use the variable as the base of action. Exponential functions have the variable in the exponent (hence, the name), and logarithmic functions have the variable in the argument (there'll be no disagreement there — I explain arguments later in this chapter)!

Here, you practice using these new functions and simplify expressions that have exponential and logarithmic terms and factors in them. Solving equations of exponential and logarithmic functions takes some different types of twists and turns. Lastly, you get to graph exponential and logarithmic functions, which requires knowing where and how the C-shaped curve comes from the function equation. Read on.

Evaluating e-Expressions and Powers of e

When mathematicians and scientists assigned the letter e to represent the number 2.71828..., little did they know that there'd be e-mail, eBay, eMall, and so on. Look what they started! In the world of mathematics, the letter e is a constant value — slightly smaller than the number 3. Exponential and logarithmic functions are based mainly on the constant e, although people use other bases on occasion. The usual exponential rules apply for all the constant bases.

Here's a quick review of the rules you need:

▸ $b^x \cdot b^y = b^{x+y}$

▸ $b^x \div b^y = b^{x-y}$

▸ $(b^x)^y = b^{xy}$

▸ $b^0 = 1$

Q. Given the exponential function $f(x) = 5^{x+1}$, determine $f(-1)$, $f(0)$, and $f(2)$

A. $f(-1) = 1$; $f(0) = 5$; $f(2) = 125$. Substituting the input value for x in each case, $f(-1) = 5^{-1+1} = 5^0 = 1$, $f(0) = 5^{0+1} = 5^1 = 5$, and $f(2) = 5^{2+1} = 5^3 = 125$.

Q. Given the exponential function $f(x) = xe^{x+1}$, determine $f(-1)$, $f(0)$, and $f(2)$

A. $f(-1) = -1$; $f(0) = 0$; $f(2) = 2e^3$. Substituting the input value for x in each case, $f(-1) = -1e^{-1+1} = -1 \cdot e^0 = -1 \cdot 1 = -1$, $f(0) = 0 \cdot e^{0+1} = 0 \cdot e^1 = 0$, and $f(2) = 2 \cdot e^{2+1} = 2e^3$. You usually leave powers of e just as they are; don't bother finding a decimal approximation.

---

1. Given the exponential function $f(x) = 2^{x-1}$, determine $f(0)$, $f(1)$, and $f(2)$

Solve It

2. Given the exponential function $f(x) = 3^x(1 - 2 \cdot 3^x)$, determine $f(0)$, $f(1)$, and $f(2)$

Solve It

---

3. Given the exponential function $f(x) = e^x - 2e^{2x}$, determine $f(0)$ and $f(1)$

Solve It

4. Given the exponential function $f(x) = (e^x + 2)^2 - (e^x - 2)^2$, determine $f(0)$, $f(1)$, and $f(2)$

Solve It

---

5. Given the exponential function $f(x) = 2(e^{x+1})^2$, determine $f(0)$, $f(1)$, and $f(-1)$

Solve It

6. Given the exponential function $f(x) = x^e - e^x$, determine $f(0)$, $f(1)$, and $f(2)$

Solve It

Solving Exponential Equations

In the real world, people use exponential equations to show how something grows or decays. These little wonders can help you figure out how long radiation is going to stick around, estimate the age of an artifact from an archeological dig, or determine how many bacteria will be teeming in a petri dish in a couple hours. Or, as the next section shows you, they can help you track the cash in your bank account.

An *exponential equation* has an exponential expression in it. In an exponential equation, you may have 2^x or e^x or some other base with a variable as its exponent. In general, to solve an exponential equation, you want to adjust your equation so you can set two numbers that have the same base equal to one another, or you want to factor a quadratic-like trinomial (see Chapter 2) or some binomial so you can apply the multiplication property of zero.

Q. Solve for x in $4^{2x+1} = 8^{3x+4}$

A. $x = -2$. First change the 4 and 8 to powers of 2 so your bases match: $(2^2)^{2x+1} = (2^3)^{3x+4}$. Next, using the property of exponents that lets you raise a power to a power, you can write $2^{4x+2} = 2^{9x+12}$. Set the two exponents equal to one another, and solve the linear equation: $4x + 2 = 9x + 12$; $-5x = 10$; $x = -2$. In the original equation, if you replace the x with -2, you get $4^{-3} = 8^{-2}$, or $\frac{1}{4^3} = \frac{1}{8^2}$; $\frac{1}{64} = \frac{1}{64}$.

Q. Solve for x in $3^{2x} - 12(3^x) + 27 = 0$

A. $x = 1$ or $x = 2$. This equation is quadratic-like. Replace each 3^x with y and think of the equation as having the format $y^2 - 12y + 27 = (y-3)(y-9) = 0$. Then you can factor the original equation as $(3^x - 3)(3^x - 9) = 0$. Using the multiplication property of zero, you set each factor equal to zero and solve for x. When $3^x - 3 = 0$, $3^x = 3^1$, giving you $x = 1$. When $3^x - 9 = 0$, $3^x = 9 = 3^2$, so $x = 2$.

7. Solve for x in $2^x = 4^{x+1}$

Solve It

8. Solve for x in $9^{x-5} = \frac{1}{27^x}$

Solve It

9. Solve for x in $e^{4x-2} = e^{x+7}$

Solve It

10. Solve for x in $2^{2x} - 5 \cdot 2^x + 4 = 0$

Solve It

11. Solve for x in $e^x - 1 = 0$

Solve It

12. Solve for x in $xe^x + 5e^x = 0$

Solve It

Making Cents: Applying Compound Interest and Continuous Compounding

When you deposit your inheritance in an account that earns 4.5% interest, compounded quarterly, your money then earns interest on the deposit, and the interest earns interest, and the interest on the interest earns interest — well, I'm sure you get the point here. Compound interest is a very powerful instrument and the basis of the saving and borrowing that you do in your life.

The formula for determining the total amount of money you accumulate in an account after t years is $A = P\left(1 + \frac{r}{n}\right)^{nt}$, where A is the total amount of money accumulated, P is the principal (initial deposit), r is the rate of interest (written as a decimal), and n is the number of times each year that the interest is computed.

The formula for determining the total amount of money accumulated when it's compounded *continuously* is $A = Pe^{rt}$. People use the continuously compounding formula at times for convenience of figuring and also in other financial formulas that deal with the present and future value of an investment. Essentially, in continuous compounding, the n, or number of times you compound, is infinity — every little part of every second.

Q. Determine the total amount of money accumulated after 10 years if you deposit your inheritance of $50,000 in an account that earns 4.5% interest, compounded quarterly

A. **$78,218.84.** Using the compound interest formula, you get $A = 50,000\left(1 + \dfrac{0.045}{4}\right)^{4\,(10)} = 50,000(1.01125)^{40} \approx 78,218.84$. Your money grows to almost $80,000.

Q. Determine the total amount of money accumulated after 10 years if you deposit your inheritance of $50,000 in an account that earns 4.5% interest, compounded continuously

A. **$78,415.61.** Using the compound interest formula, you get $A = 50,000e^{0.045\,(10)} = 50,000e^{0.45} \approx 78,415.61$. Your money grows to almost $80,000, just as it does with the quarterly compounding. The difference is less than $200.

13. Determine the total amount of money accumulated after 20 years if you deposit $1,000,000 in an account that earns 5.25% interest, compounded monthly

Solve It

14. Determine the total amount of money accumulated after 2 years and 6 months if you deposit $25,000 in an account that earns 5% interest, compounded daily (use 365 days)

Solve It

15. Determine the total amount of money accumulated after 20 years if you deposit $1,000,000 in an account that earns 5.25% interest, compounded continuously

Solve It

16. Determine the total amount of money accumulated after 2 years and 6 months if you deposit $25,000 in an account that earns 5% interest, compounded continuously

Solve It

Checking Out the Properties of Logarithms

With logarithmic functions, numbers rise or fall quickly and then even out. Mathematicians, scientists, and others use logs to determine just how bad that earthquake was, just how loud that music coming from the car next to you is, and just how long ago that person became dearly departed. Logarithms work for all these situations and more. They aren't easy to use, so calculators or charts are necessary, but logarithms tell it like it is — or was.

A logarithmic function is the inverse of some exponential function. Here's the relationship between a logarithmic function with base b and an exponential function with base b:

$$y = \log_b x \leftrightarrow b^y = x$$

Mathematicians say that the log function $f(x) = \log_b x$ has a *base* of b and an *argument* x. When you see $\log_b x$, the output of the function is an exponent: This formula is saying, "The answer is the exponent that you put on b to get the x." Logarithms have some very handy properties — all linked to their respective exponential partner.

The most frequently used logarithmic bases are base 10 and base e. Base 10 logarithms are called the *common* logs, and base e logarithms are the *natural* logs. If you have a scientific or graphing calculator, you probably have a *log* button and an *ln* button. No subscripts are shown — they're just understood to be there. The *log* button is shorthand for base 10 or common logs, and the *ln* button is for the natural logs. I show the following rules for logs with base b (any base, including 10) and for natural logs.

Laws of Logarithms	For Log Base b	For Natural Log
Log of a product	$\log_b xy = \log_b x + \log_b y$	$\ln xy = \ln x + \ln y$
Log of a quotient	$\log_b \frac{x}{y} = \log_b x - \log_b y$	$\ln \frac{x}{y} = \ln x - \ln y$
Log of a power	$\log_b x^n = n\log_b x$	$\ln x^n = n\ln x$
Log of 1	$\log_b 1 = 0$	$\ln 1 = 0$
Log of the base	$\log_b b = 1$	$\ln e = 1$
Log of the reciprocal	$\log_b \frac{1}{y} = -\log_b y$	$\ln \frac{1}{y} = -\ln y$
Change of bases	$\log_b x = \log x \div \log b$	$\log_b x = \ln x \div \ln b$

Note: You don't need the change of bases rule for the problems in this book, but this rule can really come in handy if you want to use your calculator for logarithms with bases other than 10 or e.

People use the various logarithm rules to make the logarithmic expressions simpler and easier to handle when solving equations, doing higher mathematics, and using some scientific applications.

Q. Simplify the logarithmic expression using the laws of logarithms: $\log_4 \frac{1}{16}$

A. −2. You first apply the rule for the log of a reciprocal and then rewrite the 16 as a power of 4. (The number 16 is also a power of 2, but you want to have the base of the logarithm and the base of the number match so you can take advantage of another property of logs.) Use the rule for the log of a power to bring the 2 in front of the log. Then use the rule for the log of the base to end up multiplying the −2 by 1:

$$\log_4 \frac{1}{16} = -\log_4 16 = -\log_4 4^2 = -2\log_4 4 = -2(1) = -2.$$

Q. Show how the expression $\log_6 \frac{48}{x}$ is equivalent to $1 + \log_6 8 - \log_6 x$ using the laws of logarithms

A. First use the rule for the log of a quotient to write the logarithm as $\log_6 48 - \log_6 x$. Next, factor the number 48 into 6 times 8 to get $\log_6 6 + \log_6 8 - \log_6 x$. Finally, use the rule for the log of the base to change the first term to a 1.

---

17. Determine which of the following is equivalent to $\log 45$

   a. $\log 3 + \log 15$

   b. $\log 40 + \log 5$

   c. $5 \log 9$

   d. $45 \log 1$

Solve It

---

18. Determine which of the following is equivalent to $\log \frac{x}{2}$

   a. $\frac{\log x}{2}$

   b. $\log x + \log 2$

   c. $\log x - 2$

   d. $\log x - \log 2$

Solve It

---

19. Determine which of the following is equivalent to $\log_a \frac{1}{3}$

   a. $\log_a 1 + \log_a 3$

   b. $-\log_a 3$

   c. $-3 \log_a 1$

   d. $\frac{1}{\log_a 3}$

Solve It

---

20. Determine which of the following is equivalent to $\log_{10} 10$

   a. 10

   b. 0

   c. 1

   d. 100

Solve It

---

21. Determine which of the following is equivalent to $\log_6 1$

   a. 6

   b. 0

   c. 1

   d. -6

Solve It

---

22. Determine which of the following is equivalent to $\log 8$

   a. $\log 4^2$

   b. $3 \log 2$

   c. $2 \log 3$

   d. $4 \log 2$

Solve It

Presto-Chango: Expanding and Contracting Expressions with Log Functions

The logarithmic expression $\log_2 \dfrac{4x^3 \sqrt{9x+1}}{(3x^2+7)^5(x-5)}$ is pretty complicated with its numerator, denominator, radical, and powers. The beauty of logarithms is that the rules governing them allow you to pull apart and simplify these rather intricately constructed terms. For instance, using the rules of logarithms, you can write the expression as $2 + 3\log_2 x + \frac{1}{2}\log_2(9x+1) - 5\log_2(3x^2+7) - \log_2(x-5)$. Yes, you have five separate terms now instead of the one, but each of the terms is much simpler — all the logarithms have arguments that are polynomials.

In other mathematics courses, such as calculus, you expand the logarithms in order to have an easier time when doing derivatives. Then, in a later study of math, you take the expanded format and change it back into a single logarithm. You can make all these spectacular changes using the laws of logarithms, which I cover earlier in this chapter, in the section "Checking out the Properties of Logarithms."

Q. Show how to expand the logarithm
$\log_2 \dfrac{4x^3 \sqrt{9x+1}}{(3x^2+7)^5(x-5)}$ to produce $2 + 3\log_2 x + \frac{1}{2}\log_2(9x+1) - 5\log_2(3x^2+7) - \log_2(x-5)$, the answer given in the section intro

A. First use the rule for the log of a quotient to rewrite the logarithm as $\log_2 4x^3 \sqrt{9x+1} - \log_2(3x^2+7)^5(x-5)$. Next, use the rule for the log of a product to split up the factors in the two terms. Be sure to distribute the negative sign over the two terms that come from the second term (the one after the minus sign): $\log_2 4 + \log_2 x^3 + \log_2 \sqrt{9x+1} - \log_2(3x^2+7)^5 - \log_2(x-5)$. Now change the 4 to a power of 2 and the radical to a fractional power: $\log_2 2^2 + \log_2 x^3 + \log_2(9x+1)^{1/2} - \log_2(3x^2+7)^5 - \log_2(x-5)$. Next, use the rule for the log of a power to bring the 2, 3, ½, and 5 to the front of their respective log terms: $2\log_2 2 + 3\log_2 x + \frac{1}{2}\log_2(9x+1) - 5\log_2(3x^2+7) - \log_2(x-5)$. Finally, change the $\log_2 2$ to a 1, using the rule for the log of the base. Multiply for the final result: $2 + 3\log_2 x + \frac{1}{2}\log_2(9x+1) - 5\log_2(3x^2+7) - \log_2(x-5)$.

Q. Rewrite the terms with logs as a single log expression: $4\log_7(x-2) + 5\log_7(x^2+1) - 3\log_7 x - \log_7(x^5-12) - \frac{1}{2}\log_7(x^3+2x+1)$

A. $\log_7 \dfrac{(x-2)^4 (x^2+1)^5}{x^3(x^5-12)\sqrt{x^3+2x+1}}$. First, move all the coefficients from in front of the logs to their positions as exponents on the arguments, using the rule for the log of a power: $\log_7(x-2)^4 + \log_7(x^2+1)^5 - \log_7 x^3 - \log_7(x^5-12) - \log_7(x^3+2x+1)^{1/2}$. Next, write the first two terms as a product of logs and the last three terms as the product of logs after factoring the negative sign out of each: $\log_7(x-2)^4(x^2+1)^5 - \log_7 x^3(x^5-12)(x^3+2x+1)^{1/2}$. The last three terms are all negative, so their product ends up in the denominator. Finally, write the expression as a quotient, using the rule for the log of a quotient.

23. Expand $\log \dfrac{x^4(x-2)^3}{\sqrt{x+3}}$ using the rules for logarithms

Solve It

24. Expand $\ln \sqrt{\dfrac{x^5(x-2)^3}{(3x+1)^7}}$ using the rules for logarithms

Solve It

25. Rewrite $2\log(x-3) + 3\log(2x+1) - 4\log(x+2) - \frac{1}{2}\log(x-5)$ as a single logarithm using the rules for logarithms

Solve It

26. Rewrite $3\ln(x-5) - 2\ln(x+4) + 3\ln(x+5) - 2\ln(x-4)$ as a single logarithm using the rules for logarithms

Solve It

Solving Logarithmic Equations

A *logarithmic* equation is one with a log in it. No, I don't mean an oak or birch log, but working with a logarithmic equation can be as easy as falling off that piece of fallen tree! You may think I'm joking, but this falling off business is pretty close to the truth.

The best way to solve a logarithmic equation is to get off it — move away from that format. Change a log equation to either an algebraic format or an exponential equation so you can solve it:

> ✔ **Change the log equation $\log_b x = \log_b y$ to $x = y$.** Then solve the algebraic equation. Notice that only one log term is on each side and that the bases are the same. You set the two arguments equal to one another.

> ✔ **Change the log equation $\log_b x = y$ to $b^y = x$.** Then solve the exponential equation. Again, notice that the equation has only one log term, that the base of that logarithm becomes the base of the exponential expression, and that the exponential expression (b^y) is set equal to the argument (x).

Whenever you change the format of an equation to make it easier to solve, check for extraneous (false) solutions by inserting your answers in the original equation. Even though not all answers work, changing the format is still the preferred method for solving most logarithmic equations.

Q. Solve the equation $\log_4(x-5) + \log_4(x+3) = \log_4(5x+3)$

A. $x = 9$. Before changing the equation to an algebraic format, you have to combine the two terms on the left. For you to use the rule that gets rid of the logs, the equation can have only one log term on each side. Using the rule for the log of a product, you rewrite the equation as $\log_4[(x-5)(x+3)] = \log_4(5x+3)$. Now you can set the two *arguments* of the logarithms equal to one another to get $(x-5)(x+3) = 5x+3$. This equation is quadratic. Solve it by first multiplying the two binomials together on the left and then setting the equation equal to 0. The equation $x^2 - 2x - 15 = 5x + 3$ becomes $x^2 - 7x - 18 = 0$. Factor the quadratic into $(x-9)(x+2) = 0$. This quadratic equation has two solutions: $x = 9$ or $x = -2$. The first solution, $x = 9$, is a valid solution of the original log equation. The equation is true when you replace all the x's in the original equation with 9. You get $\log_4(9-5) + \log_4(9+3) = \log_4(5 \cdot 9 + 3)$; $\log_4(4) + \log_4(12) = \log_4(48)$; $\log_4(4 \cdot 12) = \log_4(48)$. The other solution, $x = -2$, doesn't work in the log equation. You get $\log_4(-2-5) + \log_4(-2+3) = \log_4[5(-2)+3]$, which results in negative arguments in the first and last logarithmic terms. The arguments of logarithms are always positive. The solution $x = -2$ is extraneous.

Q. Solve the equation $\log_4(x^2 + 15) = 3$

A. $x = \pm7$. Rewrite the equation as an exponential equation. You get $4^3 = x^2 + 15$. Simplifying, the equation becomes $64 = x^2 + 15$. Subtracting 15 from each side, $x^2 = 49$. Taking the square root of each side, $x = \pm7$. Both the 7 and −7 work; because they get squared, the argument doesn't become negative.

27. Solve for x in $\log_3(x+3) + \log_3(2x-1) = \log_3(7x+1)$

Solve It

28. Solve for x in $\log_8(3x-1) - \log_8(x-3) = \log_8(x+2)$

Solve It

29. Solve for x in $\log_5(2x+3) + \log_5 x = 1$

Solve It

30. Solve for x in $\log(x+8) + 1 = \log(2-3x)$

Solve It

31. Solve for x in $\ln(3x + 5) + \ln(2x + 3) = \ln(x + 3)$

Solve It

32. Solve for x in $\ln(x + 1) - \ln x = 2$

Solve It

They Ought to Be in Pictures: Graphing Exponential and Logarithmic Functions

Exponential and logarithmic functions are inverses of one another. Consequently, the graphs of exponential and logarithmic functions are mirror images of one another with respect to the line $y = x$. In Figure 9-1, you see the graph of $y = 2^x$ and $y = \log_2 x$ and how they're symmetric with respect to one another over that diagonal line.

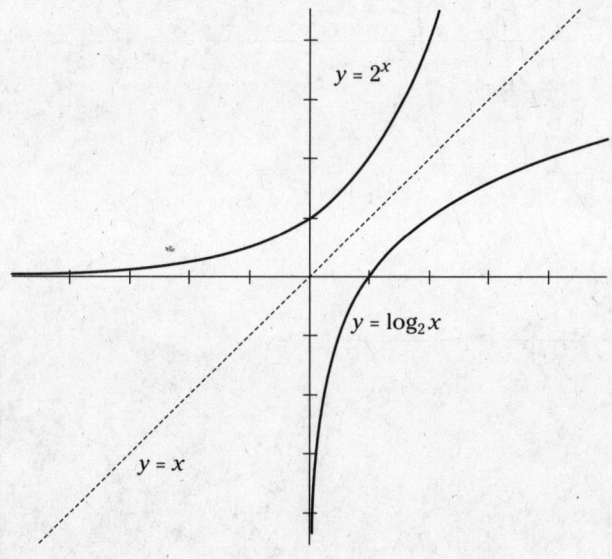

$y = 2^x$

$y = \log_2 x$

$y = x$

Figure 9-1:
Exponential and log graphs have soft C-shapes.

The basic exponential function rises from left to right, getting steeper the farther you move to the right. The basic logarithmic function also rises from left to right, but the curve gets flatter as you move to the right.

You can make the curves fall instead of rise if you use a negative sign. Negating the x in the exponential function makes it fall from left to right, and negating the log in the logarithmic function makes it fall instead of rise. The basic curves can also be made to slide left, right, up, or down by adding or subtracting constants. Here's how you can control the movement:

✔ **Up:** Add a constant to the function.

✔ **Down:** Subtract a constant from the function.

✔ **Left:** Add a constant to the *x*-term (exponent or argument).

✔ **Right:** Subtract a constant from the *x*-term (exponent or argument).

Q. Sketch the graphs of $y = 3^x$, $y = 3^x + 2$, $y = 3^{x-2}$, and $y = 3^{-x}$ all on the same set of axes

A. Adding 2 to the function $y = 3^x$ raises the basic graph by two units. Subtracting 2 from the *x* in the exponent part moves the basic graph two units to the right. Negating the *x* in the exponent makes the graph fall as the curve moves from left to right. Take a look at Figure 9-2, which shows the curves with these corresponding letters: (A) $y = 3^x$, (B) $y = 3^x + 2$, (C) $y = 3^{x-2}$, and (D) $y = 3^{-x}$.

Q. Sketch the graphs of $y = \ln x$, $y = -3 + \ln x$, $y = \ln (x + 3)$, and $y = 3 - \ln x$ all on the same set of axes

A. Subtracting 3 from the basic graph lowers it by three units. Adding 3 to the *x* value moves the graph three units to the left. Adding 3 and negating the *ln* raises the basic graph by three units and makes it fall as you read from left to right. In Figure 9-3, the graphs correspond to the following letters: (A) $y = \ln x$, (B) $y = -3 + \ln x$, (C) $y = \ln (x + 3)$, and (D) $y = 3 - \ln x$.

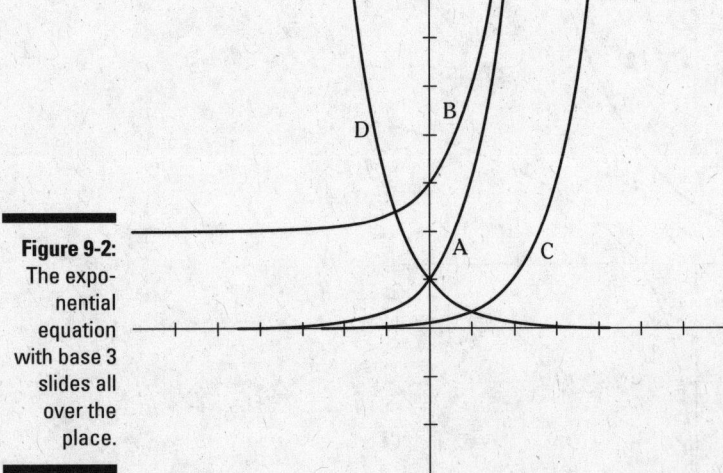

Figure 9-2:
The exponential equation with base 3 slides all over the place.

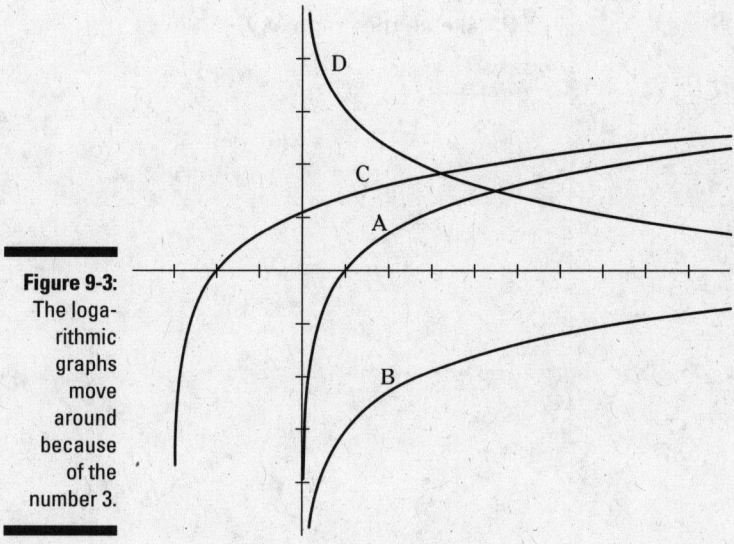

Figure 9-3:
The logarithmic graphs move around because of the number 3.

33. Sketch the graph of $y = 2^{x-3}$

Solve It

34. Sketch the graph of $y = e^{x+2}$

Solve It

35. Sketch the graph of $y = \ln (x - 1)$

Solve It

36. Sketch the graph of $y = 2 + \ln x$

Solve It

37. Sketch the graph of $y = 4^x - 4$

Solve It

38. Sketch the graph of $y = 4 + \log (x + 4)$

Solve It

Answers to Problems on Exponential and Logarithmic Functions

The following are the answers to the practice problems presented earlier in this chapter.

1 Given the exponential function $f(x) = 2^{x-1}$, determine $f(0)$, $f(1)$, and $f(2)$. The answer is $f(0) = \frac{1}{2}$; $f(1) = 1$; $f(2) = 2$.

Replacing the x's with the input values, you get $f(0) = 2^{0-1} = 2^{-1} = \frac{1}{2}$, $f(1) = 2^{1-1} = 2^0 = 1$, and $f(2) = 2^{2-1} = 2^1 = 2$.

2 Given the exponential function $f(x) = 3^x(1 - 2 \cdot 3^x)$, determine $f(0)$, $f(1)$, and $f(2)$. The answer is $f(0) = -1$; $f(1) = -15$; $f(2) = -153$.

Replacing the x's with the input values, you get $f(0) = 3^0(1 - 2 \cdot 3^0) = 1(1 - 2) = 1(-1) = -1$, $f(1) = 3^1(1 - 2 \cdot 3^1) = 3(1 - 6) = 3(-5) = -15$, and $f(2) = 3^2(1 - 2 \cdot 3^2) = 9(1 - 2 \cdot 9) = 9(1 - 18) = 9(-17) = -153$.

3 Given the exponential function $f(x) = e^x - 2e^{2x}$, determine $f(0)$ and $f(1)$. The answer is $f(0) = -1$; $f(1) = e - 2e^2$.

Replacing the x's with input values, you get $f(0) = e^0 - 2e^{2(0)} = 1 - 2(1) = -1$ and $f(1) = e^1 - 2e^{2(1)} = e - 2e^2$.

4 Given the exponential function $f(x) = (e^x + 2)^2 - (e^x - 2)^2$, determine $f(0)$, $f(1)$, and $f(2)$. The answer is $f(0) = 8$; $f(1) = 8e$; $f(2) = 8e^2$.

Replacing the x's with input values, you have $f(0) = (e^0 + 2)^2 - (e^0 - 2)^2 = (1 + 2)^2 - (1 - 2)^2 = 9 - 1 = 8$. For $f(1)$, you get $(e^1 + 2)^2 - (e^1 - 2)^2$. Now expand each binomial and simplify: $(e^2 + 4e + 4) - (e^2 - 4e + 4) = e^2 + 4e + 4 - e^2 + 4e - 4 = 8e$.

Instead of squaring the binomials and simplifying, you can factor the expression as the difference of two squares. That's how I handle the last part of the problem.

The difference of two squares factors into the difference and sum of the roots: $a^2 - b^2 = (a - b)(a + b)$.

Here, $f(2) = (e^2 + 2)^2 - (e^2 - 2)^2 = [(e^2 + 2) - (e^2 - 2)][(e^2 + 2) + (e^2 - 2)] = [e^2 + 2 - e^2 + 2][e^2 + 2 + e^2 - 2] = [4][2e^2] = 8e^2$.

5 Given the exponential function $f(x) = 2(e^{x+1})^2$, determine $f(0)$, $f(1)$, and $f(-1)$. The answer is $f(0) = 2e^2$; $f(1) = 2e^4$; $f(-1) = 2$.

Replacing the x's with input values, you get $f(0) = 2(e^{0+1})^2 = 2(e^1)^2 = 2e^2$, $f(1) = 2(e^{1+1})^2 = 2(e^2)^2 = 2e^4$, and $f(-1) = 2(e^{-1+1})^2 = 2(e^0)^2 = 2(1)^2 = 2$.

6 Given the exponential function $f(x) = x^e - e^x$, determine $f(0)$, $f(1)$, and $f(2)$. The answer is $f(0) = -1$; $f(1) = 1 - e$; $f(2) = 2^e - e^2$.

Replacing the x's with input values, you get $f(0) = 0^e - e^0 = 0 - 1 = -1$, $f(1) = 1^e - e^1 = 1 - e$, and $f(2) = 2^e - e^2$.

7 Solve for x in $2^x = 4^{x+1}$. The answer is $x = -2$.

First, rewrite the 4 on the right as 2^2 and simplify the power to get $2^x = 2^{2(x+1)} = 2^{2x+2}$. Now set the two exponents equal to one another: $x = 2x + 2$. Solving for x, $-x = 2$; $x = -2$.

8 Solve for x in $9^{x-5} = \frac{1}{27^x}$. The answer is $x = 2$.

Both the 9 and 27 are powers of 3, so rewrite the terms and simplify the exponents:

$$\left(3^2\right)^{x-5} = \frac{1}{\left(3^3\right)^x}$$

$$3^{2x-10} = \frac{1}{3^{3x}} = 3^{-3x}$$

Setting the exponents equal to one another and solving, $2x - 10 = -3x;\ 5x = 10;\ x = 2$.

9 Solve for x in $e^{4x-2} = e^{x+7}$. The answer is **$x = 3$.**

The bases are the same, so set the exponents equal to one another: $4x - 2 = x + 7;\ 3x = 9;\ x = 3$.

10 Solve for x in $2^{2x} - 5 \cdot 2^x + 4 = 0$. The answer is **$x = 0, 2$.**

This problem is a quadratic-like trinomial that you can factor into the product of two binomials. Let $y = 2^x$. For a pattern, think of the trinomial $y^2 - 5y + 4$, which factors into $(y - 4)(y - 1)$. Factoring the trinomial with the exponential terms, you get $(2^x - 1)(2^x - 4) = 0$. Setting the first factor equal to 0, you get $2^x - 1 = 0;\ 2^x = 1;\ x = 0$. With the second factor, you calculate $2^x - 4 = 0$; $2^x = 4$. The 4 is a power of 2, so write the equation as $2^x = 2^2$; thus, $x = 2$. Both solutions work.

11 Solve for x in $e^x - 1 = 0$. The answer is **$x = 0$.**

Rewrite the equation as $e^x = 1$. Think of 1 as being equal to e^0; thus, $e^x = e^0;\ x = 0$.

12 Solve for x in $xe^x + 5e^x = 0$. The answer is **$x = -5$.**

The binomial factors — the common factor is e^x. So $e^x(x + 5) = 0$. Using the multiplication property of zero, you set each of the factors equal to 0 and solve for x. But $e^x = 0$ has no solution. No power of e is equal to 0, so this factor doesn't yield a solution. However, $x + 5 = 0$ when $x = -5$.

13 Determine the total amount of money accumulated after 20 years if you deposit $1,000,000 in an account that earns 5.25% interest, compounded monthly. The answer is **$2,851,114.02.**

In 20 years, the money almost triples. Using the formula for compound interest, you write

$$A = 1,000,000\left(1 + \frac{0.0525}{12}\right)^{12\,(20)}$$

$$= 1,000,000\,(1.004375)^{240}$$

$$\approx 2,851,114.02$$

14 Determine the total amount of money accumulated after 2 years and 6 months if you deposit $25,000 in an account that earns 5% interest, compounded daily (use 365 days). The answer is **$28,328.47.**

Six months is half a year, so write 2 years, 6 months as 2.5 years. Use the compound interest formula:

$$A = 25,000\left(1 + \frac{0.05}{365}\right)^{365\,(2.5)}$$

$$\approx 25,000\,(1.000136986)^{912.5}$$

$$\approx 28,328.47$$

15 Determine the total amount of money accumulated after 20 years if you deposit $1,000,000 in an account that earns 5.25% interest, compounded continuously. The answer is **$2,857,651.12.**

Using the formula for compounding continuously, $A = 1,000,000e^{0.0525\,(20)} = 1,000,000e^{1.05} \approx 2,857,651.12$. You get a difference of over $6,000 between compounding monthly and compounding continuously (refer to problem 13 for the calculation that involves monthly compounding).

16 Determine the total amount of money accumulated after 2 years and 6 months if you deposit $25,000 in an account that earns 5% interest, compounded continuously. The answer is **$28,328.71.**

Using the formula for compounding continuously, $A = 25,000e^{0.05\,(2.5)} = 1,000,000e^{0.125} \approx 28,328.71$. When you compare this answer to the result in problem 14, you don't find much difference between compounding continuously and compounding daily.

17 Determine which of the following is equivalent to log 45. The answer is **a. log 3 + log 15.**

Because 45 is the product of 3 · 15, you can write log 45 = log (3 · 15) and then apply the rule for the log of a product.

18 Determine which of the following is equivalent to $\log \frac{x}{2}$. The answer is **d. log x – log 2.**

The expression involves a quotient, so apply the rule for the log of a quotient.

19 Determine which of the following is equivalent to $\log_a \frac{1}{3}$. The answer is **b. $-\log_a 3$.**

Two different approaches to this problem are available. You can apply the rule for the log of a quotient and write the expression as $\log_a 1 - \log_a 3$ and then replace $\log_a 1$ with 0, using the rule for the log of 1. Simplifying $0 - \log_a 3$, you get $-\log_a 3$. The easier approach is to just know the rule for the log of a reciprocal and apply it. The easier or simpler way isn't always the most obvious, so either way is really acceptable.

20 Determine which of the following is equivalent to $\log_{10} 10$. The answer is **c. 1.**

Just apply the rule for the log of the base. Technically, you don't have to show the base when the base is 10. You're working with the common logarithm.

21 Determine which of the following is equivalent to $\log_6 1$. The answer is **b. 0.**

This problem follows the rule for the log of 1.

22 Determine which of the following is equivalent to log 8. The answer is **b. 3log 2.**

First change the 8 to a power of 2, and then apply the rule for the log of a power. You get log 8 = log 2^3 = 3log 2.

23 Expand $\log \frac{x^4(x-2)^3}{\sqrt{x+3}}$ using the rules for logarithms. The answer is **4log x + 3log (x – 2) – ½ log (x + 3).**

First rewrite the log using the rule for the log of a quotient to get $\log x^4(x-2)^3 - \log \sqrt{x+3}$. Now use the rule for the log of a product on the first term, and then change the radical to a fractional power: $\log x^4 + \log (x-2)^3 - \log (x+3)^{1/2}$. Use the rule for the log of a power to rewrite each term without an exponent.

24 Expand $\ln \sqrt{\frac{x^5(x-2)^3}{(3x+1)^7}}$ using the rules for logarithms. The answer is $\frac{5}{2} \ln x + \frac{3}{2} \ln (x-2) - \frac{7}{2} \ln (3x+1)$.

Change the radical to an exponent, and then apply the rule for the log of a power, moving the ½ in front of the ln. Then apply the rule for a quotient; be sure to use a grouping symbol so that each term is multiplied by the ½ factor in the front: $\frac{1}{2} \ln \frac{x^5(x-2)^3}{(3x+1)} = \frac{1}{2}\left[\ln x^5(x-2)^3 - \ln (3x+1)^7\right]$. Now apply the rule for a product, and then move the powers in front of the respective lns. Lastly, distribute the ½ through all the terms:

$$= \frac{1}{2}\left[\ln x^5 + \ln (x-2)^3 - \ln (3x+1)^7\right]$$

$$= \frac{1}{2}\left[5\ln x + 3\ln (x-2) - 7\ln (3x+1)\right]$$

$$= \frac{5}{2}\ln x + \frac{3}{2}\ln (x-2) - \frac{7}{2}\ln (3x+1)$$

25 Rewrite $2\log (x-3) + 3\log (2x+1) - 4\log (x+2) - \frac{1}{2}\log (x-5)$ as a single logarithm using the rules for logarithms. The answer is $\log \dfrac{(x-3)^2(2x+1)^3}{(x+2)^4\sqrt{x-5}}$.

First rewrite each term with its multiplier changed to an exponent. Then group the first two terms and rewrite them using the rule for the log of a product. Factor the negative sign out of the second two terms and apply the rule for a product again. Lastly, use the rule for a quotient and change the binomial with the ½ exponent into a binomial under a radical:

$$\log (x-3)^2 + \log (2x+1)^3 - \log (x+2)^4 - \log (x-5)^{1/2}$$

$$= \log \left[(x-3)^2(2x+1)^3\right] - \left[\log (x+2)^4 + \log (x-5)^{1/2}\right]$$

$$= \log \left[(x-3)^2(2x+1)^3\right] - \left[\log (x+2)^4(x-5)^{1/2}\right]$$

$$= \log \frac{(x-3)^2(2x+1)^3}{(x+2)^4(x-5)^{1/2}} = \log \frac{(x-3)^2(2x+1)^3}{(x+2)^4\sqrt{x-5}}$$

26 Rewrite $3\ln (x-5) - 2\ln (x+4) + 3\ln (x+5) - 2\ln (x-4)$ as a single logarithm using the rules for logarithms. The answer is $\ln \dfrac{(x^2-25)^3}{(x^2-16)^2}$.

Rewrite each term by moving the multiplier in front to its place as an exponent. Next, rearrange the terms, grouping the positive and negative, and factoring the negative sign out of the negative terms. Apply the rule for the log of a product for the two groupings, and then apply the rule for the log of a quotient. Notice that the binomials in the numerator have the same power, as do the binomials in the denominator. The binomials are the sum and difference of the same two numbers, so you can write them as the difference of squares:

$$\ln (x-5)^3 - \ln (x+4)^2 + \ln (x+5)^3 - \ln (x-4)^2$$

$$= \left[\ln (x-5)^3 + \ln (x+5)^3\right] - \left[\ln (x+4)^2 + \ln (x-4)^2\right]$$

$$= \ln \left[(x-5)^3(x+5)^3\right] - \ln \left[(x+4)^2(x-4)^2\right]$$

$$= \ln \frac{(x-5)^3(x+5)^3}{(x+4)^2(x-4)^2} = \ln \frac{(x^2-25)^3}{(x^2-16)^2}$$

27 Solve for x in $\log_3(x+3) + \log_3(2x-1) = \log_3(7x+1)$. The answer is $x = 2$.

First use the rule for the log of a product to rewrite the left side of the equation, giving you $\log_3[(x+3)(2x-1)] = \log_3(7x+1)$. This equation is now in the form $\log x = \log y$, so you can set the two arguments equal to one another and solve the algebraic equation: $(x+3)(2x-1) = 7x+1$. Multiply on the left. Then set the equation equal to 0 and factor. The equation $2x^2 + 5x - 3 = 7x + 1$ becomes $2x^2 - 2x - 4 = 0$. The polynomial factors into $2(x^2 - x - 2) = 2(x-2)(x+1) = 0$. The two solutions of the quadratic equation are $x = 2$ and $x = -1$. The answer -1 doesn't work, because it gives you logs of negative numbers in the original equation. Thus, the solution $x = -1$ is extraneous.

28 Solve for x in $\log_8(3x-1) - \log_8(x-3) = \log_8(x+2)$. The answer is $x = 5$.

Apply the rule for the log of a quotient to rewrite the equation as $\log_8 \dfrac{3x-1}{x-3} = \log_8(x+2)$. Then set the two arguments equal to one another and multiply each side by $x-3$:

$$\frac{3x-1}{x-3} = x + 2$$

$$(x-3)\frac{3x-1}{x-3} = (x+2)(x-3)$$

$$3x - 1 = x^2 - x - 6$$

Setting the equation equal to 0, you get $0 = x^2 - 4x - 5 = (x - 5)(x + 1)$. The two solutions of the quadratic are $x = 5$ and $x = -1$, but only the $x = 5$ works — the other is extraneous.

29 Solve for x in $\log_5(2x + 3) + \log_5 x = 1$. The answer is **$x = 1$.**

Use the rule for the log of a product to write the equation as $\log_5[(2x + 3)x] = 1$. Now change this format to $5^1 = (2x + 3)x$. Distribute the x on the right, and then set the equation equal to 0, giving you $5 = 2x^2 + 3x$; $0 = 2x^2 + 3x - 5$. The quadratic factors: $0 = (2x + 5)(x - 1)$. The two solutions are $x = -\frac{5}{2}$ and $x = 1$, but only the $x = 1$ works — the other is extraneous.

30 Solve for x in $\log(x + 8) + 1 = \log(2 - 3x)$. The answer is **$x = -6$.**

First subtract $\log(x + 8)$ from each side, and then apply the rule for the log of a quotient. Rewrite the equation in exponential format; remember that $\log x$ is a common logarithm — the base is 10:

$$1 = \log(2 - 3x) - \log(x + 8)$$

$$1 = \log\frac{2 - 3x}{x + 8}$$

$$10^1 = \frac{2 - 3x}{x + 8}$$

Now multiply each side of the equation by $x + 8$ and solve for x:

$$10(x + 8) = 2 - 3x$$

$$10x + 80 = 2 - 3x$$

$$13x = -78$$

$$x = -6$$

31 Solve for x in $\ln(3x + 5) + \ln(2x + 3) = \ln(x + 3)$. The answer is **$x = -1$.**

Rewrite the left side using the rule for the log of a product: $\ln[(3x + 5)(2x + 3)] = \ln(x + 3)$. Then use the rule where two logs are equal to create the algebraic equation: $(3x + 5)(2x + 3) = x + 3$. Multiply on the left; then set the equation equal to 0 and factor: $6x^2 + 19x + 15 = x + 3$; $6x^2 + 18x + 12 = 0$; $6(x^2 + 3x + 2) = 6(x + 1)(x + 2) = 0$. The two solutions are $x = -1$ and $x = -2$. The solution $x = -1$ works, but the solution $x = -2$ is extraneous.

32 Solve for x in $\ln(x + 1) - \ln x = 2$. The answer is **$x = \dfrac{1}{e^2 - 1}$.**

Rewrite the left side using the rule for the log of a quotient. Then write the exponential form of the equation.

The base of the logarithm designated by $\ln$ is base e.

$$\ln\frac{x + 1}{x} = 2$$

$$e^2 = \frac{x + 1}{x}$$

Multiply each side by x, and then solve for x:

$$e^2 x = x + 1$$

$$e^2 x - x = 1$$

$$x(e^2 - 1) = 1$$

$$x = \frac{1}{e^2 - 1}$$

33 Sketch the graph of $y = 2^{x-3}$.

The y-intercept is $\left(0, \dfrac{1}{8}\right)$. The graph of this function is the basic exponential graph of $y = 2^x$ moved three units to the right.

34 Sketch the graph of $y = e^{x+2}$.

The y-intercept is $(0, e^2)$ which is about $(0, 7.39)$. The graph of this function is the basic exponential graph of $y = e^x$ moved two units to the left.

35 Sketch the graph of $y = \ln(x - 1)$.

The x-intercept is $(2, 0)$. The graph of this function is the basic logarithmic graph of $y = \ln x$ moved one unit to the right.

36 Sketch the graph of $y = 2 + \ln x$.

The x-intercept is $\left(\frac{1}{e^2}, 0\right)$ which is about $(0.14, 0)$. The graph of this function is the basic logarithmic graph of $y = \ln x$ moved two units up.

37 Sketch the graph of $y = 4^x - 4$.

The two intercepts are (0, –3) and (1, 0). The graph of this function is the basic exponential graph of $y = 4^x$ lowered by four units.

38 Sketch the graph of $y = 4 + \log(x + 4)$.

The two intercepts are $\left(0, 4 + \log 4\right)$ and (–3.9999, 0), which are approximately (0, 4.6) and (–4,0). You get the y-intercept by setting x equal to 0, giving you $y = 4 + \log 4$. You solve for the x-intercept by letting $y = 0$ and solving $0 = 4 + \log(x + 4)$, which becomes $\log(x + 4) = -4$; $10^{-4} = x + 4$; $x = -4 + 10^{-4}$. The graph of this function is the basic logarithmic graph of $y = \log x$ moved up four units and left four units.

Part III
Conics and Systems of Equations

In this part . . .

Not all paths are straight, and you don't always have to come *full circle*. But in this part, you can. The conic sections investigate ovals and various paths that can take you way beyond the ordinary graph — to the U-shaped parabola, the two-part hyperbola, and even that pizza lover's delight, the circle. Systems of equations sometimes even have coefficients that form square and rectangular figures. Be open to the opportunities in this part.

Chapter 10

Any Way You Slice It: Conic Sections

∙ ∙

In This Chapter

▶ Writing equations of parabolas, circles, ellipses, and hyperbolas in standard form

▶ Identifying centers, foci, and axes of conic sections

▶ Sketching parabolas, circles, ellipses, and hyperbolas

∙ ∙

Chapter 6 contains information on quadratic functions, which you graph as U-shaped curves called *parabolas*. Here, you see that not all parabolas represent functions. When a parabola opens to the left or right, you get more than one *y*-value for many *x*-values, and that just doesn't work with functions. This chapter introduces parabolas and some other conic sections that don't exactly fit the function mold. All parabolas are conic sections, but not all parabolas are functions — it depends on how they lie.

A *conic section,* or *conic,* is represented by a slice of a cone (or really, two cones stacked on top of each other, like an hourglass). How you slice the cone — at what angle and how big a chunk you take — determines what kind of conic section you produce. The four types of conic sections are the parabola, circle, ellipse, and hyperbola. Here's how they compare:

✔ **Parabola:** You get this U-shaped curve when you take a slice parallel to one of the sides of a cone.

✔ **Circle:** A circle appears when you cut a cone straight across, parallel to the base.

✔ **Ellipse:** If you slice through a cone at a slant, you get an ellipse, or oval.

✔ **Hyperbola:** Slicing straight down through both cones gives you a hyperbola. It looks like two U-shaped curves facing away from each other.

In this chapter, you use the standard forms of equations of conics to recognize which type of conic section an equation represents. You also find the defining characteristics of the conic sections and use these traits to help you graph the conics.

Putting Equations of Parabolas in Standard Form

Parabola equations come in two standard forms to separate the functions opening upward or downward from curves that open sideways. The standard forms tell you what the parabola looks like — its general width or narrowness, in which direction it opens, and where the vertex (turning point) of the graph is.

Here are the two standard forms for the equations of parabolas (and what they tell you):

Equation	Direction the Parabola Opens	Location of the Vertex
$(y - k)^2 = 4a(x - h)$	Right if a is positive, left if a is negative	(h, k)
$(x - h)^2 = 4a(y - k)$	Up if a is positive, down if a is negative	(h, k)

The multiplier $4a$ is a constant that tells you how steep or wide the parabola is. You can use that information in the next section, "Shaping Up: Determining the Focus and Directrix of a Parabola."

EXAMPLE

Q. Write the equation of the parabola $x^2 - 16x - 4y + 52 = 0$ in standard form to determine its vertex and in which direction it opens

A. $(x - 8)^2 = 4(1)(y + 3)$; **vertex: (8, –3); opens upward.** First rewrite the equation to isolate the x-terms: Leave the two terms with x's on the left, and get the other two terms on the right by adding $4y$ and subtracting 52 from both sides. You get $x^2 - 16x = 4y - 52$. Next, complete the square (go to Chapter 2 if you need to know more about this technique) on the left side of the equation. Be sure to add the 64 to both sides of the equation to keep it balanced; $x^2 - 16x + 64 = 4y - 52 + 64$ becomes $(x - 8)^2 = 4y + 12$. Now factor out the 4 from the terms on the right to get $(x - 8)^2 = 4(y + 3)$. The vertex of the parabola is (8, –3). The parabola opens upward, because the x term is squared and the multiplier on the right is positive. The $4a$ part of the standard form is actually $4(1)$ — the a value is 1. (You need to know information about the a in the next section.)

Q. Write the equation of the parabola $2y^2 + 28y + x + 97 = 0$ in standard form to determine its vertex and in which direction it opens

A. $(y + 7)^2 = 4\left(-\dfrac{1}{8}\right)(x - 1)$; **vertex: (1, –7); opens left.** First rewrite the equation, leaving the two terms with y's on the left and moving the others to the right. You get $2y^2 + 28y = -x - 97$. Factor the 2 out of the terms on the left before completing the square; $2(y^2 + 14y) = -x - 97$ becomes $2(y^2 + 14y + 49) = -x - 97 + 98$. Notice that you have to add 98 to the right, because the 49 that you added to complete the square is multiplied by 2. Simplifying and factoring, you have $2(y + 7)^2 = -x + 1$. Now factor –1 from each term on the right, and then divide both sides by 2: $2(y + 7)^2 = -1(x - 1)$; $(y + 7)^2 = -\dfrac{1}{2}(x - 1)$. The vertex of the parabola is at (1, –7), and the parabola opens to the left. The coefficient on the right, $-\dfrac{1}{2}$, is written $4\left(-\dfrac{1}{8}\right)$ to put it in the $4a$ form.

1. Write the equation of the parabola $x^2 - 6x - y + 4 = 0$ in standard form to determine its vertex and in which direction it opens

Solve It

2. Write the equation of the parabola $2x^2 + 8x + y + 3 = 0$ in standard form to determine its vertex and in which direction it opens

Solve It

3. Write the equation of the parabola $3y^2 - 12y - x + 9 = 0$ in standard form to determine its vertex and in which direction it opens

Solve It

4. Write the equation of the parabola $y^2 - x + 4 = 0$ in standard form to determine its vertex and in which direction it opens

Solve It

5. Write the equation of the parabola $y^2 + y + x = 0$ in standard form to determine its vertex and in which direction it opens

Solve It

6. Write the equation of the parabola $4x^2 - x - y + 1 = 0$ in standard form to determine its vertex and in which direction it opens

Solve It

Shaping Up: Determining the Focus and Directrix of a Parabola

A *parabola* is technically defined as all the points that are the same distance from a fixed point called the *focus* and a fixed line called the *directrix*. Both elements are related to the parabola's axis of symmetry. The directrix is always perpendicular to the parabola's axis of symmetry, and the vertex of the parabola lies on the axis of symmetry and is exactly halfway between the focus and the directrix. Figure 10-1 shows you a parabola with all these features identified.

You can identify the focus and directrix (as well as the vertex and axis of symmetry) of a parabola from the standard form of the parabola's equation:

Standard Form	Focus	Directrix	Vertex	Axis of Symmetry
$(y - k)^2 = 4a(x - h)$	$(h + a, k)$	$x = h - a$	(h, k)	$y = k$
$(x - h)^2 = 4a(y - k)$	$(h, k + a)$	$y = k - a$	(h, k)	$x = h$

So what can foci do for you in the real world? Well, if you're trying to concentrate the sun's rays with a parabolic mirror so you can set a fleet of Roman ships on fire, you have to know where your target should be, right? (A pretty cool idea in theory, if not in practice. We don't know whether Archimedes pulled off the attack back in 212 BC, but the concept does work for harnessing solar energy in power plants!)

Figure 10-1:
The focus is inside the parabola.

axis of symmetry

distance to focus

parabola

distance to directrix

focus

vertex   directrix

Q. Find the vertex, focus, directrix, and axis of symmetry for the parabola $x^2 = -8y$

A. **Vertex: (0, 0); focus: (–2, 0); directrix: $y = 2$; axis of symmetry: $x = 0$.** The standard form for this parabola, indicating the values of h, k, and a, is written $(x - 0)^2 = 4(-2)(y - 0)$. The value of a is –2. The vertex of the parabola is at (0, 0), and the focus is at $(0 - 2, 0) = (-2, 0)$. The directrix is $y = 0 - (-2)$ or $y = 2$, and the axis of symmetry is $x = 0$.

Q. Find the vertex, focus, directrix, and axis of symmetry for the parabola $(y - 2)^2 = 24(x + 3)$

A. **Vertex: (–3, 2); focus: (3, 2); directrix: $x = -9$; axis of symmetry: $y = -2$.** The standard form for this parabola, indicating the values of h, k, and a, is written $(y - 2)^2 = 4(6)(x + 3)$. (**Note:** In other problems, you may have to complete the square to put the equation in standard form. See the preceding section and Chapter 2 for more information.) The value of a is 6. The vertex of the parabola is at (–3, 2), and the focus is at $(-3 + 6, 2) = (3, 2)$. The directrix is $x = -3 - 6$ or $x = -9$, and the axis of symmetry is $y = -2$.

7. Find the vertex, focus, directrix, and axis of symmetry for the parabola $x^2 = 16y$

Solve It

8. Find the vertex, focus, directrix, and axis of symmetry for the parabola $y^2 = -100x$

Solve It

9. Find the vertex, focus, directrix, and axis of symmetry for the parabola $8x^2 - 32x + 38 - y = 0$

Solve It

10. Find the vertex, focus, directrix, and axis of symmetry for the parabola $y^2 - 2y + 4x + 13 = 0$

Solve It

Back to the Drawing Board: Sketching Parabolas

A parabola is a smooth, U-shaped curve, and you can sketch a parabola using all the valuable information gleaned from its equation in the standard form. The vertex, focus, directrix, and axis of symmetry are items you use in the sketch. You can also solve for intercepts and other random points if you need more information.

The value of *a* in the standard form of the parabola determines how wide or steep the parabola is and what direction it faces:

✔ When *a* is positive, the parabola opens upward or to the right.

✔ When *a* is negative, the parabola opens downward or to the left.

✔ The bigger the absolute value of *a*, the wider or flatter the parabola. As the absolute value of *a* gets larger, the focus and directrix get farther apart.

✔ When *a* is a number between –1 and +1 (when its absolute value is a proper fraction but not 0), the parabola steepens — the focus and directrix get pretty close together.

Refer to the preceding section for info on finding the vertex, focus, directrix, and axis of symmetry. (I cover finding intercepts of quadratic functions in Chapter 2.)

Q. Sketch the graph of $(y - 3)^2 = 12(x + 1)$. Label the vertex, focus, directrix, and axis of symmetry.

A. As Figure 10-2 shows, the vertex is $(-1, 3)$; the value of *a* is 3; the focus is $(-1 + 3, 3) = (2, 3)$; the directrix is $x = -1 - 3 = -4$; and the axis of symmetry is $y = 3$. The parabola opens to the right, because *a* is positive.

Q. Sketch the graph of $(x - 5)^2 = -(y - 3)$. Label the vertex, focus, directrix, and axis of symmetry.

A. Check out Figure 10-3. The vertex is $(5, 3)$; the value of *a* is $-\frac{1}{4}$; the focus is $\left(5, 3 - \frac{1}{4}\right) = \left(5, 2\frac{3}{4}\right)$; the directrix is $y = 3 - \left(-\frac{1}{4}\right) = 3\frac{1}{4}$; and the axis of symmetry is $x = 5$. The parabola opens downward, because *a* is negative.

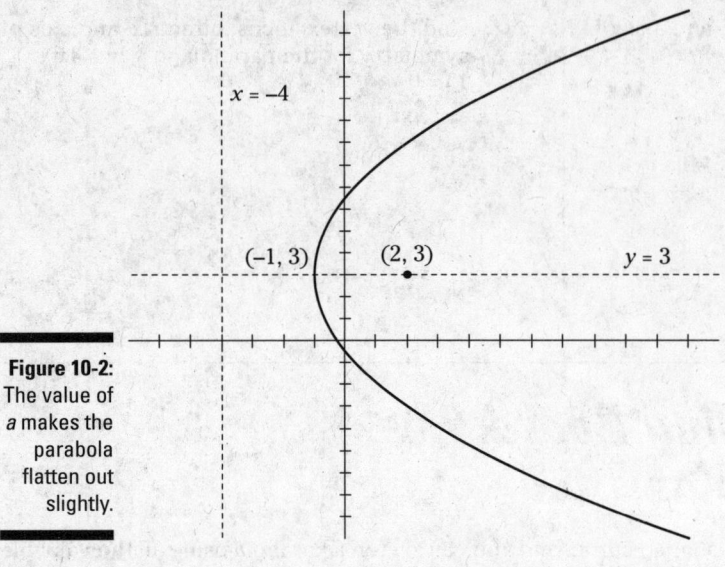

Figure 10-2:
The value of *a* makes the parabola flatten out slightly.

Figure 10-3:
The value of *a* is relatively small, so the graph steepens.

11. Sketch the graph of $(x - 4)^2 = 12(y + 1)$. Label the vertex, focus, directrix, and axis of symmetry.

Solve It

12. Sketch the graph of $(y - 1)^2 = 8(x - 5)$. Label the vertex, focus, directrix, and axis of symmetry.

Solve It

13. Sketch the graph of $4y^2 + 24y + 16x + 33 = 0$. Label the vertex, focus, directrix, and axis of symmetry.

Solve It

14. Sketch the graph of $2x^2 + 8x + 11 = y$. Label the vertex, focus, directrix, and axis of symmetry.

Solve It

Writing the Equations of Circles and Ellipses in Standard Form

A *circle* is all the points that are a constant distance, or *radius,* from a fixed point called its *center.*

An ellipse is an oval that's taller than wide or wider than tall, and it's based around two fixed points called *foci* (the plural of *focus*). An *ellipse* consists of all the points where the sum of the distances from a point on the ellipse to the *foci* is constant.

The foci lie along the ellipse's *major axis,* which is the longest line you can draw inside the ellipse. A *minor axis* is perpendicular to the major axis and intersects the major axis at the *center* of the ellipse. Figure 10-4a shows a circle with its center and radius (r) identified. Figure 10-4b has an ellipse with its axes, foci, and segments drawn from one point to the two foci.

Here are the standard forms for circles and ellipses:

✔ **Circle:** $(x - h)^2 + (y - k)^2 = r^2$, where

   • *(h, k)* is the center

   • *r* is the radius

✔ **Ellipse:** $\dfrac{(x - h)^2}{a^2} + \dfrac{(y - k)^2}{b^2} = 1$, where

   • *(h, k)* is the center

   • *2a* is the length of one axis and *2b* is the length of the other (the major axis corresponds to whichever is larger, *a* or *b*)

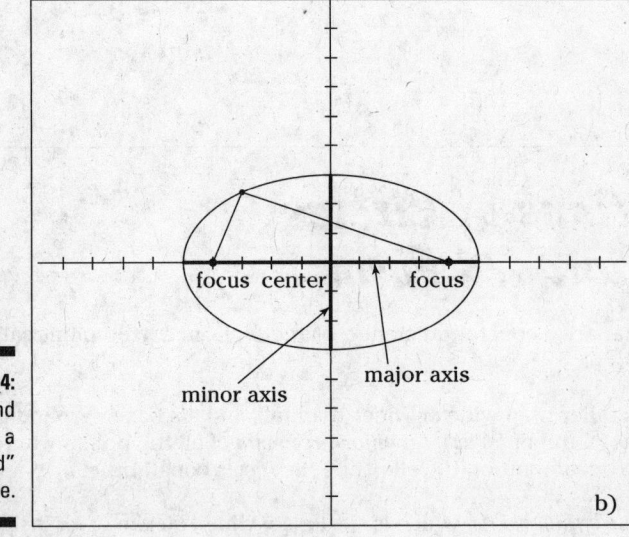

Figure 10-4:
A circle and
an ellipse, a
"squished"
circle.

Q. Write the equation of the circle $x^2 + y^2 - 12x + 24y = 45$ in standard form. Identify the center and radius of the circle.

A. $(x - 6)^2 + (y + 12)^2 = 225$; **center: (6, –12); radius: 15 units.** You have to complete the square twice (see Chapter 2). First group the x and y terms and determine what has to be added to each grouping to create a perfect square trinomial. You get $(x^2 - 12x + 36) + (y^2 + 24y + 144) = 45 + 36 + 144$. Be sure to add the same amount to each side of the equation. Factoring on the left and simplifying on the right, you get $(x - 6)^2 + (y + 12)^2 = 225$. The center of the circle is $(6, -12)$, and the radius is 15 units.

Q. Write the equation of the ellipse $25x^2 + 150x + 9y^2 + 18y + 9 = 0$ in standard form. Identify the center, the length of the major axis, and the length of the minor axis.

A. $\dfrac{(x+3)^2}{9} + \dfrac{(y+1)^2}{25} = 1$; center: $(-3, -1)$; **major axis: 10 units; minor axis: 6 units.** You have to complete the square twice to create the standard form. Factor the 25 out of the first two terms and 9 out of the second two; then subtract 9 from each side: $25(x^2 + 6x) + 9(y^2 + 2y) = -9$. To create a perfect square trinomial in the first set of parentheses, add a 9 after the $6x$ and add

225 to the right side of the equation (the 25 multiplies the 9, also). To create a perfect square trinomial in the second set of parentheses, add 1 after the $2y$ and 9 to the right side. Factoring and simplifying, you have $25(x+3)^2 + 9(y+1)^2 = 225$. Now divide each term by 225 and simplify to get the standard form. The center of the ellipse is $(-3, -1)$. The major axis is 10 units long. You get this answer by taking the square root of the larger of the two denominators, the 25, and doubling it. The minor axis is 6 units long — the square root of 9 is 3, and twice 3 is 6. This ellipse is taller than it is wide, because the larger denominator is under the y-values.

15. Write the equation of the circle $x^2 + y^2 - 6x + 8y = 11$ in standard form. Identify the center and radius of the circle.

Solve It

16. Write the equation of the circle $x^2 + y^2 + 2x - 10y + 17 = 0$ in standard form. Identify the center and radius of the circle.

Solve It

17. Write the equation of the ellipse $9x^2 + 4y^2 - 90x - 16y + 205 = 0$ in standard form. Identify the center and the lengths of the two axes.

Solve It

18. Write the equation of the ellipse $x^2 + 16y^2 + 4x - 32y + 19 = 0$ in standard form. Identify the center and the lengths of the two axes.

Solve It

Determining Foci and Vertices of Ellipses

The *foci* of an ellipse are two points that lie on the major axis (the longer of the two axes). The foci are sort of anchors or determiners for the ellipse. You measure the distances from a particular point on the ellipse to each of the foci, add up the distances, and that sum is the same for any point on the ellipse.

The closer to the center the foci are, the closer the ellipse is to a circle. You can determine the distance from the center to each focus *(c)* with the formula $c^2 = a^2 - b^2$ or $c^2 = b^2 - a^2$, depending on whether *a* or *b* is larger — you want the difference to be positive (in other words, you want the absolute value of $a^2 - b^2$). The a^2 and b^2 are the values in the denominator of the standard form for the ellipse, $\dfrac{(x-h)^2}{a^2} + \dfrac{(y-k)^2}{b^2} = 1$. They represent half the length of the axes. The value of *c* comes out to be $\pm c$ when you take the square root, and those numbers give you the distances of the foci on either side of the center of the ellipse.

The *vertices* of an ellipse are the two endpoints of the major axis. To find these points, you add and subtract the value of *a* or *b* — whichever one corresponds to the major axis — from the coordinates of the center *(h, k)* of the ellipse. Do the addition and subtraction to the *x* part of the coordinate if the major axis is horizontal, and do the computations to the *y* part of the coordinate if the major axis is vertical.

Q. Find the foci and vertices of the ellipse
$$\frac{(x+3)^2}{49} + \frac{(y-5)^2}{625} = 1.$$

A. **Foci: (–3, 29), (–3, –19); vertices: (–3, 30), (–3, –20).** The foci lie on the major axis, and the major axis is the vertical axis in this case — the larger denominator is under the *y*-values. Using $c^2 = b^2 - a^2$, you get $c^2 = 625 - 49 = 576$. So $c = \pm 24$. To find

the foci, add and subtract 24 from the *y*-value of the center of the ellipse, (–3, 5). The foci are at (–3, 29) and (–3, –19). The vertices are on the same axis and are 25 units (*b* units) above and below the center. The vertices are at (–3, 30) and (–3, –20). With the foci so close to the vertices, you have a very tall and narrow ellipse. (Check out the graph of this ellipse in the second example of the next section.)

---

19. Find the foci and vertices of the ellipse
$$\frac{(x+6)^2}{16} + \frac{(y-2)^2}{25} = 1$$

Solve It

20. Find the foci and vertices of the ellipse
$$\frac{(x-1)^2}{625} + \frac{(y-7)^2}{576} = 1$$

Solve It

21. Find the foci and vertices of the ellipse
$\dfrac{x^2}{169} + \dfrac{y^2}{144} = 1$

Solve It

22. Find the foci and vertices of the ellipse
$100x^2 + 36y^2 = 3{,}600$

Solve It

Rounding Out Your Sketches: Circles and Ellipses

Sketching a circle is really rather simple — if you have the standard form of its equation (see "Writing the Equations of Circles and Ellipses in Standard Form"). You find the center of the circle and determine its radius. Then you mark the center on your graph paper; find points that are the distance of the radius above, below, to the right, and to the left of the center; and draw in the rest of the circle.

Sketching an ellipse isn't all that much harder than drawing a circle. You find the center, the vertices (the endpoints of the major axis), and then the two endpoints of the minor axis. The ellipse is sort of like a circle, except the horizontal and vertical distances are different from one another. Again, you need the equation of the ellipse in its standard form.

Q. Sketch the circle $9(x - 1)^2 + 9(y + 4)^2 = 49$, indicating the center and radius on the graph

A. First divide each term by 9 to put the equation of the circle in its standard form. The center of the circle is at (1, –4), and the radius is $\dfrac{7}{3}$ — just a little over 2 units (see Figure 10-5). Mark the center and then estimate the distance about 2 units above, below, left, and right. Fill in the rest.

Q. Sketch the ellipse $\dfrac{(x + 3)^2}{49} + \dfrac{(y - 5)^2}{625} = 1$, indicating the center, endpoints of the minor axis, and vertices.

A. This ellipse is the same as the one in the example in "Determining Foci and Vertices of Ellipses." The center is at (–3, 5), and the vertices are at (–3, 30) and (–3, –20). The minor axis is 14 units long; it goes 7 units on either side of the center, so the endpoints are at (4, 5) and (–10, 5). Mark the center and endpoints of the axes. Then fill in the rest (see Figure 10-6).

Figure 10-5:
The circle stays in the third and fourth quadrants.

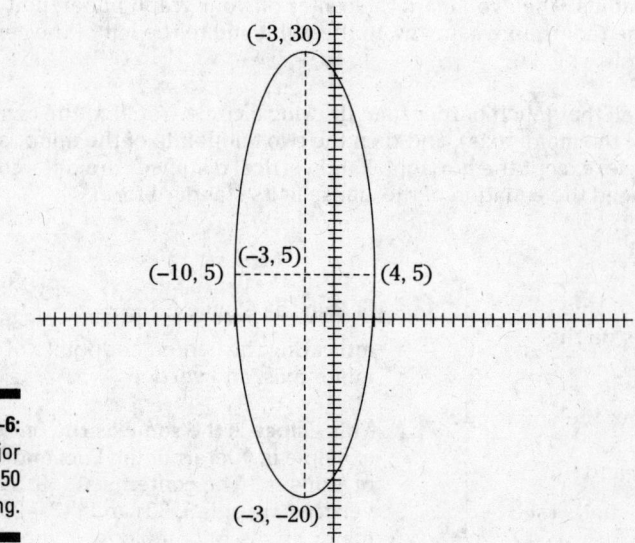

Figure 10-6:
The major axis is 50 units long.

23. Sketch the circle $4x^2 + 4y^2 = 1$, indicating the center and radius on the graph

Solve It

24. Sketch the ellipse $x^2 + 9y^2 = 81$, indicating the center, endpoints of the minor axis, and vertices

Solve It

25. Sketch the circle $x^2 + y^2 - 10x + 2y = 23$, indicating the center and radius on the graph

Solve It

26. Sketch the ellipse $4x^2 + 25y^2 - 8x - 50y = 71$, indicating the center, endpoints of the minor axis, and vertices

Solve It

Hyperbola: Standard Equations and Foci

Of the four types of conic sections, the hyperbola is the only conic that seems a bit disconnected. The graph of a hyperbola is two separate curves seeming to face away from one another. The standard forms for the equation of hyperbolas are

$$\frac{(x-h)^2}{a^2} - \frac{(y-k)^2}{b^2} = 1 \text{ or } \frac{(y-k)^2}{b^2} - \frac{(x-h)^2}{a^2} = 1$$

Notice that these formulas look just like the equation for the ellipse except for the minus sign between the two fractions.

Two forms of the standard equation exist; the form with the *x*-term in front is for hyperbolas that open to the left and right, and the form with the *y*-term in front is for hyperbolas that open upward and downward. The center of the hyperbola is the same old *(h, k)*, as in the circles and ellipses.

You measure distances from the *foci* of a hyperbola to a point on the hyperbola. The *difference* between the distances (in the ellipse it's the *sum*) is always the same for any point on the hyperbola. Solve for the foci with $c^2 = a^2 + b^2$, and let $\pm c$ be the distance from the center to the foci, either vertically or horizontally (depending on the equation, which tells you whether the hyperbola opens up and down or left and right).

Q. Find the standard form of the hyperbola $16x^2 - 9y^2 = 144$. Then give the coordinates of the center and the coordinates of the foci.

A. $\frac{x^2}{9} - \frac{y^2}{16} = 1$; center: **(0, 0)**; foci: **(-5, 0), (5, 0)**. Divide each side of the equation by 144, and you get the standard form. The hyperbola opens left and right, because the x term appears first in the standard form. The center of the hyperbola is $(0, 0)$, the origin. To find the foci, solve for c with $c^2 = a^2 + b^2 = 9 + 16 = 25$. The value of c is ± 5. Counting 5 units to the left and right of the center, the coordinates of the foci are $(-5, 0)$ and $(5, 0)$.

Q. Find the standard form of the hyperbola $576(y - 5)^2 - 49(x - 3)^2 = 28,224$. Then give the coordinates of the center and the coordinates of the foci.

A. $\frac{(y - 5)^2}{49} - \frac{(x - 3)^2}{576} = 1$; center: **(3, 5)**; foci: **(3, 30), (3, -20)**. Divide each side of the equation by 28,224 (yes, the number is huge, but the fractions reduce very nicely) to get the standard form. The hyperbola opens upward and downward, because the y term appears first in the standard form. The center of the hyperbola is $(3, 5)$. To find the foci, solve for c with $c^2 = a^2 + b^2 = 49 + 576 = 625$. The value of c is ± 25. Counting 25 units upward and downward from the center, the coordinates of the foci are $(3, 30)$ and $(3, -20)$.

27. Find the standard form of the hyperbola $3x^2 - 18y^2 = 18$. Then give the coordinates of the center and the coordinates of the foci.

Solve It

28. Find the standard form of the hyperbola $25y^2 - 144x^2 = 3,600$. Then give the coordinates of the center and the coordinates of the foci.

Solve It

29. Find the standard form of the hyperbola $9x^2 - 4y^2 - 54x + 16y + 29 = 0$. Then give the coordinates of the center and the coordinates of the foci.

Solve It

30. Find the standard form of the hyperbola $4y^2 - x^2 + 56y - 4x + 188 = 0$. Then give the coordinates of the center and the coordinates of the foci.

Solve It

Determining the Asymptotes and Intercepts of Hyperbolas

Hyperbolas have two diagonal, intersecting asymptotes (literal guide*lines*) that help you with the general shape of the two curves. As with asymptotes of rational functions, these lines aren't really a part of the curve (see Chapter 8 for info on asymptotes). The asymptotes are helpful when graphing a hyperbola because they show lines that the curves approach but don't cross.

Solve the equation $\dfrac{(x-h)^2}{a^2} = \dfrac{(y-k)^2}{b^2}$ for y to find the equations of the two asymptotes. You get this equation by setting the standard form equal to 0 instead of 1 and then finding the square root of each side, cross-multiplying, and finally writing the two linear equations of the asymptotes.

Hyperbolas have intercepts where they cross the x-axis or y-axis. If the center of the hyperbola is the origin, then the graph has x-intercepts or y-intercepts but not both. To solve for x-intercepts, set y equal to 0 and solve for x. To solve for y-intercepts, let x equal 0 and solve for y.

Q. Find the asymptotes and intercepts of the hyperbola $\dfrac{(x-6)^2}{4} - \dfrac{(y+3)^2}{9} = 1$.

A. **Asymptotes:** $y = \dfrac{3}{2}x - 12$, $y = -\dfrac{3}{2}x + 3$; **y-intercepts:** $(0, -3 + 6\sqrt{2})$, $(0, -3 - 6\sqrt{2})$; **x-intercepts:** $(6 + 2\sqrt{2}, 0)$, $(6 - 2\sqrt{2}, 0)$.

First, solve the equation $\dfrac{(x-6)^2}{4} = \dfrac{(y+3)^2}{9}$ by taking the square root of each side, cross-multiplying, and simplifying:

$$\pm\sqrt{\dfrac{(x-6)^2}{4}} = \sqrt{\dfrac{(y+3)^2}{9}}$$

$$\pm\dfrac{x-6}{2} = \dfrac{y+3}{3}$$

$$\pm\dfrac{3(x-6)}{2} = y+3$$

$$-3 \pm \left(\dfrac{3}{2}x - 9\right) = y$$

$$y = \dfrac{3}{2}x - 12 \text{ or } y = -\dfrac{3}{2}x + 3$$

You have to put the $\pm$ sign on only one side of the equation. (Putting it on both sides just creates four equations, two of which are repeats.) To find the y-intercepts, let x be 0 and solve for y:

$$\dfrac{(0-6)^2}{4} - \dfrac{(y+3)^2}{9} = 1$$

$$9 - \dfrac{(y+3)^2}{9} = 1$$

$$8 = \dfrac{(y+3)^2}{9}$$

$$72 = (y+3)^2$$

$$\pm\sqrt{72} = \sqrt{(y+3)^2}$$

$$\pm 6\sqrt{2} = y+3$$

$$y = -3 \pm 6\sqrt{2}$$

The y-intercepts are approximately at $(0, 5.485)$ and $(0, -11.485)$. To solve for the x-intercepts, let the y in the equation be 0 and solve for x:

$$\dfrac{(x-6)^2}{4} - \dfrac{(0+3)^2}{9} = 1$$

$$\dfrac{(x-6)^2}{4} - 1 = 1$$

$$\dfrac{(x-6)^2}{4} = 2$$

$$(x-6)^2 = 8$$

$$\sqrt{(x-6)^2} = \pm\sqrt{8}$$

$$x - 6 = \pm 2\sqrt{2}$$

$$x = 6 \pm 2\sqrt{2}$$

The two x-intercepts are approximately at $(8.828, 0)$ and $(3.172, 0)$.

31. Find the asymptotes and intercepts of the hyperbola $\dfrac{x^2}{100} - \dfrac{y^2}{64} = 1$

Solve It

32. Find the asymptotes and intercepts of the hyperbola $\dfrac{y^2}{169} - \dfrac{x^2}{25} = 1$

Solve It

33. Find the asymptotes and intercepts of the hyperbola $\dfrac{(x-8)^2}{16} - \dfrac{(y+10)^2}{25} = 1$

Solve It

34. Find the asymptotes and intercepts of the hyperbola $\dfrac{y^2}{169} - \dfrac{(x+5)^2}{144} = 1$

Solve It

Sketching the Hyperbola

The intercepts, foci, asymptotes, and center of a hyperbola are all important characteristics of this conic, but you can make the actual sketch with just the center and a carefully placed rectangle:

- ✔ The center of the rectangle is the center of the hyperbola.

- ✔ The width of the rectangle is twice the square root of the value under the x's in the standard equation; the height of the rectangle is twice the square root of the number under the y's in the equation.

- ✔ The asymptotes are extended diagonals of the rectangle.

- ✔ The vertices of the hyperbola are at the midpoints of the rectangle's sides — either on the left and right sides or on the top and bottom, depending on which ways the hyperbola faces.

Q. Sketch the hyperbola $\dfrac{(x+4)^2}{4} - \dfrac{(y-3)^2}{64} = 1$

A. The center of the hyperbola is (–4, 3). Insert a rectangle around that center by counting two units to the left and two units to the right of the center (these are the midpoints of the sides of the rectangle). Then count eight units up and eight units down from the center to find the other two midpoints. Lightly sketch in the rectangle.

For the asymptotes, draw the diagonals of the rectangle and extend them. See Figure 10-7a for those first steps in the drawing. The hyperbola opens to the right and left, because the x-term comes first in the standard form. Starting with the midpoints on the sides of the rectangle, sketch in the hyperbola, having it get closer to the asymptote as it gets farther from the center. See Figure 10-7b for the entire sketch.

Preliminary pieces of the graph

a)

Final graph

Figure 10-7:
The rectangle is 4 units wide and 16 units high. The asymptotes and rectangle aren't really a part of the hyperbola.

b)

35. Sketch the hyperbola $\dfrac{(x-1)^2}{1} - \dfrac{(y+3)^2}{16} =$

Solve It

36. Sketch the hyperbola $\dfrac{(y-6)^2}{625} - \dfrac{(x-2)^2}{49} =$

Solve It

37. Sketch the hyperbola $x^2 - 9y^2 = 9$

Solve It

38. Sketch the hyperbola $9y^2 - 25x^2 + 54y + 150x = 369$

Solve It

Answers to Problems on Conic Sections

The following are the answers to the practice problems presented earlier in this chapter.

1 Write the equation of the parabola $x^2 - 6x - y + 4 = 0$ in standard form to determine its vertex and in which direction it opens. The answer is **equation: $(x - 3)^2 = y + 5$; vertex: (3, –5); opens upward.**

Complete the square on the left by moving the y and 4 to the right side and adding 9 to each side of the equation. Factor and simplify. The parabola opens upward, because the value of $4a$, the multiplier on the right, is +1.

2 Write the equation of the parabola $2x^2 + 8x + y + 3 = 0$ in standard form to determine its vertex and in which direction it opens. The answer is **equation: $(x + 2)^2 = -\frac{1}{2}(y - 5)$; vertex: (–2, 5); opens downward.**

Move the y and 3 to the right, and then factor 2 out of the two x terms to get $2(x^2 + 4x) = -y + 5$. Complete the square in the parentheses, and add 8 to the right side. Simplify and factor to get $2(x + 2)^2 = -1(y - 5)$. Divide each side by 2. The parabola opens downward, because the value of $4a$, the multiplier on the right, is $-\frac{1}{2}$.

3 Write the equation of the parabola $3y^2 - 12y - x + 9 = 0$ in standard form to determine its vertex and in which direction it opens. The answer is **equation: $\left(y - 2\right)^2 = \frac{1}{3}(x + 3)$; vertex: (–3, 2); opens to the right.**

Move the x and 9 to the right. Then factor 3 out of the two y terms to get $3(y^2 - 4y) = x - 9$. When you complete the square, you have to add 12 on the right. Simplifying and factoring, you get $3(y - 2)^2 = x + 3$. Divide each side by 3. The parabola opens to the right, because the value of $4a$, the multiplier on the right, is positive.

4 Write the equation of the parabola $y^2 - x + 4 = 0$ in standard form to determine its vertex and in which direction it opens. The answer is **equation: $y^2 = x - 4$; vertex: (4, 0); opens to the right.**

Add x and subtract 4 from each side, and you have the standard form. Think of the equation as $(y - 0)^2 = x - 4$. The parabola opens to the right, because the value of $4a$ is +1.

5 Write the equation of the parabola $y^2 + y + x = 0$ in standard form to determine its vertex and in which direction it opens. The answer is **equation: $\left(y + \frac{1}{2}\right)^2 = -1\left(x - \frac{1}{4}\right)$; vertex: $\left(\frac{1}{4}, -\frac{1}{2}\right)$; opens to the left.**

Subtract x from each side. Then complete the square and add $\frac{1}{4}$ to the right side. Factor –1 out of the terms on the right. The –1 is the value of $4a$, so the parabola opens to the left.

6 Write the equation of the parabola $4x^2 - x - y + 1 = 0$ in standard form to determine its vertex and in which direction it opens. The answer is **equation: $\left(x - \frac{1}{8}\right)^2 = \frac{1}{4}\left(y - \frac{15}{16}\right)$; vertex: $\left(\frac{1}{8}, \frac{15}{16}\right)$; opens upward.**

Move the y and 1 to the right. Then factor 4 out of the two x terms. Complete the square, adding $\frac{1}{16}$ to the right. Factor and simplify. The parabola opens upward, because the value of $4a$ is $\frac{1}{4}$.

7 Find the vertex, focus, directrix, and axis of symmetry for the parabola $x^2 = 16y$. The answer is **vertex: (0, 0); focus: (0, 4); directrix: $y = -4$; axis of symmetry: $x = 0$.**

When the equation is written $x^2 = 4(4)y$, you see that the value of a is 4.

8 Find the vertex, focus, directrix, and axis of symmetry for the parabola $y^2 = -100x$. The answer is **vertex: (0, 0); focus: (–25, 0); directrix: $x = 25$; axis of symmetry: $y = 0$.**

When the equation is written $y^2 = 4(-25)x$, you see that the value of a is –25.

9 Find the vertex, focus, directrix, and axis of symmetry for the parabola $8x^2 - 32x + 38 - y = 0$.

The answer is **vertex: (2, 6); focus:** $\left(2, 6\frac{1}{32}\right)$; **directrix:** $y = 5\frac{31}{32}$; **axis of symmetry:** $x = 2$.

The standard form for this equation is

$$(x - 2)^2 = \frac{1}{8}(y - 6)$$
$$= 4\left(\frac{1}{32}\right)(y - 6)$$

when a is equal to $\frac{1}{32}$. You get the standard form by using completing the square. Add y and subtract 38 from each side to get the equation $8x^2 - 32x = y - 38$. Factor 8 from the terms on the left, giving you $8(x^2 - 4x) = y - 38$. Complete the square, adding 32 to the right. $8(x^2 - 4x + 4) = y - 38 + 32$ factors into $8(x - 2)^2 = y - 6$. Finally, divide each side by 8 to get the standard form.

10 Find the vertex, focus, directrix, and axis of symmetry for the parabola $y^2 - 2y + 4x + 13 = 0$. The answer is **vertex: (–3, 1); focus: (–4, 1); directrix: $x = -1$; axis of symmetry: $y = 1$.**

The standard form for this equation is $(y - 2)^2 = -4(x + 3)$. The value of a is –1. Use completing the square to solve for the standard form. First subtract $4x$ and 13 from each side to get $y^2 - 2y = -4x - 13$. Now complete the square on the left, adding 1 to each side: $y^2 - 2y + 1 = -4x - 12$. Factor both sides, and you get $(y - 1)^2 = -4(x + 3)$. Divide each side by –4 to finalize the form.

11 Sketch the graph of $(x - 4)^2 = 12(y + 1)$. Label the vertex, focus, directrix, and axis of symmetry.

The value of a is 3, so the parabola is relatively wide.

12 Sketch the graph of $(y - 1)^2 = 8(x - 5)$. Label the vertex, focus, directrix, and axis of symmetry.

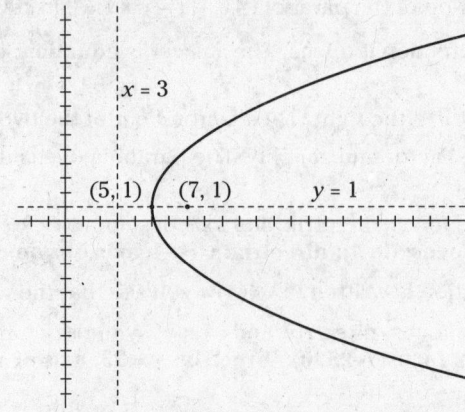

The value of a is 2, so the parabola widens out a bit.

13 Sketch the graph of $4y^2 + 24y + 16x + 33 = 0$. Label the vertex, focus, directrix, and axis of symmetry.

The standard form of the equation is $\left(y + 3\right)^2 = -4\left(x - \frac{3}{16}\right)$.

14 Sketch the graph of $2x^2 + 8x + 11 = y$. Label the vertex, focus, directrix, and axis of symmetry.

The standard form of the equation is $\left(x + 2\right)^2 = \frac{1}{2}\left(y - 3\right)$.

15 Write the equation of the circle $x^2 + y^2 - 6x + 8y = 11$ in standard form. Identify the center and radius of the circle. The answer is **equation: $(x - 3)^2 + (y + 4)^2 = 36$; center: (3, –4); radius: 6.**

Group the x terms and y terms and complete the square in each grouping: $(x^2 - 6x) + (y^2 + 8y) = 11$; $(x^2 - 6x + 9) + (y^2 + 8y + 16) = 11 + 9 + 16 = 36$. Factor the perfect square trinomials to get the standard form.

16 Write the equation of the circle $x^2 + y^2 + 2x - 10y + 17 = 0$ in standard form. Identify the center and radius of the circle. The answer is **equation: $(x + 1)^2 + (y - 5)^2 = 9$; center: (–1, 5); radius: 3.**

Group the x terms and y terms, and move the 17 to the right. Complete the square in each grouping: $(x^2 + 2x) + (y^2 - 10y) = -17$; $(x^2 + 2x + 1) + (y^2 - 10y + 25) = -17 + 1 + 25 = 9$. Factor the perfect square trinomials to get the standard form.

17 Write the equation of the ellipse $9x^2 + 4y^2 - 90x - 16y + 205 = 0$ in standard form. Identify the center and the lengths of the two axes. The answer is **equation:** $\dfrac{(x-5)^2}{4} + \dfrac{(y-2)^2}{9} = 1$; **center: (5, 2); major axis: 6 units; minor axis: 4 units.**

Group the x's together and the y's together; subtract 205 from each side. Factor 9 out of the x terms and 4 out of the y terms. You have $9(x^2 - 10x) + 4(y^2 - 4y) = -205$. Complete the square in each set of parentheses. Be sure to multiply what you insert into the parentheses by the factor outside. Then add that product to the other side: $9(x^2 - 10x + 25) + 4(y^2 - 4y + 4) = -205 + 225 + 16 = 36$. Factor the perfect square trinomials and divide each term by 36. The length of the major axis is twice the square root of 9: $2(3) = 6$. The length of the minor axis is twice the square root of 4: $2(2) = 4$.

18 Write the equation of the ellipse $x^2 + 16y^2 + 4x - 32y + 19 = 0$ in standard form. Identify the center and the lengths of the two axes. The answer is **equation:** $\dfrac{(x+2)^2}{1} + \dfrac{(y-1)^2}{\frac{1}{16}} = 1$; **center: (–2, 1); major axis: 2 units; minor axis: $\frac{1}{2}$ unit.**

Group the x's together and the y's together; subtract 19 from each side. Factor 16 out of the y terms. You have $(x^2 + 4x) + 16(y^2 - 2y) = -19$. Complete the square in each set of parentheses. Be sure to multiply what you insert into the second set of parentheses by the factor outside. Then add that product to the other side: $(x^2 + 4x + 4) + 16(y^2 - 2y + 1) = -19 + 4 + 16 = 1$. Factor the perfect square trinomials. Set the first term over 1. Multiply the numerator and denominator of the second term by $\frac{1}{16}$. The length of the major axis is twice the square root of 1: $2(1) = 2$. The length of the minor axis is twice the square root of $\frac{1}{16}$: $2\left(\frac{1}{4}\right) = \frac{1}{2}$.

19 Find the foci and vertices of the ellipse $\dfrac{(x+6)^2}{16} + \dfrac{(y-2)^2}{25} = 1$. The answer is **foci: (–6, 5), (–6, –1); vertices: (–6, 7), (–6, –3).**

Solving for c, $c^2 = 25 - 16 = 9$; $c = \pm 3$. The major axis runs vertically, because the larger denominator is under the y terms. Add and subtract 3 from the y-coordinate of the center, (–6, 2), to get the coordinates of the foci. Add and subtract 5, the value of b, from the y-coordinate of the center to get the coordinates of the vertices.

20 Find the foci and vertices of the ellipse $\dfrac{(x-1)^2}{625} + \dfrac{(y-7)^2}{576} = 1$. The answer is **foci: (8, 7), (–6, 7); vertices: (26, 7), (–24, 7).**

Solving for c, $c^2 = 625 - 576 = 49$; $c = \pm 7$. The major axis runs horizontally, because the larger denominator is under the x terms. Add and subtract 7 from the x-coordinate of the center, (1, 7), to get the coordinates of the foci. Add and subtract 25, the value of a, from the x-coordinate of the center to get the coordinates of the vertices.

21 Find the foci and vertices of the ellipse $\dfrac{x^2}{169} + \dfrac{y^2}{144} = 1$. The answer is **foci: (5, 0), (–5, 0); vertices: (13, 0), (–13, 0).**

Solving for c, $c^2 = 169 - 144 = 25$; $c = \pm 5$. The major axis runs horizontally, because the larger denominator is under the x terms. Add and subtract 5 from the x-coordinate of the center, (0, 0), to get the coordinates of the foci. Add and subtract 13, the value of a, from the x-coordinate of the center to get the coordinates of the vertices.

22 Find the foci and vertices of the ellipse $100x^2 + 36y^2 = 3,600$. The answer is **foci: (0, 8), (0, –8); vertices: (0, 10), (0, –10).**

The standard form of the equation is $\dfrac{x^2}{36} + \dfrac{y^2}{100} = 1$. Solving for c, $c^2 = 100 - 36 = 64$; $c = \pm 8$. The major axis runs vertically, because the larger denominator is under the y terms. Add and subtract 8 from the y-coordinate of the center, (0, 0), to get the coordinates of the foci. Add and subtract 10, the value of b, from the y-coordinate of the center to get the coordinates of the vertices.

23 Sketch the circle $4x^2 + 4y^2 = 1$, indicating the center and radius on the graph.

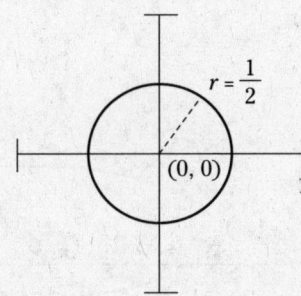

Divide each term in the equation by 4 to get the standard form, $x^2 + y^2 = \frac{1}{4}$. The radius is $\frac{1}{2}$.

24 Sketch the ellipse $x^2 + 9y^2 = 81$, indicating the center, endpoints of the minor axis, and vertices.

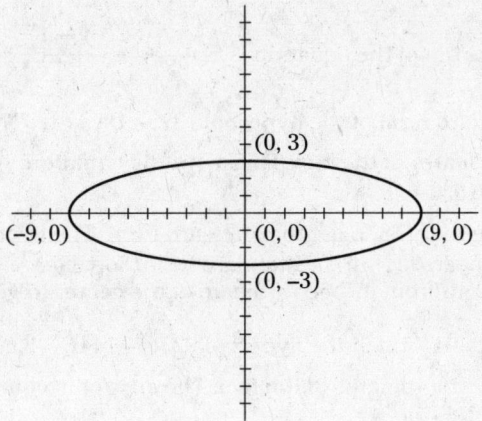

Divide each term in the equation by 81 to get the standard form, $\frac{x^2}{81} + \frac{y^2}{9} = 1$.

25 Sketch the circle $x^2 + y^2 - 10x + 2y = 23$, indicating the center and radius on the graph.

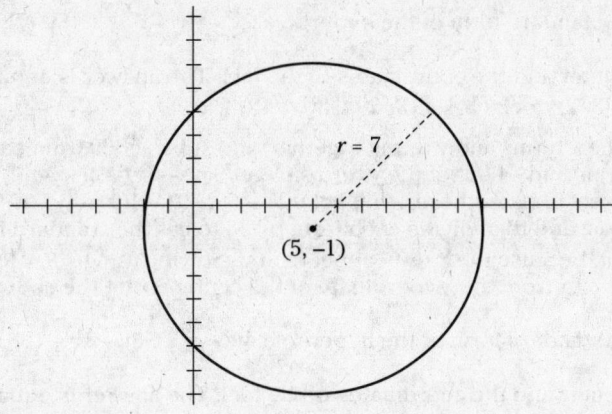

The standard form of the equation is $(x - 5)^2 + (y + 1)^2 = 49$.

26 Sketch the ellipse $4x^2 + 25y^2 - 8x - 50y = 71$, indicating the center, endpoints of the minor axis, and vertices.

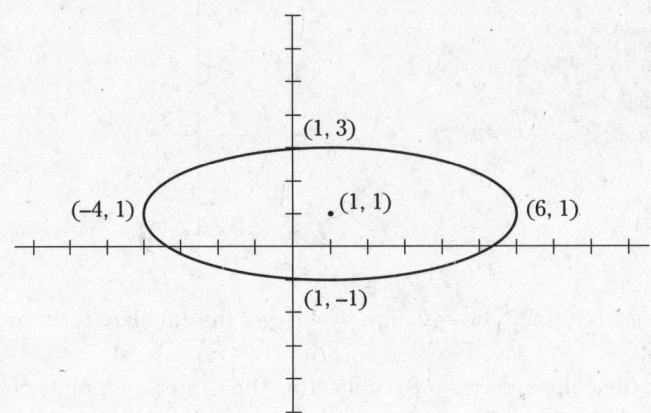

The standard form of the equation is $\dfrac{(x-1)^2}{25} + \dfrac{(y-1)^2}{4} = 1$.

27 Find the standard form of the hyperbola $3x^2 - 18y^2 = 18$. Then give the coordinates of the center and the coordinates of the foci. The answer is **equation:** $\dfrac{x^2}{6} - \dfrac{y^2}{1} = 1$; **center: (0, 0); foci:** $\left(\sqrt{7},0\right), \left(-\sqrt{7},0\right)$.

Divide each term by 18 to get the standard form. The hyperbola opens left and right, because the x term appears first in the standard form. Solving $c^2 = 6 + 1 = 7$, you find that $c = \pm\sqrt{7}$. Add and subtract c to/from the x-coordinate of the center to get the coordinates of the foci.

28 Find the standard form of the hyperbola $25y^2 - 144x^2 = 3{,}600$. Then give the coordinates of the center and the coordinates of the foci. The answer is **equation:** $\dfrac{y^2}{144} - \dfrac{x^2}{25} = 1$; **center: (0, 0); foci: (0, 13), (0, –13).**

Divide each term by 3,600 to get the standard form. The hyperbola opens upward and downward, because the y term appears first in the standard form. Solving $c^2 = 144 + 25 = 169$, you find that $c = \pm13$. Add and subtract c to/from the y-coordinate of the center to get the coordinates of the foci.

29 Find the standard form of the hyperbola $9x^2 - 4y^2 - 54x + 16y + 29 = 0$. Then give the coordinates of the center and the coordinates of the foci. The answer is **equation:** $\dfrac{(x-3)^2}{4} - \dfrac{(y-2)^2}{9} = 1$; **center: (3, 2); foci:** $\left(3 + \sqrt{13}, 2\right), \left(3 - \sqrt{13}, 2\right)$.

Group the x terms and y terms together, and subtract 29 from each side. Then factor 9 out of the x terms and –4 out of the y terms to get $9(x^2 - 6x) - 4(y^2 - 4y) = -29$. Completing the square in both sets of parentheses, you get $9(x^2 - 6x + 9) - 4(y^2 - 4y + 4) = -29 + 81 - 16 = 36$. Factor the trinomials, and then divide each term by 36 to get the standard form. The hyperbola opens left and right, because the x term appears first. Solving $c^2 = 4 + 9 = 13$, you get $c = \pm\sqrt{13}$. Add and subtract c to/from the x-coordinate of the center to get the coordinates of the foci.

30 Find the standard form of the hyperbola $4y^2 - x^2 + 56y - 4x + 188 = 0$. Then give the coordinates of the center and the coordinates of the foci. The answer is **equation:** $\dfrac{(y+7)^2}{1} - \dfrac{(x+2)^2}{4} = 1$; **center: (–2, –7); foci:** $\left(-2, -7 + \sqrt{5}\right), \left(-2, -7 - \sqrt{5}\right)$.

Group the y terms and x terms together, and subtract 188 from each side. Then factor 4 out of the y terms and –1 out of the x terms to get $4(y^2 + 14y) - (x^2 + 4x) = -188$. Completing the square in both sets of parentheses, you get $4(y^2 + 14y + 49) - (x^2 + 4x + 4) = -188 + 196 - 4 = 4$. Factor the trinomials, and then divide each term by 4 to get the standard form. The hyperbola opens upward and downward, because the y term appears first. Solving $c^2 = 1 + 4 = 5$, you find that $c = \pm\sqrt{5}$. Add and subtract c to/from the y-coordinate of the center to get the coordinates of the foci.

31 Find the asymptotes and intercepts of the hyperbola $\dfrac{x^2}{100} - \dfrac{y^2}{64} = 1$. The answer is **asymptotes:** $y = \dfrac{4}{5}\,x,\ y = -\dfrac{4}{5}\,x$; **intercepts: (10, 0), (–10, 0).**

Set the two fractions equal to one another. Then solve for y, multiplying each side by 64 and taking the square root of each side to get the two equations of the asymptotes. Find the x-intercepts by letting $y = 0$ and solving for x. The curve has no y-intercepts.

32 Find the asymptotes and intercepts of the hyperbola $\dfrac{y^2}{169} - \dfrac{x^2}{25} = 1$. The answer is **asymptotes:** $y = \dfrac{13}{5}\,x,\ y = -\dfrac{13}{5}\,x$; **intercepts: (0, 13), (0, –13).**

Set the two fractions equal to one another. Then solve for y, multiplying each side by 169 and taking the square root of each side to get the two equations of the asymptotes. Find the y-intercepts by letting $x = 0$ and solving for y. There are no x-intercepts.

33 Find the asymptotes and intercepts of the hyperbola $\dfrac{(x-8)^2}{16} - \dfrac{(y+10)^2}{25} = 1$. The answer is **asymptotes:** $y = \dfrac{5}{4}\,x - 20,\ y = -\dfrac{5}{4}\,x$; **intercepts:** $\left(0,\ -10 + 5\sqrt{3}\right),\ \left(0,\ -10 - 5\sqrt{3}\right),\ \left(8 + 4\sqrt{5}, 0\right)$ $\left(8 - 4\sqrt{5}, 0\right).$

Set the two fractions equal to one another. Then solve for y, multiplying each side by 25 and taking the square root of each side. Finally, subtract 10 from each side and simplify to get the two equations of the asymptotes. Determine the y-intercepts by letting $x = 0$ and solving for y. Find the x-intercepts by letting $y = 0$ and solving for x.

34 Find the asymptotes and intercepts of the hyperbola $\dfrac{y^2}{169} - \dfrac{(x+5)^2}{144} = 1$. The answer is **asymptotes:** $y = \dfrac{13}{12}\,x + \dfrac{65}{12},\ y = -\dfrac{13}{12}\,x - \dfrac{65}{12}$; **intercepts:** $\left(0,\ \dfrac{169}{12}\right),\ \left(0,\ -\dfrac{169}{12}\right).$

Set the two fractions equal to one another. Then solve for y, multiplying each side by 169 and taking the square root of each side to get the two equations of the asymptotes. Find the y-intercepts by letting $x = 0$ and solving for y. There are no x-intercepts.

35 Sketch the hyperbola $\dfrac{(x-1)^2}{1} - \dfrac{(y+3)^2}{16} = 1.$

The center is at (1, –3), and the hyperbola opens left and right.

36   Sketch the hyperbola $\dfrac{(y-6)^2}{625} - \dfrac{(x-2)^2}{49} = 1$.

The center is at $(2, 6)$, and the hyperbola opens upward and downward.

37   Sketch the hyperbola $x^2 - 9y^2 = 9$.

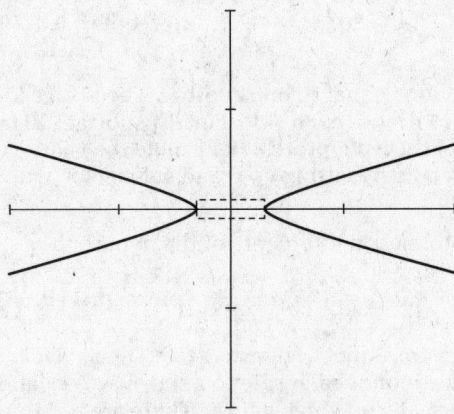

The standard form of the equation is $\dfrac{x^2}{9} - \dfrac{y^2}{1} = 1$. The center is at $(0, 0)$, and the hyperbola opens left and right.

38   Sketch the hyperbola $9y^2 - 25x^2 + 54y + 150x = 369$.

The standard form of the equation is $\dfrac{(y-3)^2}{25} - \dfrac{(x-3)^2}{9} = 1$. The center is at $(3, 3)$, and the hyperbola opens upward and downward.

Chapter 11

Solving Systems of Linear Equations

Linear equations have variables that don't exceed the power of one. For example, the equation $2x + 3y - z = 0$ is linear. A *system* of linear equations can contain any number of equations and any number of variables. But the only systems that have the possibility of having a solution are those where you have at least as many equations as variables. When looking for solutions of systems of equations, you try to get one numerical value for x, one for y, one for z, and so on.

In this chapter, you use substitution and elimination to solve the linear systems. You also use Cramer's Rule as an alternative when fractional answers get nasty. Finally, you *decompose fractions,* which means breaking fractions into simpler fractions. Linear equations come in very handy when decomposition is desired (bet you never expected to read that decomposition is desirable!).

Solving Two Linear Equations Algebraically

The solution of a system of two linear equations consists of the values of x and y that make both of the equations true — at the same time. Graphically, the solution is the point where the two lines intersect. The two most frequently used methods for solving systems of linear equations are elimination and substitution:

▶ **Elimination (also called add-subtract):** This method involves adding the two equations together — or multiples of the two equations — so that in the sum, the coefficient on one of the variables becomes 0. That variable drops out (is eliminated), so you can solve for the other variable. Then you plug the solution back into one of the original equations and solve for the variable you eliminated.

▶ **Substitution:** This method has you set one of the equations equal to x or y. You can then substitute the *equivalent* of the variable from one equation for that variable in the other equation. You end up with a single-variable equation, which you can solve. Then plug that answer into one of the original equations and solve for the other variable.

You can use either method to solve linear systems, and you choose one over the other if a method seems to work better in a particular system (substitution works best if the coefficient on one of the variables is 1 or –1). The following examples show the same system of equations solved using both methods.

Q. Use elimination to solve for the common solution in the two equations: $x + 3y = 4$ and $2x + 5y = 5$

A. **$x = -5$, $y = 3$.** Multiply each term in the first equation by –2 (you get $-2x - 6y = -8$) and then add the terms in the two equations together. You choose the number –2 as a multiplier because it makes the coefficient of the x term in the first equation equal to –2, while the coefficient on x in the second equation is 2. The numbers –2 and 2 are opposites, so adding the equations together *eliminates* the x term:

$$-2x - 6y = -8$$
$$\underline{2x + 5y = 5}$$
$$-y = -3$$

Now solve $-y = -3$ for y, and you get $y = 3$. Put 3 in for y in the first original equation, and you have $x + 3(3) = 4$; $x + 9 = 4$; $x = -5$. The solution is $x = -5$, $y = 3$, also written as the ordered pair $(-5, 3)$. You can also solve for the x-value by putting the 3 into the second equation — you get the same result.

Q. Use substitution to solve for the common solution in the two equations: $x + 3y = 4$ and $2x + 5y = 5$

A. **$x = -5$, $y = 3$.** To use substitution, select a variable in one of the equations with a coefficient of 1 or –1. The only variable that qualifies in this system is x in the first equation. Solve for x in terms of y in that equation. You get $x = 4 - 3y$. Substitute that equivalent of x into the second equation. The second equation becomes $2(4 - 3y) + 5y = 5$. Solve that equation for y: $8 - 6y + 5y = 5$; $8 - y = 5$; $-y = -3$; $y = 3$. That answer should look familiar. Substitute the 3 into $x + 3y = 4$ to get x: $x + 3(3) = 4$; $x + 9 = 4$; $x = -5$.

1. Solve for the common solution in the two equations: $5x - 3y = 7$ and $2x + 3y = 7$

Solve It

2. Solve for the common solution in the two equations: $8x - 3y = 41$ and $3x + 2y = 6$

Solve It

3. Solve for the common solution in the two equations: $4x + 5y = 11$ and $y = 2x + 5$

Solve It

4. Solve for the common solution in the two equations: $3x - 4y = 10$ and $x - 3y = 10$

Solve It

Using Cramer's Rule to Defeat Unruly Fractions

Cramer's Rule — named for the mathematician Gabriel Cramer — is a way of solving a system of linear equations by focusing on just the coefficients and constants and ignoring (well, not really) the variables. You write the linear equations with the variables in exactly the same order, and then you can just leave the variables out of the computations. The technique is really rather slick, but you have to be organized and remember what you put where.

Cramer's Rule works on systems of two linear equations — whether they're unruly or not. You prefer to use elimination or substitution when the systems add together nicely or substitute one into the other (see the preceding section). This rule comes in handy when the coefficients of the variables are large or have no factors in common. Cramer is most appreciated when the answers are big, unwieldy fractions.

Essentially, Cramer's Rule involves a bunch of cross-products. You arrange squares of numbers, multiply across the diagonals, and find the difference to solve the systems of equations.

Writing a system of two linear equations as

$$a_1 x + b_1 y = c_1$$
$$a_2 x + b_2 y = c_2$$

let d represent the denominator, let n_x represent the x numerator, and let n_y represent the y numerator. The squares of numbers corresponding to these values are

$$d = \begin{vmatrix} a_1 & b_1 \\ a_2 & b_2 \end{vmatrix}, \; n_x = \begin{vmatrix} c_1 & b_1 \\ c_2 & b_2 \end{vmatrix}, \text{ and } n_y = \begin{vmatrix} a_1 & c_1 \\ a_2 & c_2 \end{vmatrix}.$$ Use these squares to form the cross-products:

$$d = a_1 b_2 - b_1 a_2$$
$$n_x = c_1 b_2 - b_1 c_2$$
$$n_y = a_1 c_2 - c_1 a_2$$

The solution of the system of equations is then $x = \dfrac{n_x}{d} = \dfrac{c_1 b_2 - b_1 c_2}{a_1 b_2 - b_1 a_2}, \; y = \dfrac{n_y}{d} = \dfrac{a_1 c_2 - c_1 a_2}{a_1 b_2 - b_1 a_2}.$

The formulas for Cramer's Rule aren't really as complicated as they may appear at first glance. When you're solving for x, you get the square of numbers for the numerator by replacing the x coefficients with the constants. And likewise, when you're solving for y, you get the square of numbers for the numerator by replacing the y coefficients with the constants.

Q. Solve the system of equations $12x + 13y = 10$ and $4x + 9y = 12$ using Cramer's Rule

A. $x = -\dfrac{33}{28}, y = \dfrac{13}{7}$. The three squares of numbers and their cross products are

$$d = \begin{vmatrix} 12 & 13 \\ 4 & 9 \end{vmatrix} = 12(9) - 13(4) = 108 - 52 = 56$$

$$n_x = \begin{vmatrix} 10 & 13 \\ 12 & 9 \end{vmatrix} = 10(9) - 13(12) = 90 - 156 = -66$$

$$n_y = \begin{vmatrix} 12 & 10 \\ 4 & 12 \end{vmatrix} = 12(12) - 10(4) = 144 - 40 = 104$$

To solve for x, divide n_x by d to get $\dfrac{-66}{56} = -\dfrac{33}{28}$. To solve for y, divide n_y by d to get $\dfrac{104}{56} = \dfrac{13}{7}$. These fractions are perfectly good numbers. They just don't cooperate as well with the elimination or substitution methods.

5. Solve the system of equations $16x + 5y = 4$ and $13x + 3y = -1$ using Cramer's Rule

Solve It

6. Solve the system of equations $5x + 7y = 11$ and $4x + 6y = 7$ using Cramer's Rule

Solve It

7. Solve the system of equations $3x - 8y = 9$ and $2x - 5y = 4$ using Cramer's Rule

Solve It

8. Solve the system of equations $6x - 9y = 8$ and $12x + 7y = 4$ using Cramer's Rule

Solve It

A Third Variable: Upping the Systems to Three Linear Equations

When working with systems of equations, you can solve for only one variable at a time. So if a third linear equation comes along (bringing of course, its variable z), well, three's a crowd. However, you can easily deal with all the variables as long as you address each in turn.

You solve systems of three (or more) linear equations using the elimination method ("Solving Two Linear Equations Algebraically," earlier in this chapter, explains how the basic method works):

1. **Starting with three equations, eliminate one variable to create two equations with the two remaining variables.**

   Pair the first equation with the second, the second with the third, or the first with the third to eliminate one of the variables. Then choose a different pairing and eliminate the same variable.

2. **From those two new equations, eliminate a second variable so you can solve for the one that remains.**

3. **Substitute back into the other equations to find the values of the other variables.**

   Plug the first variable you solved for into one of the two-variable equations you found in step 1. Then solve for the third variable by plugging the known values into one of the original equations.

You may more easily understand this process with an example, so that's where I go next. (You can also solve systems of equations with matrices. See Chapter 14 for details.)

Q. Find the common solution of the system of equations $x + 5y - 2z = 2$, $4x + 3y + 2z = 2$, and $3x - 3y - 5z = 38$

A. **$x = 4$, $y = -2$, $z = -4$ — also written as the ordered triple (4, -2, -4).** You can choose to eliminate any of the three variables, but there's usually a good-better-best-worse-worst decision to make. In this problem, the best choice is to eliminate the x variable. The x variable has the only coefficient of 1 in all of the equations. You look for a 1 or -1 or for multiples of the same number in the coefficients of a single variable.

Do two pairings of elimination. Multiply the first equation by -4 and add it to the second equation:

$$-4x - 20y + 8z = -8$$
$$\underline{4x + 3y + 2x = 2}$$
$$-17y + 10z = -6$$

For the second pairing, multiply the first equation by -3 and add it to the third equation:

$$-3x - 15y + 6z = -6$$
$$\underline{3x - 3y - 5z = 38}$$
$$-18y + z = 32$$

Then add the two equations that result (after multiplying the second equation by -10 so you can eliminate the z's):

$$-17y + 10z = -6$$
$$\underline{180y - 10z = -320}$$
$$163y = -326$$

Divide each side of the equation by 163 to get $y = -2$. Replace the y in $-18y + z = 32$ with the -2, and you get $-18(-2) + z = 32$; $36 + z = 32$; $z = -4$. Now take the values for y and z and put them into any of the original equations (I'd choose the first) to solve for x. You get $x + 5(-2) - 2(-4) = 2$; $x - 10 + 8 = 2$; $x - 2 = 2$; $x = 4$.

9. Find the common solution of the system of equations $3x + 4y - z = 7$, $2x - 3y + 3z = 5$, and $x + 5y - 2z = 0$

Solve It

10. Find the common solution of the system of equations $8x + 3y - 2z = -2$, $x - 3y + 4z = -13$, and $6x + 4y - z = -3$

Solve It

11. Find the common solution of the system of equations $x + y - 4z = 6$, $4x - 5y + 2z = 24$, and $3x - 8y - 3z = 5$

Solve It

12. Find the common solution of the system of equations $6x + y - 4z = 11$, $3x - y - 3z = 6$, and $9x + 4y + z = 5$

Solve It

A Line by Any Other Name: Writing Generalized Solution Rules

A system of linear equations has a single numerical solution — if one exists. Parallel lines never cross, so if the equations represent parallel lines, there's no solution.

But a system such as $x - 3y = 7$ and $2x = 6y + 14$ has an infinite number of common solutions. The points $(4, -1)$, $(7, 0)$, and $(-2, -3)$ all work — to name a few. This situation occurs when one equation is actually a multiple of the other; the two equations are just two different ways of saying the same thing. Graphically, these equations represent the same line. A generalized solution, or a rule that creates all the possible solutions, is $(3y + 7, y)$. This answer is in a point form or ordered pair, *(x, y),* and it shows that you pick a number for y and then find the x by multiplying y by 3 and adding 7.

The same type of situation can also occur with three or more linear equations — where there's an infinite number of solutions, all bound by a relationship tied to one of the variables. You find these types of solutions when elimination completely wipes out one of the equations. When that happens, you solve for each of the variables in terms of the others.

Q. Solve the system of equations $x - 2y = 5$ and $8x = 16y + 40$

A. **(2y + 5, y).** Rewrite the system using the same format for each equation so you can apply elimination. You get $x - 2y = 5$ and $8x - 16y = 40$. Multiply the terms in the first equation by -8 and add them to the second equation. You get

$$\begin{aligned} -8x + 16y &= -40 \\ \underline{8x - 16y} &= \underline{40} \\ 0 + 0 &= 0 \end{aligned}$$

Zero always equals 0. This is your signal that this system has an infinite number of solutions. Solve the first equation for x to get $x = 2y + 5$. You write your solution as $(2y + 5, y)$. The answer means that if you substitute any number for y, you get the corresponding value of x. All these solutions work in both of the original equations.

Q. Solve the system of equations $4x - 3y + z = 3$, $x - 4y - z = 15$, and $2x + 5y + 3z = -27$

A. $\left(x, \dfrac{5x - 18}{7}, \dfrac{-13x - 33}{7}\right)$. Eliminate the z's. Add the first and second equations together to get $5x - 7y = 18$. Next, multiply the second equation by 3 and add it to the third:

$$\begin{aligned} 3x - 12y - 3z &= 45 \\ \underline{2x + 5y + 3z} &= \underline{-27} \\ 5x - 7y &= 18 \end{aligned}$$

You get the same equation as you did when adding the first two equations together. Combining the two resulting equations just gets you $0 = 0$. So solve for either x or y in terms of the other, and then solve for z in terms of that same variable. Solving for y in terms of x, you get $5x - 18 = 7y$; $y = \dfrac{5x - 18}{7}$. Now, to solve for z in terms of x, substitute for y in the first equation:

$$4x - 3\left(\frac{5x - 18}{7}\right) + z = 3$$

$$z = 3 - 4x + \frac{15x - 54}{7} = \frac{21 - 28x}{7} + \frac{15x - 54}{7}$$

$$z = \frac{-13x - 33}{7}$$

The ordered triple representing the solution is $\left(x, \frac{5x - 18}{7}, \frac{-13x - 33}{7}\right)$. For instance, if you let $x = -2$, you get $y = -4$ and $z = -1$. The possibilities are endless.

---

13. Solve the system of equations: $x - y = 3$ and $2x = 2y + 6$

Solve It

14. Solve the system of equations: $3y - x = 7$ and $2x + 14 = 6y$

Solve It

15. Solve the system of equations: $x - y = 11$, $5x + 3z = 60$, and $5y + 3z = 5$

Solve It

16. Solve the system of equations: $x + 2y + z = 8$, $3x - 2y + 4z = 1$, and $x - 6y + 2z = -15$

Solve It

Decomposing Fractions Using Systems

To add or subtract fractions, you first find a common denominator. Then, when you perform the operation, you combine the numerators and leave the denominators the same.

You use the same principle for the reverse process: taking a fraction apart to see what was added or subtracted to get that result. The process is called *decomposition of fractions* or *partial fractions,* and it uses systems of equations to determine the solution. (Really, decomposition is a pretty nifty trick. I could introduce it to you with lots of letters representing constants, but I'd prefer not to muddy the waters! Take a look at the example to see how it's done.)

Q. Find the two fractions whose sum or difference is equal to $\dfrac{12x - 3}{2x^2 - 7x - 4}$

A. $\dfrac{2}{(2x + 1)} + \dfrac{5}{(x - 4)}$. Here how to get the answer, step by step.

1. **Factor the denominator.**

   The denominator of the fraction factors into $(2x + 1)(x - 4)$, so there are two fractions — one with a denominator of $2x + 1$ and another with a denominator of $x - 4$ — whose sum or difference is this fraction.

2. **Set the original fraction equal to A over the first binomial plus B over the second binomial.**

   Write $\dfrac{12x - 3}{2x^2 - 7x - 4} = \dfrac{A}{2x + 1} + \dfrac{B}{x - 4}$. The goal is to find out what A and B are equal to.

3. **On the right, add the two fractions together by writing each with the common denominator; set the numerator equal to the numerator of the original problem.**

   Your common denominator is $(2x + 1)(x - 4)$:

   $$\frac{A}{2x + 1} + \frac{B}{x - 4} = \frac{A(x - 4)}{(2x + 1)(x - 4)} + \frac{B(2x + 1)}{(x - 4)(2x + 1)}$$
   $$= \frac{A(x - 4) + B(2x + 1)}{(2x + 1)(x - 4)}$$

   The numerator of the original problem must be equal to the numerator of the sum, so write $12x - 3 = A(x - 4) + B(2x + 1)$.

   Tip: You can do step 3 by simply multiplying both sides of the equation by the least common denominator (use the factored denominator from step 1).

4. **On the right, distribute the A and the B; factor out the x's and group the remaining terms.**

   $12x - 3 = Ax - 4A + 2Bx + B$

   $12x - 3 = (A + 2B)x + (-4A + B)$

5. **Use these groupings to set up a system of equations; solve for A and B.**

   ✔ The coefficient of x is equal to 12 on the left and *(A + 2B)* on the right, so $12 = A + 2B$.

   ✔ The constants are equivalent to –3 and $(-4A + B)$, so $-3 = -4A + B$.

   Solve the system of equations $12 = A + 2B$ and $-3 = -4A + B$ using elimination. Multiply the first equation by 4 and add it to the second equation:

   $$\begin{aligned} 48 &= 4A + 8B \\ -3 &= -4A + B \\ \hline 45 &= \phantom{-4A + {}}9B \\ 5 &= B \end{aligned}$$

   Substituting $B = 5$ into the first equation, $12 = A + 2(5)$; $12 = A + 10$; $A = 2$.

6. **Plug the A and B into your equation from step 2.**

   Replacing the A with 2 and the B with 5 in the fractions, you get $\dfrac{12x - 3}{2x^2 - 7x - 4} = \dfrac{2}{2x + 1} + \dfrac{5}{x - 4}$. The fraction is decomposed, but it certainly doesn't stink!

17. Find the two fractions whose sum or difference is equal to $\dfrac{10x + 5}{2x^2 + x - 3}$

Solve It

18. Find the two fractions whose sum or difference is equal to $\dfrac{2x - 22}{x^2 - x - 12}$

Solve It

19. Find the two fractions whose sum or difference is equal to $\dfrac{20 - x}{2x^2 + 19x + 35}$

Solve It

20. Find the two fractions whose sum or difference is equal to $\dfrac{7x}{x^2 - x - 12}$

Solve It

Answers to Problems on Systems of Equations

The following are the answers to the practice problems presented earlier in this chapter.

1 Solve for the common solution in the two equations: $5x - 3y = 7$ and $2x + 3y = 7$. The answer is **$x = 2, y = 1$.**

The coefficients of the y terms are opposites of one another, so when you add the two equations together, you get $7x = 14$; $x = 2$. Replace the x with 2 in the first equation: $5(2) - 3y = 7$; $10 - 3y = 7$, $-3y = -3$; $y = 1$.

2 Solve for the common solution in the two equations: $8x - 3y = 41$ and $3x + 2y = 6$. The answer is **$x = 4, y = -3$.**

Multiply the terms in the first equation by 2 and the terms in the second equation by 3. As a result, you end up adding $-6y$ and $6y$ together, which eliminates the y terms when you add the two equations. You get $25x = 100$; $x = 4$. Replace the x with 4 in the second equation: $3(4) + 2y = 6$; $12 + 2y = 6$; $2y = -6$; $y = -3$.

3 Solve for the common solution in the two equations: $4x + 5y = 11$ and $y = 2x + 5$. The answer is **$x = -1, y = 3$.**

The second equation is already solved for y. Substitute the equivalent of y from the second equation into the first equation to get $4x + 5(2x + 5) = 11$. Distribute and simplify: $4x + 10x + 25 = 11$; $14x + 25 = 11$; $14x = -14$; $x = -1$. Replace the x with -1 in the second equation: $y = 2(-1) + 5 = 3$.

4 Solve for the common solution in the two equations: $3x - 4y = 10$ and $x - 3y = 10$. The answer is **$x = -2, y = -4$.**

Solve for x in the second equation to get $x = 3y + 10$. Substitute the equivalent of x into the first equation, and you get $3(3y + 10) - 4y = 10$. Distribute and simplify: $9y + 30 - 4y = 10$; $5y = -20$; $y = -4$. Now replace the y in the second equation to get $x - 3(-4) = 10$; $x + 12 = 10$; $x = -2$.

5 Solve the system of equations $16x + 5y = 4$ and $13x + 3y = -1$ using Cramer's Rule. The answer is **$x = -1, y = 4$.**

$$x = \frac{n_x}{d} = \frac{\begin{vmatrix} 4 & 5 \\ -1 & 3 \end{vmatrix}}{\begin{vmatrix} 16 & 5 \\ 13 & 3 \end{vmatrix}} = \frac{4(3) - 5(-1)}{16(3) - 5(13)} = \frac{12 + 5}{48 - 65} = \frac{17}{-17} = -1$$

$$y = \frac{n_y}{d} = \frac{\begin{vmatrix} 16 & 4 \\ 13 & -1 \end{vmatrix}}{\begin{vmatrix} 16 & 5 \\ 13 & 3 \end{vmatrix}} = \frac{16(-1) - 4(13)}{16(3) - 5(13)} = \frac{-16 - 52}{48 - 65} = \frac{-68}{-17} = 4$$

6 Solve the system of equations $5x + 7y = 11$ and $4x + 6y = 7$ using Cramer's Rule. The answer is **$x = \frac{17}{2}, y = -\frac{9}{2}$.**

$$x = \frac{n_x}{d} = \frac{\begin{vmatrix} 11 & 7 \\ 7 & 6 \end{vmatrix}}{\begin{vmatrix} 5 & 7 \\ 4 & 6 \end{vmatrix}} = \frac{11(6) - 7(7)}{5(6) - 7(4)} = \frac{66 - 49}{30 - 28} = \frac{17}{2}$$

$$y = \frac{n_y}{d} = \frac{\begin{vmatrix} 5 & 11 \\ 4 & 7 \end{vmatrix}}{\begin{vmatrix} 5 & 7 \\ 4 & 6 \end{vmatrix}} = \frac{5(7) - 11(4)}{5(6) - 7(4)} = \frac{35 - 44}{30 - 28} = \frac{-9}{2}$$

7 Solve the system of equations $3x - 8y = 9$ and $2x - 5y = 4$ using Cramer's Rule. The answer is $x = -13$, $y = -6$.

$$x = \frac{n_x}{d} = \frac{\begin{vmatrix} 9 & -8 \\ 4 & -5 \end{vmatrix}}{\begin{vmatrix} 3 & -8 \\ 2 & -5 \end{vmatrix}} = \frac{9(-5) - (-8)(4)}{3(-5) - (-8)(2)} = \frac{-45 + 32}{-15 + 16} = \frac{-13}{1} = -13$$

$$y = \frac{n_y}{d} = \frac{\begin{vmatrix} 3 & 9 \\ 2 & 4 \end{vmatrix}}{\begin{vmatrix} 3 & -8 \\ 2 & -5 \end{vmatrix}} = \frac{3(4) - 9(2)}{3(-5) - (-8)(2)} = \frac{12 - 18}{-15 + 16} = \frac{-6}{1} = -6$$

8 Solve the system of equations $6x - 9y = 8$ and $12x + 7y = 4$ using Cramer's Rule. The answer is $x = \frac{46}{75}$, $y = -\frac{12}{25}$.

$$x = \frac{n_x}{d} = \frac{\begin{vmatrix} 8 & -9 \\ 4 & 7 \end{vmatrix}}{\begin{vmatrix} 6 & -9 \\ 12 & 7 \end{vmatrix}} = \frac{8(7) - (-9)(4)}{6(7) - (-9)(12)} = \frac{56 + 36}{42 + 108} = \frac{92}{150} = \frac{46}{75}$$

$$y = \frac{n_y}{d} = \frac{\begin{vmatrix} 6 & 8 \\ 12 & 4 \end{vmatrix}}{\begin{vmatrix} 6 & -9 \\ 12 & 7 \end{vmatrix}} = \frac{6(4) - (8)(12)}{6(7) - (-9)(12)} = \frac{24 - 96}{42 + 108} = \frac{-72}{150} = -\frac{12}{25}$$

9 Find the common solution of the system of equations $3x + 4y - z = 7$, $2x - 3y + 3z = 5$, and $x + 5y - 2z = 0$. The answer is $x = 4$, $y = -2$, $z = -3$.

Eliminate x's by multiplying the third equation by -3 and adding it to the first equation; you get $-11y + 5z = 7$. Then eliminate x's in another combination by multiplying the original third equation by -2 and adding it to the second equation; you get $-13y + 7z = 5$. Use Cramer's Rule on these two resulting equations (see the section "Using Cramer's Rule to Defeat Unruly Fractions"):

$$y = \frac{n_y}{d} = \frac{\begin{vmatrix} 7 & 5 \\ 5 & 7 \end{vmatrix}}{\begin{vmatrix} -11 & 5 \\ -13 & 7 \end{vmatrix}} = \frac{49 - 25}{-77 + 65} = \frac{24}{-12} = -2$$

$$z = \frac{n_z}{d} = \frac{\begin{vmatrix} -11 & 7 \\ -13 & 5 \end{vmatrix}}{\begin{vmatrix} -11 & 5 \\ -13 & 7 \end{vmatrix}} = \frac{-55 + 91}{-77 + 65} = \frac{36}{-12} = -3$$

Now substitute -2 for y and -3 for z in the original third equation to solve for x. You get $x + 5(-2) - 2(-3) = 0$; $x - 10 + 6 = 0$; $x - 4 = 0$; $x = 4$.

10 Find the common solution of the system of equations $8x + 3y - 2z = -2$, $x - 3y + 4z = -13$, and $6x + 4y - z = -3$. The answer is $x = -1$, $y = 0$, $z = -3$.

Eliminate z's by multiplying the second equation by -8 and adding it to the first equation to get $17x + 3y = -17$. Then eliminate z's in another combination by multiplying the original second equation by -6 and adding it to the third equation; you get $25x + 13y = -25$. Use Cramer's Rule on these two resulting equations:

$$x = \frac{n_x}{d} = \frac{\begin{vmatrix} -17 & 3 \\ -25 & 13 \end{vmatrix}}{\begin{vmatrix} 17 & 3 \\ 25 & 13 \end{vmatrix}} = \frac{-221 + 75}{221 - 75} = \frac{-146}{146} = -1$$

$$y = \frac{n_y}{d} = \frac{\begin{vmatrix} 17 & -17 \\ 25 & -25 \end{vmatrix}}{\begin{vmatrix} 17 & 3 \\ 25 & 13 \end{vmatrix}} = \frac{-425 + 425}{221 - 75} = \frac{0}{146} = 0$$

Now substitute $x = -1$ and $y = 0$ into the original third equation to get $6(-1) + 4(0) - z = -3$; $-6 - z = -3$; $-z = 3$; $z = -3$.

11 Find the common solution of the system of equations $x + y - 4z = 6$, $4x - 5y + 2z = 24$, and $3x - 8y - 3z = 5$. The answer is $\boldsymbol{x = 8, y = 2, z = 1}$.

Eliminate y's by multiplying the first equation by 5 and adding it to the second equation to get $9x - 18z = 54$. Divide the terms in this equation by 9 to get $x - 2z = 6$, and then solve for x so you can do a substitution: $x = 2z + 6$. Now multiply the original first equation by 8 and add it to the third equation to get $11x - 35z = 53$. Substitute $2z + 6$ in for the x to get $11(2z + 6) - 35z = 53$; $22z + 66 - 35z = 53$; $-13z + 66 = 53$; $-13z = -13$; $z = 1$. The value of x is then $x = 2(1) + 6 = 8$. Replace the x with 8 and the z with 1 in the first original equation to get $8 + y - 4(1) = 6$; $y + 4 = 6$; $y = 2$.

12 Find the common solution of the system of equations $6x + y - 4z = 11$, $3x - y - 3z = 6$, and $9x + 4y + z = 5$. The answer is $\boldsymbol{x = \frac{1}{3};\ y = 1;\ z = -2}$.

Eliminate y's by adding the first and second equations together to get $9x - 7z = 17$. Now multiply the second equation by 4 and add it to the third equation to get $21x - 11z = 29$. Use Cramer's Rule on these two resulting equations:

$$x = \frac{n_x}{d} = \frac{\begin{vmatrix} 17 & -7 \\ 29 & -11 \end{vmatrix}}{\begin{vmatrix} 9 & -7 \\ 21 & -11 \end{vmatrix}} = \frac{-187 + 203}{-99 + 147} = \frac{16}{48} = \frac{1}{3}$$

$$z = \frac{n_z}{d} = \frac{\begin{vmatrix} 9 & 17 \\ 21 & 29 \end{vmatrix}}{\begin{vmatrix} 9 & -7 \\ 21 & -11 \end{vmatrix}} = \frac{261 - 357}{-99 + 147} = \frac{-96}{48} = -2$$

Substitute the values of x and z back into the first equation to get $6\left(\frac{1}{3}\right) + y - 4(-2) = 11$; $2 + y + 8 = 11$; $y = 1$.

13 Solve the system of equations: $x - y = 3$ and $2x = 2y + 6$. The answer is $\boldsymbol{(x, x - 3)}$ **or** $\boldsymbol{(y + 3, y)}$.

Rewrite the second equation as $2x - 2y = 6$. Multiplying the terms in the first equation by -2 and adding them to the second equation, you get $0 = 0$. This system has an infinite number of solutions, so write the general rule for the solutions by solving for either y or x. Solving the first equation for y, you get $y = x + 3$, so the general rule as an ordered pair is $(x, x - 3)$; after you choose a value for x, you can subtract 3 from that number to get y. For instance, $(5, 2)$, $(3, 0)$, and $(-4, -7)$ are all ordered pairs using this rule. If you, instead, chose to solve for x, you get $x = y + 3$, and the general rule as an ordered pair is $(y + 3, y)$. Letting $y = 2$, 0, or -7, you get the same ordered pairs listed before.

14 Solve the system of equations: $3y - x = 7$ and $2x + 14 = 6y$. The answer is $(3y - 7, y)$ or $\left(x, \dfrac{x + 7}{3}\right)$.

Rewrite the second equation as $-6y + 2x = -14$. Then multiply the terms in the first equation by 2 and add the equations together. You get $0 = 0$, an indication that multiple solutions exist. Solving the first equation for x, you get $x = 3y - 7$, so the ordered pair for the general solution is $(3y - 7, y)$. After you choose a value for y, you can solve for x using the rule in the first coordinate position. For instance, if $y = 6$, $x = 11$. If you choose to solve for y instead of x, you see that your general rule has a fraction. That's not bad — it's just not as nice to work with.

15 Solve the system of equations: $x - y = 11$, $5x + 3z = 60$, and $5y + 3z = 5$. The answer is $\left(x, x - 11, \dfrac{60 - 5x}{3}\right)$ or $\left(y + 11, y, \dfrac{5 - 5y}{3}\right)$ or $\left(\dfrac{60 - 3z}{5}, \dfrac{5 - 3z}{5}, z\right)$.

Multiply the terms in the third equation by -1 and add them to the second equation to get $5x - 5y = 55$. Divide each term by 5, and the result is the same as the first equation, so this system has an infinite number of solutions. Solving for y yields $y = x - 11$. Replace the y in the third equation with $x - 1$, and you get $5(x - 11) + 3z = 5$; $5x - 55 + 3z = 5$; $5x + 3z = 60$. Solve for z by subtracting $5x$ from each side and dividing each term by 3. After you choose a value for x, you can solve for y and z. For instance, if $x = 15$, then $y = 4$, and $z = -5$. You can see that the answers work in the other general solutions, too.

16 Solve the system of equations: $x + 2y + z = 8$, $3x - 2y + 4z = 1$, and $x - 6y + 2z = -15$. The answer is $\left(\dfrac{9 - 5z}{4}, \dfrac{23 + z}{8}, z\right)$ or $\left(x, \dfrac{31 - x}{10}, \dfrac{9 - 4x}{5}\right)$ or $\left(31 - 10y, y, \dfrac{40y - 115}{5}\right)$.

Add the first two equations together to get $4x + 5z = 9$. Then multiply the terms in the first equation by 3 and add them to the third equation. Once again, you get $4x + 5z = 9$. This system has an infinite number of solutions. Solve for x by subtracting $5z$ from each side and dividing by 4. Then substitute this equivalent of x into the first equation and simplify (first multiply each term by 4 and then combine like terms): $\dfrac{9 - 5z}{4} + 2y + z = 8$; $9 - 5z + 8y + 4z = 32$; $-z + 8y = 23$. Now add z to each side and divide by 8: $8y = 23 + z$; $y = \dfrac{23 + z}{8}$. Letting $z = 5$, you get $x = -4$ and $y = \dfrac{7}{2}$. You can also write the general solution in terms of x or y, as shown.

17 Find the two fractions whose sum or difference is equal to $\dfrac{10x + 5}{2x^2 + x - 3}$. The answer is $\dfrac{4}{2x + 3} + \dfrac{3}{x - 1}$.

The denominator of the fraction factors into $(2x + 3)(x - 1)$. Write the equation involving the original fraction and the sum of the two fractions with unknown numerators: $\dfrac{10x + 5}{2x^2 + x - 3} = \dfrac{A}{2x + 3} + \dfrac{B}{x - 1}$. Multiply each side of the equation by the common denominator, and you're left with just the numerators:

$$(2x + 3)(x - 1)\frac{10x + 5}{(2x + 3)(x - 1)} = \frac{A}{2x + 3}(2x + 3)(x - 1) + \frac{B}{x - 1}(2x + 3)(x - 1)$$

$$10x + 5 = A(x - 1) + B(2x + 3)$$

Simplify the equation on the right by distributing, rearranging, and factoring:

$$10x + 5 = Ax - A + 2Bx + 3B$$

$$= Ax + 2Bx - A + 3B$$

$$= (A + 2B)x + (-A + 3B)$$

Equating the coefficients of x, you get $10 = A + 2B$. Equating the constants, you get $5 = -A + 3B$. Add the first of these new equations to the second to get $15 = 5B$; $B = 3$. Replace the B in either of the two new equations and solve for A: $A = 4$.

18 Find the two fractions whose sum or difference is equal to $\dfrac{2x-22}{x^2-x-12}$. The answer is $\dfrac{-2}{x-4}+\dfrac{4}{x+3}$.

The denominator of the fraction factors into $(x-4)(x+3)$. Write the equation involving the original fraction and the sum of the two fractions with unknown numerators: $\dfrac{2x-22}{x^2-x-12}=\dfrac{A}{x-4}+\dfrac{B}{x+3}$. Multiply each side of the equation by the common denominator, and you're left with just the numerators:

$$(x-4)(x+3)\dfrac{2x-22}{(x-4)(x+3)}=\dfrac{A}{x-4}(x-4)(x+3)+\dfrac{B}{x+3}(x-4)(x+3)$$

Simplify the equation on the right by distributing, rearranging, and factoring:

$$2x-22=Ax+3A+Bx-4B$$
$$=Ax+Bx+3A-4B$$
$$=(A+B)x+(3A-4B)$$

Equating the coefficients of x, you get $2=A+B$. Equating the constants, you get $-22=3A-4B$. Multiply the first of these new equations by 4 and add it to the second to get $-14=7A$; $A=-2$. Substitute -2 for A in the first equation, and you get that $B=4$.

19 Find the two fractions whose sum or difference is equal to $\dfrac{20-x}{2x^2+19x+35}$. The answer is $\dfrac{5}{2x+5}-\dfrac{3}{x+7}$.

The denominator of the fraction factors into $(2x+5)(x+7)$. Write the equation involving the original fraction and the sum of the two fractions with unknown numerators: $\dfrac{20-x}{2x^2+19x+35}=\dfrac{A}{2x+5}+\dfrac{B}{x+7}$. Multiply each side of the equation by the common denominator, and you're left with just the numerators. Then simplify the equation on the right by distributing, rearranging, and factoring:

$$20-x=Ax+7A+2Bx+5B$$
$$=Ax+2Bx+7A+5B$$
$$-x+20=(A+2B)x+(7A+5B)$$

Equating the coefficients of x, you get $-1=A+2B$. Equating the constants, you get $20=7A+5B$. Multiply the first of these new equations by -7 and add it to the second to get $-9B=27$; $B=-3$. Substitute -3 for B in the first equation, and you get $A=5$.

20 Find the two fractions whose sum or difference is equal to $\dfrac{7x}{x^2-x-12}$. The answer is $\dfrac{4}{x-4}+\dfrac{3}{x+3}$.

The denominator of the fraction factors into $(x-4)(x+3)$. Write the equation involving the original fraction and the sum of the two fractions with unknown numerators:

$$\dfrac{7x}{x^2-x-12}=\dfrac{A}{x-4}+\dfrac{B}{x+3}$$

Multiply each side of the equation by the common denominator, and you're left with just the numerators. Then simplify the equation on the right by distributing, rearranging, and factoring:

$$7x=Ax+3A+Bx-4B$$
$$=Ax+Bx+3A-4B$$
$$=(A+B)x+(3A-4B)$$

Equating the coefficients of x, you get $7=A+B$. Equating the constants gives you $0=3A-4B$. Multiply the first of these new equations by 4 and add it to the second to get $7A=28$; $A=4$. Substitute 4 for A in the first equation, and you get that $B=3$.

Chapter 12

Solving Systems of Non-Linear Equations and Inequalities

In This Chapter

▶ Figuring out where lines and parabolas meet

▶ Solving systems of equations involving circles and parabolas

▶ Finding where exponential equations intersect

▶ Solving systems of inequalities with graphs

The graph of a line and a parabola may intersect in two points or one point or not at all. So the solution of a system of equations involving both linear and quadratic equations, whose graphs are a line and a parabola, may have two solutions, one solution, or none at all. (Don't you just love ambiguity?) The number of possible solutions increases with the powers on the equations.

In this chapter, you deal with the multiple solutions you find when working with systems that mix lines, parabolas, and circles. You also dabble with systems of inequalities to find a *feasible region* or area where all the points in the area are solutions of the system. Read on to find the methods to this madness.

Finding the Intersections of Lines and Parabolas

A line can cut through a parabola in two points, or it may just be tangent to the parabola and touch it at one point. And then, sadly, a line and a parabola may never meet. When solving systems of equations involving lines and parabolas, you usually use the substitution method — solving for x or y in the equation of the line and substituting into the equation of the parabola.

Sometimes the equations lend themselves to elimination — when adding the equations (or multiples of the equations) together eliminates one of the variables entirely because its coefficient becomes 0. Elimination works only occasionally, but substitution always works. (For more information on the basics of substitution and elimination, see Chapter 11.)

Q. Find the common solution(s) in the equations $y = -5x^2 + 12x + 3$ and $8x + y = 18$

A. The points of intersection are **(1, 10)**, **(3, –6)**. Here's another way to write this solution: When $x = 1$, $y = 10$, and when $x = 3$, $y = -6$. To find these solutions, rewrite the equation of the line as $y = 18 - 8x$. Replace the y in the equation of the parabola with its equivalent to get $18 - 8x = -5x^2 + 12x + 3$. Move all the terms to the left and combine like terms, giving you $5x^2 - 20x + 15 = 0$. Divide each term by 5 and then factor, which gives you the equation $5(x^2 - 4x + 3) = 5(x - 3)(x - 1) = 0$. Using the multiplication property of zero (in order for a product to equal 0, one of the factors must be 0), you know that $x = 3$ or $x = 1$. Substitute those values back into the equation of the line to get the corresponding y-values.

Tip: Always substitute back into the equation with the lower exponents. You can avoid creating extraneous solutions.

Q. Find the common solution(s) in the equations $y = x^2 - 4x$ and $2x + y + 1 = 0$

A. **(1, –3).** Solve for y in the equation of the line to get $y = -2x - 1$. Substitute this value into the equation of the parabola to get $-2x - 1 = x^2 - 4x$. Moving the terms to the right and simplifying, $0 = x^2 - 2x + 1 = (x - 1)^2$. The only solution is $x = 1$. Replacing x with 1 in the equation of the line, you find that $y = -3$. The line is tangent to the parabola at the point of intersection, which is why this problem has only one solution.

1. Find the common solution(s) in the equations $y = x^2 + 4x + 7$ and $3x - y + 9 = 0$

Solve It

2. Find the common solution(s) in the equations $y = 4x^2 - 8x - 3$ and $4x + y = 5$

Solve It

3. Find the common solution(s) in the equations $x = 2y^2 - y$ and $x + 3y = 12$

Solve It

4. Find the common solution(s) in the equations $x = y^2 + 5y + 6$ and $x - 3y = 6$

Solve It

Crossing Curves: Finding the Intersections of Parabolas and Circles

When a parabola and circle intersect, the possibilities for their meeting are many and varied. The two curves can intersect in as many as four different points, or maybe three, or just two or even just one point. Keep your options open and be alert for as many common solutions as possible (right — up to four). And, yes, the system may have no solution at all. The curves may miss each other completely. I guess some things just aren't meant to be.

For these problems, you usually turn to substitution. However, you don't have to set one of the equations equal to x or y by itself. You may solve an equation for $4x$ or $(y-3)^2$ or some other term that appears in the other equation. As long as the terms match, you can swap one value for the other.

Q. Find the common solutions of the circle $(x-2)^2 + (y-2)^2 = 4$ and the parabola $2y = x^2 - 4x + 4$

A. **(0, 2), (2, 0), (4, 2).** Rewrite the equation of the parabola as $2y = (x-2)^2$. Next, replace the first term in the first equation with $2y$ and simplify: $2y + (y-2)^2 = 4$; $2y + y^2 - 4y + 4 = 4$; $y^2 - 2y = 0$. Factor the terms on the left to get $y(y-2) = 0$. So $y = 0$ or $y = 2$. Letting $y = 0$ in the equation of the parabola, you get $2(0) = x^2 - 4x + 4$, or $0 = (x-2)^2$. So when $y = 0$, $x = 2$. Next, let $y = 2$ in the parabola equation. You get $2(2) = x^2 - 4x + 4$; $4 = x^2 - 4x + 4$; $0 = x^2 - 4x$. Factoring, $0 = x(x-4)$, so $x = 0$ or $x = 4$. When $y = 2$, $x = 0$ or 4.

Q. Find the common solutions of $x^2 + y^2 = 100$ and $y^2 + 6x = 100$.

A. **(0, 10), (0, –10), (6, 8), (6, –8).** Solving the second equation for y^2, you get $y^2 = 100 - 6x$. Replace the y^2 in the first equation with its equivalent to get $x^2 + 100 - 6x = 100$. Simplifying and factoring, the equation becomes $x^2 - 6x = x(x-6) = 0$. So $x = 0$ or $x = 6$. Replacing x with 0 in the equation of the parabola, $y^2 = 100$; $y = \pm 10$. Replacing x with 6 in the equation of the parabola, $y^2 + 36 = 100$; $y^2 = 64$; $y = \pm 8$.

5. Find the common solutions of $x^2 + y^2 = 25$ and $x^2 + 4y = 25$

Solve It

6. Find the common solutions of $x^2 + y^2 = 9$ and $5x^2 - 6y = 18$

Solve It

7. Find the common solutions of $x^2 + y^2 = 5$ and $y^2 = -4x$

Solve It

8. Find the common solutions of $x^2 + y^2 = 9$ and $y = x^2 + 11$

Solve It

Appealing to a Higher Power: Dealing with Exponential Systems

You can solve systems of exponential equations algebraically when the bases of the exponential terms are the same number or when *obvious* (don't you hate that word in mathematics?) solutions pop out because of the simple natures of the equations involved. If the bases match, you can simply set the exponents equal to each other. When an algebraic solution isn't available, then a good graphing calculator or computer program can find the solution — which usually includes lots of decimal values and/or logarithmic functions.

This section deals with the types of problems that you can solve algebraically (or *simply*). Of course, the most simple is just plugging in a number that you're pretty sure works. But that method can be time consuming if you have to plug away a lot — reserve it for the sure thing. Refer to Chapter 9 if you need to know more about solving equations that involve exponential terms.

Q. Find the common solutions of $y = 2^{x^2 + 6x}$ and $y = 16^{x+2}$

A. $(2, 65{,}536)$, $\left(-4, \dfrac{1}{256}\right)$. You want the bases to match, so first change the exponential term in the second equation to a power of 2. It becomes $y = (2^4)^{x+2} = 2^{4x+8}$. Setting the two y-values of the two different equations equal to one another, you get $2^{x^2 + 6x} = 2^{4x+8}$. Now set the two exponents equal to each other: $x^2 + 6x = 4x + 8$. Moving all the terms to the left and factoring, $x^2 + 2x - 8 = (x + 4)(x - 2) = 0$. The solutions of this quadratic equation are $x = -4$ or $x = 2$. Replace the x with -4 in either of the original equations, and you get $y = \dfrac{1}{256}$. Replace x with 2 in either equation, and you get $y = 65{,}536$.

Q. Find the common solutions of $y = 3^{x+1}$ and $y = 2x + 3$

A. $(0, 3)$, $(-1, 1)$. A graphing calculator would show you an exponential curve rising from left to right and a line appearing to cut through the curve in two places near the y-axis. You'd have to zoom in closely to see the two points of intersection. These equations were chosen carefully so that the answers are integers. If you evaluate the two functions for a few values, you can determine the solutions with minimal computation. Let $x = 0$ in the first equation, and you get $y = 3^{0+1} = 3$. Let $x = 0$ in the second equation, and you get $y = 2(0) + 3 = 3$. A solution! Let $x = -1$ in the first equation, and you get $y = 3^{-1+1} = 3^0 = 1$. Let $x = -1$ in the second equation, and you get $y = 2(-1) + 3 = -2 + 3 = 1$. These are the only two solutions.

9. Find the common solution(s) of $y = 3^{x-1}$ and $y = 9^x$

Solve It

10. Find the common solution(s) of $y = 8^{2-x}$ and $y = 4^{x^2-x}$

Solve It

11. Find the common solution(s) of $y = 2^x$ and $y = 1 - x$

Solve It

12. Find the common solution(s) of $y = e^{x^2}$ and $y = e$

Solve It

Solving Systems of Inequalities

An inequality such as $x > 2$ has an infinite number of solutions. Any number bigger than 2 is a solution. The inequality $y > 4$ also has an infinite number of solutions. Put the two inequalities together and solve for all the ordered pairs *(x, y)* that make both inequalities true at the same time, and you have lots and lots of points, such as (3, 5), (4, 6), (3, 8), and so on.

The best way to represent all the solutions of this system of inequalities is with a picture. You show a graph that has the upper right-hand corner all shaded in — everything to the right of $x = 2$ and everything above $y = 4$. This drawing method is essentially how you solve any system of inequalities: Shade in everything (all the points) shared by the inequalities in the system.

Here's how the process works:

1. **If the inequalities aren't already in their proper forms, rewrite them so you can more easily graph the curves (yes, straight lines are types of curves).**

   This step often means isolating the *y* (or *x*) on one side of the inequality so you can plot some points (solving for *y* can give you slope-intercept form of a line — see Chapter 4). If you're working with a circle or ellipse, graphing may be easier if you write the inequality in its standard form (see the equations in Chapter 10 for details). Remember to reverse the inequality sign whenever you divide by a negative number.

2. **Pretend the inequality signs are equal signs, and graph the curves.**

   Use dashed lines if the equation says > or <; use solid lines if it includes ≥ or ≤.

3. Try a test point to determine whether a region contains solutions to one of the inequalities.

Select a *test point,* a random point that's clearly on one side of a curve or the other (I suggest something simple, like the origin). Replace the x and y in the chosen inequality with the test coordinates and see whether the statement is true.

4. Lightly shade in the region that makes the statement true; repeat steps 3 and 4 for the other curves.

Your solution for the system of equations is where the shading overlaps.

EXAMPLE

Q. Determine the solution of the system of inequalities $y \geq x + 4$ and $x + y + 2 > 0$

A. Sketch in the graphs of the lines $y = x + 4$ and $x + y + 2 = 0$. Use a solid line for the first line, because the inequality has *greater than or equal to,* which means you have to include all the points on the line, too. The graph of the second line should be a dotted or dashed line to show that you don't include the points on the line in the solution. Figure 12-1 shows the graphs of the lines and their point of intersection $(-3, 1)$. You find that point of intersection using either elimination or substitution (see Chapter 11 if you need more info on finding solutions of systems of linear equations).

The two intersecting lines divide the graphing plane into four different regions. You need to determine which region contains the solutions of the system. All the points that satisfy the inequality $y \geq x + 4$ are to the left of the line. You can tell that by just trying a test point, such as $(0, 0)$, which is clearly below the line. Replace the x and y in the inequality with 0's, and you get $0 \geq 0 + 4$. That's not true, so the point $(0, 0)$ isn't in the solution of the inequality. The answers must be on the other side of the line. Lightly shade in the area above that line. Trying that same test point in the second inequality, you get $0 + 0 + 2 > 0$, which is true. So the point $(0, 0)$ is a solution of the second inequality — as are all the points to the right of the line. Shade in the right side of that second line. The two shaded areas overlap in the upper region of the graph, and the shaded area is darker there. All the points in that region are solutions of the system of inequalities.

Figure 12-1:
The darker shading shows the points in the solution to this system of inequalities.

$y = x + 4$

$x + y + 2 = 0$

$(-3, 1)$

13. Determine the solution of the system of inequalities: $x + y \leq 10$ and $2x + y > 18$

Solve It

14. Determine the solution of the system of inequalities: $3x + 5y \leq 15$, $x \geq 0$, and $y \geq 0$

Solve It

15. Determine the solution of the system of inequalities: $2x - y \leq 22$, $x \leq 2$, and $y \leq 0$

Solve It

16. Determine the solution of the system of inequalities: $y \geq 2x^2 - 3x + 4$ and $x \geq y - 4$

Solve It

17. Determine the solution of the system of inequalities: $2x^2 + y \leq 8$ and $y + 4 \geq x^2$

Solve It

18. Determine the solution of the system of inequalities: $x^2 + y^2 \leq 25$ and $x + 2y \geq 5$

Solve It

Answers to Problems on Solving Systems of Non-Linear Equations and Inequalities

The following are the answers to the practice problems presented earlier in this chapter.

1 Find the common solution(s) in the equations $y = x^2 + 4x + 7$ and $3x - y + 9 = 0$. The answer is **(–2, 3), (1, 12)**.

Solve for y in the second equation (you get $y = 3x + 9$), and substitute that into the equation of the parabola: $3x + 9 = x^2 + 4x + 7$. Move all the terms to the right and factor the equation: $0 = x^2 + x - 2 = (x + 2)(x - 1)$. So $x = -2$ or 1. Letting $x = -2$ in the equation of the line, $3(-2) - y + 9 = 0$; $-6 - y + 9 = 0$; $-y = -3$; $y = 3$. And when $x = 1$ in the equation of the line, $3(1) - y + 9 = 0$; $3 - y + 9 = 0$; $-y = -12$; $y = 12$.

When solving for the second coordinate in the solution of a system of equations, use the simpler equation — the one with the smaller exponents — to avoid introducing extraneous solutions.

2 Find the common solution(s) in the equations $y = 4x^2 - 8x - 3$ and $4x + y = 5$. The answer is **(–1, 9), (2, –3)**.

Solve for y in the second equation ($y = 5 - 4x$) and substitute the equivalent of y into the equation of the parabola: $5 - 4x = 4x^2 - 8x - 3$. Move all the terms to the right and factor the equation: $0 = 4x^2 - 4x - 8 = 4(x^2 - x - 2) = 4(x + 1)(x - 2)$. Using the multiplication property of zero, you find that $x = -1$ or $x = 2$. When $x = -1$ in the equation of the line, $4(-1) + y = 5$; $-4 + y = 5$; $y = 9$. And substituting $x = 2$ in the equation of the line, $4(2) + y = 5$; $8 + y = 5$; $y = -3$.

3 Find the common solution(s) in the equations $x = 2y^2 - y$ and $x + 3y = 12$. The answer is **(21, –3), (6, 2)**.

Solve for x in the second equation ($x = 12 - 3y$) and substitute that value into the equation of the parabola to get $12 - 3y = 2y^2 - y$. Move all the terms to the right and factor the equation: $0 = 2y^2 + 2y - 12 = 2(y^2 + y - 6) = 2(y + 3)(y - 2)$. The solutions are $y = -3$ or $y = 2$. Substitute –3 for y in the equation of the line, and you get $x + 3(-3) = 12$; $x - 9 = 12$; $x = 21$. Next, substituting 2 for y, $x + 3(2) = 12$; $x + 6 = 12$; $x = 6$.

4 Find the common solution(s) in the equations $x = y^2 + 5y + 6$ and $x - 3y = 6$. The answer is **(6, 0), (0, –2)**.

Solve for x in the second equation ($x = 3y + 6$), and substitute that value into the equation of the parabola to get $3y + 6 = y^2 + 5y + 6$. Move all the terms to the right and factor the equation: $0 = y^2 + 2y = y(y + 2)$. The solutions are $y = 0$ or $y = -2$. Substitute 0 for y in the equation of the line, and you determine that $x - 3(0) = 6$; $x - 0 = 6$; $x = 6$. Next, substituting –2 for y, $x - 3(-2) = 6$; $x + 6 = 6$; $x = 0$.

5 Find the common solutions of $x^2 + y^2 = 25$ and $x^2 + 4y = 25$. The answer is **(5, 0), (–5, 0), (3, 4), (–3, 4)**.

Solve the second equation for x^2 (you get $x^2 = 25 - 4y$) and replace the x^2 in the first equation with its equivalent. The new equation reads $25 - 4y + y^2 = 25$. Simplifying, you get $y^2 - 4y = 0$. This equation factors into $y(y - 4) = 0$. The two solutions are $y = 0$ and $y = 4$. Go back to the second equation, the equation of the parabola, because it has only one squared term (it has lower exponents, so choosing this equation lets you avoid extraneous solutions). Replace the y in that equation with 0 to get $x^2 + 4(0) = 25$; $x^2 = 25$. That equation has two solutions: $x = 5$ or $x = -5$. Now, going back and replacing the y with 4 in that equation of the parabola, $x^2 + 4(4) = 25$; $x^2 + 16 = 25$; $x^2 = 9$. This equation also has two solutions: $x = 3$ or $x = -3$. Pairing up the y's and their respective x's, you get the four different solutions. The circle and parabola intersect in four distinct points.

6 Find the common solutions of $x^2 + y^2 = 9$ and $5x^2 - 6y = 18$. The answer is $\left(\frac{12}{5}, \frac{9}{5}\right)$, $\left(-\frac{12}{5}, \frac{9}{5}\right)$, **(0, -3).**

Eliminate the x terms: Multiply the terms of the first equation by -5 (which gives you $-5x^2 - 5y^2 = -45$) and add the two equations together. The resulting equation is $-5y^2 - 6y = -27$. Rewrite the equation, setting it equal to 0, and factor. You get $0 = 5y^2 + 6y - 27 = (5y - 9)(y + 3)$. Use the multiplication property of zero to solve for the two solutions of this equation. Replacing the y in the second equation (the equation of the parabola) with $\frac{9}{5}$, you get $5x^2 - 6\left(\frac{9}{5}\right) = 18$; $5x^2 - \frac{54}{5} = 18$; $5x^2 = \frac{144}{5}$. Then divide both sides of the equation by 5 and take the square root of each side: $x^2 = \frac{144}{25}$; $x = \pm\frac{12}{5}$. To find the other solution, let the y in the equation of the parabola be equal to -3. You get $5x^2 - 6(-3) = 18$; $5x^2 + 18 = 18$; $5x^2 = 0$. So $x = 0$. The circle and parabola intersect or touch in three distinct points.

7 Find the common solutions of $x^2 + y^2 = 5$ and $y^2 = -4x$. The answer is **(-1, 2), (-1, -2).**

Replace the y^2 in the first equation with $-4x$. Rewrite the equation $x^2 - 4x = 5$ as $x^2 - 4x - 5 = 0$, which factors into $(x - 5)(x + 1) = 0$. The two solutions of this equation are $x = 5$ or $x = -1$. Replace the x with 5 in the second equation, and you get the equation $y^2 = -4(5) = -20$. But $y^2 = -20$ doesn't have any real solution. So when $x = 5$, there isn't a common solution. You can use $x = 5$ in the equation of the circle, but you can't use $x = 5$ in the equation of the parabola. Now let $x = -1$, and you get $y^2 = -4(-1) = 4$; $y^2 = 4$. The two solutions of this equation are 2 and -2. The circle and parabola intersect only twice.

8 Find the common solutions of $x^2 + y^2 = 9$ and $y = x^2 + 11$. **No solution.**

Solve for x^2 in the second equation and replace the x^2 in the first equation with $y - 11$. You get $y - 11 + y^2 = 9$; $y^2 + y - 20 = 0$. Factor, and you find that $(y + 5)(y - 4) = 0$. The solutions are $y = -5$ or $y = 4$. When you replace the y in the parabola with -5, you get $-5 = x^2 + 11$; $-16 = x^2$. This equation has no real solution. You don't fare any better letting $y = 4$. You get $4 = x^2 + 11$; $x^2 = -7$. This entire system has no solution. The circle and parabola never cross or touch.

9 Find the common solution(s) of $y = 3^{x-1}$ and $y = 9^x$. The answer is $\left(-1, \frac{1}{9}\right)$.

Set y equal to y to get $3^{x-1} = 9^x$. Change the 9 to 3^2 and simplify: $3^{x-1} = (3^2)^x = 3^{2x}$. Now that the bases are the same, you can set the two exponents equal to one another and solve for x: $x - 1 = 2x$; $x = -1$. Replacing the x with -1 in $y = 3^{x-1}$, you get $y = 3^{-1-1} = 3^{-2} = \frac{1}{9}$.

10 Find the common solution(s) of $y = 8^{2-x}$ and $y = 4^{x^2-x}$. The answer is $\left(\frac{3}{2}, 2\sqrt{2}\right)$, **(-2, 4,096).**

First set y equal to y, change the 8 and 4 to powers of 2, and simplify the equation: $8^{2-x} = 4^{x^2-x}$; $(2^3)^{2-x} = (2^2)^{x^2-x}$; $2^{6-3x} = 2^{2x^2-2x}$. The bases are the same, so set the exponents equal to one another: $6 - 3x = 2x^2 - 2x$ becomes $0 = 2x^2 + x - 6$. Factoring, you get $0 = (2x - 3)(x + 2)$. When $x = \frac{3}{2}$, $y = 8^{2-3/2} = 8^{1/2} = \sqrt{8} = 2\sqrt{2}$, and when $x = -2$, $y = 8^{2-(-2)} = 8^4 = 4{,}096$.

11 Find the common solution(s) of $y = 2^x$ and $y = 1 - x$. The answer is **(0, 1).**

The first equation is an exponential that rises steadily as the x-values increase. The graph of the second equation is a line that falls steadily from left to right. They intersect at a single point. With some careful selections of points, you can quickly determine their single common solution, (0, 1). Replacing the x with 0 in the exponential gives you $y = 2^0 = 1$. And replacing the x with 0 in the line gives you $y = 1 - 0 = 1$.

12 Find the common solution(s) of $y = e^{x^2}$ and $y = e$. The answer is **(1, e), (–1, e).**

The exponential function is positive for all values of x that you input. And the line is horizontal with a y-intercept of (0, 1). If you replace the x in the exponential with 1, you get $y = e^1$ or $y = e$. The same thing occurs when you replace the x with –1; the square of –1 is also 1, so you get the same y-value.

13 Determine the solution of the system of inequalities: $x + y \leq 10$ and $2x + y > 18$.

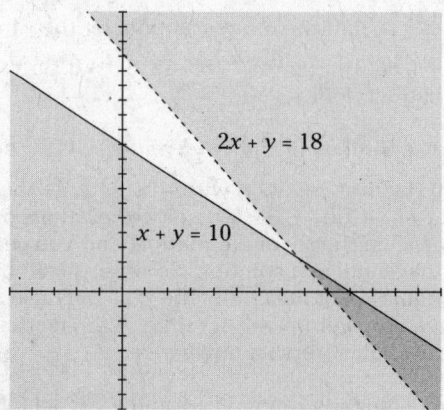

Draw the line $x + y = 10$ (in slope-intercept form, $y = -x + 10$) as a solid line, and draw a dashed line for $2x + y = 18$ (or $y = -2x + 18$). The test point (0, 0) works for the inequality $x + y \leq 10$, so you shade to the left of that line. The test point (0, 0) doesn't satisfy the inequality $2x + y > 18$, so you don't want that point included in its solution; Therefore, shade to the right of that line. The intersection of the two shaded areas moves downward between the intersection of the two lines.

14 Determine the solution of the system of inequalities: $3x + 5y \leq 15$, $x \geq 0$, and $y \geq 0$.

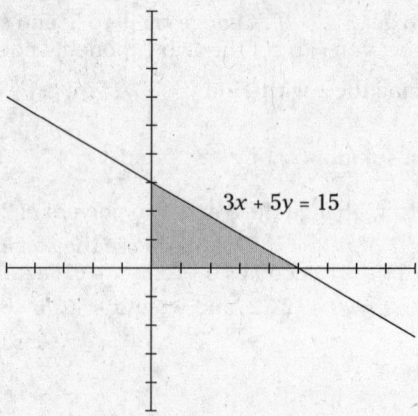

The lines are all solid, because the inequalities all have an *or equal to* in them. The $x \geq 0$ and $y \geq 0$ indicate that all the points in the solution are in the first quadrant — where the x- and y-coordinates are positive. The test point $(0, 0)$ works for the inequality $3x + 5y \leq 15$, so the shaded area is below and to the left of the line $3x + 5y = 15$ (in slope-intercept form, $y = -\frac{3}{5}x + 3$).

15 Determine the solution of the system of inequalities: $2x - y \leq 22$, $x \leq 2$, and $y \leq 0$.

The inequality $x \leq 2$ includes all the points to the left of the line $x = 2$, and the inequality $y \leq 0$ includes all the points below the x-axis. Rewrite the inequality $2x - y \leq 22$ in slope-intercept form by subtracting $2x$ from each side and then multiplying each side by -1. Don't forget to reverse the inequality. You get $y \geq 2x - 22$. The test point $(0, 0)$ works for $y \geq 2x - 22$ (you draw the solid line $y = 2x - 22$), so the test point is in the shaded area. The solutions for that inequality are to the left of and above that line. The points that meet all the qualifications are below the x-axis, left of $x = 2$, and above the diagonal line.

16 Determine the solution of the system of inequalities: $y \geq 2x^2 - 3x + 4$ and $x \geq y - 4$.

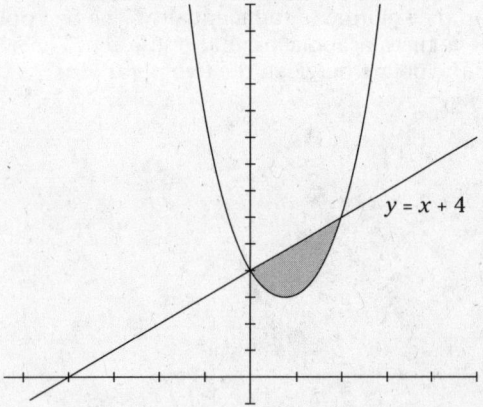

The inequality $y \geq 2x^2 - 3x + 4$ doesn't include the test point $(0, 0)$, so the points under the parabola don't work for the solution. All the points above or inside the parabola are part of the solution. The test point $(0, 0)$ does work for $x \geq y - 4$ (in slope-intercept form, the line is $y = x + 4$), so all the points below that line are in its solution. The solution to this system consists of the points beneath the line and inside the parabola.

17 Determine the solution of the system of inequalities: $2x^2 + y \leq 8$ and $y + 4 \geq x^2$.

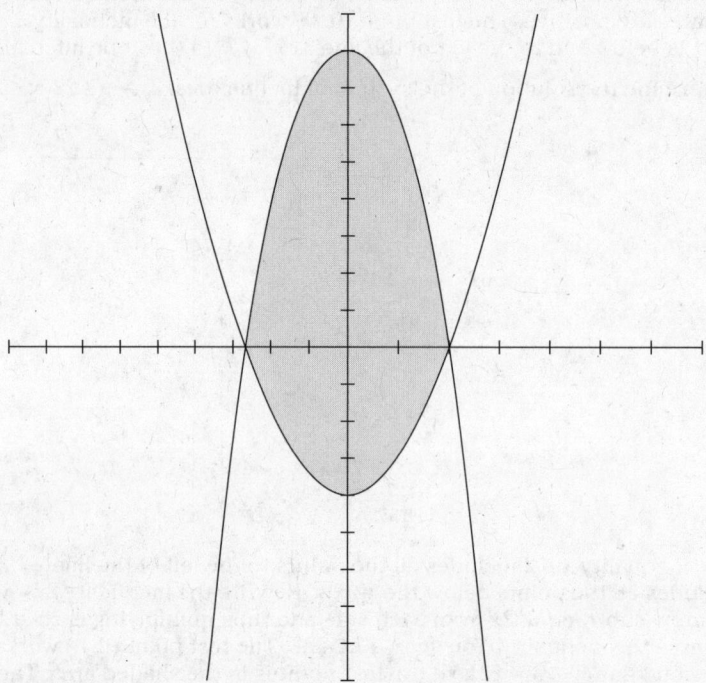

To graph the parabolas, solve the equations for y in terms of x so you can plot points, or convert the parabola equations to standard form (see Chapter 10). The equations are $y = -2x^2 + 8$ and $y = x^2 - 4$. Use these for graphing, but go back to the inequalities to apply the test point. The test point $(0, 0)$ does work for the inequality $2x^2 + y \leq 8$, so all the points below that parabola are in the solution of the inequality. The test point $(0, 0)$ also works in $y + 4 \geq x^2$, so all the points above that parabola are in the solution. The solution of the system of inequalities consists of all the points between the two parabolas.

18   Determine the solution of the system of inequalities: $x^2 + y^2 \leq 25$ and $x + 2y \geq 5$.

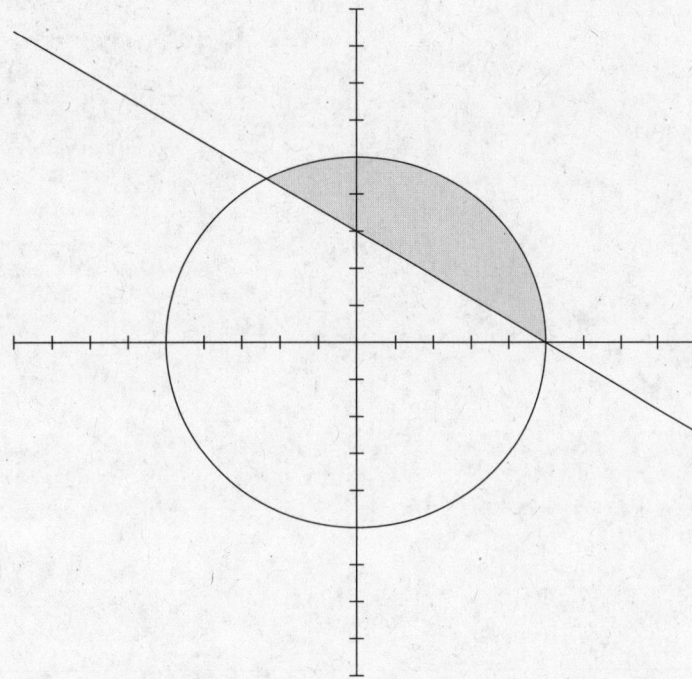

You pretty much have a circle equation in standard form, and it tells you that the circle's center is at the origin and that the radius is 5 (see Chapter 10). Draw the circle and the line (in slope-intercept form, $y = -\frac{1}{2}x + \frac{5}{2}$) as solid lines. The test point $(0, 0)$ works for the inequality involving the circle, so any point inside the circle satisfies this inequality. The test point $(0, 0)$ doesn't work for the inequality involving the line, so only points above the line are in the solution. The intersection of the two inequalities consists of the points in the circle that are above the line.

Part IV
Other Good Stuff: Lists, Arrays, and Imaginary Numbers

"It's nothing. One of my removable discontinuities is stuck."

In this part . . .

*T*he key to being successful in business — and in mathematics — is using your *imagination,* not being *set* in your ways, and choosing a good *sequence* of events to create a *series* of successes. There — you can't always come up with a way to connect imaginary numbers, sequences and series, and sets of numbers, but they're all here, in this part, for your sampling pleasure.

Chapter 13

Getting More Complex with Imaginary Numbers

· ·

In This Chapter

▶ Keeping an eye on the complex *i* and its powers

▶ Operating on complex numbers: Addition, subtraction, and multiplication

▶ Complex division: Multiplying by conjugates

▶ Solving equations for complex solutions

· ·

The real numbers satisfied mathematicians and other scientists for centuries, but mathematicians eventually came across equations that couldn't be solved and problems that couldn't be answered without wandering into the realm of imaginary numbers.

Imaginary numbers involve the symbol *i*, which represents $\sqrt{-1}$. Put imaginary numbers together with real numbers, and you get *complex* numbers. The complex number system consists of numbers of the form $a + bi$, with *a* and *b* representing real numbers.

In this chapter, you investigate the powers of *i*, perform operations on complex numbers, write the answers in the standard form, and finally, solve equations such as $x^2 = -16$. Until the advent of imaginary numbers, equations such as this had no solution at all. Good thing they have solutions today, though, because some scientists and engineers use imaginary numbers on a regular basis to solve real-world problems. And, hey, if you understand complex numbers, you may be well on your way to that dream career in quantum mechanics!

Simplifying Powers of i

Performing operations on complex numbers requires multiplying by *i* and simplifying powers of *i*. By definition, $i = \sqrt{-1}$, so $i^2 = -1$. If you want i^3, you compute it by writing $i^3 = i^2 \cdot i = -1 \cdot i = -i$. Also, $i^4 = i^2 \cdot i^2 = (-1)(-1) = 1$. And then the values of the powers start repeating themselves, because $i^5 = i$, $i^6 = -1$, $i^7 = -i$, and $i^8 = 1$. So what do you do if you want a higher power, such as i^{345}, or something else pretty high up there? You don't want to have to write out all the powers up to i^{345} using the pattern (not when you could be whitewater rafting or cleaning your room or watching the Cubs win the World Series!). Instead, use the following rule:

To compute the value of a power of *i*, determine whether the power is a multiple of 4, one more than a multiple of 4, two more than, or three more than a multiple of 4. Then apply the following:

▸ $i^{4n} = 1$

▸ $i^{4n+1} = i$

▸ $i^{4n+2} = -1$

▸ $i^{4n+3} = -i$

Q. Simplify the following: i^{444}, $i^{3,003}$, $i^{54,321}$, and $i^{111,002}$

A. $i^{444} = 1$; $i^{3,003} = -i$; $i^{54,321} = i$; $i^{111,002} = -1$.
Writing the power of i as a multiple of 4 and what's left over (you know, the remainder), you get $i^{444} = i^{4(111)} = 1$, $i^{3,003} = i^{4(750)+3} = -i$, $i^{54,321} = i^{4(13,580)+1} = i$, and $i^{111,002} = i^{4(27,750)+2} = -1$.

---

1. Simplify the following: i^{45}

Solve It

2. Simplify the following: i^{60}

Solve It

---

3. Simplify the following: $i^{4,007}$

Solve It

4. Simplify the following: $i^{2,002}$

Solve It

Not Quite Brain Surgery: Doing Operations on Complex Numbers

A complex number has the standard form $a + bi$, where a and b are real numbers. You can add, subtract, and multiply complex numbers using the same algebraic rules as those for real numbers and then simplify the final answer so it's in the standard form.

For the most part, the i works just like any other variable (except you can simplify powers of i, as I mention in the preceding section). In general, you combine all real numbers, change all powers of i to 1, –1, i, or –i, and then combine all terms with i's in them. Division of complex numbers is something else, so division gets its own section — the next section in this chapter.

EXAMPLE

Q. Given the two complex numbers $3 + 4i$ and $-5 + 2i$, perform three separate operations: add them, subtract the second from the first, and multiply them together

A. **Add: $-2 + 6i$; subtract: $8 + 2i$; multiply: $-23 - 14i$.** Adding, you have $(3 + 4i) +$ $(-5 + 2i)$. Rearrange the terms so that the like terms are together. You get $3 - 5 + 4i +$ $2i = -2 + 6i$. Subtracting, you have to negate each term in the second complex number: $(3 + 4i) - (-5 + 2i) = 3 + 4i + 5 - 2i = 3 + 5 +$ $4i - 2i = 8 + 2i$. To multiply the complex numbers, first use the FOIL method (*First, Outer, Inner, Last* — see Chapter 1): $(3 + 4i)(-5 + 2i) = -15 + 6i - 20i + 8i^2 = -15 - 14i + 8i^2$. However, this answer isn't in standard form; you have to simplify the squared term. Remember that $i^2 = -1$, so write the answer as $-15 - 14i + 8(-1) =$ $-15 - 14i - 8 = -23 - 14i$.

---

5. Compute: $(2 + 3i) + (-3 + 4i)$

Solve It

6. Compute: $(-6 + i) - (3 - 2i)$

Solve It

---

7. Compute: $(-8 + 6i) - 4(-1 + i)$

Solve It

8. Compute: $(-3 + 2i)(4 - 9i)$

Solve It

---

9. Compute: $(-6 + 3i)(6 + 3i)$

Solve It

10. Compute: $(2 - 3i)^2$

Solve It

"Dividing" Complex Numbers with a Conjugate

Mathematicians (that's you) can add, subtract. and multiply complex numbers. Technically, you can't divide complex numbers — in the traditional sense. You divide complex numbers by writing the division problem as a fraction and then multiplying the numerator and denominator by a conjugate.

The *conjugate* of the complex number $a + bi$ is $a - bi$.

The product of $(a + bi)(a - bi)$ is $a^2 + b^2$. How does that happen? Where's the i? Look at the steps in the multiplication: $(a + bi)(a - bi) = a^2 - abi + abi - b^2i^2 = a^2 - b^2(-1) = a^2 + b^2$, which is a real number — with no complex part. So when you need to divide one complex number by another, you multiply the numerator and denominator of the problem by the conjugate of the denominator. This step creates a real number in the denominator of the answer, which allows you to write the answer in the standard form of a complex number.

Q. Divide $10 + 5i$ by $4 - 3i$

A. **1 + 2i.** Express the division as a fraction. Multiply both the numerator and denominator of the fraction $\dfrac{10 + 5i}{4 - 3i}$ by $4 + 3i$. You get $\dfrac{10 + 5i}{4 - 3i} \cdot \dfrac{4 + 3i}{4 + 3i} = \dfrac{40 + 50i + 15i^2}{16 + 9} = \dfrac{40 - 15 + 50i}{25} = \dfrac{25 + 50i}{25}$. Writing the answer in standard form, you get $1 + 2i$.

---

11. Divide $40 - 20i$ by $3 + i$

Solve It

12. Divide $5 + 10i$ by $2 - i$

Solve It

---

13. Divide $20 - 10i$ by $-3 + 4i$

Solve It

14. Divide $20i$ by $2 + 6i$

Solve It

Solving Equations with Complex Solutions

You often come across equations that have no real solutions — or equations that have the potential for many more real solutions than they actually have. For instance, the equation $x^2 + 1 = 0$ has no real solutions. If you write it as $x^2 = -1$ and try to take the square root of each side, you run into trouble. Not until you have the imaginary numbers can you write that the solution of this equation is $x = \pm i$. The equation has two complex solutions.

An example of an equation without enough real solutions is $x^4 - 81 = 0$. This equation factors into $(x^2 - 9)(x^2 + 9) = 0$. The two real solutions of this equation are 3 and –3. The two complex solutions are $3i$ and $-3i$.

To solve for the complex solutions of an equation, you use factoring, the square root property for solving quadratics, and the quadratic formula. Refer to Chapter 2 for a rundown of these techniques.

Q. Find all the roots, real and complex, of the equation $x^3 - 2x^2 + 25x - 50 = 0$

A. $x = 2, 5i, -5i$. First factor the equation to get $x^2(x - 2) + 25(x - 2) = (x - 2)(x^2 + 25) = 0$. Using the multiplication property of zero, you determine that $x - 2 = 0$; $x = 2$. You also get $x^2 + 25 = 0$; $x^2 = -25$. Take the square root of each side, and $x = \pm\sqrt{-25}$. Simplify the radical, using the equivalence for i, and the complex solutions are $\pm\sqrt{-25} = \pm\sqrt{25}\sqrt{-1} = \pm 5i$. The real root is 2, and the imaginary roots are $5i$ and $-5i$.

Q. Find all the roots, real and imaginary, of the equation $5x^2 - 8x + 5 = 0$

A. $x = 0.4 + 0.6i, 0.4 - 0.6i$. The quadratic doesn't factor, so you use the quadratic formula:

$$x = \frac{-(-8) \pm \sqrt{(-8)^2 - 4(5)(5)}}{2(5)}$$
$$= \frac{8 \pm \sqrt{64 - 100}}{10}$$
$$= \frac{8 \pm \sqrt{-36}}{10}$$
$$= \frac{8 \pm 6i}{10} = \frac{4}{10} \pm \frac{6}{10}i$$

The only two solutions are complex: $0.4 + 0.6i$ and $0.4 - 0.6i$.

---

15. Find all the roots, real and imaginary, of $x^2 + 9 = 0$

Solve It

16. Find all the roots, real and imaginary, of $x^2 + 4x + 7 = 0$

Solve It

17. Find all the roots, real and imaginary, of $5x^2 + 6x + 3 = 0$

Solve It

18. Find all the roots, real and imaginary, of $x^4 + 12x^2 - 64 = 0$

Solve It

Answers to Problems on Complex Numbers

The following are the answers to the practice problems presented earlier in this chapter.

1 Simplify the following: i^{45}. The answer is **i**.

Rewrite the term as $i^{4(11)+1} = i$.

2 Simplify the following: i^{60}. The answer is **1**.

Rewrite the term as $i^{4(15)} = 1$.

3 Simplify the following: $i^{4,007}$. The answer is **$-i$**.

Rewrite the term as $i^{4(1,001)+3} = -i$.

4 Simplify the following: $i^{2,002}$. The answer is **-1**.

Rewrite the term as $i^{4(5,00)+2} = -1$.

5 Compute: $(2 + 3i) + (-3 + 4i)$. The answer is **$-1 + 7i$**.

Group the real parts and imaginary parts and simplify: $2 - 3 + 3i + 4i = -1 + 7i$.

6 Compute: $(-6 + i) - (3 - 2i)$. The answer is **$-9 + 3i$**.

Distribute the negative sign over the second complex number; then group the real parts and imaginary parts and simplify: $-6 + i - 3 + 2i = -6 - 3 + i + 2i = -9 + 3i$.

7 Compute: $(-8 + 6i) - 4(-1 + i)$. The answer is **$-4 + 2i$**.

Distribute the -4 over the second complex number; then group the real parts and imaginary parts and simplify: $-8 + 6i + 4 - 4i = -8 + 4 + 6i - 4i = -4 + 2i$.

8 Compute: $(-3 + 2i)(4 - 9i)$. The answer is **$6 + 35i$**.

Multiply the two binomials together using FOIL. Replace the i^2 with -1. Then group the real parts and imaginary parts and simplify: $-12 + 27i + 8i - 18i^2 = -12 + 27i + 8i - 18(-1) = -12 + 18 + 27i + 8i = 6 + 35i$.

9 Compute: $(-6 + 3i)(6 + 3i)$. The answer is **-45**.

Multiply the two binomials together using FOIL. Combine the two opposite terms to get 0. Replace the i^2 with -1 and simplify: $-36 - 18i + 18i + 9i^2 = -36 + 9(-1) = -36 - 9 = -45$.

10 Compute: $(2 - 3i)^2$. The answer is **$-5 - 12i$**.

Multiply the two binomials together using FOIL. Replace the i^2 with -1. Then group the real parts and imaginary parts and simplify: $4 - 6i - 6i + 9i^2 = 4 - 6i - 6i + 9(-1) = 4 - 9 - 6i - 6i = -5 - 12i$.

11 Divide $40 - 20i$ by $3 + i$. The answer is **$10 - 10i$**.

Write the problem as a fraction. Then multiply both the numerator and denominator by the conjugate of the denominator, $3 - i$. Simplify, and write the final answer in $a + bi$ form:

$$\frac{40 - 20i}{3 + i} \cdot \frac{3 - i}{3 - i} = \frac{120 - 40i - 60i + 20i^2}{9 + 1} = \frac{120 - 20 - 100i}{10} = \frac{100 - 100i}{10}.$$

12 Divide $5 + 10i$ by $2 - i$. The answer is **$5i$**.

Write the problem as a fraction. Then multiply both the numerator and denominator by the conjugate of the denominator, $2 + i$. Simplify, and write the final answer in $a + bi$ form:

$$\frac{5 + 10i}{2 - i} \cdot \frac{2 + i}{2 + i} = \frac{10 + 5i + 20i + 10i^2}{4 + 1} = \frac{10 - 10 + 25i}{5} = \frac{0 + 25i}{5}$$

13 Divide $20 - 10i$ by $-3 + 4i$. The answer is **$-4 - 2i$.**

Write the problem as a fraction. Then multiply both the numerator and denominator by the conjugate of the denominator, $-3 - 4i$. Simplify, and write the final answer in $a + bi$ form:
$$\frac{20 - 10i}{-3 + 4i} \cdot \frac{-3 - 4i}{-3 - 4i} = \frac{-60 - 80i + 30i + 40i^2}{9 + 16} = \frac{-60 - 40 - 50i}{25} = \frac{-100 - 50i}{25}$$

14 Divide $20i$ by $2 + 6i$. The answer is **$3 + i$.**

Write the problem as a fraction. Then multiply both the numerator and denominator by the conjugate of the denominator, $2 - 6i$. Simplify, and write the final answer in $a + bi$ form:
$$\frac{20i}{2 + 6i} \cdot \frac{2 - 6i}{2 - 6i} = \frac{40i - 120i^2}{4 + 36} = \frac{120 + 40i}{40}.$$

15 Find all the roots, real and imaginary, of $x^2 + 9 = 0$. The answer is **$x = 3i, -3i$.**

Add -9 to each side to get $x^2 = -9$. Take the square root of each side. Then simplify the expression using i for the negative under the radical: $x = \pm\sqrt{9} = \pm\sqrt{9}\sqrt{-1} = \pm3i$.

16 Find all the roots, real and imaginary, of $x^2 + 4x + 7 = 0$. The answer is **$x = -2 + i\sqrt{3}$ and $-2 - i\sqrt{3}$.**

Use the quadratic formula to solve for x. Simplify the expression using i for the negative under the radical:
$$x = \frac{-4 \pm \sqrt{4^2 - 4(1)(7)}}{2(1)}$$
$$= \frac{-4 \pm \sqrt{16 - 28}}{2}$$
$$= \frac{-4 \pm \sqrt{-12}}{2} = \frac{-4 \pm \sqrt{4}\sqrt{3}\sqrt{-1}}{2}$$
$$= \frac{-4 \pm 2i\sqrt{3}}{2} = -2 \pm i\sqrt{3}$$

17 Find all the roots, real and imaginary, of $5x^2 + 6x + 3 = 0$. The answer is **$x = -\frac{3}{5} + \frac{\sqrt{6}}{5}i$ and $-\frac{3}{5} - \frac{\sqrt{6}}{5}i$.**

Use the quadratic formula to solve for x. Simplify the expression using i for the negative under the radical:
$$x = \frac{-6 \pm \sqrt{6^2 - 4(5)(3)}}{2(5)}$$
$$= \frac{-6 \pm \sqrt{36 - 60}}{10}$$
$$= \frac{-6 \pm \sqrt{-24}}{10} = \frac{-4 \pm \sqrt{4}\sqrt{6}\sqrt{-1}}{10}$$
$$= \frac{-6 \pm 2i\sqrt{6}}{10} = \frac{-6}{10} \pm \frac{2i\sqrt{6}}{10} = -\frac{3}{5} \pm i\frac{\sqrt{6}}{5}$$

18 Find all the roots, real and imaginary, of $x^4 + 12x^2 - 64 = 0$. The answer is **$x = 2, -2, 4i, -4i$.**

Factor the left side: $(x^2 + 16)(x^2 - 4) = (x^2 + 16)(x - 2)(x + 2) = 0$. Obtain the two real roots by setting $x - 2$ and $x + 2$ equal to 0. When $x^2 + 16 = 0$, you find that $x^2 = -16$. Taking the square root of each side and using i for the -1 under the radical gives you the two imaginary roots.

Chapter 14

Making Headway with Matrices

In algebra, a *matrix* is a rectangular array of numbers, not a computer-generated universe inhabited by Keanu Reeves and Laurence Fishburne. An array of numbers is a quick, down-and-dirty way of expressing a lot of information without going into a lot of detail. In the real world, matrices are a way of organizing business elements or transportation capabilities into a format that you can operate on and analyze.

In this chapter, you add, subtract, multiply, and divide (well, sort of divide) matrices — when allowed. I cover the rules governing operations on matrices. People also use matrices to solve large systems of linear equations with ease and efficiency, so you get to practice that type of solving technique firsthand.

Describing Dimensions and Types of Matrices

Matrices are rectangular arrays of numbers, usually denoted as being matrices by having the numbers surrounded by a bracket. The numbers in a matrix are its *elements*. Matrices are classified in several ways, the most immediately apparent being their *dimension,* or size.

The dimension of a matrix is written $r \times c$, where r is the number of rows in the matrix and c is the number of columns. The dimension may be 3×2 or 4×7 or 1×5 or some other combination of numbers. Here are the types of matrices:

- A *column* matrix has one column and any number of rows.

- A *row* matrix has one row and any number of columns.

- A *zero* matrix is quite predictable — all its elements are 0s.

- A *square* matrix is just what the name suggests: The number of rows and columns is the same.

- An *identity* matrix is a square matrix in which the elements along the diagonal that runs from the upper left-hand corner to the lower right corner are all 1s, with all the other elements being 0s. An identity matrix is one of the more important matrices because it shows up in finding inverse matrices and in solving systems of equations (which I cover later in this chapter).

You name matrices using capital letters to distinguish one matrix from another.

Q. Identify the dimension of each matrix and determine any special classifications:

$$A = \begin{bmatrix} 2 & -3 & 0 & 4 \\ -1 & 1 & 0 & 5 \end{bmatrix}, B = \begin{bmatrix} 1 & 0 \\ 0 & 1 \end{bmatrix}, C = \begin{bmatrix} -1 \\ 0 \\ 2 \end{bmatrix}$$

A. List the number of rows first and then columns:

- ✔ **Matrix A: 2 × 4; no other special classification**
- ✔ **Matrix B: 2 × 2; square matrix, identity matrix**
- ✔ **Matrix C: 3 × 1; column matrix**

---

1. Identify the dimension of the matrix and determine any special classifications

$$\begin{bmatrix} 1 & 0 & 0 & 0 \\ 0 & 1 & 0 & 0 \\ 0 & 0 & 1 & 0 \\ 0 & 0 & 0 & 1 \end{bmatrix}$$

Solve It

2. Identify the dimension of the matrix and determine any special classifications

$$\begin{bmatrix} 0 & 0 & 0 & 0 & 0 & 0 \end{bmatrix}$$

Solve It

3. Identify the dimension of the matrix and determine any special classifications

$$\begin{bmatrix} 4 & 3 \\ 0 & -1 \\ -1 & 7 \\ 5 & 5 \\ -3 & 0 \end{bmatrix}$$

Solve It

4. Identify the dimension of the matrix and determine any special classifications

$$\begin{bmatrix} 1 & 0 \\ 1 & 6 \end{bmatrix}$$

Solve It

Adding, Subtracting, and Doing Scalar Multiplication on Matrices

As long as matrices have the exact same dimensions, you can add them to or subtract them from one another. The dimensions have to be the same, because each element has to pair up with its counterpart in the other matrix; you perform the operation on those elements.

Scalar multiplication of matrices means multiplying every element in a particular matrix by some number. Because only one matrix is involved in the operation, the dimension doesn't matter in scalar multiplication — any dimension works.

Q. Given the matrices $A = \begin{bmatrix} -3 & 2 \\ 0 & 4 \\ 1 & -5 \\ 6 & -1 \end{bmatrix}$, $B = \begin{bmatrix} -4 & -2 \\ 3 & 3 \\ 1 & -5 \\ 8 & 2 \end{bmatrix}$ perform the following operations: $A+B$, $2B$, and $3A - 4B$

A. $A + B = \begin{bmatrix} -3 & 2 \\ 0 & 4 \\ 1 & -5 \\ 6 & -1 \end{bmatrix} + \begin{bmatrix} -4 & -2 \\ 3 & 3 \\ 1 & -5 \\ 8 & 2 \end{bmatrix} = \begin{bmatrix} -3+(-4) & 2+(-2) \\ 0+3 & 4+3 \\ 1+1 & -5+(-5) \\ 6+8 & -1+2 \end{bmatrix} = \begin{bmatrix} -7 & 0 \\ 3 & 7 \\ 2 & -10 \\ 14 & 1 \end{bmatrix}$

$2B = 2\begin{bmatrix} -4 & -2 \\ 3 & 3 \\ 1 & -5 \\ 8 & 2 \end{bmatrix} = \begin{bmatrix} 2(-4) & 2(-2) \\ 2(3) & 2(3) \\ 2(1) & 2(-5) \\ 2(8) & 2(2) \end{bmatrix} = \begin{bmatrix} -8 & -4 \\ 6 & 6 \\ 2 & -10 \\ 16 & 4 \end{bmatrix}$

$3A - 4B = 3\begin{bmatrix} -3 & 2 \\ 0 & 4 \\ 1 & -5 \\ 6 & -1 \end{bmatrix} - 4\begin{bmatrix} -4 & -2 \\ 3 & 3 \\ 1 & -5 \\ 8 & 2 \end{bmatrix}$

$= \begin{bmatrix} 3(-3) & 3(2) \\ 3(0) & 3(4) \\ 3(1) & 3(-5) \\ 3(6) & 3(-1) \end{bmatrix} + \begin{bmatrix} -4(-4) & -4(-2) \\ -4(3) & -4(3) \\ -4(1) & -4(-5) \\ -4(8) & -4(2) \end{bmatrix}$

$= \begin{bmatrix} -9 & 6 \\ 0 & 12 \\ 3 & -15 \\ 18 & -3 \end{bmatrix} + \begin{bmatrix} 16 & 8 \\ -12 & -12 \\ -4 & 20 \\ -32 & -8 \end{bmatrix} = \begin{bmatrix} 7 & 14 \\ -12 & 0 \\ -1 & 5 \\ -14 & -11 \end{bmatrix}$

5. If $A = \begin{bmatrix} 2 & 3 & 0 \\ 4 & 1 & 5 \end{bmatrix}$ and $B = \begin{bmatrix} -1 & -2 & 0 \\ 1 & 0 & 2 \end{bmatrix}$, find $A - B$

Solve It

6. If $A = \begin{bmatrix} -3 & 6 \\ 2 & 2 \end{bmatrix}$ and $B = \begin{bmatrix} -1 & -2 \\ -3 & 2 \end{bmatrix}$, find $4A + 2B$

Solve It

7. If $A = \begin{bmatrix} 4 & -2 & 3 \\ 1 & 6 & -5 \\ 4 & 4 & -2 \end{bmatrix}$ and $B = \begin{bmatrix} -4 & 2 & -3 \\ -1 & -6 & 5 \\ -4 & -4 & 2 \end{bmatrix}$,

find $A + B$

Solve It

8. If $A = \begin{bmatrix} 3 & 0 & 0 \\ 0 & 0 & 7 \\ 0 & 5 & 1 \end{bmatrix}$ and $B = \begin{bmatrix} 1 & 0 & 0 \\ 0 & 1 & 0 \\ 0 & 0 & 1 \end{bmatrix}$,

find $9A - B$

Solve It

Trying Times: Multiplying Matrices by Each Other

You can multiply two matrices together only if the number of *columns* in the first matrix is equal to the number of *rows* in the second matrix. The resulting matrix has a dimension determined by the number of rows in the first matrix and the number of columns in the second matrix.

If matrix A has dimension $m \times n$ and matrix B has dimension $p \times q$, then the product $A \cdot B$ is possible only if $n = p$. Furthermore, the dimensions of the matrix formed from $A \cdot B$ (the answer matrix) are $m \times q$.

To multiply the matrices together, you sum up products. The element in the first row, first column of the answer matrix is the result of multiplying each element in the first *row* of the first matrix times the corresponding elements in the first *column* of the second matrix — and finding the sum of all those products. You have to find n products (where n is the number of columns in the first matrix and rows in the second) and add them together just to put a single number in the answer matrix.

One more example of how the numbers match up: The element in the third row, fourth column of the answer matrix is the result of multiplying all the elements in the third row of the first matrix times the fourth column of the second matrix — and summing up the products. Check out an example.

Q. Find the product $A \cdot B$ if $A = \begin{bmatrix} 2 & 0 \\ -1 & 3 \\ 0 & 1 \end{bmatrix}$ and $B = \begin{bmatrix} -2 & 1 \\ 1 & 0 \end{bmatrix}$.

A. **The resulting matrix is** $\begin{bmatrix} -4 & 2 \\ 5 & -1 \\ 1 & 0 \end{bmatrix}$. Matrix A is 3×2, and matrix B is 2×2. The number of columns in A matches the number of rows in B, so you can multiply the matrices together. The resulting matrix is 3×2. The following are the sums of the products you need to produce the elements in the answer matrix:

	Column 1	Column 2
Row 1	First row of A times first column of B $2(-2) + 0(1) = -4 + 0 = -4$	First row of A times second column of B $2(1) + 0(0) = 2 + 0 = 2$
Row 2	Second row of A times first column of B $-1(-2) + 3(1) = 2 + 3 = 5$	Second row of A times second column of B $-1(1) + 3(0) = -1 + 0 = -1$
Row 3	Third row of A times first column of B $0(-2) + 1(1) = 0 + 1 = 1$	Third row of A times second column of B $0(1) + 1(0) = 0 + 0 = 0$

9. Find the product of the two matrices

$$A = \begin{bmatrix} 2 & 1 \\ 0 & 4 \end{bmatrix} \text{ and } B = \begin{bmatrix} 3 & -2 & 4 \\ 1 & 0 & 2 \end{bmatrix}$$

Solve It

10. Find the product of the two matrices

$$C = \begin{bmatrix} 1 & 0 \\ 4 & 3 \\ 2 & -2 \end{bmatrix} \text{ and } A = \begin{bmatrix} 2 & 1 \\ 0 & 4 \end{bmatrix}$$

Solve It

11. Find the product of the two matrices

$$B = \begin{bmatrix} 3 & -2 & 4 \\ 1 & 0 & 2 \end{bmatrix} \text{ and } C = \begin{bmatrix} 1 & 0 \\ 4 & 3 \\ 2 & -2 \end{bmatrix}$$

Solve It

12. Find the product of the two matrices

$$A = \begin{bmatrix} 2 & 1 \\ 0 & 4 \end{bmatrix} \text{ and } I = \begin{bmatrix} 1 & 0 \\ 0 & 1 \end{bmatrix}$$

Solve It

The Search for Identity: Finding Inverse Matrices

The inverse of matrix A is matrix A^{-1}. Two matrices are *inverses* if their product is an identity matrix (see "Describing Dimensions and Types of Matrices," earlier in this chapter). Only square matrices have inverses, but not all square matrices have inverses. So in terms of matrices, having an inverse is a pretty special characteristic.

The matrices $B = \begin{bmatrix} 4 & 5 \\ 5 & 6 \end{bmatrix}$ and $B^{-1} = \begin{bmatrix} -6 & 5 \\ 5 & -4 \end{bmatrix}$ are inverses of one another. In general, matrix multiplication is *not* commutative — you can't change the order of multiplication and expect to get the same answer. However, with inverses, the order you multiply them in doesn't matter — you get the same result, and that result is an identity matrix:

$$B \cdot B^{-1} = \begin{bmatrix} 4 & 5 \\ 5 & 6 \end{bmatrix} \cdot \begin{bmatrix} -6 & 5 \\ 5 & -4 \end{bmatrix} = \begin{bmatrix} -24+25 & 20-20 \\ -30+30 & 25-24 \end{bmatrix} = \begin{bmatrix} 1 & 0 \\ 0 & 1 \end{bmatrix}$$

$$B^{-1} \cdot B = \begin{bmatrix} -6 & 5 \\ 5 & -4 \end{bmatrix} \cdot \begin{bmatrix} 4 & 5 \\ 5 & 6 \end{bmatrix} = \begin{bmatrix} -24+25 & -30+30 \\ 20-20 & 25-24 \end{bmatrix} = \begin{bmatrix} 1 & 0 \\ 0 & 1 \end{bmatrix}$$

If you have a matrix and want to find its inverse (if it has one), you can use a method in which you put your matrix and an identity matrix of the same dimensions side by side in one big matrix. Then you change the elements in your original square matrix until they look like an identity matrix by performing operations such as multiplying and adding rows. The operations you can perform are exactly those you use when solving systems of equations: You can add two rows together to eliminate one of the numbers (when you add opposites), or you can multiply a row by some number and add it to another row. The result is the inverse of the original matrix sitting alongside the identity matrix. Here are some guidelines on making the transformation:

✔ Start with the first column on the left and work your way to the right; finish work on one column before moving to the next.

✔ When you start work on a new column, get the 1 in its proper place before dealing with the 0s (1s should start in the upper left and go down along a diagonal). Figure out what number you can multiply an element by to get a product of 1 (find the element's reciprocal) and then multiply the entire row by that number.

✔ Change numbers to 0s using the elimination process (see Chapter 11). Basically, figure out what number you'd have to add to an element to get a sum of 0. Multiply one of the other rows by that number and add the two rows together.

If you're working on the first column, multiply row one by that magic elimination number; the second column, multiply row two; third column, row three; and so on.

In the examples, I show you the step-by-step process for finding the inverse of a square matrix (plus a slick shortcut for 2×2 matrices).

Q. Find the inverse of matrix $A = \begin{bmatrix} 1 & 0 & 4 \\ -2 & 1 & -6 \\ 3 & 1 & 9 \end{bmatrix}$

A. Use the following steps.

1. **Write the original matrix and an identity matrix all in one array:** $\begin{bmatrix} 1 & 0 & 4 & \vdots & 1 & 0 & 0 \\ -2 & 1 & -6 & \vdots & 0 & 1 & 0 \\ 3 & 1 & 9 & \vdots & 0 & 0 & 1 \end{bmatrix}$

2. **Change the first column of the original matrix so it looks like that of an identity matrix.**

You don't have to do anything to the first row, because you want that 1 in the upper left-hand corner. However, you do need the other two elements to be 0s. The second element is –2, so to

get a sum of 0, you have to add 2 to it. Add 2 times row one to row two: $\begin{bmatrix} 1 & 0 & 4 & \vdots & 1 & 0 & 0 \\ 0 & 1 & 2 & \vdots & 2 & 1 & 0 \\ 3 & 1 & 9 & \vdots & 0 & 0 & 1 \end{bmatrix}$

The third element in this column is 3. To change it to 0, multiply each element in row one by –3

and add the result to row three; this sum becomes your new row three: $\begin{bmatrix} 1 & 0 & 4 & \vdots & 1 & 0 & 0 \\ 0 & 1 & 2 & \vdots & 2 & 1 & 0 \\ 0 & 1 & -3 & \vdots & -3 & 0 & 1 \end{bmatrix}$

3. **Make the matrix's second column look like the second column of an identity matrix.**

First you need a 1 in the second row, second column. Luckily, your other calculations already put a 1 there. You already have a 0 above it, so you only need to work on putting a 0 in the third row. Add −1

times row two to row three, and write the sum as the new row three:

$$\begin{bmatrix} 1 & 0 & 4 & \vdots & 1 & 0 & 0 \\ 0 & 1 & 2 & \vdots & 2 & 1 & 0 \\ 0 & 0 & -5 & \vdots & -5 & -1 & 1 \end{bmatrix}$$

4. **Make the third column of the matrix look like that of an identity matrix.**

You need a 1 in the third row, third column. This element is currently −5, so you need to divide

every element in that row by −5 (multiply by −⅕):

$$\begin{bmatrix} 1 & 0 & 4 & \vdots & 1 & 0 & 0 \\ 0 & 1 & 2 & \vdots & 2 & 1 & 0 \\ 0 & 0 & 1 & \vdots & 1 & 0.2 & -0.2 \end{bmatrix}.$$

The other terms need to be 0s, so use the elimination method: Add −4 times row three to row one:

$$\begin{bmatrix} 1 & 0 & 0 & \vdots & -3 & -0.8 & 0.8 \\ 0 & 1 & 2 & \vdots & 2 & 1 & 0 \\ 0 & 0 & 1 & \vdots & 1 & 0.2 & -0.2 \end{bmatrix}$$ Add −2 times row three to row two: $$\begin{bmatrix} 1 & 0 & 0 & \vdots & -3 & -0.8 & 0.8 \\ 0 & 1 & 0 & \vdots & 0 & 0.6 & 0.4 \\ 0 & 0 & 1 & \vdots & 1 & 0.2 & -0.2 \end{bmatrix}.$$

5. **You now have an identity matrix to the left. The matrix on the right is the inverse of the**

original matrix: $A^{-1} = \begin{bmatrix} -3 & -0.8 & 0.8 \\ 0 & 0.6 & 0.4 \\ 1 & 0.2 & -0.2 \end{bmatrix}.$

Q. Find the inverse of the matrix $\begin{bmatrix} 5 & 8 \\ 7 & 11 \end{bmatrix}$

A. The process I outline in the preceding example works for any size square matrix. But you can use a neater, quicker way to find the inverse of a 2×2 matrix.

The inverse of matrix $\begin{bmatrix} a & b \\ c & d \end{bmatrix}$ is $\begin{bmatrix} \dfrac{d}{m} & -\dfrac{b}{m} \\ -\dfrac{c}{m} & \dfrac{a}{m} \end{bmatrix}$, where $m = ad - bc$, the difference of the cross-products. Of course, if the difference is 0, the matrix has no inverse. Basically, you just swap the elements a and d, negate the elements b and c, and divide each element by the difference of the cross-products.

For this particular matrix, find the difference of the cross-products: $m = 5(11) - 8(7) = 55 - 56 = -1$. Now switch the 5 and 11 and negate both the 8 and the 7. Then divide each element by −1:

$$\begin{bmatrix} 11 & -8 \\ -7 & 5 \end{bmatrix} \rightarrow \begin{bmatrix} \dfrac{11}{-1} & \dfrac{-8}{-1} \\ \dfrac{-7}{-1} & \dfrac{-5}{-1} \end{bmatrix} \rightarrow \begin{bmatrix} -11 & 8 \\ 7 & -5 \end{bmatrix}$$

The inverse of the original matrix is $\begin{bmatrix} -11 & 8 \\ 7 & -5 \end{bmatrix}.$

13. Find the inverse of the matrix $\begin{bmatrix} 3 & 2 \\ 7 & 5 \end{bmatrix}$

Solve It

14. Find the inverse of the matrix $\begin{bmatrix} -4 & -5 \\ 7 & 9 \end{bmatrix}$

Solve It

15. Find the inverse of the matrix $\begin{bmatrix} 6 & 3 \\ 4 & 2 \end{bmatrix}$

Solve It

16. Find the inverse of the matrix $\begin{bmatrix} -8 & 4 \\ 5 & -2 \end{bmatrix}$

Solve It

17. Find the inverse of the matrix $\begin{bmatrix} 1 & 2 & 3 \\ 0 & 2 & 4 \\ -1 & 4 & -1 \end{bmatrix}$

Solve It

18. Find the inverse of the matrix $\begin{bmatrix} 1 & 2 & -3 \\ 3 & 5 & -7 \\ -1 & -3 & 6 \end{bmatrix}$

Solve It

Using Matrices to Solve Systems of Equations

One very nice application of matrices is using them to solve systems of linear equations. This application is especially helpful when you have large systems with lots of variables, as well as when you have a graphing calculator or computer program to do all the computations for you.

When solving systems of equations using matrices, the system has to have the same number of equations as it has variables. To solve systems of linear equations using matrices, do the following:

1. **Write all the equations so the variables are in the same order, and set the equations equal to the constants.**

2. **Create a *coefficient* matrix made up of all the coefficients of the equations and a *constant* matrix made up of all the constants.**

3. **Find the solution of the system by multiplying the *inverse* of the coefficient matrix times the constant matrix.**

Here's an example to show the process in more detail.

Q. Solve the system of equations using matrices:
$$\begin{aligned} x - y - z &= 7 \\ -x + 2y - 3z &= -7 \\ x - 2y + z &= 9 \end{aligned}$$

A. **(2, –4, –1); or $x = 2$, $y = –4$, $z = –1$.** You first create the coefficient matrix: $\begin{bmatrix} 1 & -1 & -1 \\ -1 & 2 & -3 \\ 1 & -2 & 1 \end{bmatrix}$

Then find the inverse of that matrix using the techniques from the preceding section. The inverse of the coefficient matrix is $\begin{bmatrix} 2 & -1.5 & -2.5 \\ 1 & -1 & -2 \\ 0 & -0.5 & -0.5 \end{bmatrix}$.

Now create the *constant* matrix: $\begin{bmatrix} 7 \\ -7 \\ 9 \end{bmatrix}$.

Multiply the coefficient matrix times the constant matrix: $\begin{bmatrix} 2 & -1.5 & -2.5 \\ 1 & -1 & -2 \\ 0 & -0.5 & -0.5 \end{bmatrix} \cdot \begin{bmatrix} 7 \\ -7 \\ 9 \end{bmatrix} = \begin{bmatrix} 2 \\ -4 \\ -1 \end{bmatrix}$. (Refer to "Trying Times: Multiplying Matrices by Each Other," earlier in the chapter, if you need help.) You see the solutions in the resulting matrix, going down, in the order in which you wrote the coefficients.

19. Solve the system of equations using matrices: $3x - 2y = 17$ and $-4x + 3y = -23$

Solve It

20. Solve the system of equations using matrices: $-6x + 5y = 8$ and $-4x + 3y = 6$

Solve It

21. Solve the system of equations using matrices: $x - y - z = -1$, $2x + 3y + z = 0$, and $3y + 2z = 2$

Solve It

22. Solve the system of equations using matrices: $2x + y - z = 9$, $x + 2y - z = 6$, and $x - 2y + z = 0$

Solve It

Answers to Problems on Matrices

The following are the answers to the practice problems presented earlier in this chapter.

1 Identify the dimension of the matrix and determine any special classifications: $\begin{bmatrix} 1 & 0 & 0 & 0 \\ 0 & 1 & 0 & 0 \\ 0 & 0 & 1 & 0 \\ 0 & 0 & 0 & 1 \end{bmatrix}$. The answer is **4 × 4; square, identity.**

The elements in the diagonal running from left to right are all 1s, and the rest of the elements are 0s, making this array an identity matrix.

2 Identify the dimension of the matrix and determine any special classifications: $\begin{bmatrix} 0 & 0 & 0 & 0 & 0 & 0 \end{bmatrix}$. The answer is **1 × 6; row, zero.**

This matrix has only one row, and all the elements are 0s.

3 Identify the dimension of the matrix and determine any special classifications: $\begin{bmatrix} 4 & 3 \\ 0 & -1 \\ -1 & 7 \\ 5 & 5 \\ -3 & 0 \end{bmatrix}$. The answer is **5 × 2; no special characteristics.**

This matrix has five rows and two columns.

4 Identify the dimension of the matrix and determine any special classifications: $\begin{bmatrix} 1 & 0 \\ 1 & 6 \end{bmatrix}$. The answer is **2 × 2; square.**

The number of rows and columns is the same.

5 If $A = \begin{bmatrix} 2 & 3 & 0 \\ 4 & 1 & 5 \end{bmatrix}$ and $B = \begin{bmatrix} -1 & -2 & 0 \\ 1 & 0 & 2 \end{bmatrix}$, find $A - B$. The answer is $\begin{bmatrix} 3 & 5 & 0 \\ 3 & 1 & 3 \end{bmatrix}$.

Subtract each element in the second matrix from the corresponding elements in the first matrix: $\begin{bmatrix} 2-(-1) & 3-(-2) & 0-0 \\ 4-1 & 1-0 & 5-2 \end{bmatrix} = \begin{bmatrix} 3 & 5 & 0 \\ 3 & 1 & 3 \end{bmatrix}$.

6 If $A = \begin{bmatrix} -3 & 6 \\ 2 & 2 \end{bmatrix}$ and $B = \begin{bmatrix} -1 & -2 \\ -3 & 2 \end{bmatrix}$, find $4A + 2B$. The answer is $\begin{bmatrix} -14 & 20 \\ 2 & 12 \end{bmatrix}$.

Multiply each element in the first matrix by 4 and all those in the second matrix by 2. Then add the respective elements together: $4\begin{bmatrix} -3 & 6 \\ 2 & 2 \end{bmatrix} + 2\begin{bmatrix} -1 & -2 \\ -3 & 2 \end{bmatrix} = \begin{bmatrix} -12+(-2) & 24+(-4) \\ 8+(-6) & 8+4 \end{bmatrix} = \begin{bmatrix} -14 & 20 \\ 2 & 12 \end{bmatrix}$

7 If $A = \begin{bmatrix} 4 & -2 & 3 \\ 1 & 6 & -5 \\ 4 & 4 & -2 \end{bmatrix}$ and $B = \begin{bmatrix} -4 & 2 & -3 \\ -1 & -6 & 5 \\ -4 & -4 & 2 \end{bmatrix}$, find $A + B$. The answer is $\begin{bmatrix} 0 & 0 & 0 \\ 0 & 0 & 0 \\ 0 & 0 & 0 \end{bmatrix}$.

Add all the elements in the first matrix to their counterparts in the second matrix. Because each element is the opposite of what you add it to, the result is a zero matrix: $\begin{bmatrix} 4+(-4) & -2+2 & 3+(-3) \\ 1+(-1) & 6+(-6) & -5+5 \\ 4+(-4) & 4+(-4) & -2+2 \end{bmatrix} = \begin{bmatrix} 0 & 0 & 0 \\ 0 & 0 & 0 \\ 0 & 0 & 0 \end{bmatrix}$.

8 If $A = \begin{bmatrix} 3 & 0 & 0 \\ 0 & 0 & 7 \\ 0 & 5 & 1 \end{bmatrix}$ and $B = \begin{bmatrix} 1 & 0 & 0 \\ 0 & 1 & 0 \\ 0 & 0 & 1 \end{bmatrix}$, find $9A - B$. The answer is $\begin{bmatrix} 26 & 0 & 0 \\ 0 & -1 & 63 \\ 0 & 45 & 8 \end{bmatrix}$.

Multiply each element in the first matrix by 9. Then subtract all the elements in the second

matrix from those results: $9\begin{bmatrix} 3 & 0 & 0 \\ 0 & 0 & 7 \\ 0 & 5 & 1 \end{bmatrix} - \begin{bmatrix} 1 & 0 & 0 \\ 0 & 1 & 0 \\ 0 & 0 & 1 \end{bmatrix} = \begin{bmatrix} 27-1 & 0-0 & 0-0 \\ 0-0 & 0-1 & 63-0 \\ 0-0 & 45-0 & 9-1 \end{bmatrix} = \begin{bmatrix} 26 & 0 & 0 \\ 0 & -1 & 63 \\ 0 & 45 & 8 \end{bmatrix}$.

9 Find the product of the two matrices $A = \begin{bmatrix} 2 & 1 \\ 0 & 4 \end{bmatrix}$ and $B = \begin{bmatrix} 3 & -2 & 4 \\ 1 & 0 & 2 \end{bmatrix}$. The answer is $\begin{bmatrix} 7 & -4 & 10 \\ 4 & 0 & 8 \end{bmatrix}$.

The product of a 2×2 matrix times a 2×3 matrix is a 2×3 matrix. Find the sums of all the row-to-column products: $\begin{bmatrix} 2(3)+1(1) & 2(-2)+1(0) & 2(4)+1(2) \\ 0(3)+4(1) & 0(-2)+4(0) & 0(4)+4(2) \end{bmatrix} = \begin{bmatrix} 7 & -4 & 10 \\ 4 & 0 & 8 \end{bmatrix}$.

10 Find the product of the two matrices $C = \begin{bmatrix} 1 & 0 \\ 4 & 3 \\ 2 & -2 \end{bmatrix}$ and $A = \begin{bmatrix} 2 & 1 \\ 0 & 4 \end{bmatrix}$. The answer is $\begin{bmatrix} 2 & 1 \\ 8 & 16 \\ 4 & -6 \end{bmatrix}$.

The product of a 3×2 matrix times a 2×2 matrix is a 3×2 matrix. Find the sums of all the row-to-column products: $\begin{bmatrix} 1(2)+0(0) & 1(1)+0(4) \\ 4(2)+3(0) & 4(1)+3(4) \\ 2(2)+(-2)(0) & 2(1)+(-2)(4) \end{bmatrix} = \begin{bmatrix} 2 & 1 \\ 8 & 16 \\ 4 & -6 \end{bmatrix}$.

11 Find the product of the two matrices $B = \begin{bmatrix} 3 & -2 & 4 \\ 1 & 0 & 2 \end{bmatrix}$ and $C = \begin{bmatrix} 1 & 0 \\ 4 & 3 \\ 2 & -2 \end{bmatrix}$. The answer is $\begin{bmatrix} 3 & -14 \\ 5 & -4 \end{bmatrix}$.

The product of a 2×3 matrix times a 3×2 matrix is a 2×2 matrix. Find the sums of all the row-to-column products: $\begin{bmatrix} 3(1)+(-2)(4)+4(2) & 3(0)+(-2)(3)+4(-2) \\ 1(1)+0(4)+2(2) & 1(0)+0(3)+2(-2) \end{bmatrix} = \begin{bmatrix} 3 & -14 \\ 5 & -4 \end{bmatrix}$.

12 Find the product of the two matrices $A = \begin{bmatrix} 2 & 1 \\ 0 & 4 \end{bmatrix}$ and $I = \begin{bmatrix} 1 & 0 \\ 0 & 1 \end{bmatrix}$. The answer is $\begin{bmatrix} 2 & 1 \\ 0 & 4 \end{bmatrix}$.

The product of a 2×2 matrix times a 2×2 matrix is a 2×2 matrix. Find the sums of all the row-to-column products. The second matrix is an identity matrix, which is why the result is actually the first, original matrix: $\begin{bmatrix} 2(1)+1(0) & 2(0)+1(1) \\ 0(1)+4(0) & 0(0)+4(1) \end{bmatrix} = \begin{bmatrix} 2 & 1 \\ 0 & 4 \end{bmatrix}$.

13 Find the inverse of the matrix $\begin{bmatrix} 3 & 2 \\ 7 & 5 \end{bmatrix}$. The answer is $\begin{bmatrix} 5 & -2 \\ -7 & 3 \end{bmatrix}$.

The difference between the cross-products, m, is $15 - 14 = 1$. Reverse the 3 and 5, and negate the 2 and 7. Because you're dividing by 1, the new entries don't change.

14 Find the inverse of the matrix $\begin{bmatrix} -4 & -5 \\ 7 & 9 \end{bmatrix}$. The answer is $\begin{bmatrix} -9 & -5 \\ 7 & 4 \end{bmatrix}$.

The difference between the cross-products, m, is $-36 - (-35) = -1$. Reverse the -4 and 9. Then negate the -5 to make it 5 and the 7 to make it -7. Dividing every term by -1, all the signs change — even the two that you changed already (they change back).

15 Find the inverse of the matrix $\begin{bmatrix} 6 & 3 \\ 4 & 2 \end{bmatrix}$. **No inverse.**

The difference between the cross-products, m, is $12 - 12 = 0$. This result means that this matrix has no inverse.

16 Find the inverse of the matrix $\begin{bmatrix} -8 & 4 \\ 5 & -2 \end{bmatrix}$. The answer is $\begin{bmatrix} \frac{1}{2} & 1 \\ \frac{5}{4} & 2 \end{bmatrix}$.

The difference between the cross-products, m, is $16 - 20 = -4$. Reverse the -8 and -2. Then change the 4 and 5 to negative numbers. Finally, divide each number by -4 and simplify.

17 Find the inverse of the matrix $\begin{bmatrix} 1 & 2 & 3 \\ 0 & 2 & 4 \\ -1 & 4 & -1 \end{bmatrix}$. The answer is $\begin{bmatrix} \frac{9}{10} & -\frac{7}{10} & -\frac{1}{10} \\ \frac{1}{5} & -\frac{1}{10} & \frac{1}{5} \\ -\frac{1}{10} & \frac{3}{10} & -\frac{1}{10} \end{bmatrix}$.

Write the matrix and a 3×3 identity matrix next to it to make one big array.

Change the first column so it looks like an identity column: The element in the first row is already a 1 and the element in the second row is already a 0, so you just have to change the -1 to a 0. Add row one to row three:

$$\begin{bmatrix} 1 & 2 & 3 & : & 1 & 0 & 0 \\ 0 & 2 & 4 & : & 0 & 1 & 0 \\ -1 & 4 & -1 & : & 0 & 0 & 1 \end{bmatrix} \rightarrow \begin{bmatrix} 1 & 2 & 3 & : & 1 & 0 & 0 \\ 0 & 2 & 4 & : & 0 & 1 & 0 \\ 0 & 6 & 2 & : & 1 & 0 & 1 \end{bmatrix}.$$

Change the second column so it looks like an identity column: The element in the second row, second column needs to be a 1, but it's currently a 2. Divide row two by 2. You need 0s above and below the 1, so you then multiply row two by -2 and add that answer to row one. Multiply row two by -6 and add it to row three:

$$\begin{bmatrix} 1 & 2 & 3 & : & 1 & 0 & 0 \\ 0 & 1 & 2 & : & 0 & \frac{1}{2} & 0 \\ 0 & 6 & 2 & : & 1 & 0 & 1 \end{bmatrix} \rightarrow \begin{bmatrix} 1 & 0 & -1 & : & 1 & -1 & 0 \\ 0 & 1 & 2 & : & 0 & \frac{1}{2} & 0 \\ 0 & 0 & -10 & : & 1 & -3 & 1 \end{bmatrix}.$$

Change the third column so it looks like an identity column: The -10 in the third row, third column is changed to a 1 by dividing row three by -10. Then get 0s above the 1 by adding row three to row one. Multiply row three by -2 and add it to row two:

$$\begin{bmatrix} 1 & 0 & -1 & : & 1 & -1 & 0 \\ 0 & 1 & 2 & : & 0 & \frac{1}{2} & 0 \\ 0 & 0 & 1 & : & -\frac{1}{10} & \frac{3}{10} & -\frac{1}{10} \end{bmatrix} \rightarrow \begin{bmatrix} 1 & 0 & 0 & : & \frac{9}{10} & -\frac{7}{10} & -\frac{1}{10} \\ 0 & 1 & 0 & : & \frac{1}{5} & -\frac{1}{10} & \frac{1}{5} \\ 0 & 0 & 1 & : & -\frac{1}{10} & \frac{3}{10} & -\frac{1}{10} \end{bmatrix}.$$

18 Find the inverse of the matrix $\begin{bmatrix} 1 & 2 & -3 \\ 3 & 5 & -7 \\ -1 & -3 & 6 \end{bmatrix}$. The answer is $\begin{bmatrix} -9 & 3 & -1 \\ 11 & -3 & 2 \\ 4 & -1 & 1 \end{bmatrix}$.

Write the matrix and a 3×3 identity matrix next to it.

The first row, first column already has an element of 1, so you just have to get 0s below that 1. Multiply row one by -3 and add it to row two. Add row one to row three:

$$\begin{bmatrix} 1 & 2 & -3 & : & 1 & 0 & 0 \\ 3 & 5 & -7 & : & 0 & 1 & 0 \\ -1 & -3 & 6 & : & 0 & 0 & 1 \end{bmatrix} \rightarrow \begin{bmatrix} 1 & 2 & -3 & : & 1 & 0 & 0 \\ 0 & -1 & 2 & : & -3 & 1 & 0 \\ 0 & -1 & 3 & : & 1 & 0 & 1 \end{bmatrix}.$$

Change the second column so it looks like an identity column: The element along the diagonal is currently a -1, so multiply row two by -1. You then want 0s above and below the 1 in the second row, second column, so multiply row two by -2 and add it to row one. Then add row two to row three:

$$\begin{bmatrix} 1 & 2 & -3 & : & 1 & 0 & 0 \\ 0 & 1 & -2 & : & 3 & -1 & 0 \\ 0 & -1 & 3 & : & 1 & 0 & 1 \end{bmatrix} \rightarrow \begin{bmatrix} 1 & 0 & 1 & : & -5 & 2 & 0 \\ 0 & 1 & -2 & : & 3 & -1 & 0 \\ 0 & 0 & 1 & : & 4 & -1 & 1 \end{bmatrix}.$$

The third row, third column has a 1 for an element (after all the work on the first two columns). To get 0s above that 1, multiply row three by –1 and add it to to row one. Then multiply row three by 2 and add it to row two:

$$\begin{bmatrix} 1 & 0 & 0 & : & -9 & 3 & -1 \\ 0 & 1 & 0 & : & 11 & -3 & 2 \\ 0 & 0 & 1 & : & 4 & -1 & 1 \end{bmatrix}.$$

19 Solve the system of equations using matrices: $3x - 2y = 17$ and $-4x + 3y = -23$. The answer is **$x = 5, y = -1$.**

Write the coefficient matrix and find its inverse. (Refer to the section on "The Search for Identity: Finding Inverse Matrices" for info on that technique.) Then multiply that inverse by the constant matrix. (You can find how to multiply matrices in the section "Trying Times: Multiplying Matrices by Each Other.") Here's how this problem plays out:

$$\begin{bmatrix} 3 & -2 \\ -4 & 3 \end{bmatrix}^{-1} = \begin{bmatrix} 3 & 2 \\ 4 & 3 \end{bmatrix}$$

$$\begin{bmatrix} 3 & 2 \\ 4 & 3 \end{bmatrix} \cdot \begin{bmatrix} 17 \\ -23 \end{bmatrix} = \begin{bmatrix} 5 \\ -1 \end{bmatrix}$$

20 Solve the system of equations using matrices: $-6x + 5y = 8$ and $-4x + 3y = 6$. The answer is **$x = -3, y = -2$.**

Write the coefficient matrix and find its inverse. Then multiply that inverse by the constant matrix:

$$\begin{bmatrix} -6 & 5 \\ -4 & 3 \end{bmatrix}^{-1} = \begin{bmatrix} \frac{3}{2} & -\frac{5}{2} \\ 2 & -3 \end{bmatrix}$$

$$\begin{bmatrix} \frac{3}{2} & -\frac{5}{2} \\ 2 & -3 \end{bmatrix} \cdot \begin{bmatrix} 8 \\ 6 \end{bmatrix} = \begin{bmatrix} -3 \\ -2 \end{bmatrix}$$

21 Solve the system of equations using matrices: $x - y - z = -1$, $2x + 3y + z = 0$, and $3y + 2z = 2$. The answer is **$x = 1, y = -2, z = 4$.**

Write the coefficient matrix and find its inverse. Then multiply that inverse by the constant matrix:

$$\begin{bmatrix} 1 & -1 & -1 \\ 2 & 3 & 1 \\ 0 & 3 & 2 \end{bmatrix}^{-1} = \begin{bmatrix} 3 & -1 & 2 \\ -4 & 2 & -3 \\ 6 & -3 & 5 \end{bmatrix}$$

$$\begin{bmatrix} 3 & -1 & 2 \\ -4 & 2 & -3 \\ 6 & -3 & 5 \end{bmatrix} \cdot \begin{bmatrix} -1 \\ 0 \\ 2 \end{bmatrix} = \begin{bmatrix} 1 \\ -2 \\ 4 \end{bmatrix}$$

22 Solve the system of equations using matrices: $2x + y - z = 9$, $x + 2y - z = 6$, and $x - 2y + z = 0$. The answer is **$x = 3, y = 0, z = -3$.**

Write the coefficient matrix and find its inverse. Then multiply that inverse by the constant matrix:

$$\begin{bmatrix} 2 & 1 & -1 \\ 1 & 2 & -1 \\ 1 & -2 & 1 \end{bmatrix}^{-1} = \begin{bmatrix} 0 & 0.5 & 0.5 \\ -1 & 1.5 & 0.5 \\ -2 & 2.5 & 1.5 \end{bmatrix}$$

$$\begin{bmatrix} 0 & 0.5 & 0.5 \\ -1 & 1.5 & 0.5 \\ -2 & 2.5 & 1.5 \end{bmatrix} \cdot \begin{bmatrix} 9 \\ 6 \\ 0 \end{bmatrix} = \begin{bmatrix} 3 \\ 0 \\ -3 \end{bmatrix}$$

Chapter 15

Going Out of Sequence with Sequences and Series

A *sequence* is a list of items all separated by commas (milk, sugar, eggs . . . okay, so it's generally a list of numbers). Mathematical sequences usually have a rule determining what number comes at what position in the list. A *series* is the sum of the numbers in a sequence — or that list. You add up a few or many or all the terms in a sequence.

In this chapter, you see arithmetic (pronounced air-ith-*met*-ick) and geometric sequences — some of the more commonly used sequences. You also form many other types of sequences by just writing a rule using various mathematical operations. Finally, you figure out how to add the terms in a sequence using rules that apply to series.

Writing the Terms of a Sequence

The sequence 1, 3, 5, 7, 9, ... consists of the odd positive integers. Notice that I stop listing the numbers at 9, hoping that you can determine the pattern from those few terms. The three dots, called an *ellipsis,* mean "and so on and so on." You often see the terms of a sequence written in *braces:* {1, 3, 5, 7, 9, ...}; this mathematical notation is standard for sequences. Even graphing calculators recognize the braces as designating the list of numbers in a sequence.

You can also describe sequences using a rule. You let *n* represent a counting number and write {2*n* – 1}. This rule gives you all the terms in a sequence when you replace the *n* with each of the counting numbers, one after the other.

Q. Write the first six terms of the sequence: {$n^3 - 2n + 1$}

A. **0, 5, 22, 57, 116, and 205.** When you replace *n* with 1, you get $1^3 - 2(1) + 1 = 1 - 2 + 1 = 0$. When you replace *n* with 2, you get $2^3 - 2(2) + 1 = 8 - 4 + 1 = 5$. And so on — or should I use . . .?

Q. Find a rule that describes how to find all the terms in the sequence: 2, –6, 10, –14, 18, –22, ...

A. $\{(-1)^{n+1}(4n-2)\}$. You first notice that the terms alternate in sign between positive and negative. If you disregard the signs, the numbers by themselves (their *absolute values*) go up by 4 units each time. You deal with these two features separately. First, to make numbers go up by 4 each time, multiply an n by 4, giving you $4n$. But when you let $n = 1$, you get 4, and when it's 2, you get 8. These numbers are too big by

2, so subtract 2, and you get $4n - 2$. That takes care of the absolute value: $\{4n - 2\} = \{2, 6, 10, 14, 18, 22, ...\}$. You deal with the alternating part by multiplying the numbers by powers of –1. When –1 is raised to an odd power, it's equal to –1. When you raise –1 to an even power, the number is equal to +1. When $n = 1$, you want a positive number in this sequence, so use $(-1)^{n+1}$ as a multiplier; write the sequence rule as $\{(-1)^{n+1}(4n-2)\}$. As n changes, the power on the –1 alternates between even and odd, giving you the + and – signs for the terms.

---

1. Write the first six terms of the sequence: $\{2n^2 - 5n + 3\}$

Solve It

2. Write the first six terms of the sequence: $\left\{n! + \dfrac{(n+1)!}{2^n}\right\}$

Note: The factorial operation, $n!$, is defined as follows: $n! = n(n-1)(n-2)\cdots 3 \cdot 2 \cdot 1$. In other words, you multiply the number by every positive integer smaller than that number. (For practice working with factorials, see Chapter 16.)

Solve It

---

3. Write the first six terms of the sequence: $\{(-1)^{n+1}3^n\}$

Solve It

4. Write the first six terms of the sequence: $\{(-1)^{n+1}n^2 + (-1)^n n^3\}$

Solve It

5. Write the next two terms of this sequence, and then write a rule for determining all the terms of the sequence: $\{1, -4, 9, -16, 25, -36, ...\}$

Solve It

6. Write the next two terms of this sequence, and then write a rule for determining all the terms of the sequence: $\left\{\frac{1}{2}, \frac{1}{4}, \frac{1}{8}, \frac{1}{16}, \frac{1}{32}, \frac{1}{64}, ...\right\}$

Solve It

Differences and Multipliers: Working with Special Sequences

Two very special types of sequences are the arithmetic sequences and geometric sequences:

✔ **Arithmetic:** These sequences always have terms that are a constant difference apart. For instance, the sequence $\{3, 8, 13, 18, 23, ...\}$ is an arithmetic sequence with terms that you can find by adding the *difference,* 5, to the preceding term.

Arithmetic sequences have a general term of $a_n = a_{n-1} + d$, where you can find the next term by adding the difference, d, to the preceding term. To find a particular term in the sequence — the nth term — you can also use the formula $a_n = a_1 + (n-1)d$, where a_1 is the first term in the sequence and d is the difference.

✔ **Geometric:** These sequences have terms that are a constant multiplier or ratio apart. For instance, the sequence $\{1, 3, 9, 27, 81, ...\}$ is a geometric sequence with terms that you can find by multiplying the *ratio,* 3, by the preceding term.

Geometric sequences have a general term of $g_n = rg_{n-1}$, where you can find the next term by multiplying the ratio, r, by the preceding term. To find a particular term in the sequence — the nth term — you can also use the formula $g_n = g_1(r^{n-1})$, where g_1 is the first term in the sequence and r is the ratio.

Q. Find the 100th and 101st terms in the sequence beginning with –45, –43, –41, –39, ... and also find a formula for all the terms in the sequence

A. **Terms: 153 and 155; formula: $\{2n - 47\}$.** This sequence is arithmetic, because you can see a common difference of 2 between each pair of terms. The first term in the

sequence is –45. To find a formula, use $a_n = a_1 + (n-1)d$, replacing the a_1 with –45 and d with 2. Solving for the general term, $a_n = -45 + (n-1)2 = -45 + 2n - 2 = 2n - 47$. Using the formula and substituting in the 100 for a_n, you get $a_{100} = -45 + (100 - 1)2 = -45 + (99)2 = -45 + 198 = 153$. The 101st term is the next term, so just add 2 to 153 to get 155.

7. Find the 10th, 11th, and 12th terms of the arithmetic sequence whose first term is –4 and whose difference between the terms is 3

Hint: Find the formula first.

Solve It

8. Find the 10th, 11th, and 12th terms of the geometric sequence whose first term is 512 with a ratio of $\frac{3}{2}$

Hint: Find the formula first.

Solve It

9. Find the next three terms of the sequence beginning with 9, 15, 21, 27, ... and write a formula for all the terms

Solve It

10. Find the next three terms of the sequence beginning with 1, –5, 25, –125, ... and write a formula for all the terms

Solve It

Backtracking: Constructing Recursively Defined Sequences

A *recursively* defined sequence is one in which the successive terms are built from two or more of the preceding terms. One of the best-known recursively defined sequences is the Fibonacci Sequence: 1, 1, 2, 3, 5, 8, 13, 21, 34, ... — each term after the first two is the sum of the preceding two terms.

To define this sequence with the proper notation, you say that $a_1 = 1$, $a_2 = 1$, and $a_n = a_{n-2} + a_{n-1}$. The subscript notation says that a_{n-2} is two terms before the term being created and that a_{n-1} is the term right before the one being created.

Q. Determine the first eight terms of the sequence where $a_1 = 3$, $a_2 = 5$, and $a_n = 2a_{n-2} - a_{n-1}$

A. **3, 5, 1, 9, –7, 25, –39, 89.** You already have the 3 and 5. The next term is $a_3 = 2a_{3-2} - a_{3-1} = 2a_1 - a_2 = 2(3) - 5 = 6 - 5 = 1$. The fourth term is $a_4 = 2a_2 - a_3 = 2(5) - 1 = 10 - 1 = 9$. The fifth term is $a_5 = 2a_3 - a_4 = 2(1) - 9 = 2 - 9 = -7$. The sixth term is $a_6 = 2a_4 - a_5 = 2(9) - (-7) = 18 + 7 = 25$. The seventh term is $a_7 = 2a_5 - a_6 = 2(-7) - 25 = -14 - 25 = -39$. And the eighth term is $a_8 = 2a_6 - a_7 = 2(25) - (-39) = 50 + 39 = 89$.

11. Determine the first six terms of the sequence where $a_1 = 2$, $a_2 = 3$, and $a_n = 2a_{n-2} - 3a_{n-1}$

Solve It

12. Determine the first six terms of the sequence where $a_1 = 4$, $a_2 = 1$, and $a_n = a_{n-2} - (a_{n-1})^2$

Solve It

13. Determine the first six terms of the sequence where $a_1 = 2$, $a_2 = 4$, $a_3 = 7$, and $a_n = 3a_{n-3} + a_{n-1}$

Solve It

14. You're offered a job with a salary of $10,000 the first year and $20,000 the second year. After the second year, your salary will be the sum of the previous year's salary and 10 percent of the salary two years before that year. The formula for the salaries forms a sequence that's recursively defined. Find the first six years' salary amounts, using $a_1 = 10,000$, $a_2 = 20,000$, and $a_n = 0.1a_{n-2} + a_{n-1}$.

Solve It

Using Summation Notation

A sequence is a list of things — usually numbers — and a series is the sum of a list of numbers. You can say, "Find the sum of the first six numbers in the sequence that has the general rule $a_n = n^2 - 3n + 4$." Or you can use *summation notation,* which tells you the same thing in a more cryptic (but quicker) fashion.

The summation notation for finding the sum of the first six terms of the sequence $a_n = n^2 - 3n + 4$ is $\sum_{n=1}^{6}(n^2 - 3n + 4)$. The big Σ stands for *summation* — it's the capital letter sigma in the Greek alphabet. The values below and above the sigma tell you where to start ($n = 1$) and stop (6), and the rule after the sigma, in the parentheses, tells you how to form the numbers to be added up. In this case, you add up $2 + 2 + 4 + 8 + 14 + 22$ to get 52.

Q. Find the sum of the series $\sum_{n=1}^{5} (n^2 + 3^n - 3)$

A. **403.** Using the rule to list the terms to be added up, you get $(1 + 3 - 3) + (4 + 9 - 3) + (9 + 27 - 3) + (16 + 81 - 3) + (25 + 243 - 3) = 1 + 10 + 33 + 94 + 265 = 403$.

Q. Find the sum of the series $\sum_{n=1}^{50} \left(\frac{1}{2n-1} - \frac{1}{2n+1} \right)$

A. $\frac{100}{101}$. At first, this problem may seem to be a bit daunting, with 50 terms to deal with and all those fractions. But after you list a few terms, you may see a pattern that can help you:

$$\left(\frac{1}{1} - \frac{1}{3} \right) + \left(\frac{1}{3} - \frac{1}{5} \right) + \left(\frac{1}{5} - \frac{1}{7} \right) + \left(\frac{1}{7} - \frac{1}{9} \right) + \cdots$$

$$= 1 + \left(-\frac{1}{3} + \frac{1}{3} \right) + \left(-\frac{1}{5} + \frac{1}{5} \right) + \left(-\frac{1}{7} + \frac{1}{7} \right) + \left(-\frac{1}{9} + \cdots \right)$$

The second term in each grouping is the opposite of the first term in the next grouping. All the opposites sum to 0. So only the first term and the last term will be left. You can jump right to the final sum:

$$= 1 + \left(-\frac{1}{3} + \frac{1}{3} \right) + \left(-\frac{1}{5} + \frac{1}{5} \right) + \cdots + \left(-\frac{1}{97} + \frac{1}{97} \right) + \left(-\frac{1}{99} + \frac{1}{99} \right) - \frac{1}{101}$$

$$= 1 - \frac{1}{101} = \frac{101}{101} - \frac{1}{101} = \frac{100}{101}$$

---

15. Find the sum of the series $\sum_{n=1}^{4} (3^{n+1} - 4)$

Solve It

16. Find the sum of the series $\sum_{n=3}^{6} (n^2 - 2n - 1)$

Solve It

---

17. Find the sum of the series $\sum_{n=0}^{5} \left(\frac{n}{(n+1)^2} \right)$

Solve It

18. Find the sum of the series $\sum_{n=0}^{4} \left(\frac{n!}{n+1} \right)$

Note: $n!$ means $n(n-1)(n-2) \cdots 3 \cdot 2 \cdot 1$. Note that by definition, $0! = 1$. For more info on factorial operations, see Chapter 16.

Solve It

Finding Sums with Special Series

The sums of the terms in sequences have many practical applications, from performing calculations in the world of finance and banking to counting the seats in an ampitheater (or, naturally, impressing your crush by determining the value of *e* to 100 decimal places). If the number of terms in the sequence is very large, the summing can get pretty awful, and the sums can get enormous.

Many commonly used sequences have series that have formulas for their sums. Algebraic and geometric series, for instance, have rules for their sums. Also, series consisting of terms that are squares of numbers or cubes of numbers have formulas for their sums. Refer to Chapter 18 if you want to see some of the special formulas that I use in this section.

Some formulas for the sums of series are in Figure 15-1.

Sequence Type	Rule for Sequence Terms	Formula for Sum of Terms
Arithmetic	$a_n = a_1 + (n-1)d$	$S_n = \frac{n}{2}[2a_1 + (n-1)d] = \frac{n}{2}[a_1 + a_n]$
Geometric	$g_n = g_1 + (r)^{n-1}$	$S_n = \frac{g_1(1-r^n)}{1-r}$
Converging geometric	$g_n = g_1(r)^{n-1}, |r|<1$	$S_\infty = \frac{g_1}{1-r}$
Odd numbers	$a_n = 2n-1$	$S_n = n^2$

Figure 15-1: Sequences and their sum formulas.

Q. Find the sum of the series 2 + 5 + 8 + 11 + 14 + ... + 89

A. **1,365.** The terms in this series come from an arithmetic sequence of terms. The common difference is 3. You can use either formula to find the sum, but in each, you need to know how many terms are in the list. Counting them one by one is a bit tedious and is likely to produce an error. Instead, create the rule for the general term. The terms are three apart, so you need 3n in your rule. Subtract 1 from a multiple of 3, and you get the terms. The rule is 3n − 1. Set that rule equal to 89 to solve for *n*. If 3n − 1 = 89, then 3n = 90; n = 30. This series contains 30 terms. Now, using the number of terms and the first and last terms in the formula, $S_n = \frac{30}{2}(2+89) = 15(91) = 1,365.$

Q. Find the sum of the series $1 + \frac{1}{4} + \frac{1}{16} + \frac{1}{64} + \frac{1}{256} + ...$; this problem asks for the sum of an infinite number of terms

A. $\frac{4}{3}$. The terms in this series come from a geometric sequence of terms in which the ratio is a proper fraction — its absolute value is less than 1. You an use the general sum formula for a geometric series if you're adding up a certain number of terms — 10 or 20 or 50 of them. But you can also find the sum of *all* of them when the ratio of the geometric series is a proper fraction. The terms in a *converging* geometric series get so very small that their sum never exceeds a particular number — the series has a limit (see Chapter 8 for details on limits). Using the formula for an infinite sum on this series, you put the first term, 1, in the numerator and the ratio, ¼, in the denominator: $S_\infty = \frac{1}{1-\frac{1}{4}} = \frac{1}{\frac{3}{4}} = \frac{4}{3}.$

Q. Find the sum of the series $\dfrac{1}{4^5} + \dfrac{1}{4^6} + \dfrac{1}{4^7} + \dfrac{1}{4^8} + \ldots$

A. $\dfrac{1}{768}$. One way to approach this problem is to find the sum of the infinite series starting with a first term, $\dfrac{1}{4^5} = \dfrac{1}{1,024}$, and using the formula where the ratio is ¼. You get $\dfrac{\dfrac{1}{1,024}}{1-\dfrac{1}{4}} = \dfrac{\dfrac{1}{1,024}}{\dfrac{3}{4}} = \dfrac{1}{\underset{256}{1,024}} \cdot \dfrac{\overset{1}{4}}{3} = \dfrac{1}{768}$.

Another approach is to subtract the first four terms of a series with a ratio of ¼ from the infinite series. As you see, the answer is the same:

$$\sum_{n=0}^{\infty}\left(\frac{1}{4^n}\right) - \sum_{n=0}^{4}\left(\frac{1}{4^n}\right) = \frac{4}{3} - \frac{1\left(1-\left(\frac{1}{4}\right)^4\right)}{1-\frac{1}{4}} = \frac{4}{3} - \frac{\left(1-\frac{1}{1,024}\right)}{\frac{3}{4}}$$

$$= \frac{4}{3} - \frac{\dfrac{1,023}{1,024}}{\dfrac{3}{4}} = \frac{4}{3} - \frac{\overset{341}{1,023}}{\underset{256}{1,024}} \cdot \frac{\overset{1}{4}}{\overset{1}{3}}$$

$$= \frac{4}{3} - \frac{341}{256} = \frac{4(256) - 3(341)}{768}$$

$$= \frac{1,024 - 1,023}{768} = \frac{1}{768}$$

This technique doesn't look like a particularly good way to do the problem, but keep in mind that sometimes it's easier to subtract a few terms from the entire sum.

---

19. Find the sum of the series $1 + 2 + 3 + 4 + \ldots + 500$

Solve It

20. Find the sum of the series $1 + 2 + 4 + 8 + 16 + \ldots + 512$

Solve It

21. Find the sum of the series $-3 + 9 - 27 + 81 - 243 + \ldots + 3^{10}$

Solve It

22. Find the sum of the series $1 + 3 + 5 + 7 + 9 + \ldots + 99$

Solve It

23. Find the sum of the series $1 + \frac{2}{3} + \frac{4}{9} + \frac{8}{27} + \ldots$

Solve It

24. Find the sum of the series $101 + 103 + 105 + \ldots + 1{,}001$

Solve It

Answers to Problems on Sequences and Series

The following are the answers to the practice problems presented earlier in this chapter.

1 Write the first six terms of the sequence: $\{2n^2 - 5n + 3\}$. The answer is **0, 1, 6, 15, 28, 45.**

Replace the n in the rule with 1 to get $2(1)^2 - 5(1) + 3 = 2 - 5 + 3 = 0$. Repeat the process, letting $n = 2, 3, 4, 5$, and 6 to determine the rest of the terms.

2 Write the first six terms of the sequence: $\left\{n! + \dfrac{(n+1)!}{2^n}\right\}$. The answer is **2, 3.5, 9, 31.5, 142.5, 798.75.**

Replace the n in the rule with 1 to get $1! + \dfrac{(1+1)!}{2^1} = 1 + \dfrac{2!}{2} = 1 + \dfrac{2}{2} = 2$. Repeat the process, letting $n = 2, 3, 4, 5$, and 6 to determine the rest of the terms.

3 Write the first six terms of the sequence: $\{(-1)^{n+1}3^n\}$. The answer is **3, –9, 27, –81, 243, –729.**

Replace the n in the rule with 1 to get $(-1)^{1+1}3^1 = 1 \cdot 3 = 3$. Repeat the process, letting $n = 2, 3, 4, 5$, and 6 to determine the rest of the terms. Notice that even powers of –1 result in a positive term and that odd powers give you a negative term.

4 Write the first six terms of the sequence: $\{(-1)^{n+1}n^2 + (-1)^n n^3\}$. The answer is **0, 4, –18, 48, –100, 180.**

Replace the n in the rule with 1 to get $(-1)^{1+1}1^2 + (-1)^1 1^3 = 1 \cdot 1 + (-1) \cdot 1 = 1 - 1 = 0$. Repeat the process, letting $n = 2, 3, 4, 5$, and 6 to determine the rest of the terms.

5 Write the next two terms of this sequence, and then write a rule for determining all the terms of the sequence: $\{1, -4, 9, -16, 25, -36, \ldots\}$. The answer is **49, –64; $\{(-1)^{n+1} n^2\}$.**

The sequence consists of the squares of the numbers 1, 2, 3, 4, and so on with the signs alternating on the squares. When the square is an odd number, the sign is positive, so the exponent has to be even. You accomplish this adjustment by adding 1 to the number of the term in the exponent.

6 Write the next two terms of this sequence, and then write a rule for determining all the terms of the sequence: $\left\{\dfrac{1}{2}, \dfrac{1}{4}, \dfrac{1}{8}, \dfrac{1}{16}, \dfrac{1}{32}, \dfrac{1}{64}, \ldots\right\}$. The answer is $\dfrac{1}{128}, \dfrac{1}{256}; \left\{\dfrac{1}{2^n}\right\}$.

The sequence consists of powers of ½ — in other words, powers of 2 in the denominator of a fraction. The powers shown are $2^1 = 2$, $2^2 = 4$, $2^3 = 8$, $2^4 = 16$, $2^5 = 32$, and $2^6 = 64$.

7 Find the 10th, 11th, and 12th terms of the arithmetic sequence whose first term is –4 and whose difference between the terms is 3. The answer is **23, 26, 29.**

Using $a_n = a_1 + (n - 1)d$, replace a_1 with –4 and d with 3 to get $a_n = -4 + (n - 1)3 = -4 + 3n - 3 = 3n - 7$. The rule for the sequence is $\{3n - 7\}$. Replace the n with 10 to get 23 for the 10th term. Then just add 3 for the 11th and 3 more for the 12th term.

8 Find the 10th, 11th, and 12th terms of the geometric sequence whose first term is 512 with a ratio of $\dfrac{3}{2}$. The answer is **19,683, $\dfrac{59,049}{2}$, $\dfrac{177,147}{4}$.**

Using $g_n = g_1(r^{n-1})$, replace g_1 with 512 and r with $\dfrac{3}{2}$ to get $g_n = 512\left(\dfrac{3}{2}\right)^{n-1}$. Replacing the n with 10, you get $g_{10} = 512\left(\dfrac{3}{2}\right)^{10-1} = 2^9\left(\dfrac{3}{2}\right)^9 = \dfrac{2^9}{1} \cdot \dfrac{3^9}{2^9} = \dfrac{3^9}{1} = 19{,}683$. Multiply by the ratio to get the 11th term and then by the ratio, again, to get the 12th term.

9 Find the next three terms of the sequence beginning with 9, 15, 21, 27, ... and write a formula for all the terms. The answer is **33, 39, 45; {6n + 3}.**

The sequence is arithmetic, because you can see a common difference between the terms: 6. The first term is 9, so using $a_n = a_1 + (n - 1)d$, you determine that $a_n = 9 + (n - 1)6 = 9 + 6n - 6 = 6n + 3$. Use this formula to find the next three terms — or just add 6 onto the 27, then 6 more and 6 more.

10 Find the next three terms of the sequence beginning with 1, –5, 25, –125, ... and write a formula for all the terms. The answer is **625, –3,125, 15,625; $g_n = (-1)^{n+1}(5)^{n-1}$.**

This sequence is geometric, because you can see a constant multiple of –5 between the consecutive terms. Another way of writing the rule is $g_n = (-1)^{n-1}(5)^{n-1} = (-5)^{n-1}$. Use this rule to find the next three terms, or just multiply the –125 by –5 and then by –5 again and again.

11 Determine the first six terms of the sequence where $a_1 = 2$, $a_2 = 3$, and $a_n = 2a_{n-2} - 3a_{n-1}$. The answer is **2, 3, –5, 21, –73, 261.**

The rule has you perform operations on two consecutive terms to construct the next. Multiply the first 2 by 2 and subtract 3 times the 3 to get 2(2) – 3(3) = 4 – 9 = –5. For the fourth term, multiply the 3 by 2 and subtract 3 times the –5 to get 2(3) – 3(–5) = 6 + 15 = 21. Continue with the pattern to get the rest of the terms.

12 Determine the first six terms of the sequence where $a_1 = 4$, $a_2 = 1$, and $a_n = a_{n-2} - (a_{n-1})^2$. The answer is **4, 1, 3, –8, –61, –3,713.**

The rule has you perform operations on two consecutive terms to construct the next. Take the first term and subtract the square of the second from it. The third term is $4 - (1)^2 = 3$, the fourth term is $1 - (3)^2 = 1 - 9 = -8$, and so on.

13 Determine the first six terms of the sequence where $a_1 = 2$, $a_2 = 4$, $a_3 = 7$, and $a_n = 3a_{n-3} + a_{n-1}$. The answer is **2, 4, 7, 13, 25, 46.**

The rule has you perform operations on the first and third of three consecutive terms. You multiply the first term by 3 and add the third term to that product. The fourth term is 3(2) + 7 = 6 + 7 = 13. The fifth term is 3(4) + 13 = 12 + 13 = 25. The sixth term is 3(7) + 25 = 21 + 25 = 46.

14 You're offered a job with a salary of $10,000 the first year and $20,000 the second year. After the second year, your salary will be the sum of the previous year's salary and 10 percent of the salary two years before that year. The formula for the salaries forms a sequence that's recursively defined. Find the first six years' salary amounts, using $a_1 = 10,000$, $a_2 = 20,000$, and $a_n = 0.1a_{n-2} + a_{n-1}$. The answer is **$10,000; $20,000; $21,000; $23,000; $25,100; $27,400.**

The rule has you perform operations on two consecutive salaries to determine the next. Multiply the first salary by 0.1 and add that amount to the second salary. The third year, the salary is 0.1(10,000) + 20,000 = 1,000 + 20,000 = 21,000. The fourth year, the salary is 0.1(20,000) + 21,000 = 2,000 + 2,1000 = 23,000. And so on.

15 Find the sum of the series $\sum_{n=1}^{4} (3^{n+1} - 4)$. The answer is **344.**

The sum of the four terms is (5) + (27 – 4) + (81 – 4) + (243 – 4) = 5 + 23 + 77 + 239 = 344.

16 Find the sum of the series $\sum_{n=3}^{6} (n^2 - 2n - 1)$. The answer is **46.**

Notice that the terms start with $n = 3$. The sum of the four terms is (9 – 6 – 1) + (16 – 8 – 1) + (25 – 10 – 1) + (36 – 12 – 1) = 2 + 7 + 14 + 23 = 46.

17 Find the sum of the series $\sum_{n=0}^{5}\left(\dfrac{n}{(n+1)^2}\right)$. The answer is $\dfrac{343}{450}$.

Notice that the terms start with $n = 0$. The sum of the six terms is $0 + \dfrac{1}{4} + \dfrac{2}{9} + \dfrac{3}{16} + \dfrac{4}{25} + \dfrac{5}{36} = \dfrac{900 + 800 + 675 + 144 + 225}{3,600} = \dfrac{686}{900} = \dfrac{343}{450}$.

18 Find the sum of the series $\sum_{n=0}^{4}\left(\dfrac{n!}{n+1}\right)$. The answer is $\dfrac{127}{15}$.

You see that the terms start with $n = 0$. Also, remember that $0! = 1$, by definition. The sum of the five terms is $1 + \dfrac{1}{2} + \dfrac{2}{3} + \dfrac{3}{2} + \dfrac{24}{5} = \dfrac{30 + 15 + 20 + 45 + 144}{30} = \dfrac{254}{30} = \dfrac{127}{15}$.

19 Find the sum of the series $1 + 2 + 3 + 4 + \ldots + 500$. The answer is **125,250.**

The series is arithmetic with a first term of 1, a last term of 500, and a total of 500 terms. Using the formula, $S_n = \dfrac{n}{2}[a_1 + a_n] = \dfrac{500}{2}[1 + 500] = 250[501] = 125,250$.

20 Find the sum of the series $1 + 2 + 4 + 8 + 16 + \ldots + 512$. The answer is **1,023.**

The series is geometric with a first term of 1, a ratio of 2, and a tenth term of 512. Using the formula, $S_n = \dfrac{g_1(1 - r^n)}{1 - r} = \dfrac{1(1 - 2^{10})}{1 - 2} = \dfrac{1 - 1,024}{-1} = \dfrac{-1,023}{-1} = 1,023$. You determine that 512 is the tenth term by solving $1(2^{x-1}) = 512$ for x.

21 Find the sum of the series $-3 + 9 - 27 + 81 - 243 + \ldots + 3^{10}$. The answer is **−132,861.**

The series is geometric with a first term of −3, a ratio of −3, and a total of 11 terms. Using the formula, $S_n = \dfrac{g_1(1 - r^n)}{1 - r} = \dfrac{-3(1 - (-3)^{11})}{1 - (-3)} = \dfrac{-3(1 + 177,147)}{4} = \dfrac{-531,444}{4} = -132,861$

22 Find the sum of the series $1 + 3 + 5 + 7 + 9 + \ldots + 99$. The answer is **2,500.**

The series is arithmetic with a first term of 1, a difference of 2, and 50 terms in the list. To determine the number of terms, you solve $2n - 1 = 99$ and find that $n = 50$. You can use the formula for the sum of the arithmetic series, or you can use the rule for the sum of n odd integers, which is n^2.

23 Find the sum of the series $1 + \dfrac{2}{3} + \dfrac{4}{9} + \dfrac{8}{27} + \ldots$ The answer is **3.**

The series is geometric with a ratio of $\dfrac{2}{3}$. Use the formula for an infinite series to get $S_\infty = \dfrac{1}{1 - \dfrac{2}{3}} = \dfrac{1}{\dfrac{1}{3}} = 3$.

24 Find the sum of the series $101 + 103 + 105 + \ldots + 1,001$. The answer is **248,501.**

The series is arithmetic with a first term of 101, a difference of 2, and a total of 451 terms. You find the number of terms by solving $1,001 = 2n + 99$ for n. You can use the formula for the sum of the arithmetic series, or you can find n^2 and then subtract the terms from 1 through 99. Using the formula for the arithmetic series, $S_n = \dfrac{n}{2}[a_1 + a_n] = \dfrac{451}{2}[101 + 1,001] = \dfrac{451}{2}[1,102] = 248,501$.

Chapter 16

Everything You Ever Wanted to Know about Sets and Counting

A *set* is a mathematical entity that consists of a collection or group of objects. For instance, a set named A may consist of the first six prime numbers. You write A = {2, 3, 5, 7, 11, 13}. The set notation looks a bit like the notation you use to write the terms of a sequence (see Chapter 15), but with sets, the elements don't have to be in any particular order. You can just as easily write A = {3, 13, 11, 7, 5, 2}. The two sets here are the same.

In addition, sets don't require a mathematical rule for their elements. You can have B = {horse, dog, 5, blue}. The only thing the four elements in set B have in common is that they're in the same set. (Much like a family, a set can be a real mishmash.)

In this chapter, you become familiar with set notation and ways to perform operations such as union and intersection on sets. Then you count the number of elements in sets using the multiplication property, permutations, and combinations. I cover the Binomial Theorem, which is directly related to combinations, as well.

Writing the Elements of a Set from Rules or Patterns

You can describe the elements in a set in several different ways, but you usually want to choose the method that's quickest and most efficient and/or clearest to the reader. The two main methods for describing a set are roster and rule (or set-builder).

A *roster* is a list of the elements in a set. When the set doesn't include many elements, then this description works fine. If the set contains a lot of elements, you can use an ellipses (…) if the pattern is obvious (a nasty word in mathematics). A rule works well when you find lots and lots of elements in the set.

Q. Use roster and rule notation to describe the set F, which consists of all the positive multiples of 5 that are less than 50

A. **Roster notation: F = {5, 10, 15, 20, 25, 30, 35, 40, 45}; rule notation: F = {$x \mid x = 5n$, where $n \, \varepsilon \, Z$ and $0 < n < 10$}.** You read the rule as "Set F consists of all x's such that x is equal to five times n, where n is an integer and n is a number between zero and ten." (Remember, the symbol Z stands for integers.) Of course, you don't have to write the elements in the set in order. You can just as easily write F = {10, 20, 30, 40, 45, 35, 25, 15, 5}.

Q. Write the elements of set C in roster form if C = {$x \mid x = a^2$ and $x = b^3$, where $0 < a, b < 30$}

A. **C = {1, 64, 729}.** The rule for C is that x has to be a perfect square *and* a perfect cube. The roots of x (a and b) are positive numbers less than 30. The best way to approach this problem is to find all the squares of the numbers from 1 to 30 and then determine which are cubes: C = {1, 64, 729}. You get these elements because $1 = 1^2 = 1^3$, $64 = 8^2 = 4^3$, and $729 = 27^2 = 9^3$.

1. Write set A using roster notation if A = {$x \mid x$ is odd, $x = 7n$, $0 < x < 70$}

Solve It

2. Write set B using roster notation if B = {$x \mid x = n^2 - 1$, $0 < n \leq 10$}

Solve It

3. Write set C using a rule if C = {11, 21, 31, 41, 51, 61}

Solve It

4. Write set D using a rule if D = {1, 5, 9, 13, 17, 21, ...}

Solve It

Get Together: Combining Sets with Unions, Intersections, and Complements

Sets of elements can be combined or changed by using set operations. Much like addition or subtraction of real numbers, set operations are strictly defined to do something to the sets involved. The set operations are union, intersection, and complement:

✔ The *union* of two sets A and B is denoted A ∪ B and asks for all the elements in sets A and B — all of them together (without repeating any elements that they share).

✔ The *intersection* of the two sets A and B is denoted A ∩ B and asks for all the elements that A and B have in common. If the two sets have nothing in common, then your answer is the *empty set* or *null set,* written : { } or ∅.

✔ The *complement* of a set A, denoted A', asks for all the elements that *aren't* in the set but are in the universal set. The *universal set* is everything under consideration at the time. For instance, if you're working on sets that contain the letters of the English alphabet, then the universal set is all 26 letters.

Q. Given the sets A = {2, 4, 6, 8}, B = {4, 8, 16, 24, 32}, C = {3, 6, 9, 12, 15, 18, 21}, and the universal set is U = {x | 0 < x ≤ 32, x ε Z} = {1, 2, 3, 4, ... , 31, 32}, then find A ∪ B, A ∩ B, B ∩ C, and C'

A. The union of A and B, A ∪ B, consists of all the elements in A and B both, so **A ∪ B = {2, 4, 6, 8, 16, 24, 32}.** Notice that the 4 and 8 aren't repeated. The intersection of A and B, A ∩ B, consists of all the elements the two sets share, so **A ∩ B = {4, 8}.** The intersection of B and C, B ∩ C, is all the elements that the two sets share, but the two sets have nothing in common, so **B ∩ C = ∅,** the empty set. The complement of set C consists of everything that's in the universal set that's *not* in set C, so **C' = {1, 2, 4, 5, 7, 8, 10, 11, 13, 14, 16, 17, 19, 20, 22, 23, 24, 25, 26, 27, 28, 29, 30, 31, 32}.**

Q. Given the sets D = {0, 1, 2, 3, 4}, E = {0, 2, 6, 10}, F = {1, 3, 6, 10}, and U = {0, 1, 2, 3, ... , 10}, find the sets (D ∩ F)' and (D ∪ E) ∩ (E ∪ F)

A. The parentheses in these set operation problems work the same way as parentheses in algebraic expressions — you perform what's inside the parentheses first. To find the complement of the intersection of sets D and F, (D ∩ F)', you first find the intersection D ∩ F = {1, 3}; then, referring back to the universal set, U, you find the complement: **(D ∩ F)' = {0, 2, 4, 5, 6, 7, 8, 9, 10}.** The complement is everything *except* the 1 and 3 in the intersection. To find the intersection of the two unions, first find the two unions: D ∪ E = {0, 1, 2, 3, 4, 6, 10}, and E ∪ F = {0, 1, 2, 3, 6, 10}. You write the intersection — what the two results have in common — as **(D ∪ E) ∩ (E ∪ F) = {0, 1, 2, 3, 6, 10}.**

5. Given the sets A = {0, 2, 4, 6, 8, ... , 20}, B = {0, 5, 10, 15, 20}, C = {7, 11, 17}, and the universal set U = {0, 1, 2, 3, 4, ... , 20}, find A ∩ B

Solve It

6. Given the sets A = {0, 2, 4, 6, 8, ... , 20}, B = {0, 5, 10, 15, 20}, C = {7, 11, 17}, and the universal set U = {0, 1, 2, 3, 4, ... , 20}, find A ∪ B

Solve It

7. Given the sets A = {0, 2, 4, 6, 8, ... , 20}, B = {0, 5, 10, 15, 20}, C = {7, 11, 17} and the universal set U = {0, 1, 2, 3, 4, ... , 20}, find B ∩ C

Solve It

8. Given the sets A = {0, 2, 4, 6, 8, ... , 20}, B = {0, 5, 10, 15, 20}, C = {7, 11, 17}, and the universal set U = {0, 1, 2, 3, 4, ... , 20}, find A'

Solve It

9. Given the sets A = {0, 2, 4, 6, 8, ... , 20}, B = {0, 5, 10, 15, 20}, C = {7, 11, 17}, and the universal set U = {0, 1, 2, 3, 4, ... , 20}, find (A ∪ C)'

Solve It

10. Given the sets A = {0, 2, 4, 6, 8, ... , 20}, B = {0, 5, 10, 15, 20}, C = {7, 11, 17}, and the universal set U = {0, 1, 2, 3, 4, ... , 20}, find (A ∪ B) ∩ (B ∪ C)

Solve It

Multiplication Countdowns: Simplifying Factorial Expressions

Sets of elements have special operations to combine them or change them. (The preceding section covers all those operations, if you need to refer back.) Another operation that's used with sets (but that isn't exclusive to sets) is *factorial*, denoted by the exclamation point. You use the factorial operation to count the number of elements in the union, intersection, or complement of sets. Factorials appear in the formulas you use to count the elements in sets that are really large. The upcoming sections "Counting on Permutations When Order Matters" and "Mixing It Up with Combinations" show you how the factorial works in the formulas.

The factorial operation, $n!$, is defined as $n! = n(n-1)(n-2)(n-3) \cdots 4 \cdot 3 \cdot 2 \cdot 1$. In other words, you multiply the number n, being operated upon, by every positive integer smaller than n. Some values of $n!$ are: $1! = 1$, $2! = 2$, $3! = 6$, $4! = 24$, $5! = 120$, $6! = 720$, and so on. You see that they're getting pretty big pretty fast.

One other factorial value that you need is $0! = 1$. You may think that's a typo. Nope. By definition, 0 factorial is equal to 1. It's one of those quirky things that mathematicians declare and make everyone use so that answers to problems come out right. People wanted the formulas for counting to be consistent for all the numbers used.

Simplifying factorials isn't difficult, but it isn't as easy as you may think at first glance. To simplify $\frac{6!}{3!}$, you can't just reduce the 6 and the 3. You have to look at all the factors involved in each factorial operation. Write out the factorials, and you get $\frac{6!}{3!} = \frac{6 \cdot 5 \cdot 4 \cdot 3 \cdot 2 \cdot 1}{3 \cdot 2 \cdot 1}$. Now reduce the like factors and simplify: $\frac{6!}{3!} = \frac{6 \cdot 5 \cdot 4 \cdot \cancel{3} \cdot \cancel{2} \cdot \cancel{1}}{\cancel{3} \cdot \cancel{2} \cdot \cancel{1}} = 6 \cdot 5 \cdot 4 = 120$.

Q. Simplify the factorial expression: $\frac{18!}{3!15!}$

A. **816.** First, write out the expansions of the factorials. But wait! (Notice that despite the exclamation point, the factorial doesn't work on the word *wait*.) Instead of writing out all the factors of 15!, just write 18! as $18 \cdot 17 \cdot 16 \cdot 15!$. You choose to stop with

the 15 because of the 15! in the denominator. The 15! terms will cancel out, so don't bother to write out all those identical terms in both numerator and denominator: $\frac{18!}{3!15!} = \frac{18 \cdot 17 \cdot 16 \cdot 15!}{3 \cdot 2 \cdot 1 \cdot 15!}$. Now divide out any other common factors and simplify: $\frac{18!}{3!15!} = \frac{\overset{3}{\cancel{18}} \cdot 17 \cdot 16 \cdot \cancel{15!}}{\cancel{3} \cdot \cancel{2} \cdot 1 \cdot \cancel{15!}} = 3 \cdot 17 \cdot 16 = 816$.

---

11. Simplify the expression $\frac{8!}{4!}$

Solve It

12. Simplify the expression $\frac{52!}{50!}$

Solve It

---

13. Simplify the expression $\frac{5!}{2!3!}$

Solve It

14. Simplify the expression $\frac{20!}{5!15!}$

Solve It

Checking Your Options: Using the Multiplication Property

When you want to count up how many things are in a set, you have quite a few options. When the set contains too many elements to count accurately, you look for some sort of pattern or rule to help out. In this section, you practice the multiplication property. In the next two sections, you count the elements using permutations and combinations. Each method has its place in the world of mathematical counting. You can count on it!

If you can do task one in m_1 ways, task two in m_2 ways, task three in m_3 ways, and so on, then you can perform all the tasks in a total of $m_1 \cdot m_2 \cdot m_3 \cdots$ ways.

Q. How many ways can you fly from San Francisco to New York City, stopping in Denver, Chicago, and Buffalo, if the Web site offers four ways to fly from from San Francisco to Denver, six ways to fly from Denver to Chicago, two ways to fly from Chicago to Buffalo, and three ways to fly from Buffalo to New York City?

A. **144.** Multiply $4 \cdot 6 \cdot 2 \cdot 3 = 144$. This method doesn't tell you what all the routes are; it just tells you how many are possible so you know when you've listed all of them. (Better get to work on that.)

Q. How many ways can one write a password if the first symbol has to be a digit from 1 to 9; the second, third, and fourth symbols have to be letters of the English alphabet; and the last symbol has to be from the set {!, @, #, $, %, ^, &, *, +}?

A. **1,423,656.** You multiply $9 \cdot 26 \cdot 26 \cdot 26 \cdot 9 = 1,423,656$. This system allows a lot of passwords, but most institutions make you use eight or more characters, which makes the number possible grow even more.

15. If you have to take one class in each subject, how many different course loads can you create if you have a choice of four math classes, three history classes, eight English classes, and five science classes?

Solve It

16. How many different ice cream sundaes can you create if you have a choice of five ice cream flavors, three sauces, and five sprinkled toppings if you choose one of each type?

Solve It

17. How many different automobiles can you order if you have a choice of six colors, four interiors, two trim options, three warranties, and two types of seats?

Solve It

18. How many different dinners can you order if you have a choice of twelve appetizers, eight entrees, five potatoes, six desserts, and a choice of soup or salad?

Solve It

Counting on Permutations When Order Matters

Permutations involve taking a specific number of items from an available group or set and seeing how many different ways the items can be selected and then arranged. For instance, if you choose three letters from the set {a, r, s, t} and arrange them as many ways as possible, you get the arrangements of {a, r, s}: *ars, asr, ras, rsa, sar,* and *sra;* the arrangements of {a, r, t}: *art, atr, rat, rta, tar,* and *tra;* and the arrangements of {a, s, t}: *ast, ats, sat, sta, tas,* and *tsa;* and the arrangements of {r, s, t}: *rst, rts, srt, str, trs,* and *tsr.* The number of permutations is 24.

Note: If you lose track of how to figure out all the "words," you can list all these arrangements using a *tree.* You can find a discussion of trees in *Algebra II For Dummies* (Wiley) or in other algebra books. Without a tree, just figure out a way to do a systematic listing.

You can find the number of permutations of *n* things taken *r* at a time with the formula $P(n,r) = \frac{n!}{(n-r)!}$. The *n* is the grouping or set that you're pulling items from. The *r* is how many of those items you're taking at a time. The notation *P(n, r)* is standard notation for indicating permutations.

Q. How many permutations are possible if you choose three letters from a set of four? (This problem is the one I describe at the beginning of this section.)

A. **24.** Using the formula, $P(4,3) = \frac{4!}{(4-3)!} = \frac{4!}{1!} = \frac{4 \cdot 3 \cdot 2 \cdot 1}{1} = 24$

Q. How many arrangements are possible if you choose any three letters from the English alphabet and then take any three digits from the digits 0–9 and use them for a password (assume that none are repeated)?

A. **11,232,000.** First use the formula to find the number of arrangements of letters: $P(26,3) = \frac{26!}{(26-3)!} = \frac{26!}{23!} = \frac{26 \cdot 25 \cdot 24 \cdot 23!}{23!} = 26 \cdot 25 \cdot 24 = 15{,}600$. Then find the number of arrangements of digits: $P(10,3) = \frac{10!}{(10-3)!} = \frac{10!}{7!} = \frac{10 \cdot 9 \cdot 8 \cdot 7!}{7!} = 10 \cdot 9 \cdot 8 = 720$. Finally, using the multiplication from the preceding section, "Checking Your Options: Using the Multiplication Property," multiply the two answers together: $15{,}600 \cdot 720 = 11{,}232{,}000$.

19. How many arrangements ("words") are possible using three of the letters from the word *stare?*

Solve It

20. How many arrangements ("words") are possible using all five of the letters from the word *stare?*

Solve It

21. How many different license plates can you form using three letters from the English alphabet (except the letters O and I and not repeating any), two digits from the digits 1–9 (without repetition), and then two letters chosen from the first seven letters in the alphabet (not two of the same letter)?

Solve It

22. You have six blue books, five red books, and ten green books, and you decide to put four of each color on a bookshelf, keeping the same colors together. How many arrangements are possible?

Solve It

Mixing It Up with Combinations

Combinations are another way of counting items. This time, you select some items from a larger group, but you don't care what order they come in. For instance, if you select 6 lottery numbers from a listing of 54, which number was chosen first or second or third doesn't matter — just that you chose them.

You can find the number of combinations or ways to choose *r* elements from a set containing *n* elements with the formula $C(n,r) = \binom{n}{r} = \frac{n!}{r!(n-r)!}$. The two different notations are commonly used to indicate that you're finding the number of combinations. Notice that this formula looks very much like the formula for permutations. The extra factor in the denominator makes the total number smaller.

Q. Find the number of ways (combinations) of selecting 6 lottery numbers from 54 choices.

A. **25,827,165.** (At $1 per entry, no wonder your chances are so slim.) Here's the math:

$$C(54,6) = \binom{54}{6} = \frac{54!}{6!(54-6)!} = \frac{54!}{6!\,48!} = \frac{54 \cdot 53 \cdot 52 \cdot 51 \cdot 50 \cdot 49 \cdot 48!}{6 \cdot 5 \cdot 4 \cdot 3 \cdot 2 \cdot 1 \cdot 48!}$$

$$= \frac{\overset{9}{54} \cdot 53 \cdot \overset{13}{52} \cdot \overset{17}{51} \cdot \overset{5}{50} \cdot 49 \cdot 48!}{6 \cdot 5 \cdot 4 \cdot 3 \cdot 2 \cdot 1 \cdot 48!} = 9 \cdot 53 \cdot 13 \cdot 17 \cdot 5 \cdot 49$$

$$= 25,827,165$$

Each of the factors in the denominator divides into a factor in the numerator, so the result is always a whole number.

Q. At a restaurant, the specialty platter includes a choice of any two of the five entrees and any three of the ten sides. How many different platters are possible?

A. **1,200.** Use $C(5, 2)$ to choose the entrees and $C(10, 3)$ to choose the sides. Then multiply the two results together using the multiplication property from earlier in this chapter:

$$C(5,2) \cdot C(10,3) = \frac{5!}{2!\,3!} \cdot \frac{10!}{3!\,7!} = \frac{5 \cdot 4 \cdot 3!}{2 \cdot 1 \cdot 3!} \cdot \frac{10 \cdot 9 \cdot 8 \cdot 7!}{3 \cdot 2 \cdot 1 \cdot 7!}$$

$$= \frac{5 \cdot \overset{2}{4}}{\overset{}{2} \cdot 1} \cdot \frac{\overset{5}{10} \cdot \overset{3}{9} \cdot 8}{\overset{}{3} \cdot \overset{}{2} \cdot 1} = 5 \cdot 2 \cdot 5 \cdot 3 \cdot 8 = 1,200$$

You find that 1,200 different platters are possible.

---

23. You can invite 9 of your 20 best friends to your birthday party. How many different combinations of friends are possible?

Solve It

24. A lottery has you choose 8 numbers from 60. How many different winning combinations are possible?

Solve It

25. You're selecting a committee of eight. Four are to be chosen from a group of ten men, and four are to be chosen from a group of twenty women. How many different committees can you have?

Solve It

26. You're going to sample 5 fruit cups out of the total shipment of 400. If those 5 cups are okay, then you can assume that the whole shipment is okay. How many different ways can you choose 5 fruit cups out of 400?

Solve It

Raising Binomials to Powers: Investigating the Binomial Theorem

A *binomial* is a mathematical expression that has two terms. In algebra, people frequently raise binomials to powers to complete computations.

The *Binomial Theorem* says that if a and b are real numbers and n is a positive integer, then

$$(a + b)^n = a^n + na^{n-1}b^1 + \frac{n(n-1)}{2}a^{n-2}b^2 + \cdots + na^1b^{n-1} + b^n$$

$$= \binom{n}{0}a^n + \binom{n}{1}a^{n-1}b^1 + \cdots + \binom{n}{k}a^{n-k}b^k + \cdots + \binom{n}{n-1}a^1b^{n-1} + \binom{n}{n}b^n$$

You can see the rule here, in the second line, in terms of the coefficients that are created using combinations. The powers on a start with n and decrease until the power is 0 in the last term. That's why you don't see an a in the last term — it's really a^0. The powers on b increase from b^0 until the last term, where it's b^n. Notice that the power of b matches k in the combination.

Q. Use the Binomial Theorem to determine the coefficients of the expansion of $(x + y)^5$. Then write the expansion.

A. $x^5 + 5x^4y + 10x^3y^2 + 10x^2y^3 + 5xy^4 + y^5$. Using the format from the Binomial Theorem, the coefficients of the expansion are $\binom{5}{0}, \binom{5}{1}, \binom{5}{2}, \binom{5}{3}, \binom{5}{4}, \binom{5}{5} = 1, 5, 10, 10, 5, 1$. Place those coefficients along a horizontal line, leaving room after each coefficient for the powers of x and y. You have something that looks like: 1 5 10 10 5 1. Now write decreasing powers of x after each coefficient, giving you $1x^5$ $5x^4$ $10x^3$ $10x^2$ $5x^1$ 1. The last power of x, the x^0, is just a 1, so you don't write it into the term. Now write increasing powers of y after each x term and on the last term, and add the terms together (the first power is 0, so you don't need a y after the first x term). You now get $1x^5 + 5x^4y^1 + 10x^3y^2 + 10x^2y^3 + 5x^1y^4 + 1y^5$. Notice that the sum of the exponents on each term is 5.

27. Expand $(x + y)^6$ using the Binomial Theorem

Solve It

28. Expand $(x - y)^4$ using the Binomial Theorem

Solve It

29. Expand $(x + 1)^5$ using the Binomial Theorem

Solve It

30. Expand $(x - 2)^3$ using the Binomial Theorem

Solve It

Answers to Problems on Sets and Counting

The following are the answers to the practice problems presented earlier in this chapter.

1 Write set A using roster notation if A = {x | x is odd, $x = 7n$, $0 < x < 70$}. The answer is **{7, 21, 35, 49, 63}**.

According to the rule, you want numbers that are odd, multiples of 7, and between 0 and 70.

2 Write set B using roster notation if B = {x | $x = n^2 - 1$, $0 < n \leq 10$}. The answer is **{0, 3, 8, 15, 24, 35, 48, 63, 80, 99}**.

The numbers in this set are all 1 less than a perfect square — and the perfect squares are all integers between 0 and 10.

3 Write set C using a rule if C = {11, 21, 31, 41, 51, 61}. The answer is **C = {x | $x = 10n + 1$, $1 \leq n \leq 6$}**.

Another way of writing the rule is C = {x | $x = 10n + 11$, $0 \leq n \leq 5$}. Notice that in this second rule, the values of n are different — they begin and end in different places — and the constant 11 is different. What's alike in these two rules is that the n is multiplied by 10, keeping the terms 10 units apart.

4 Write set D using a rule if D = {1, 5, 9, 13, 17, 21, ...}. The answer is **D = {x | $x = 4n + 1$, $n \geq 0$}**.

The rule allows the set to be infinite — the number of terms has no end.

5 Given the sets A = {0, 2, 4, 6, 8, ... , 20}, B = {0, 5, 10, 15, 20}, C = {7, 11, 17}, and the universal set U = {0, 1, 2, 3, 4, ... , 20}, find A ∩ B. The answer is **{0, 10, 20}**.

The sets A and B share only these three elements.

6 Given the sets A = {0, 2, 4, 6, 8, ... , 20}, B = {0, 5, 10, 15, 20}, C = {7, 11, 17}, and the universal set U = {0, 1, 2, 3, 4, ... , 20}, find A ∪ B. The answer is **{0, 2, 4, 5, 6, 8, 10, 12, 14, 15, 16, 18, 20}**.

The union of A and B contains everything from A — all the even numbers from 0 to 20 — and everything from B — the multiples of 5 from 0 to 20. Essentially, you just list the even numbers and insert the 5 and 15 from set B. The 0 and 10 and 20 are already accounted for.

7 Given the sets A = {0, 2, 4, 6, 8, ... , 20}, B = {0, 5, 10, 15, 20}, C = {7, 11, 17} and the universal set U = {0, 1, 2, 3, 4, ... , 20}, find B ∩ C. The answer is **∅ or { }**.

The solution is the empty set, because sets B and C have no terms in common.

8 Given the sets A = {0, 2, 4, 6, 8, ... , 20}, B = {0, 5, 10, 15, 20}, C = {7, 11, 17}, and the universal set U = {0, 1, 2, 3, 4, ... , 20}, find A'. The answer is **{1, 3, 5, 7, ... , 19}**.

The complement of set A, the even numbers from 0 to 20, is a set that has all the odd numbers between 0 and 20.

9 Given the sets A = {0, 2, 4, 6, 8, ... , 20}, B = {0, 5, 10, 15, 20}, C = {7, 11, 17}, and the universal set U = {0, 1, 2, 3, 4, ... , 20}, find (A ∪ C)'. The answer is **{1, 3, 5, 9, 13, 15, 19}**.

You first find the union of sets A and C: A ∪ C = {0, 2, 4, 6, 7, 8, 10, 11, 12, 14, 16, 17, 18, 20}. The complement of that union is everything in the universal set, U, that isn't in the set A ∪ C. So (A ∪ C)' = {1, 3, 5, 9, 13, 15, 19}.

10 Given the sets A = {0, 2, 4, 6, 8, ... , 20}, B = {0, 5, 10, 15, 20}, C = {7, 11, 17}, and the universal set U = {0, 1, 2, 3, 4, ... , 20}, find (A ∪ B) ∩ (B ∪ C). The answer is **{0, 5, 10, 15, 20}**.

First find the union A ∪ B and the other union B ∪ C. A ∪ B = {0, 2, 4, 5, 6, 8, 10, 12, 14, 15, 16, 18, 20}, and B ∪ C = {0, 5, 7, 10, 11, 15, 17, 20}. The intersection of these two resulting sets is (A ∪ B) ∩ (B ∪ C) = {0, 5, 10, 15, 20} = B. The result is the original set B.

11 Simplify the expression $\frac{8!}{4!}$. The answer is **1,680**.

Expand the numerator, and leave the denominator as 4!. Then reduce and simplify: $\frac{8!}{4!} = \frac{8 \cdot 7 \cdot 6 \cdot 5 \cdot 4!}{4!} = 8 \cdot 7 \cdot 6 \cdot 5 = 1680$.

12 Simplify the expression $\frac{52!}{50!}$. The answer is **2,652**.

Expand the numerator, and leave the denominator as 50!. Then reduce and simplify: $\frac{52!}{50!} = \frac{52 \cdot 51 \cdot 50!}{50!} = 52 \cdot 51 = 2652$.

13 Simplify the expression $\frac{5!}{2!\,3!}$. The answer is **10**.

Expand the numerator and the first factor in the denominator. Reduce the common factors and simplify: $\frac{5!}{2!\,3!} = \frac{5 \cdot 4 \cdot 3!}{2 \cdot 1 \cdot 3!} = \frac{5 \cdot 4}{2 \cdot 1} = 5 \cdot 2 = 10$.

14 Simplify the expression $\frac{20!}{5!\,15!}$. The answer is **15,504**.

Expand the numerator and the first factor in the denominator. Reduce the common factors and simplify: $\frac{20!}{5!\,15!} = \frac{20 \cdot 19 \cdot 18 \cdot 17 \cdot 16 \cdot 15!}{5 \cdot 4 \cdot 3 \cdot 2 \cdot 1 \cdot 15!} = \frac{20 \cdot 19 \cdot 18 \cdot 17 \cdot 16}{5 \cdot 4 \cdot 3 \cdot 2 \cdot 1} = 19 \cdot 3 \cdot 17 \cdot 16 = 15,504$.

15 If you have to take one class in each subject, how many different course loads can you create if you have a choice of four math classes, three history classes, eight English classes, and five science classes? The answer is **480**.

Multiply: $4 \cdot 3 \cdot 8 \cdot 5 = 480$.

16 How many different ice cream sundaes can you create if you have a choice of five ice cream flavors, three sauces, and five sprinkled toppings if you choose one of each type? The answer is **75**.

Multiply: $5 \cdot 3 \cdot 5 = 75$.

17 How many different automobiles can you order if you have a choice of six colors, four interiors, two trim options, three warranties, and two types of seats? The answer is **288**.

Multiply: $6 \cdot 4 \cdot 2 \cdot 3 \cdot 2 = 288$.

18 How many different dinners can you order if you have a choice of twelve appetizers, eight entrees, five potatoes, six desserts, and a choice of soup or salad? The answer is **5,760**.

Multiply: $12 \cdot 8 \cdot 5 \cdot 6 \cdot 2 = 5,760$. Don't forget that *soup or salad* is two choices for that selection.

19 How many arrangements ("words") are possible using three of the letters from the word *stare?* The answer is **60**.

Use the permutation formula $P(5, 3)$. Simplifying, $P(5,3) = \frac{5!}{(5-3)!} = \frac{5 \cdot 4 \cdot 3 \cdot 2!}{2!} = 5 \cdot 4 \cdot 3 = 60$.

20 How many arrangements ("words") are possible using all five of the letters from the word *stare?* The answer is **120**.

Use the permutation formula $P(5, 5)$. Simplifying, $P(5,5) = \frac{5!}{(5-5)!} = \frac{5 \cdot 4 \cdot 3 \cdot 2 \cdot 1}{0!} = \frac{5 \cdot 4 \cdot 3 \cdot 2 \cdot 1}{1} = 5! = 120$.

21 How many different license plates can you form using three letters from the English alphabet (except the letters O and I and not repeating any), two digits from the digits 1–9 (without repetition), and then two letters chosen from the first seven letters in the alphabet (not two of the same letter)? The answer is **36,723,456**.

Use three different permutations all multiplied together. For the first three letters, use $P(24, 3)$. The two digits use $P(9, 2)$. And the last two letters use $P(7, 2)$:

$$P(24, 3) \cdot P(9, 2) \cdot P(7, 2) = \frac{24!}{(24-3)!} \cdot \frac{9!}{(9-2)!} \cdot \frac{7!}{(7-2)!}$$

$$= \frac{24!}{21!} \cdot \frac{9!}{7!} \cdot \frac{7!}{5!} = \frac{24!}{21!} \cdot \frac{9!}{5!}$$

$$= \frac{24 \cdot 23 \cdot 22 \cdot 21!}{21!} \cdot \frac{9 \cdot 8 \cdot 7 \cdot 6 \cdot 5!}{5!}$$

$$= 24 \cdot 23 \cdot 22 \cdot 9 \cdot 8 \cdot 7 \cdot 6 = 36,723,456$$

22 You have six blue books, five red books, and ten green books, and you decide to put four of each color on a bookshelf, keeping the same colors together. How many arrangements are possible? The answer is **1,306,368,000**.

Use four different permutations all multiplied together. For the blue books, use $P(6, 4)$; for the red books, use $P(5, 4)$; and for the green books, use $P(10, 4)$. You then have to account for what order the three colors are going to be in. Use $P(3, 3)$. The books will be ordered within their colors and the groups of colors will be ordered. Whew! Here's what the work looks like:

$$P(6, 4) \cdot P(5, 4) \cdot P(10, 4) \cdot P(3, 3) = \frac{6!}{(6-4)!} \cdot \frac{5!}{(5-4)!} \cdot \frac{10!}{(10-4)!} \cdot \frac{3!}{(3-3)!}$$

$$= \frac{6!}{2!} \cdot \frac{5!}{1!} \cdot \frac{10!}{6!} \cdot \frac{3!}{0!}$$

$$= \frac{5! \cdot 10! \cdot 3!}{2! \cdot 1 \cdot 1}$$

$$= \frac{5! \cdot 10! \cdot 3 \cdot 2!}{2!} = 5! \, 10! \, 3!$$

$$= 120 \cdot 3,628,800 \cdot 3$$

$$= 1,306,368,000$$

23 You can invite 9 of your 20 best friends to your birthday party. How many different combinations of friends are possible? The answer is **167,960**.

This is a combination of 20 items in which you choose 9, $C(20, 9)$. Solving,

$$C(20, 9) = \frac{20!}{9!(20-9)!} = \frac{20!}{9! \, 11!}$$

$$= \frac{20 \cdot 19 \cdot 18 \cdot 17 \cdot {}^{2}16 \cdot {}^{5}15 \cdot {}^{2}14 \cdot 13 \cdot {}^{2}12 \cdot 11!}{9 \cdot 8 \cdot 7 \cdot 6 \cdot 5 \cdot 4 \cdot 3 \cdot 2 \cdot 1 \cdot 11!}$$

$$= 19 \cdot 17 \cdot 2 \cdot 5 \cdot 2 \cdot 13 \cdot 2 = 167,960$$

You can reduce the fraction in lots of ways — you have many combinations of the factors. But in any case, you should end up with all 1s in the denominator and numbers to multiply in the numerator.

24 A lottery has you choose 8 numbers from 60. How many different winning combinations are possible? The answer is **2,558,620,845**.

It would be a wonder if anyone ever won this lottery! This problem involves a combination of 60 numbers, choosing 8. Solving,

$$C(60,8) = \frac{60!}{8!(60-8)!} = \frac{60!}{8!\,52!}$$

$$= \frac{60 \cdot 59 \cdot \overset{29}{58} \cdot 57 \cdot \overset{}{56} \cdot 55 \cdot \overset{9}{54} \cdot 53 \cdot 52!}{8 \cdot 7 \cdot 6 \cdot 5 \cdot 4 \cdot 3 \cdot 2 \cdot 1 \cdot 52!}$$

$$= 59 \cdot 29 \cdot 57 \cdot 55 \cdot 9 \cdot 53 = 2{,}558{,}620{,}845$$

25 You're selecting a committee of eight. Four are to be chosen from a group of ten men, and four are to be chosen from a group of twenty women. How many different committees can you have? The answer is **1,017,450.**

Multiply $C(10, 4)$ times $C(20, 4)$. You're simply multiplying the number of ways to choose four men times the number of ways to choose four women:

$$C(10,4) \cdot C(20,4) = \frac{10!}{4!(10-4)!} \cdot \frac{20!}{4!(20-4)!}$$

$$= \frac{10 \cdot 9 \cdot 8 \cdot 7 \cdot 6!}{4 \cdot 3 \cdot 2 \cdot 1 \cdot 6!} \cdot \frac{20 \cdot 19 \cdot 18 \cdot 17 \cdot 16!}{4 \cdot 3 \cdot 2 \cdot 1 \cdot 16!}$$

$$= \frac{10 \cdot \overset{3}{9} \cdot 8 \cdot 7}{4 \cdot 3 \cdot 2} \cdot \frac{\overset{5}{20} \cdot 19 \cdot \overset{3}{18} \cdot 17}{4 \cdot 3 \cdot 2}$$

$$= 10 \cdot 3 \cdot 7 \cdot 5 \cdot 19 \cdot 3 \cdot 17 = 1{,}017{,}450$$

26 You're going to sample 5 fruit cups out of the total shipment of 400. If those 5 cups are okay, then you can assume that the whole shipment is okay. How many different ways can you choose 5 fruit cups out of 400? The answer is **83,218,600,080.**

Solve $C(400, 5)$:

$$\frac{400!}{5!(400-5)!} = \frac{400!}{5!\,395!} = \frac{\overset{10}{400} \cdot \overset{133}{399} \cdot 398 \cdot 397 \cdot 396 \cdot 395!}{5 \cdot 4 \cdot 3 \cdot 2 \cdot 1 \cdot 395!}$$

$$= 10 \cdot 133 \cdot 398 \cdot 397 \cdot 396$$

$$= 83{,}218{,}600{,}080$$

Some scientific calculators have trouble with a number this large and may go into scientific notation for an answer. You may see something like 8.32186×10^{10}.

27 Expand $(x + y)^6$ using the Binomial Theorem. The answer is $x^6 + 6x^5y + 15x^4y^2 + 20x^3y^3 + 15x^2y^4 + 6xy^5 + y^6$.

Using the Binomial Theorem, the coefficients are $\binom{6}{0}, \binom{6}{1}, \binom{6}{2}, \binom{6}{3}, \binom{6}{4}, \binom{6}{5}, \binom{6}{6} = 1, 6, 15, 20,$ 15, 6, 1. First write the coefficients, leaving space for the decreasing powers of x: 1 6 15 20 15 6 1. Now write in the decreasing powers of x: $1x^6$ $6x^5$ $15x^4$ $20x^3$ $15x^2$ $6x$ 1. Put in increasing powers of y and add the terms together: $x^6 + 6x^5y + 15x^4y^2 + 20x^3y^3 + 15x^2y^4 + 6xy^5 + y^6$.

28 Expand $(x - y)^4$ using the Binomial Theorem. The answer is $x^4 - 4x^3y + 6x^2y^2 - 4xy^3 + 1$.

Using the Binomial Theorem, the coefficients are $\binom{4}{0}, \binom{4}{1}, \binom{4}{2}, \binom{4}{3}, \binom{4}{4} = 1, 4, 6, 4, 1$. Write out the coefficients, leaving space for the decreasing powers of x: 1 4 6 4 1. Now write in the decreasing powers of x: $1x^4$ $4x^3$ $6x^2$ $4x$ 1. The y term is negative, so the powers of y have to include the negative sign when you write in the increasing powers of y: $1x^4(-y)^0$ $4x^3(-y)^1$ $6x^2(-y)^2$ $4x(-y)^3$ $1(-y)^4$. The even powers of y turn out to be positive, and the odd powers are negative. This situation makes the terms in the expansion alternate from positive to negative and back again: $1x^4 - 4x^3y + 6x^2y^2 - 4xy^3 + 1y^4$.

29 Expand $(x + 1)^5$ using the Binomial Theorem. The answer is $x^5 + 5x^4 + 10x^3 + 10x^2 + 5x + 1$.

Notice that there are no y terms in this expansion. Powers of 1 have been multiplied through, and of course, powers of 1 are equal to 1. Using the Binomial Theorem, the coefficients are $\binom{5}{0}, \binom{5}{1}, \binom{5}{2}, \binom{5}{3}, \binom{5}{4}, \binom{5}{5} = 1, 5, 10, 10, 5, 1$. Write out the coefficients and insert the decreasing powers of x: $1x^5\ 5x^4\ 10x^3\ 10x^2\ 5x\ 1$. Multiplying each term by a power of 1 doesn't change it, so just put + signs between the terms.

30 Expand $(x - 2)^3$ using the Binomial Theorem. The answer is $x^3 - 6x^2 + 12x - 8$.

Using the Binomial Theorem, the coefficients are $\binom{3}{0}, \binom{3}{1}, \binom{3}{2}, \binom{3}{3} = 1, 3, 3, 1$. Write out the coefficients and the decreasing powers of x: $1x^3\ 3x^2\ 3x\ 1$. The second term in the binomial is -2. Now insert the increasing powers of -2: $1x^3(-2)^0\ 3x^2(-2)^1\ 3x(-2)^2\ 1(-2)^3$. Simplifying the powers of 2 and combining them with the coefficients, you get $1x^3(1) + 3x^2(-2) + 3x(4) + 1(-8) = 1x^3 - 6x^2 + 12x - 8$.

Part V
The Part of Tens

The 5th Wave   By Rich Tennant

"What exactly are we saying here?"

In this part . . .

In the world of mathematics, you find an uncountable number of graphs of curves and lines. This part gives you ten graphs that are the basis for a huge percentage of them. Also, you find ways of collecting numbers and adding them up — ten neat-o formulas to ease your way.

Chapter 17

Ten Basic Graphs

Graphing is one way of getting the characteristics of a function out there for everyone to see. The basic graphs are just that — basic. They're centered at the origin and aren't expanded or shrunken or jostled about. You can alter the basic graphs by performing translations to the left or right or up or down.

Chapter 4 presents the nuts and bolts of graphing, and you can refer to the earlier chapters in this book for more-specific information on how to change the looks of the basic graphs. Right now, though, sit back and take a look at some of these elegant images.

Putting Polynomials in Their Place

The graph of a polynomial function is a smooth curve that may or may not change direction, depending on its degree. The two simplest polynomials are the quadratic, $y = x^2$ (see Figure 17-1), and the cubic, $y = x^3$ (see Figure 17-2). They both go through the origin and the point (1, 1). You can find out more about quadratics and cubics in Chapters 6 and 7.

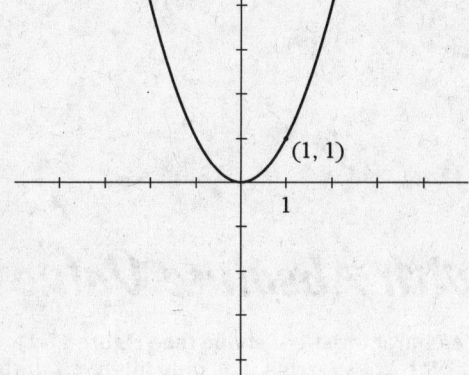

Figure 17-1:
The function $y = x^2$ has its vertex at the origin.

Figure 17-2:
The function
$y = x^3$ has
a point of
inflection
(a bend) at
the origin.

(1, 1)

1

Lining Up Front and Center

Just two points determine a unique line. This statement means that only one line can go through any two designated points.

Lines can have x- and y-intercepts — where the lines cross the axes; the *slope* of a line tells whether it rises or falls and how steeply this happens. Chapter 4 has lots of information on lines, if you want to know more. As Figure 17-3 shows, the graph of the line $y = x$ goes diagonally through the first and third quadrants. The slope is 1, and the line goes through the point (1, 1). The only intercept of this line is the origin.

(1, 1)

1

Figure 17-3:
The function
$y = x$ is a
line moving
upward
from left to
right.

Being Absolutely Sure with Absolute Value

The absolute value function $y = |x|$ has a characteristic V shape (see Figure 17-4). The V is typical of most absolute value equations with linear terms. The only intercept of this basic absolute value graph is the origin, and the function goes through the point (1, 1). Flip to Chapter 1 for more info on absolute value.

Figure 17-4:
The V opens
upward
because the
coefficient
is positive.

(1, 1)

Graphing Reciprocals of x and x²

The graphs of $y = \frac{1}{x}$ and $y = \frac{1}{x^2}$ both have a vertical asymptote of $x = 0$ and horizontal asymptote of $y = 0$ (see Figures 17-5 and 17-6). The asymptotes are actually the x- and y-axes. (You can find more information on asymptotes in Chapter 8.) Each curve goes through the point (1, 1), and each curve exhibits symmetry. As you can see in the figures, the graph of $y = \frac{1}{x}$ is symmetric with respect to the origin (a 180-degree turn gives you the same graph), and the graph of $y = \frac{1}{x^2}$ is symmetric with respect to the y-axis (it's a mirror image on either side).

(1, 1)

Figure 17-5:
The graph of
$y = \frac{1}{x}$ is in
only the first
and third
quadrants.

Figure 17-6:
The graph of
$y = \dfrac{1}{x^2}$ is
always posi-
tive (above
the y-axis).

Rooting Out Square Root and Cube Root

The graph of $y = \sqrt{x}$ starts at the origin and stays in the first quadrant (see Figure 17-7). Except for (0, 0), all the points have positive x- and y-coordinates. The curve rises gently from left to right.

Figure 17-7:
The graph of
the square
root looks
like half a
parabola.

The graph of $y = \sqrt[3]{x}$ is an odd function (see Chapter 5): It looks like twice its partner, the square root, with the square root curve spun around the origin into the third quadrant (see Figure 17-8). You can take cube roots of negative numbers, so you can find negative x- and y- values for points on this curve. Both curves go through the point (1, 1).

Figure 17-8:
The graph
of the cube
root is sym-
metric with
respect to
the origin.

(1, 1)

1

Growing Exponentially with a Graph

The graph of the exponential function $y = e^x$ is always above the x-axis (see Figure 17-9). The only intercept of this graph is the y-intercept at (0, 1). The x-axis acts as a horizontal asymptote when x is very small, and the curve grows without bound as the x-values move to the right. (Check out Chapter 9 for practice with the number e.)

Figure 17-9:
The number
e, about
2.718, raised
to the xth
power.

1

Logging In on Logarithmic Graphing

The graph of the logarithmic function $y = \ln x$ is the mirror image of its inverse function, $y = e^x$, over the line $y = x$ (see Figure 17-10). The function has one intercept, at (1, 0). The graph rises from left to right, moving from the fourth quadrant up through the first quadrant. The y-axis acts as a vertical asymptote as the values of x approach 0 — get very small. (For the scoop on logs, flip to Chapter 9.)

Figure 17-10:
The log function keeps growing more slowly all the time.

Chapter 18

Ten Special Sequences and Their Sums

- -

In This Chapter

▶ Sequences involving polynomials and formulas for their sums

▶ Describing special patterns and adding them

▶ Sequences for e and $\sin x$

- -

A sequence is a list of terms that has a formula or pattern for determining the numbers to come. A *series* is the sum of the terms in a sequence. Chapter 15 explains several types of sequences and series. In this chapter, I give you the power to add up lists of numbers — sometimes a huge number of numbers. Many sequences of numbers are used in financial and scientific formulas, and being able to add them up is essential. This chapter doesn't contain all the sum formulas possible, but the formulas here are the most common.

In Chapter 17, I tell you about basic graphs you can change through translations. The formulas in this chapter can be changed, too, but you'll have to wait until you get to a more advanced math course to get the opportunity. For now, just enjoy.

Adding As Easy As One, Two, Three

The positive integers are 1, 2, 3, 4, 5, ... , going on forever. Find the sum of the first n terms of this sequence using this formula:

$$1 + 2 + 3 + ... + n = \frac{n(n+1)}{2}$$

Thus, if you want the sum of the first 100 integers, you do the following calculation:

$$1 + 2 + 3 + ... + 100 = \frac{100(100+1)}{2} = \frac{\overset{50}{100}(101)}{\cancel{2}} = 5{,}050$$

Summing Up the Squares

The squares of the positive integers are 1, 4, 9, 16, 25, ... , n^2. To find a particular term in this sequence, you just take the square of the number of the term (the 12th term is $12^2 = 144$).

Find the sum of the first n terms using $1^2 + 2^2 + 3^2 + ... + n^2 = \dfrac{n(n+1)(2n+1)}{6}$. To add up the numbers $1 + 4 + 9 + 16 + 25 + ... + 144$, you want the sum of the first 12 squares. Using the formula,

$$1 + 4 + 9 + 16 + \ldots + 144 = 1^2 + 2^2 + 3^2 + \ldots + 12^2$$

$$= \frac{12(12 + 1)\big(2(12) + 1\big)}{6} = \frac{12(13)(25)}{6} = \frac{\overset{2}{\cancel{12}}(13)(25)}{\cancel{6}} = 2(13)(25) = 650$$

Finding the Sum of the Cubes

The cubes of the positive integers are 1, 8, 27, 64, 125, ... , n^3. The rule for the general term is n^3; you just raise the number of the term to the third power. You can find the sum of these cubes, 1^3, 2^3, 3^3, and so on, using $1^3 + 2^3 + 3^3 + \ldots + n^3 = \dfrac{n^2(n + 1)^2}{4}$. If you look closely, you can see that this formula is just the square of the formula used to find the sum of the first n integers — each factor in the formula is squared.

If you want to find the sum of $1 + 8 + 27 + 64 + 125 + 216 + \ldots + 3,375$, you first have to figure out whose cube 3,375 is. A simple calculator can find this answer for you. Or with some sleuthing, you may determine that the root has to end in a 5 for the cube to end in a 5. Seeing that $3,375 = 15^3$ doesn't take long. Here's how you find the sum:

$$1^3 + 2^3 + 3^3 + \ldots + 15^3 = \frac{15^2(15 + 1)^2}{4}$$

$$= \frac{225(256)}{4} = \frac{225\big(\overset{64}{\cancel{256}}\big)}{\cancel{4}} = 225(64) = 14,400$$

Not Being at Odds with Summing Odd Numbers

The odd positive integers are 1, 3, 5, 7, 9, ... , $(2n - 1)$. If you want the fifth odd number, the 9, you replace the n with 5 in $2n - 1$ and get $10 - 1 = 9$.

When the numbers get large, the process for finding out where a particular odd number falls in the sequence becomes even more important — you really don't want to have to add up all the numbers for the series, but you have to know how many terms are in the list. You can calculate, for instance, that the number 2,357 is the 1,179th odd number. You find this answer by solving $2n - 1 = 2,357$; $2n = 2,358$. Dividing by 2, you get $n = 1,179$, so 2,357 is the 1,179th odd number.

The formula for the sum of n odd numbers is $1 + 3 + 5 + \ldots + (2n - 1) = n^2$. To add up the odd numbers $1 + 3 + 5 + 7 + \ldots + 2,357$, you first determine how many numbers are in the list: $2n - 1 = 2,357$, so $n = 1,179$. The sum is $1,179^2 = 1,390,041$.

Evening Things Out by Adding Up Even Numbers

The even positive integers are 2, 4, 6, 8, ... , $2n$. Determine the sum of the first n even numbers with the formula $2 + 4 + 6 + 8 + \ldots + 2n = n(n + 1)$. You may notice that this formula looks somewhat like the formula for adding up the first n integers (see "Adding As Easy As One, Two, Three," earlier in this chapter). Multiply that other formula by 2, and you get the formula for the evens. Seems to make sense, even. (Ouch — horrible pun.)

Adding up the numbers $2 + 4 + 6 + 8 + ... + 500$, you determine that $n = 250$. So the sum is $500(501) = 250,500$.

Adding Everything Arithmetic

Arithmetic sequences are very predictable. The terms are always a constant difference from one another. The terms in an arithmetic sequence are $a_1, a_2, a_3, ... , a_n = a_1, a_1 + d, a_1 + 2d, a_1 + 3d, ... , a_1 + (n-1)d$. The number of times you add the common difference to the first term is 1 less than the total number of terms.

To find the sum of the terms of an arithmetic sequence, you can use one of these formulas:

$$a_1 + (a_1 + d) + (a_1 + 2d) + ... + (a_1 + (n-1)d) = \frac{n}{2}[2a_1 + (n-1)d] = \frac{n}{2}[a_1 + a_n]$$

Finding the sum using either formula requires that you know how many terms are in the sequence. Solve for n, the number of terms, using the rule for the general term of the arithmetic sequence: $a_n = a_1 + (n-1)d$.

To find the sum of $4 + 10 + 16 + 22 + 28 + ... + 304$, you recognize that the first term, a_1, is 4, the last term, a_n, is 304, and the difference between each pair of terms is 6. Solving $4 + (n-1)6 = 304$, you get $6n - 6 = 300$; $6n = 306$; $n = 51$. So the sum of those 51 terms is

$$4 + 10 + 16 + ... + 304 = \frac{51}{2}[4 + 304] = \frac{51[308]}{2} = \frac{51[\overset{154}{\cancel{308}}]}{\cancel{2}} = 7,854$$

Geometrically Speaking

Geometric sequences are almost as predictable as arithmetic sequences. The terms have a common ratio — you divide any term in a geometric sequence by the term that comes before it and you get the same ratio or multiplier, no matter which two terms you choose. The geometric sequence 1, 3, 9, 27, 81, ... has a first term of 1 and a common ratio of 3. The general term for a geometric sequence is $g_n = g_1(r^{n-1})$, where g_1 is the first term and r is the common ratio.

To add up the first n terms of a geometric sequence, use this formula:

$$g_1 + g_1 r + g_1 r^2 + ... + g_1 r^{n-1} = \frac{g_1(1 - r^n)}{1 - r}$$

Adding up $1 + 3 + 9 + 27 + ... + 6,561$, you first determine that 6,561 is 3^8. This answer makes 6,561 the ninth term in the series, because $1(3^8) = 1(3^{n-1})$, and if $8 = n - 1$, then $n = 9$. Using the formula for the sum,

$$1 + 3 + 9 + ... + 1(3^{9-1}) = \frac{1(1 - 3^9)}{1 - 3}$$

$$= \frac{1(1 - 19,683)}{-2} = \frac{-19,682}{-2} = 9,841$$

Easing into a Sum for e

The number e is approximately equal to 2.718281828459. The value for e is irrational, which means that the decimal never ends or repeats. The number, however, is oh-so-necessary in the scientific, mathematical, and financial worlds (see Chapter 9 for an application that involves your bank account).

You can determine the value for e using various formulas, one of which is the sum of a sequence of numbers. The more terms you add up in this series, the more correct decimal places for e you determine. The general term for this series is

$$e = \frac{1}{0!} + \frac{1}{1!} + \frac{1}{2!} + \frac{1}{3!} + \dots + \frac{1}{(n-1)!}$$

I don't have a handy formula for adding up the terms in this series, but a scientific calculator can do the job for a while. Here are a few of the sums:

$$\frac{1}{0!} + \frac{1}{1!} = \frac{1}{1} + \frac{1}{1} = 2$$

$$\frac{1}{0!} + \frac{1}{1!} + \frac{1}{2!} = \frac{1}{1} + \frac{1}{1} + \frac{1}{2} = 2.5$$

$$\frac{1}{0!} + \frac{1}{1!} + \frac{1}{2!} + \frac{1}{3!} = \frac{1}{1} + \frac{1}{1} + \frac{1}{2} + \frac{1}{6} \approx 2.666666667$$

$$\frac{1}{0!} + \frac{1}{1!} + \frac{1}{2!} + \frac{1}{3!} + \frac{1}{4!} = \frac{1}{1} + \frac{1}{1} + \frac{1}{2} + \frac{1}{24} \approx 2.708333333$$

$$\vdots$$

$$\frac{1}{0!} + \frac{1}{1!} + \frac{1}{2!} + \frac{1}{3!} + \dots + \frac{1}{10!} = \frac{1}{1} + \frac{1}{1} + \frac{1}{2} + \frac{1}{6} + \dots + \frac{1}{3,628,800} \approx 2.718281801$$

Signing In on the Sine

The *sine* is one of the trigonometric functions. The input (x-value) for trig functions consists of angle measures. For instance, if $y = \sin x$ (sin is the abbreviation for sine) and x is 30 degrees, then $y = \sin 30° = 0.5$.

Many ways let you determine the value of the sine of an angle, but the one that fits in this chapter is an infinite series of values. If you write the input, x, in radians instead of degrees ($360° = 2\pi$ radians), then $\sin x = x - \frac{x^3}{3!} + \frac{x^5}{5!} - \frac{x^7}{7!} + \dots + (-1)^{n-1} \frac{x^{2n-1}}{(2n-1)!}$. The more terms you use, the closer you get to the exact value.

So to find the sine of 30 degrees, change that measure to radians to get $30° \approx 0.5236$ radians ($30°$ is $\frac{1}{12}$ of a circle, and $2\pi \div 12 = \frac{\pi}{6}$ radians). Using the series,

$$\sin 0.5236 \approx 0.5236 - \frac{0.5236^3}{3!} + \frac{0.5236^5}{5!} - \frac{0.5236^7}{7!} + \dots$$

$$\approx 0.5236 - \frac{0.1435}{6} + \frac{0.0394}{120} - \frac{0.0108}{5040}$$

$$\approx 0.5236 - 0.0239 + 0.0003 - 0$$

$$= 0.5$$

In the fourth term, the number is so small that it rounds to 0 when you just want four decimal places.

Powering Up on Powers of 2

You can quickly and simply find the sum of a sequence consisting of powers of 2 by just finding the next power of 2 and subtracting 1. For instance, $1 + 2 + 4 + 8 + 16$ consists of the powers of 2 from 2^0 to 2^4. The sum is $2^5 - 1 = 32 - 1 = 31$. So the rule is $1 + 2 + 4 + 8 + ... + 2^n = 2^{n+1} - 1$.

Adding Up Fractions with Multiples for Denominators

Create a rather interesting sequence with

$$\frac{1}{1 \cdot 2} + \frac{1}{2 \cdot 3} + \frac{1}{3 \cdot 4} + ... + \frac{1}{n(n+1)}$$

where you take the number of the term times the next integer and put their product in the denominator of a fraction. The sum of the terms is equal to $\frac{n}{n+1}$. That result almost seems too easy to believe, doesn't it? Look at a few examples:

$$\frac{1}{1 \cdot 2} + \frac{1}{2 \cdot 3} + \frac{1}{3 \cdot 4} + ... + \frac{1}{n(n+1)} = \frac{n}{n+1}$$

$$\frac{1}{1 \cdot 2} + \frac{1}{2 \cdot 3} = \frac{1}{2} + \frac{1}{6} = \frac{3}{6} + \frac{1}{6} = \frac{4}{6} = \frac{2}{3}$$

$$\frac{1}{1 \cdot 2} + \frac{1}{2 \cdot 3} + \frac{1}{3 \cdot 4} = \frac{1}{2} + \frac{1}{6} + \frac{1}{12} = \frac{6}{12} + \frac{2}{12} + \frac{1}{12} = \frac{9}{12} = \frac{3}{4}$$

$$\frac{1}{1 \cdot 2} + \frac{1}{2 \cdot 3} + \frac{1}{3 \cdot 4} + \frac{1}{4 \cdot 5} = \frac{1}{2} + \frac{1}{6} + \frac{1}{12} + \frac{1}{20} = \frac{30}{60} + \frac{10}{60} + \frac{5}{60} + \frac{3}{60} = \frac{48}{60} = \frac{4}{5}$$

Index

BUSINESS, CAREERS & PERSONAL FINANCE

0-7645-5307-0

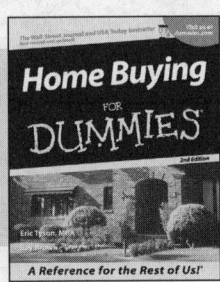

0-7645-5331-3 *†

Also available:
- Accounting For Dummies †
  0-7645-5314-3
- Business Plans Kit For Dummies †
  0-7645-5365-8
- Cover Letters For Dummies
  0-7645-5224-4
- Frugal Living For Dummies
  0-7645-5403-4
- Leadership For Dummies
  0-7645-5176-0
- Managing For Dummies
  0-7645-1771-6

- Marketing For Dummies
  0-7645-5600-2
- Personal Finance For Dummies *
  0-7645-2590-5
- Project Management For Dummies
  0-7645-5283-X
- Resumes For Dummies †
  0-7645-5471-9
- Selling For Dummies
  0-7645-5363-1
- Small Business Kit For Dummies *†
  0-7645-5093-4

HOME & BUSINESS COMPUTER BASICS

0-7645-4074-2

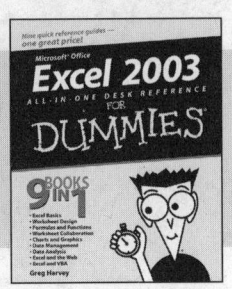

0-7645-3758-X

Also available:
- ACT! 6 For Dummies
  0-7645-2645-6
- iLife '04 All-in-One Desk Reference
  For Dummies
  0-7645-7347-0
- iPAQ For Dummies
  0-7645-6769-1
- Mac OS X Panther Timesaving
  Techniques For Dummies
  0-7645-5812-9
- Macs For Dummies
  0-7645-5656-8
- Microsoft Money 2004 For Dummies
  0-7645-4195-1

- Office 2003 All-in-One Desk Reference
  For Dummies
  0-7645-3883-7
- Outlook 2003 For Dummies
  0-7645-3759-8
- PCs For Dummies
  0-7645-4074-2
- TiVo For Dummies
  0-7645-6923-6
- Upgrading and Fixing PCs For Dummies
  0-7645-1665-5
- Windows XP Timesaving Techniques
  For Dummies
  0-7645-3748-2

FOOD, HOME, GARDEN, HOBBIES, MUSIC & PETS

0-7645-5295-3

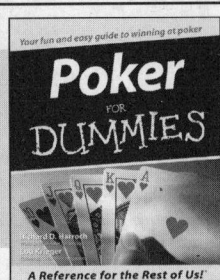

0-7645-5232-5

Also available:
- Bass Guitar For Dummies
  0-7645-2487-9
- Diabetes Cookbook For Dummies
  0-7645-5230-9
- Gardening For Dummies *
  0-7645-5130-2
- Guitar For Dummies
  0-7645-5106-X
- Holiday Decorating For Dummies
  0-7645-2570-0
- Home Improvement All-in-One
  For Dummies
  0-7645-5680-0

- Knitting For Dummies
  0-7645-5395-X
- Piano For Dummies
  0-7645-5105-1
- Puppies For Dummies
  0-7645-5255-4
- Scrapbooking For Dummies
  0-7645-7208-3
- Senior Dogs For Dummies
  0-7645-5818-8
- Singing For Dummies
  0-7645-2475-5
- 30-Minute Meals For Dummies
  0-7645-2589-1

INTERNET & DIGITAL MEDIA

0-7645-1664-7

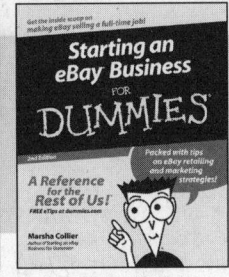

0-7645-6924-4

Also available:
- 2005 Online Shopping Directory
  For Dummies
  0-7645-7495-7
- CD & DVD Recording For Dummies
  0-7645-5956-7
- eBay For Dummies
  0-7645-5654-1
- Fighting Spam For Dummies
  0-7645-5965-6
- Genealogy Online For Dummies
  0-7645-5964-8
- Google For Dummies
  0-7645-4420-9

- Home Recording For Musicians
  For Dummies
  0-7645-1634-5
- The Internet For Dummies
  0-7645-4173-0
- iPod & iTunes For Dummies
  0-7645-7772-7
- Preventing Identity Theft For Dummies
  0-7645-7336-5
- Pro Tools All-in-One Desk Reference
  For Dummies
  0-7645-5714-9
- Roxio Easy Media Creator For Dummies
  0-7645-7131-1

* Separate Canadian edition also available
† Separate U.K. edition also available

SPORTS, FITNESS, PARENTING, RELIGION & SPIRITUALITY

0-7645-5146-9

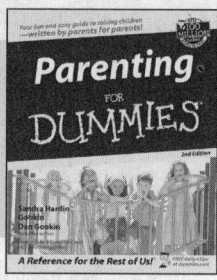

0-7645-5418-2

Also available:

- Adoption For Dummies
  0-7645-5488-3
- Basketball For Dummies
  0-7645-5248-1
- The Bible For Dummies
  0-7645-5296-1
- Buddhism For Dummies
  0-7645-5359-3
- Catholicism For Dummies
  0-7645-5391-7
- Hockey For Dummies
  0-7645-5228-7

- Judaism For Dummies
  0-7645-5299-6
- Martial Arts For Dummies
  0-7645-5358-5
- Pilates For Dummies
  0-7645-5397-6
- Religion For Dummies
  0-7645-5264-3
- Teaching Kids to Read For Dummies
  0-7645-4043-2
- Weight Training For Dummies
  0-7645-5168-X
- Yoga For Dummies
  0-7645-5117-5

TRAVEL

0-7645-5438-7

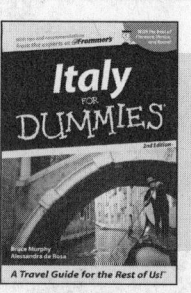

0-7645-5453-0

Also available:

- Alaska For Dummies
  0-7645-1761-9
- Arizona For Dummies
  0-7645-6938-4
- Cancún and the Yucatán For Dummies
  0-7645-2437-2
- Cruise Vacations For Dummies
  0-7645-6941-4
- Europe For Dummies
  0-7645-5456-5
- Ireland For Dummies
  0-7645-5455-7

- Las Vegas For Dummies
  0-7645-5448-4
- London For Dummies
  0-7645-4277-X
- New York City For Dummies
  0-7645-6945-7
- Paris For Dummies
  0-7645-5494-8
- RV Vacations For Dummies
  0-7645-5443-3
- Walt Disney World & Orlando For Dummies
  0-7645-6943-0

GRAPHICS, DESIGN & WEB DEVELOPMENT

0-7645-4345-8

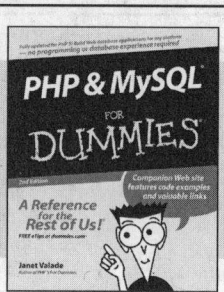

0-7645-5589-8

Also available:

- Adobe Acrobat 6 PDF For Dummies
  0-7645-3760-1
- Building a Web Site For Dummies
  0-7645-7144-3
- Dreamweaver MX 2004 For Dummies
  0-7645-4342-3
- FrontPage 2003 For Dummies
  0-7645-3882-9
- HTML 4 For Dummies
  0-7645-1995-6
- Illustrator CS For Dummies
  0-7645-4084-X

- Macromedia Flash MX 2004 For Dummies
  0-7645-4358-X
- Photoshop 7 All-in-One Desk Reference
  For Dummies
  0-7645-1667-1
- Photoshop CS Timesaving Techniques
  For Dummies
  0-7645-6782-9
- PHP 5 For Dummies
  0-7645-4166-8
- PowerPoint 2003 For Dummies
  0-7645-3908-6
- QuarkXPress 6 For Dummies
  0-7645-2593-X

NETWORKING, SECURITY, PROGRAMMING & DATABASES

0-7645-6852-3

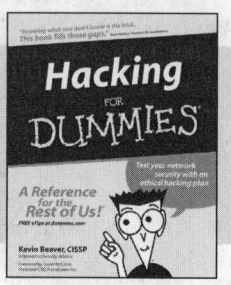

0-7645-5784-X

Also available:

- A+ Certification For Dummies
  0-7645-4187-0
- Access 2003 All-in-One Desk Reference
  For Dummies
  0-7645-3988-4
- Beginning Programming For Dummies
  0-7645-4997-9
- C For Dummies
  0-7645-7068-4
- Firewalls For Dummies
  0-7645-4048-3
- Home Networking For Dummies
  0-7645-42796

- Network Security For Dummies
  0-7645-1679-5
- Networking For Dummies
  0-7645-1677-9
- TCP/IP For Dummies
  0-7645-1760-0
- VBA For Dummies
  0-7645-3989-2
- Wireless All In-One Desk Reference
  For Dummies
  0-7645-7496-5
- Wireless Home Networking For Dummies
  0-7645-3910-8

HEALTH & SELF-HELP

0-7645-6820-5 *†

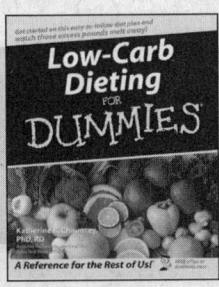

0-7645-2566-2

Also available:

Alzheimer's For Dummies
0-7645-3899-3

Asthma For Dummies
0-7645-4233-8

Controlling Cholesterol For Dummies
0-7645-5440-9

Depression For Dummies
0-7645-3900-0

Dieting For Dummies
0-7645-4149-8

Fertility For Dummies
0-7645-2549-2

Fibromyalgia For Dummies
0-7645-5441-7

Improving Your Memory For Dummies
0-7645-5435-2

Pregnancy For Dummies †
0-7645-4483-7

Quitting Smoking For Dummies
0-7645-2629-4

Relationships For Dummies
0-7645-5384-4

Thyroid For Dummies
0-7645-5385-2

EDUCATION, HISTORY, REFERENCE & TEST PREPARATION

0-7645-5194-9

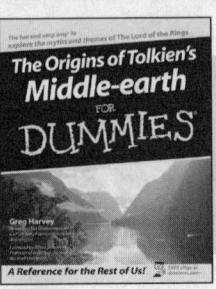

0-7645-4186-2

Also available:

Algebra For Dummies
0-7645-5325-9

British History For Dummies
0-7645-7021-8

Calculus For Dummies
0-7645-2498-4

English Grammar For Dummies
0-7645-5322-4

Forensics For Dummies
0-7645-5580-4

The GMAT For Dummies
0-7645-5251-1

Inglés Para Dummies
0-7645-5427-1

Italian For Dummies
0-7645-5196-5

Latin For Dummies
0-7645-5431-X

Lewis & Clark For Dummies
0-7645-2545-X

Research Papers For Dummies
0-7645-5426-3

The SAT I For Dummies
0-7645-7193-1

Science Fair Projects For Dummies
0-7645-5460-3

U.S. History For Dummies
0-7645-5249-X

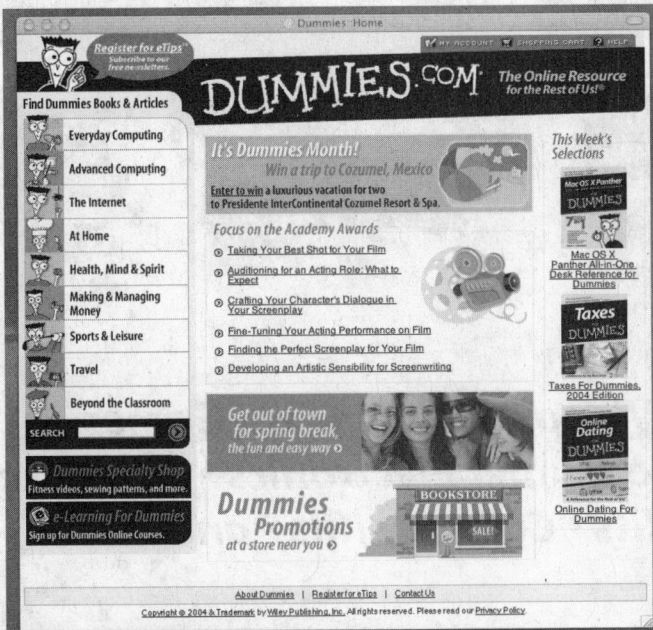

Get smart @ dummies.com®

- **Find a full list of Dummies titles**
- **Look into loads of FREE on-site articles**
- **Sign up for FREE eTips e-mailed to you weekly**
- **See what other products carry the Dummies name**
- **Shop directly from the Dummies bookstore**
- **Enter to win new prizes every month!**

* Separate Canadian edition also available
† Separate U.K. edition also available

Available wherever books are sold. For more information or to order direct: U.S. customers visit www.dummies.com or call 1-877-762-2974.
U.K. customers visit www.wileyeurope.com or call 0800 243407. Canadian customers visit www.wiley.ca or call 1-800-567-4797.